Electromagnetic Wave Propagation

Electromagnetic Wave Propagation

Donald W. Dearholt
New Mexico State University

William R. McSpadden
Battelle Memorial Institute
formerly at, New Mexico State University

McGraw-Hill Book Company

New York   St. Louis   San Francisco   Düsseldorf   Johannesburg
Kuala Lumpur   London   Mexico   Montreal   New Delhi
Panama   Rio de Janeiro   Singapore   Sydney   Toronto

Electromagnetic Wave Propagation

Copyright © 1973 by McGraw-Hill, Inc. All rights reserved. Printed in the United States of America. No part of this publication may be reproduced, stored in a retrieval system, or transmitted, in any form or by any means, electronic, mechanical, photocopying, recording, or otherwise, without the prior written permission of the publisher.

4567890 DODO 798

This book was set in Modern 8A by The Maple Press Company. The editors were Charles R. Wade, and Madelaine Eichberg; the designer was Rafael Hernandez; and the production supervisor was Thomas J. Lo Pinto. The drawings were done by Textart Services, Inc.
The printer and binder was R. R. Donnelley & Sons Company.

Library of Congress Cataloging in Publication Data

Dearholt, Donald W
   Electromagnetic wave propagation.

  1. Electromagnetic waves. 2. Electric lines. 3. Waves guides. I. McSpadden, Willian R., 1931- joint author. II. Title.
QC670.D39   530.1'41   72-5583
ISBN 0-07-016205-0

Contents

Preface

1 fundamentals of fields   1

    1.1   Introduction   1
    1.2   Coordinate Systems and Transformations   2
         The Rectangular Coordinate System   2
         The Cylindrical Coordinate System   3
         The Transformation from Rectangular to Cylindrical Coordinates   4
         The Transformation from Cylindrical to Rectangular Coordinates   10
         The Spherical Coordinate System   10
         Orthogonal Curvilinear Coordinates   13
         Table 1.2.1   Vector transformations between coordinate systems   14

	Table 1.2.2 Divergence, gradient, and curl in curvilinear coordinates	16
	Table 1.2.3 Vector identities	17
1.3	Coulomb's Law and the **E** Field	18
	Charge Density	21
	The **E** Field	22
1.4	Scalar and Vector Fields	24
	Scalar Fields	24
	Vector Fields	25
1.5	Delta Functions	26
1.6	The Gradient and the Potential Field	29
	Current Density	38
1.7	The Divergence	39
	Table 1.7.1 Gradient, divergence, curl, and scalar laplacian in the common coordinate systems	44
	Programmed Exercise 1.7.1 Divergence of the **E** field in a diode	47
1.8	Gauss' Law, the Divergence Theorem, and Poisson's Equation	49
	Gauss' Law	49
	The Divergence Theorem	54
	Poisson's Equation	55
1.9	The Curl	56
	Description of the Curl	56
	Definition of the Curl	59
	Some Properties of the Curl	60
	The Vector Components of the Curl	61
	An Alternate Form of the Curl	63
	Examples of the Curl	65
1.10	Stokes' Theorem and Properties of Vector Fields	68
	Stokes' Theorem	68
	Some Properties of Vector Fields	73
1.11	Helmholtz' Theorem and a Summary of Static Fields	75
	Combinations of the ∇ Operators	75
	Vector Potential and Poisson's Equation	76
	Solutions of the Scalar and Vector Poisson Equation	78
	Helmholtz' Theorem	80
	Summary of Static Fields	81
	Summary of Equations for Static Electromagnetic Fields	81
1.12	The Lorentz Transformation and Time-varying Fields	83
	Introduction	83

		Derivation of the Lorentz Transformation	84
		The Distortion of Distance and Time	88
		The Transformation of Velocity	91
		Programmed Exercise 1.12.1 Examples of velocity transformation	93
		Mass and Energy	94
		The Transformation of Mass and Force	96
	1.13	The Development of Maxwell's Equations	99
		The Electric and Magnetic Fields	99
		Transformation of Static Sources	104
		Maxwell's Equations	105
	Problems		112

2 time-varying fields — 115

	2.1	Postulates of Electromagnetic Theory and Their Physical Interpretations	115
		A Set of Mathematical Postulates	115
		Table 2.1.1 Mathematical postulates of electromagnetic theory	117
		Physical Interpretations	118
	2.2	The Potential Solutions of Maxwell's Equations	124
	2.3	The Wave Equation	128
	2.4	Voltage, Electromotive Force, and Potential Difference	132
	2.5	Power, Energy, and Poynting's Theorem	135
	2.6	Radiation	140
		Programmed Exercise 2.6.1 Radiation from a point source	142
	2.7	Boundary Conditions and Properties of Materials	145
		The Dielectric Tensor	145
		Boundary Conditions	148
		Table 2.7.1 Summary of the elementary field equations	154
	Problems		155

3 plane-wave propagation — 159

	3.1	Solution of the Wave Equation	159
	3.2	Traveling Waves	163
	3.3	Characteristics of a Uniform Plane Wave	166
	3.4	Plane Waves in Dissipative Media	169
		Skin Effect	173
		The Complex Poynting Vector and Power	175
	3.5	Polarization of Plane Waves	179
		Unpolarized Waves	179

		Plane-polarized Waves	179
		Elliptically Polarized Waves	180
		Circularly Polarized Waves	182
	3.6	Velocities of Propagation	182
		Group Velocity	183
	3.7	A Plane Wave Incident on a Perfect Conductor	185
		Normal Incidence	186
		Oblique Incidence	189
	3.8	A Plane Wave Incident on a Perfect Dielectric	193
		Normal Incidence	194
		Oblique Incidence	196
		Programmed Exercise 3.8.1 Undersea communication	202
	Problems		204

4 transient waves on lossless transmission lines 207

4.1	Coaxial Transmission Line	207
4.2	The Wave Equations for Voltage and Current	209
4.3	Voltage, Current, and Equivalent Circuits	216
4.4	Lines of Finite Length	222
4.5	Resistive Terminations	226
4.6	Bounce Diagrams	230
4.7	Arbitrary Waveforms	234
	Programmed Exercise 4.7.1 An ideal diode on a line	237
4.8	Nonzero Initial Conditions	238
	Programmed Exercise 4.8.1 A pulse generator	244
4.9	Discontinuities on a Transmission Line	245
4.10	Reactive Terminations	252
Problems		259

5 sinusoidal waves on lossless transmission lines 266

5.1	The Steady-state Solution from Transient Analysis	266
5.2	Steady-state Solution of the Wave Equation	268
	Table 5.2.1 Summary of the steady-state equations	272
5.3	Sinusoidal Traveling Waves and Standing Waves	273
	Sinusoidal Traveling Waves	273
	Standing Waves	275
	Programmed Exercise 5.3.1 The Ferranti effect	282
5.4	Impedance Terminations and the Gamma Plane	284
	Voltage, Current, Impedance, and the Reflection Coefficient	284
	Computer Solutions of Some Typical Problems	294

	Table 5.4.1 A lossless line with a real load	293
	Table 5.4.2 A lossless line with a complex load	294
	Table 5.4.3 A lossless line with an imaginary load	295
	Table 5.4.4 A lossless line with no load	296
	The Gamma Plane	296
5.5	The Smith Chart and the $Z\text{-}\theta$ Chart	299
	The Smith Chart	301
	The $Z\text{-}\theta$ Chart	308
5.6	Voltage, Current, and Power from the Smith Chart	309
5.7	Impedance Matching with Stub Tuners	314
	Single-stub Tuning	315
	Double-stub Tuning	318
5.8	The Quarter-wavelength Transformer	323
	Programmed Exercise 5.8.1 The exponentially tapered line	327
5.9	The Systems Approach Using the Digital Computer	329
	The Computer versus Conventional Techniques	329
	The Systems Approach	330
	Table 5.9.1 Summary of the steady-state equations	331
	The SANTLINE Computer Package	333
	Table 5.9.2 Definitions of words and element assignments for the common array TLDATA	337
	Table 5.9.3 A description of the subprograms of SANTLINE and a glossary of terms	345
5.10	Applications of the Computer	346
	A Discussion of Bandwidth	350
	Table 5.10.1 Program TLINE1	351
	Table 5.10.2 Results for the line of Fig. 5.4.7	352
	Table 5.10.3 Program double-stub bandwidth	354
	Table 5.10.4 Program QWTX	356
	Table 5.10.5 Frequency response of the quarter-wave transformer of Fig. 5.8.3a	357
	Table 5.10.6 Program QWTX2	359
	Problems	361

6   waves on dissipative lines — 366

6.1	Introduction to Dissipative Lines	366
6.2	Standing Waves	369
	Programmed Exercise 6.2.1 Measuring Z_0 and γ	372
6.3	Graphical Techniques	374
6.4	Impedance Matching	376
	Quarter-wavelength Transformers	376
	Stub Tuners	378

6.5	Computer Synthesis and Analysis Problems	381
	Table 6.5.1 Program STUBBW for Fig. 6.4.3	382
	Table 6.5.2 Results for the line of Fig. 6.4.3	383
Problems		384

7 waveguides 387

7.1	Introduction	387
7.2	Infinite Parallel Planes	388
	Phase and Group Velocities	392
	Wavelength	394
	Impedance Concepts, Power, and Energy	397
7.3	TE_{m0} Waves in Rectangular Waveguides	398
	The TE_{m0} Waves	399
7.4	A Programmed Example on Rectangular Waveguides	401
	Part 1 Helmholtz' Equation	402
	Part 2 Uniform Plane Waves	402
	Part 3 Nonuniform Plane Waves	403
	Part 4 Rectangular Waveguides	406
7.5	Waveguide Boundary-value Problems	408
7.6	Circular Waveguides	409
7.7	Resonant Cavities	417
Problems		421

Appendix A	Tables for Reference	423
Appendix B	Description of the TLDATA Array and the SANTLINE Computer Package	430
Appendix C	Charts	442

Index

Preface

Our original intention was to provide a text for a second course in electromagnetic field theory for students of electrical engineering. Since many schools require only two courses in electromagnetic theory, we planned a text with a broad theoretical base sufficiently problem-oriented to ensure that the students would work a wide variety of practical problems. At the suggestion of one of the reviewers, the development of Maxwell's equations, originally visualized as a prerequisite to the course, has been included in the text, becoming Chaps. 1 and 2. Although the information density in these chapters is quite high, they might prove adequate as the basis of an introductory course in field theory, especially at schools having a strong sequence of engineering physics. The primary usefulness of these first two chapters, however, is probably as a reference, and perhaps a second point of view, to a more advanced student. In particular, the development of Maxwell's equations from Coulomb's law and special relativity, in Chap. 1, represents a point of view not normally presented to undergraduates.

To use the text for a second fields course, it is often convenient to begin with the material in Chap. 3, in which it is assumed that Maxwell's equations are understood; and by implication, the various vector operations and the meaning of the basic partial differential equations of field theory must also be understood. Chapter 3 provides the basic concepts of electromagnetic wave propagation in free space and in space bounded by an infinite conductor or by infinite parallel dielectric layers. This study of transverse electromagnetic waves leads quite naturally to transmission lines, which seem as important as ever. Consequently, Chaps. 4, 5, and 6 cover transients on lossless lines, sinusoids on lossless lines, and sinusoids on dissipative lines, respectively. In these chapters, transmission-line phenomena are derived from electromagnetic theory rather than from an extension of circuit theory, thus providing a better understanding of the relationship between guided transverse electromagnetic waves and voltage and current. Connections with circuit theory are further strengthened by alternate derivations of some of the transmission-line equations, which are assigned in the problems for Chaps. 4 to 6.

The computational power of the digital computer has been brought to bear on a wide variety of transmission-line problems through the use of the FORTRAN program SANTLINE. A number of subroutines have been incorporated, in an attempt to minimize the programming effort required for a user of SANTLINE. The program and the various subroutines available to it are explained in detail in Sec. 5.9, with examples illustrating its use in Sec. 5.10 and 6.5. The subroutines have been chosen so that, for a small problem, the student can follow each step manually, with the aid of a Smith chart, for better understanding. To use SANTLINE, a background in basic FORTRAN is helpful but not essential. The SANTLINE program provides a means of analyzing and synthesizing transmission lines considering bandwidth requirements, and it simultaneously computes the voltage, current, and power distribution on multiple series-parallel connected lines. Thus some commonly used systems concepts can be applied to this class of problem. It is by no means necessary to teach the material on the use of SANTLINE in order to use the remainder of the text, as SANTLINE is not a prerequisite to any other material in the book. Thus, if no suitable computer is available or if the reprogramming effort is too great, ample material remains for a one-semester course at the junior or senior level.

In Chap. 7, transverse electric and transverse magnetic waves are introduced. They are derived, for parallel-plane and rectangular waveguides, from the work done in Sec. 3.7 on the oblique incidence of a plane wave on a perfect conductor. Thus transverse electric and transverse magnetic waves are initially presented as a reflection phenomenon, and this fact is emphasized in some of the early equations by the presence of the angle of incidence and by an equation relating the frequency and the angle of incidence. To conclude this chapter on guided waves (and the book), the modes of oscillation of simple cavities having a uniform cross section are considered.

The sets of problems at the end of each chapter are arranged in the same order as the topics presented in the associated chapter, and generally they become progressively more difficult throughout a problem set.

Included throughout the text at various points are programmed exercises, usually a discussion of a difficult practical problem followed by a series of carefully reasoned statements which the student should verify step by step or a series of instructions for the student to follow. The student should go through these exercises carefully to test and reinforce his concepts.

The appendixes include a set of reference tables, program listings for the SANTLINE computer package, and two types of transmission-line charts.

Donald W. Dearholt
William R. McSpadden

1
Fundamentals of Fields

1.1 INTRODUCTION

Electromagnetic field theory is one of the most fundamental subjects studied in the electrical engineering curriculum today. While there are many immediate and important applications of the theory, a primary advantage to many students is the understanding and insight into electrical phenomena it affords. Although the study is not necessarily an easy one, the student will find it very rewarding as he progresses from the most elementary concepts of fields and the development of Maxwell's equations, to the concepts of wave propagation and transmission lines, and then to waveguides and their applications.

Vector analysis is no stranger to the junior or senior electrical engineering student, although it may not be considered a friend. It is the basic language used to describe electromagnetic fields, and only through repeated exposure do we become proficient with it. From his prior work, a student is generally well prepared to handle the elementary operations of vector analysis, such as addition and subtraction of vectors and scalar and vector products (the dot product and the cross product). The junior or senior student has generally been exposed to the rectangular, cylindrical, and spherical coordinate systems, to transformations between coordinate systems, to the gradient, divergence, and curl, and often to more advanced concepts of vector analysis. Normally his exposure has not been adequate for him to feel comfortable with these concepts, although he may be able to perform the mathematical manipulations. Consequently in Chap. 1 we concentrate on those topics which often prove troublesome to the student, such as the curl. In Chap. 1 the student will also be introduced to some material that is likely to be new to him, such as Helmholtz' theorem and the special theory of relativity. Upon completion of Chap. 1, we will be able to summarize the study of electromagnetic fields and Maxwell's equations for those fields.

Throughout this text the International (or rationalized mks) System of Units is used. Table A.1.1 in Appendix A provides a summary of these units and abbreviations.

1.2 COORDINATE SYSTEMS AND TRANSFORMATIONS

THE RECTANGULAR COORDINATE SYSTEM

The rectangular, or cartesian, coordinate system consists of three mutually orthogonal axes† whose variables are denoted x, y, and z. Any point (x,y,z) in the three-dimensional space can be uniquely located by the values of the variables. The notation $\mathbf{a}_x$, $\mathbf{a}_y$, and $\mathbf{a}_z$ is reserved for unit vectors, as shown in Fig. 1.2.1.

From an abstract mathematical point of view, a vector $\mathbf{R}$ is associated with a single point. It is not a line between two points, although vectors are often introduced in elementary texts in that manner. Fortunately, the concept of a vector being associated with a single point is ideal in physical applications. For example, we consider a force on an electron as a vector $\mathbf{R}$ acting at a particular point (i.e., the point where the electron is located) in three-dimensional space.

A vector $\mathbf{R}$ at the point (x,y,z) can be written as the vector sum of three component vectors as

$$\mathbf{R}(x,y,z) = R_x \mathbf{a}_x + R_y \mathbf{a}_y + R_z \mathbf{a}_z \tag{1}$$

The magnitude, or length, of this vector is

$$R = (R_x^2 + R_y^2 + R_z^2)^{1/2} \tag{2}$$

A unit vector $\mathbf{a}_R$ in the direction of $\mathbf{R}$ can be found by dividing the vector

† All the coordinate systems used in this text will be right-handed orthogonal systems.

FUNDAMENTALS OF FIELDS

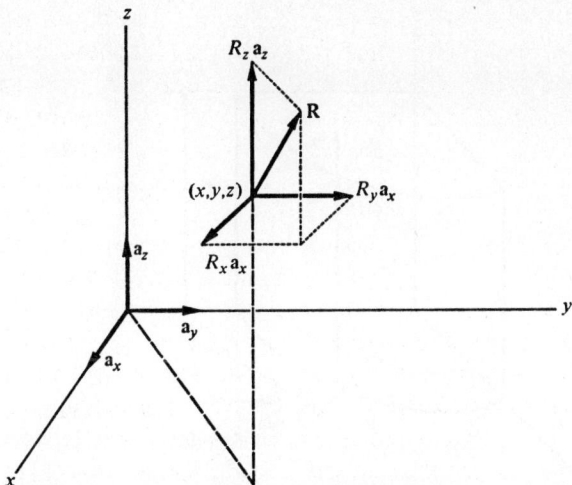

Fig. 1.2.1 The rectangular coordinate system.

by its magnitude:

$$\mathbf{a}_R = \frac{\mathbf{R}}{R} \tag{3}$$

The components R_x, R_y, and R_z of the vector usually are functions of the variables x, y, and z. The components have the same dimensional units as the vector **R**, and all the unit vectors are dimensionless.

It is assumed that the reader is familiar with the elementary properties of vectors and vector operations. Consequently, we are largely concerned with establishing a common terminology for use throughout the remainder of the text.

THE CYLINDRICAL COORDINATE SYSTEM

Any point in the cylindrical coordinate system is located by the intersection of three mutually orthogonal surfaces, which are determined by:

1. A constant radial distance r about the z axis, i.e., a cylinder
2. A constant angle ϕ, measured from the x axis in the xy plane, i.e., a vertical plane
3. A plane z = constant, i.e., a plane perpendicular to the z axis

The orthogonal unit vectors $\mathbf{a}_r$, $\mathbf{a}_\phi$, and $\mathbf{a}_z$ are normal to the surfaces at the point (r,ϕ,z), as shown in Fig. 1.2.2. At any point (r,ϕ,z) the unit vectors point in the directions in which the corresponding variables

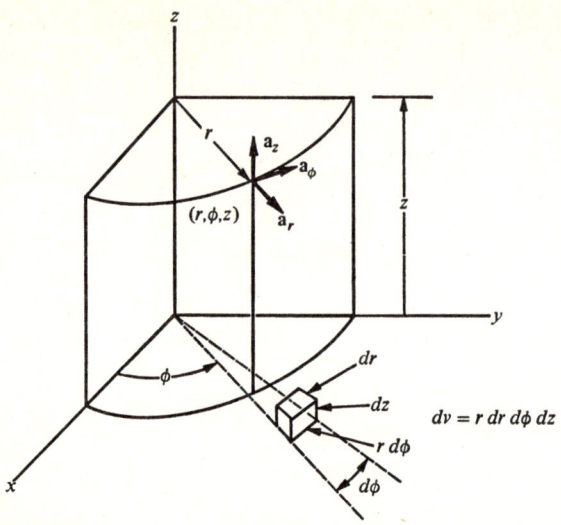

Fig. 1.2.2 The cylindrical coordinate system.

increase. Thus $\mathbf{a}_r$ points in the direction of increasing radius, $\mathbf{a}_\phi$ in the direction of increasing angle ϕ, and $\mathbf{a}_z$ in the positive z direction. An incremental volume element in cylindrical coordinates is given by

$$dv = r\, dr\, d\phi\, dz \tag{4}$$

Any vector $\mathbf{R}$ can be written in terms of the unit vectors and a set of components R_r, R_ϕ, and R_z:

$$\mathbf{R}(r,\phi,z) = R_r \mathbf{a}_r + R_\phi \mathbf{a}_\phi + R_z \mathbf{a}_z \tag{5}$$

The magnitude of $\mathbf{R}$ is

$$R = (R_r^2 + R_\phi^2 + R_z^2)^{1/2} \tag{6}$$

Thus a unit vector in the direction of $\mathbf{R}$ can be found by dividing the vector by its magnitude. In general the components R_r, R_ϕ, and R_z will each be functions of the variables r, ϕ, and z.

THE TRANSFORMATION FROM RECTANGULAR TO CYLINDRICAL COORDINATES

Often a great deal of labor can be saved if one can choose a coordinate system that "fits" the symmetries of the particular problem. Expressions that appear very complicated in one system may be quite simple when expressed in the "natural" coordinate system of the problem. However, experience is probably the best guide in choosing the natural coordinate system. Often it is convenient to set up a problem in terms

FUNDAMENTALS OF FIELDS

of rectangular coordinates and then transform it to the system that is geometrically most appropriate. Thus one must know techniques for transforming vectors from one system to another. If we are given a vector, such as a force on an electron, it has a magnitude and direction which must be the same in any coordinate system. Thus we can represent a vector **R** in either rectangular or cylindrical coordinates:

$$\mathbf{R} = R_x\mathbf{a}_x + R_y\mathbf{a}_y + R_z\mathbf{a}_z = R_r\mathbf{a}_r + R_\phi\mathbf{a}_\phi + R_z\mathbf{a}_z \tag{7}$$

If we know one set of components, say R_x, R_y, and R_z, and wish to find the other, we multiply (dot product) by the unit vectors of the unknown set. For example, from Eq. (7),

$$\mathbf{R}\cdot\mathbf{a}_r = R_x\mathbf{a}_x\cdot\mathbf{a}_r + R_y\mathbf{a}_y\cdot\mathbf{a}_r + R_z\mathbf{a}_z\cdot\mathbf{a}_r = R_r \tag{8}$$

$$\mathbf{R}\cdot\mathbf{a}_\phi = R_x\mathbf{a}_x\cdot\mathbf{a}_\phi + R_y\mathbf{a}_y\cdot\mathbf{a}_\phi + R_z\mathbf{a}_z\cdot\mathbf{a}_\phi = R_\phi \tag{9}$$

$$\mathbf{R}\cdot\mathbf{a}_z = R_x\mathbf{a}_x\cdot\mathbf{a}_z + R_y\mathbf{a}_y\cdot\mathbf{a}_z + R_z\mathbf{a}_z\cdot\mathbf{a}_z = R_z \tag{10}$$

These equations can be expressed in a more convenient form using matrices†

$$[R_x \ R_y \ R_z] \underbrace{\begin{bmatrix} \mathbf{a}_x\cdot\mathbf{a}_r & \mathbf{a}_x\cdot\mathbf{a}_\phi & \mathbf{a}_x\cdot\mathbf{a}_z \\ \mathbf{a}_y\cdot\mathbf{a}_r & \mathbf{a}_y\cdot\mathbf{a}_\phi & \mathbf{a}_y\cdot\mathbf{a}_z \\ \mathbf{a}_z\cdot\mathbf{a}_r & \mathbf{a}_z\cdot\mathbf{a}_\phi & \mathbf{a}_z\cdot\mathbf{a}_z \end{bmatrix}}_{[T_{RC}]} = [R_r \ R_\phi \ R_z] \tag{11}$$

We will call the 3 × 3 matrix the *transformation matrix* $[T_{RC}]$, where the subscripts RC designate a transformation from rectangular to cylindrical coordinates. The transformation matrix is independent of the particular vector being transformed and depends only on the particular coordinate systems. Thus it remains the same throughout any problem, regardless of how many vectors we may want to transform.

The terms within the $[T_{RC}]$ matrix can be simplified by considering each product. For example, from Fig. 1.2.3 we can see that

$$\mathbf{a}_x\cdot\mathbf{a}_r = \cos\phi \qquad \mathbf{a}_y\cdot\mathbf{a}_r = \sin\phi \qquad \mathbf{a}_z\cdot\mathbf{a}_r = 0$$

Similarly we find the other terms of the $[T_{RC}]$ matrix.

$$[T_{RC}] = \begin{bmatrix} \cos\phi & -\sin\phi & 0 \\ \sin\phi & \cos\phi & 0 \\ 0 & 0 & 1 \end{bmatrix} \tag{12}$$

† Matrix multiplication is a row-by-column multiplication. For example,

$$AB = \begin{bmatrix} a_{11} & a_{12} \\ a_{21} & a_{22} \end{bmatrix} \begin{bmatrix} b_{11} & b_{12} \\ b_{21} & b_{22} \end{bmatrix} = \begin{bmatrix} a_{11}b_{11} + a_{12}b_{21} & a_{11}b_{12} + a_{12}b_{22} \\ a_{21}b_{11} + a_{22}b_{21} & a_{21}b_{12} + a_{22}b_{22} \end{bmatrix} \neq BA$$

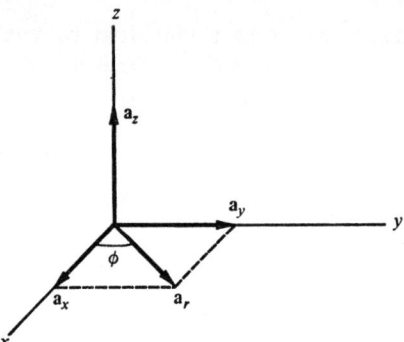

Fig. 1.2.3 The unit vectors.

Consequently the components of the transformed vector can be found, as in Eq. (11),

$$[R_r \quad R_\phi \quad R_z] = [R_x \quad R_y \quad R_z][T_{RC}] \tag{13}$$

In general the relationship is

[new coordinates] = [old coordinates][transformation matrix] (14)

The transformation matrix of Eq. (12) is a function only of the angle ϕ, because just a rotation (no translation) is required to transform from rectangular to cylindrical coordinates if both systems have the same origin.

The process can be streamlined further by postmultiplying Eq. (13) by a column matrix of the unit vectors of the cylindrical system. One then obtains the desired vector directly,

$$\mathbf{R}(r,\phi,z) = R_r \mathbf{a}_r + R_\phi \mathbf{a}_\phi + R_z \mathbf{a}_z = [R_x \quad R_y \quad R_z][T_{RC}] \begin{bmatrix} \mathbf{a}_r \\ \mathbf{a}_\phi \\ \mathbf{a}_z \end{bmatrix} \tag{15}$$

by substituting the particular values of R_x, R_y, and R_z for the given vector. However, the beginner may prefer to use Eq. (13) until he develops some confidence with the technique.

In substituting R_x, R_y, and R_z into Eqs. (13) or (15), one must replace the variables x, y, and z by their corresponding variables in the cylindrical system. These are

$$x = r \cos \phi \quad y = r \sin \phi \quad z = z \tag{16}$$

as can be seen from a study of the geometry of Fig. 1.2.2. The substitutions can be made either before or after the matrix operations are performed.

As an example, suppose we transform the vector $\mathbf{R}(x,y,z)$ to cylindrical coordinates:

$$\mathbf{R}(x,y,z) = x\mathbf{a}_x + y\mathbf{a}_y + z\mathbf{a}_z \tag{17}$$

FUNDAMENTALS OF FIELDS

This is a vector from the origin to the point (x,y,z), and is called a *position vector*. It transforms as follows:

$$\mathbf{R}(r,\phi,z) = [x \quad y \quad z][T_{RC}]\begin{bmatrix} \mathbf{a}_r \\ \mathbf{a}_\phi \\ \mathbf{a}_z \end{bmatrix}$$

$$= [r\cos\phi \quad r\sin\phi \quad z][T_{RC}]\begin{bmatrix} \mathbf{a}_r \\ \mathbf{a}_\phi \\ \mathbf{a}_z \end{bmatrix}$$

$$= [r\cos\phi \quad r\sin\phi \quad z]\begin{bmatrix} \cos\phi & -\sin\phi & 0 \\ \sin\phi & \cos\phi & 0 \\ 0 & 0 & 1 \end{bmatrix}\begin{bmatrix} \mathbf{a}_r \\ \mathbf{a}_\phi \\ \mathbf{a}_z \end{bmatrix}$$

$$= [r(\cos^2\phi + \sin^2\phi) \quad (-r+r)(\cos\phi\sin\phi) \quad z]\begin{bmatrix} \mathbf{a}_r \\ \mathbf{a}_\phi \\ \mathbf{a}_z \end{bmatrix} \quad (18)$$

$$\mathbf{R}(r,\phi,z) = r\mathbf{a}_r + 0\mathbf{a}_\phi + z\mathbf{a}_z \quad (19)$$

This vector is equivalent to $\mathbf{R}(x,y,z)$ but is expressed in cylindrical coordinates.

A word of caution is in order here. We must carefully distinguish between the *location of a point* and *a vector at that point*. One cannot correctly conclude from Eq. (19) that $\phi = 0$ because $R_\phi = 0$. Equation (19) tells us that a vector $\mathbf{R}(r,\phi,z)$, corresponding to the original vector $\mathbf{R}(x,y,z)$, is the sum of its two vector components $r\mathbf{a}_r$ and $z\mathbf{a}_z$ and that it has no component in the $\mathbf{a}_\phi$ direction. This will be true for any arbitrary point we may choose, as shown in Fig. 1.2.4. The point (r,ϕ,z) can be located, but not from Eq. (19). From Eqs. (16) we can solve for

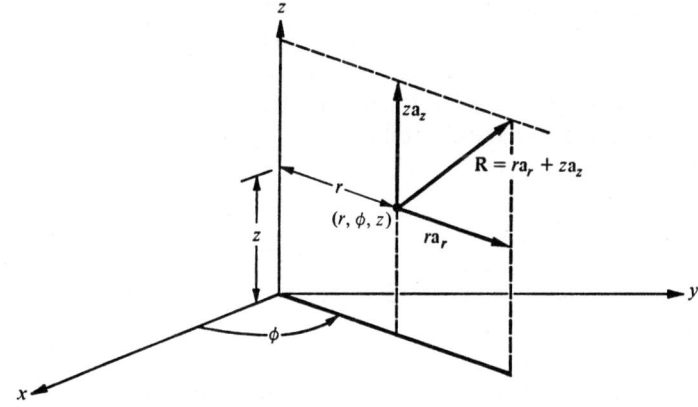

Fig. 1.2.4 The vector components of $\mathbf{R}(r,\phi,z)$.

r, ϕ, and z:

$$r = (x^2 + y^2)^{1/2} \qquad \phi = \tan^{-1}\frac{y}{x} \qquad z = z \tag{20}$$

Now the point (r,ϕ,z) can be located if we know x, y, and z.

To continue the above example, suppose we choose the point $(x = 1, y = 1, z = 1)$ which is the same as the point $(r = \sqrt{2}, \phi = 45°, z = 1)$. The corresponding vectors are

$$\mathbf{R}(x,y,z) = \mathbf{R}(1,1,1) = 1\mathbf{a}_x + 1\mathbf{a}_y + 1\mathbf{a}_z \tag{21}$$

$$\mathbf{R}(r,\phi,z) = \mathbf{R}(\sqrt{2},45°,1) = \sqrt{2}\,\mathbf{a}_r + 1\mathbf{a}_z \tag{22}$$

Both vectors have the same magnitude and direction. Alternately we could transform the vector $\mathbf{R}(1,1,1)$ from Eq. (18):

$$\mathbf{R}(r,\phi,z) = [1 \quad 1 \quad 1][T_{RC}]\begin{bmatrix}\mathbf{a}_r \\ \mathbf{a}_\phi \\ \mathbf{a}_z\end{bmatrix} \tag{23}$$

$$\mathbf{R}(r,\phi,z) = (\cos\phi + \sin\phi)\mathbf{a}_r + (\cos\phi - \sin\phi)\mathbf{a}_\phi + 1\mathbf{a}_z \tag{24}$$

The angle ϕ is not arbitrary here, since x and y have been specified. When we substitute $\phi = 45°$ in Eq. (24), we obtain Eq. (22), as we darkly suspected we might.

We have just developed a technique for *transforming* a single vector from the rectangular coordinate system to the cylindrical system. The vector remains located at the same point in space in either system. In many problems we are also concerned with the process of *translation*, whereby a free vector is moved from one point to another in the same coordinate system. When both transformation and translation are required, there is usually an "easy" way and a "hard" way to proceed.†
We illustrate both procedures.

Suppose that we are given the position vectors $\mathbf{R}_1$ and $\mathbf{R}_2$ at points $p_1 = (1,0,0)$ and $p_2 = (0,1,0)$ (in rectangular coordinates) and we want to find the vector $\mathbf{R}$ at p_1, as in Fig. 1.2.5, in cylindrical coordinates. Two possible procedures follow.

Procedure 1

1. Express $\mathbf{R}_1$ and $\mathbf{R}_2$ in rectangular coordinates.
2. *Translate* the vectors to point p_1.
3. Compute their difference: $\mathbf{R} = \mathbf{R}_2 - \mathbf{R}_1$.
4. *Transform* $\mathbf{R}$ to cylindrical coordinates using $[T_{RC}]$.

† In many problems "the law of conservation of difficulty" holds, wherein simplicity in one phase of the work causes increased difficulty in another. Here we see an exception to the rule.

FUNDAMENTALS OF FIELDS

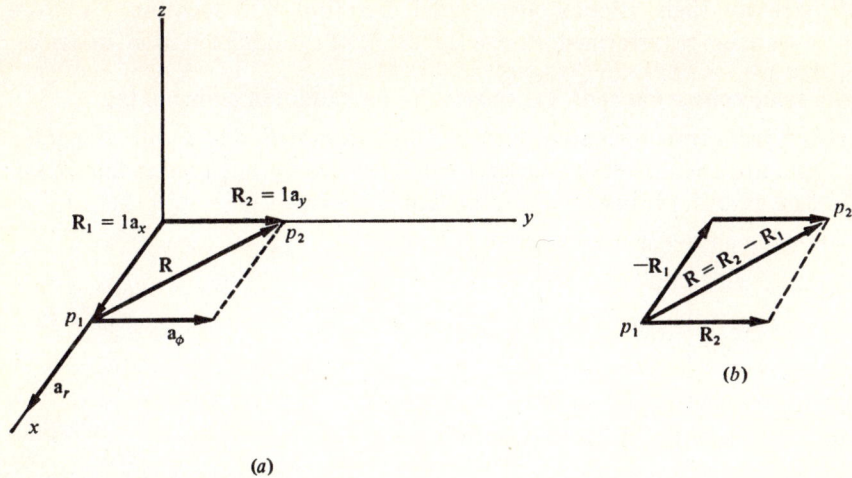

Fig. 1.2.5 Vectors and translations.

Procedure 2
1. Express R_1 and R_2 in rectangular coordinates.
2. *Transform* R_1 and R_2 to cylindrical coordinates using $[T_{RC}]$.
3. *Translate* R_1 and R_2 to point p_1.
4. Compute their difference: $R = R_2 - R_1$.

Procedure 1 is the easier of the two because (1) the translation process is done in the rectangular system, where it is simple and automatic, and (2) only one vector is transformed. However, since some subtle points arise, the student should use each procedure to solve the problem given above. The correct answer is $R(1,0°,0) = -a_r + a_\phi$.

We can draw the following conclusions from these derivations and examples:

1. A vector *location* is identified by three variables (x, y, z, or r, ϕ, z) which, in general, do not have the same numerical values as the *components* (R_x, R_y, R_z, or R_r, R_ϕ, R_z) of the vector.
2. The respective components of two vectors cannot be added or subtracted unless the vectors are located at the same point.
3. Vectors can be *translated* to a common point, but the components change under translation in any system except the rectangular one. Vectors retain their original magnitude and direction under translation.
4. Vectors which are transformed from one coordinate system to another

retain their original magnitude, direction, and location. Vectors can be transformed efficiently using the transformation matrices.

THE TRANSFORMATION FROM CYLINDRICAL TO RECTANGULAR COORDINATES

The reverse transformation matrix $[T_{CR}]$, from cylindrical to rectangular coordinates, can be derived by the same process as that used to find $[T_{RC}]$. In this case the transformation matrix $[T_{CR}]$ is†

$$[T_{CR}] = \begin{bmatrix} \frac{x}{r} & \frac{y}{r} & 0 \\ -\frac{y}{r} & \frac{x}{r} & 0 \\ 0 & 0 & 1 \end{bmatrix} \tag{25}$$

and the variable substitutions are

$$r = (x^2 + y^2)^{1/2} \qquad \phi = \tan^{-1}\frac{y}{x} \qquad z = z \tag{26}$$

Thus to transform a vector $\mathbf{R}(r,\phi,z)$ to $\mathbf{R}(x,y,z)$ in rectangular coordinates, one may use

$$\mathbf{R}(x,y,z) = [R_r \quad R_\phi \quad R_z][T_{CR}]\begin{bmatrix} \mathbf{a}_x \\ \mathbf{a}_y \\ \mathbf{a}_z \end{bmatrix} \tag{27}$$

The $[T_{CR}]$ matrix is independent of the particular vector being transformed, and so one need only substitute for the components R_r, R_ϕ, and R_z.

THE SPHERICAL COORDINATE SYSTEM

Any point in the spherical coordinate system is determined by the intersection of three mutually orthogonal surfaces (but not by the same surfaces as those of the cylindrical coordinate system). These surfaces are shown in Fig. 1.2.6, and are determined by:

1. A constant radial distance about the origin, i.e., a sphere
2. A constant angle θ measured from the z axis, i.e., a vertical cone
3. A constant angle ϕ measured from the x axis, i.e., a vertical plane

Any point p in the three-dimensional space can be written in terms of the variables r, θ, and ϕ as (r,θ,ϕ). The unit vectors, denoted $\mathbf{a}_r$, $\mathbf{a}_\theta$, and $\mathbf{a}_\phi$, point in the direction of increasing values of the variables. The

---

† The "old" variable r is included in the transformation matrix for convenience in writing the matrix. Eventually it must be replaced by the "new" variable $r = (x^2 + y^2)^{1/2}$, as in Eq. (26).

FUNDAMENTALS OF FIELDS

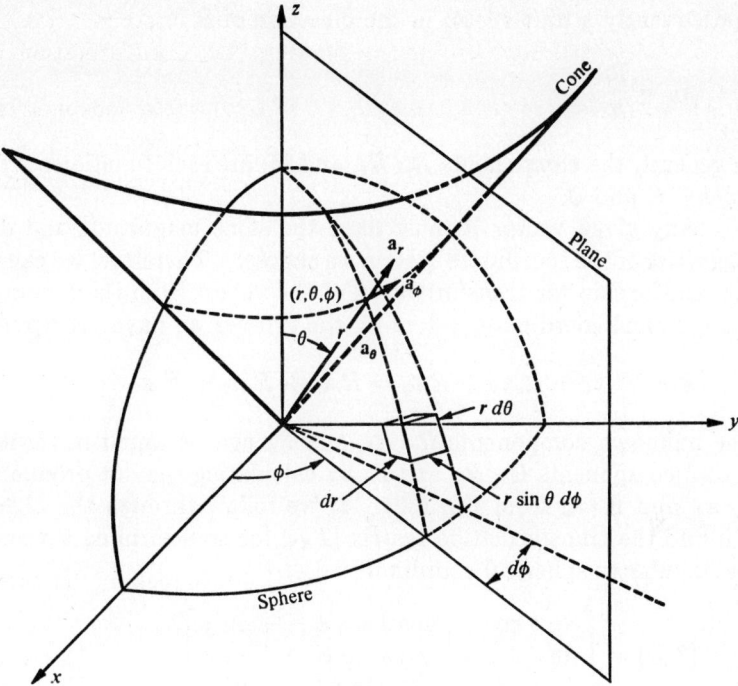

Fig. 1.2.6 The spherical coordinate system.

variable r is *not* the same in this system as in the cylindrical system, since here a constant r determines a spherical surface rather than a cylindrical one. The variable ϕ is the same in either system, but it appears as the third variable of (r,θ,ϕ) rather than as the second variable, as in the cylindrical system. Because of the way of measuring r, θ, and ϕ we have chosen, ϕ appears as the third variable in order to maintain a right-hand system; that is, $\mathbf{a}_r \times \mathbf{a}_\theta = \mathbf{a}_\phi$.

We now state some of the properties of the spherical coordinate system. The volume element, as shown in Fig. 1.2.6, is

$$dv = r^2 \sin\theta \, dr \, d\theta \, d\phi \tag{28}$$

Any vector in the three-dimensional space can be expressed as

$$\mathbf{R} = R_r \mathbf{a}_r + R_\theta \mathbf{a}_\theta + R_\phi \mathbf{a}_\phi \tag{29}$$

and has a magnitude of

$$R = (R_r^2 + R_\theta^2 + R_\phi^2)^{1/2} \tag{30}$$

Consequently a unit vector in the direction of **R** is

$$\mathbf{a}_R = \frac{\mathbf{R}}{R} \tag{31}$$

In general, the components R_r, R_θ, and R_ϕ are each functions of the variables r, θ, and ϕ.

Any given vector **R** must have the same magnitude and direction regardless of the coordinate system we choose. Therefore, we can develop the relationship for transformation of the vector from the rectangular to the spherical coordinate system by the process we have used previously:

$$\mathbf{R} = R_x\mathbf{a}_x + R_y\mathbf{a}_y + R_z\mathbf{a}_z = R_r\mathbf{a}_r + R_\theta\mathbf{a}_\theta + R_\phi\mathbf{a}_\phi \tag{32}$$

The unknown components R_r, R_θ, and R_ϕ can be found in terms of the known components R_x, R_y, and R_z by calculating the dot products $\mathbf{R} \cdot \mathbf{a}_r$, $\mathbf{R} \cdot \mathbf{a}_\theta$, and $\mathbf{R} \cdot \mathbf{a}_\phi$ from Eq. (32). If we follow through the algebra, we will find the transformation matrix $[T_{RS}]$ for transforming a vector from rectangular to spherical coordinates:

$$[T_{RS}] = \begin{bmatrix} \sin\theta\cos\phi & \cos\theta\cos\phi & -\sin\phi \\ \sin\theta\sin\phi & \cos\theta\sin\phi & \cos\phi \\ \cos\theta & -\sin\theta & 0 \end{bmatrix} \tag{33}$$

The transformation is accomplished by the matrix multiplication

$$[R_r \quad R_\theta \quad R_\phi] = [R_x \quad R_y \quad R_z][T_{RS}] \tag{34}$$

As before, we can write the vector **R** directly by postmultiplying by the unit vector matrix:

$$\mathbf{R}(r,\theta,\phi) = [R_r \quad R_\theta \quad R_\phi]\begin{bmatrix} \mathbf{a}_r \\ \mathbf{a}_\theta \\ \mathbf{a}_\phi \end{bmatrix} \tag{35}$$

The change in variables is accomplished by substituting the following values for x, y, and z:

$$x = r\sin\theta\cos\phi \quad y = r\sin\theta\sin\phi \quad z = r\cos\theta \tag{36}$$

To illustrate the transformation, let us find the vector $\mathbf{R}(r,\theta,\phi)$ corresponding to the position vector $\mathbf{R}(x,y,z)$ of

$$\mathbf{R}(x,y,z) = x\mathbf{a}_x + y\mathbf{a}_y + z\mathbf{a}_z \tag{37}$$

$$\mathbf{R}(r,\theta,\phi) = [x \quad y \quad z][T_{RS}]\begin{bmatrix} \mathbf{a}_r \\ \mathbf{a}_\theta \\ \mathbf{a}_\phi \end{bmatrix} \tag{38}$$

FUNDAMENTALS OF FIELDS

$$R(r,\theta,\phi) = [r \sin \theta \cos \phi \quad r \sin \theta \sin \phi \quad r \cos \theta]$$

$$\begin{bmatrix} \sin \theta \cos \phi & \cos \theta \cos \phi & -\sin \phi \\ \sin \theta \sin \phi & \cos \theta \sin \phi & \cos \phi \\ \cos \theta & -\sin \theta & 0 \end{bmatrix} \begin{bmatrix} a_r \\ a_\theta \\ a_\phi \end{bmatrix}$$

$$= [r \sin^2 \theta(\sin^2 \phi + \cos^2 \phi) + r \cos^2 \theta$$

$$r \sin \theta \cos \theta(\cos^2 \phi + \sin^2 \phi) - r \cos \theta \sin \theta$$

$$-r \sin \theta \sin \phi \cos \phi + r \sin \theta \sin \phi \cos \phi] \begin{bmatrix} a_r \\ a_\theta \\ a_\phi \end{bmatrix} \quad (39)$$

$$R(r,\theta,\phi) = [r \quad 0 \quad 0] \begin{bmatrix} a_r \\ a_\theta \\ a_\phi \end{bmatrix} = r a_r \quad (40)$$

Thus the vector $R(x,y,z)$ becomes simply $r a_r$ in spherical coordinates.

As with all other coordinate systems, we must take care not to confuse the location of a point (r,θ,ϕ) with the components R_r, R_θ, and R_ϕ of the vector at that point. Also we cannot add two vectors in spherical coordinates unless they are located at the same point.

The transformation from spherical to rectangular coordinates can be accomplished by the same method. This transformation matrix $[T_{SR}]$ is given in Table 1.2.1, along with the other transformation matrices and variable changes of this section.

ORTHOGONAL CURVILINEAR COORDINATES

Although there are many other coordinate systems that fit special geometries,[†] such as the paraboloidal and toroidal systems, we will not be concerned with them in this text. However, we will summarize some of the properties that are common to all orthogonal systems by introducing a generalized curvilinear coordinate system.

The orthogonal curvilinear coordinates, denoted u, v, and w, are the three variables of any three-dimensional orthogonal coordinate system; for example, $u = r$, $v = \theta$, and $w = \phi$ for the spherical system. The corresponding orthogonal unit vectors are a_u, a_v, and a_w. Then a vector in the curvilinear coordinate system is

$$R(u,v,w) = R_u a_u + R_v a_v + R_w a_w \quad (41)$$

The dimensional units of R_u, R_v, and R_w are the same as those of the vector R, and the unit vectors are dimensionless.

In general the variables u, v, and w, and their differentials du, dv, and dw, are *not* distances in the three-dimensional space. The exception

[†] P. Moon and D. Spencer, "Field Theory for Engineers," D. Van Nostrand Company, Inc., Princeton, N.J., 1961.

ELECTROMAGNETIC WAVE PROPAGATION

Table 1.2.1 Vector transformations between coordinate systems

Given a vector $\mathbf{R}(u_1,v_1,w_1)$ with components R_{u_1}, R_{v_1}, and R_{w_1} in an "old" system, the transformed vector $\mathbf{R}(u_2,v_2,w_2)$ in the "new" system is given by

$$\mathbf{R}(u_2,v_2,w_2) = [R_{u_1} \quad R_{v_1} \quad R_{w_1}][T_{12}]\begin{bmatrix} a_{u_2} \\ a_{v_2} \\ a_{w_2} \end{bmatrix}$$

where $[T_{12}]$ is the transformation matrix from system 1 to system 2. The components R_{u_1}, R_{v_1}, and R_{w_1} must be expressed in terms of the new variables u_2, v_2, and w_2.

Variable substitutions and transformation matrices

Rectangular to cylindrical	Cylindrical to rectangular
$x = r\cos\phi,\ y = r\sin\phi,\ z = z$	$r = (x^2 + y^2)^{1/2},\ \phi = \tan^{-1}\left(\dfrac{y}{x}\right),\ z = z$
$[T_{RC}] = \begin{bmatrix} \cos\phi & -\sin\phi & 0 \\ \sin\phi & \cos\phi & 0 \\ 0 & 0 & 1 \end{bmatrix}$	$[T_{CR}] = \begin{bmatrix} \dfrac{x}{r} & \dfrac{y}{r} & 0 \\ -\dfrac{y}{r} & \dfrac{x}{r} & 0 \\ 0 & 0 & 1 \end{bmatrix}$

Rectangular to spherical	Spherical to rectangular
$x = r\sin\theta\cos\phi$ $y = r\sin\theta\sin\phi$ $z = r\cos\theta$	$r = (x^2 + y^2 + z^2)^{1/2}$ $\theta = \cos^{-1}\dfrac{z}{r}$ $\phi = \tan^{-1}\dfrac{y}{x}$
$[T_{RS}] = \begin{bmatrix} \sin\theta\cos\phi & \cos\theta\cos\phi & -\sin\phi \\ \sin\theta\sin\phi & \cos\theta\sin\phi & \cos\phi \\ \cos\theta & -\sin\theta & 0 \end{bmatrix}$	$[T_{SR}] = \begin{bmatrix} \dfrac{x}{r} & \dfrac{y}{r} & \dfrac{z}{r} \\ \dfrac{xz}{r\alpha} & \dfrac{yz}{r\alpha} & -\dfrac{\alpha}{r} \\ -\dfrac{y}{\alpha} & \dfrac{x}{\alpha} & 0 \end{bmatrix}$ where $\alpha = (x^2 + y^2)^{1/2}$

Cylindrical to spherical	Spherical to cylindrical
r_c = radial variable in cylindrical system	r_s = radial variable in spherical system
$r_s = (r_c^2 + z^2)^{1/2}$	$r_c = r_s \sin\theta$
$\theta = \cos^{-1}\dfrac{z}{r_s}$	$\phi = \phi$
$\phi = \phi$	$z = r_s \cos\theta$
$[T_{CS}] = \begin{bmatrix} \sin\theta & \cos\theta & 0 \\ 0 & 0 & 1 \\ \cos\theta & -\sin\theta & 0 \end{bmatrix}$	$[T_{SC}] = \begin{bmatrix} \dfrac{r_c}{r_s} & 0 & \dfrac{z}{r_s} \\ \dfrac{z}{r_s} & 0 & -\dfrac{r_c}{r_s} \\ 0 & 1 & 0 \end{bmatrix}$

FUNDAMENTALS OF FIELDS

is the rectangular system, where x, y, and z represent distances. A differential arc length is

$$ds = [(dx)^2 + (dy)^2 + (dz)^2]^{1/2} \tag{42}$$

By replacing dx, dy, and dz by their total differentials in terms of the curvilinear variables, for example

$$dx = \frac{\partial x}{\partial u} du + \frac{\partial x}{\partial v} dv + \frac{\partial x}{\partial w} dw$$

the reader can show† that the corresponding arc length in the curvilinear system is

$$ds = [(h_1 \, du)^2 + (h_2 \, dv)^2 + (h_3 \, dw)^2]^{1/2} \tag{43}$$

The parameters h_1, h_2, and h_3, called the *scale factors*, are important. They are also called the *metric coefficients*, although some authors reserve that term for the squares of the scale factors. They are given by the equations

$$h_1 = \left[\left(\frac{\partial x}{\partial u}\right)^2 + \left(\frac{\partial y}{\partial u}\right)^2 + \left(\frac{\partial z}{\partial u}\right)^2\right]^{1/2}$$

$$h_2 = \left[\left(\frac{\partial x}{\partial v}\right)^2 + \left(\frac{\partial y}{\partial v}\right)^2 + \left(\frac{\partial z}{\partial v}\right)^2\right]^{1/2}$$

$$h_3 = \left[\left(\frac{\partial x}{\partial w}\right)^2 + \left(\frac{\partial y}{\partial w}\right)^2 + \left(\frac{\partial z}{\partial w}\right)^2\right]^{1/2} \tag{44}$$

Thus if one knows the relationship between the rectangular coordinate variables x, y, z and the u, v, w variables, one can compute the scale factors. For example, we found that for the spherical system,

$$x = r \sin \theta \cos \phi$$
$$y = r \sin \theta \sin \phi$$
$$z = r \cos \theta \tag{45}$$

and so the corresponding scale factors, from Eq. (44), are

$$h_1 = 1$$
$$h_2 = r$$
$$h_3 = r \sin \theta \tag{46}$$

Differential distances are then $h_1 \, du$, $h_2 \, dv$, and $h_3 \, dw$, and a differential

† Interpreted literally, this means the problem is long and tedious and we do not want to do it.

arc length is

$$ds = [(dr)^2 + (r\,d\theta)^2 + (r\sin\theta\,d\phi)^2]^{1/2} \tag{47}$$

When the differential distances are known, one can determine the differential vector and the surface and volume elements used in integrals

$$d\mathbf{l} = h_1\,du\,\mathbf{a}_u + h_2\,dv\,\mathbf{a}_v + h_3\,dw\,\mathbf{a}_w \quad \text{vector element} \tag{48}$$

$$dv = h_1 h_2 h_3\,du\,dv\,dw \quad \text{volume element} \tag{49}$$

Table 1.2.2 Divergence, gradient, and curl in curvilinear coordinates

Coordinate system	Variables			Unit vectors			Scale factors		
	u	v	w	$\mathbf{a}_u$	$\mathbf{a}_v$	$\mathbf{a}_w$	h_1	h_2	h_3
Cartesian	x	y	z	$\mathbf{a}_x$	$\mathbf{a}_y$	$\mathbf{a}_z$	1	1	1
Cylindrical	r	ϕ	z	$\mathbf{a}_r$	$\mathbf{a}_\phi$	$\mathbf{a}_z$	1	r	1
Spherical	r	θ	ϕ	$\mathbf{a}_r$	$\mathbf{a}_\theta$	$\mathbf{a}_\phi$	1	r	$r\sin\theta$

Vector: $\mathbf{A} = A_u\mathbf{a}_u + A_v\mathbf{a}_v + A_w\mathbf{a}_w$ \hfill (1)

Differential vector: $d\mathbf{l} = h_1\,du\,\mathbf{a}_u + h_2\,dv\,\mathbf{a}_v + h_3\,dw\,\mathbf{a}_w$ \hfill (2)

Differential volume: $dV = h_1 h_2 h_3\,du\,dv\,dw$ \hfill (3)

$$\operatorname{grad}\psi = \nabla\psi = \frac{1}{h_1}\frac{\partial\psi}{\partial u}\mathbf{a}_u + \frac{1}{h_2}\frac{\partial\psi}{\partial v}\mathbf{a}_v + \frac{1}{h_3}\frac{\partial\psi}{\partial w}\mathbf{a}_w \tag{4}$$

$$\operatorname{div}\mathbf{A} = \nabla\cdot\mathbf{A} = \frac{1}{h_1 h_2 h_3}\left(\frac{\partial}{\partial u}h_2 h_3 A_u + \frac{\partial}{\partial v}h_3 h_1 A_v + \frac{\partial}{\partial w}h_1 h_2 A_w\right) \tag{5}$$

$$\operatorname{curl}\mathbf{A} = \nabla\times\mathbf{A} = \frac{1}{h_1 h_2 h_3}\begin{vmatrix} h_1\mathbf{a}_u & h_2\mathbf{a}_v & h_3\mathbf{a}_w \\ \dfrac{\partial}{\partial u} & \dfrac{\partial}{\partial v} & \dfrac{\partial}{\partial w} \\ h_1 A_u & h_2 A_v & h_3 A_w \end{vmatrix} \tag{6}$$

Scalar laplacian:

$$\nabla^2\psi = \frac{1}{h_1 h_2 h_3}\left[\frac{\partial}{\partial u}\left(\frac{h_2 h_3}{h_1}\frac{\partial\psi}{\partial u}\right) + \frac{\partial}{\partial v}\left(\frac{h_3 h_1}{h_2}\frac{\partial\psi}{\partial v}\right) + \frac{\partial}{\partial w}\left(\frac{h_1 h_2}{h_3}\frac{\partial\psi}{\partial w}\right)\right] \tag{7}$$

Vector laplacian:

$\nabla^2\mathbf{A} = \nabla(\nabla\cdot\mathbf{A}) - \nabla\times\nabla\times\mathbf{A}$ \quad any coordinate system \hfill (8)

$\nabla^2\mathbf{A} = (\nabla^2 A_x)\mathbf{a}_x + (\nabla^2 A_y)\mathbf{a}_y + (\nabla^2 A_z)\mathbf{a}_z$ \quad rectangular coordinates only \hfill (9)

FUNDAMENTALS OF FIELDS

Table 1.2.3 Vector identities
A, B, C = vectors; a_i = unit vectors; a_n = normal unit vector; Ψ = scalar

The divergence theorem $\int_v \nabla \cdot \mathbf{A}\, dv = \oint_s \mathbf{A} \cdot \mathbf{a}_n\, ds$	(1)
Stokes' theorem $\oint_C \mathbf{A} \cdot d\mathbf{l} = \int_s \nabla \times \mathbf{A} \cdot \mathbf{a}_n\, ds$	(2)

The following equations are valid only in the rectangular coordinate system:

$$\nabla \equiv \mathbf{a}_x \frac{\partial}{\partial x} + \mathbf{a}_y \frac{\partial}{\partial y} + \mathbf{a}_z \frac{\partial}{\partial z} \tag{3}$$

$$\nabla \cdot \mathbf{A} = \mathbf{a}_x \frac{\partial A_x}{\partial x} + \mathbf{a}_y \frac{\partial A_y}{\partial y} + \mathbf{a}_z \frac{\partial A_z}{\partial z} \tag{4}$$

$$\nabla \times \mathbf{A} = \left(\frac{\partial A_z}{\partial y} - \frac{\partial A_y}{\partial z}\right)\mathbf{a}_x + \left(\frac{\partial A_x}{\partial z} - \frac{\partial A_z}{\partial x}\right)\mathbf{a}_y + \left(\frac{\partial A_y}{\partial x} - \frac{\partial A_x}{\partial y}\right)\mathbf{a}_z \tag{5}$$

$$\nabla \Psi = \frac{\partial \Psi}{\partial x}\mathbf{a}_x + \frac{\partial \Psi}{\partial y}\mathbf{a}_y + \frac{\partial \Psi}{\partial z}\mathbf{a}_z \tag{6}$$

$$\nabla^2 \Psi = \frac{\partial^2 \Psi}{\partial x^2} + \frac{\partial^2 \Psi}{\partial y^2} + \frac{\partial^2 \Psi}{\partial z^2} \tag{7}$$

$$\nabla^2 \mathbf{A} = (\nabla^2 A_x)\mathbf{a}_x + (\nabla^2 A_y)\mathbf{a}_y + (\nabla^2 A_z)\mathbf{a}_z \tag{8}$$

The following equations are independent of the coordinate system:

$\nabla \cdot \Psi \mathbf{A} = \nabla \Psi \cdot \mathbf{A} + \Psi \nabla \cdot \mathbf{A}$	(9)
$\nabla \times \Psi \mathbf{A} = \nabla \Psi \times \mathbf{A} + \Psi \nabla \times \mathbf{A}$	(10)
$\nabla \times \nabla \Psi = 0$	(11)
$\nabla \cdot (\nabla \times \mathbf{A}) = 0$	(12)
$\nabla \times \nabla \times \mathbf{A} = \nabla(\nabla \cdot \mathbf{A}) - \nabla^2 \mathbf{A}$	(13)
$\nabla \times (\mathbf{A} \times \mathbf{B}) = \mathbf{A}(\nabla \cdot \mathbf{B}) - \mathbf{B}(\nabla \cdot \mathbf{A}) - (\mathbf{A} \cdot \nabla)\mathbf{B} + (\mathbf{B} \cdot \nabla)\mathbf{A}$	(14)
$\mathbf{A} \cdot \mathbf{B} \times \mathbf{C} = \mathbf{C} \cdot \mathbf{A} \times \mathbf{B} = \mathbf{B} \cdot \mathbf{C} \times \mathbf{A}$	(15)
$\mathbf{A} \times (\mathbf{B} \times \mathbf{C}) = (\mathbf{A} \cdot \mathbf{C})\mathbf{B} - (\mathbf{A} \cdot \mathbf{B})\mathbf{C}$	(16)
$(\mathbf{A} \times \mathbf{B}) \times \mathbf{C} = (\mathbf{A} \cdot \mathbf{C})\mathbf{B} - (\mathbf{B} \cdot \mathbf{C})\mathbf{A}$	(17)

Surface elements can be constructed by multiplying any two of the differential distance elements.

When a vector operation is given in terms of curvilinear coordinates and scale factors, the equation is valid for any coordinate system. This is true for the gradient, divergence, curl, and laplacian operations. For example, the gradient in curvilinear coordinates is

$$\nabla V = \frac{1}{h_1}\frac{\partial V}{\partial u}\mathbf{a}_u + \frac{1}{h_2}\frac{\partial V}{\partial v}\mathbf{a}_v + \frac{1}{h_3}\frac{\partial V}{\partial w}\mathbf{a}_w \tag{50}$$

A substitution of the appropriate scale factors converts it to any given coordinate system.

The common operations in orthogonal curvilinear coordinates are given in Table 1.2.2 (see p. 16). Table 1.2.3 (see p. 17), vector identities, is also included for convenience.

1.3 COULOMB'S LAW AND THE E FIELD

Electric charge is a fundamental characteristic of the most elementary particles of matter. Atomic theory ascribes this characteristic to electrons and protons. But what is electric charge? In asking such a question, we are in the same situation as the philosopher of 2000 years ago, who, believing in the four-element theory,† asked "What is water?" He could answer the question only by describing the characteristics of water as he knew them. Although his answer may have included descriptions of a mirrorlike pond, or thundering surf, or perhaps taste, color, density, etc., it would be couched in terms of things he could sense and measure. A man of relatively recent generations could give a more satisfying answer by adding a description of the molecular structure and constituents of water.

At present we must describe electric charge in terms of those characteristics which we can sense and measure, for we know of no more elementary substructure of charge. The characteristics of electric charge may be stated as follows:

1. There are two, and only two, kinds of electric charge, called *positive* and *negative.*
2. They are associated with subatomic particles called *protons* and *electrons*, respectively.
3. Charge occurs in indivisible quanta or packets, and one quantum is the amount of charge associated with one electron.
4. Like charges repel and unlike attract with measurable force.

† The universe was believed to be constructed of four elements, earth, air, fire, and water.

FUNDAMENTALS OF FIELDS

5. For two point charges, the force acts along a line through their centers and is inversely proportional to the square of the distance of separation.

These are the characteristics which we can verify experimentally. Yet from these characteristics and the known laws of mechanics, we can construct a theory that will help us understand a multitude of physical phenomena.

The fourth and fifth properties given above are now called Coulomb's law, although other well-known scientists of the eighteenth century besides Coulomb arrived independently at this law.†

The mathematical statement of this law (see Fig. 1.3.1) is

$$\mathbf{F} = \frac{q_1 q_2}{4\pi\varepsilon_0 R^2} \mathbf{a}_R \qquad (1)$$

where $\mathbf{F}$ = force, N
q_1, q_2 = two point charges in free space, C
R = straight-line distance between charges, m

The unit vector $\mathbf{a}_R$ is directed along the centerline of q_1 and q_2, and ε_0 is a constant of proportionality, called the *permittivity* or the *dielectric constant of free space* ($\varepsilon_0 = 8.854 \times 10^{-12} \approx 10^{-9}/36\pi$), with units of coulombs squared per newton-meters squared, or farads per meter. This law, called an inverse-square law because force varies inversely with the square of distance, is of the same form as Newton's law of gravitation. However, unlike the gravitational force, the coulomb force of Eq. (1) may be either positive or negative according to the algebraic signs of the charges q_1 and q_2. Although Coulomb's law is an experimental law, the exponent 2 appears to be exact. Maxwell expressed the exponent as $2 \pm \delta$ and then showed that $\delta < 1/21{,}600$. More recent researchers‡ have shown that $\delta < 10^{-9}$.

As an example of the use of Coulomb's law, we will compute the electrostatic force between the electron and the proton in a hydrogen

---

† Franklin, Priestley, Cavendish, Coulomb, Poisson, and others were all concerned with this law between 1700 and 1800; see R. S. Elliott, "Electromagnetics," pp. 98–117, McGraw-Hill Book Company, New York, 1966.

‡ S. J. Plimpton and W. E. Lawton, A Very Accurate Test of Coulomb's Law of Force between Charges, *Phys. Rev.*, vol. 50, p. 1066, 1936.

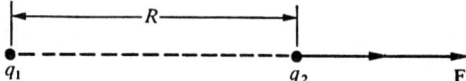

Fig. 1.3.1 Coulomb's law.

atom. Approximate values of the constants for Eq. (1) are

$$q_1 = -q_2 = -1.60 \times 10^{-19} \text{ C} \quad \text{charge of an electron}$$
$$R = 0.53 \times 10^{-10} \text{ m} \quad \text{radius of the first Bohr orbit}$$

Consequently the force is

$$\mathbf{F} = \frac{q_1 q_2}{4\pi\varepsilon_0 R^2} \mathbf{a}_R = -8.21 \times 10^{-8} \mathbf{a}_R \quad \text{N}$$

The force (of about 1.84×10^{-8} lb) tends to pull the charges together, as indicated by the negative sign. This electrostatic force must be in equilibrium with the centrifugal force for the electron to remain in a stable orbit.

Coulomb's law is valid for electrostatic forces between point charges in regions as small as atomic radii. For electrostatic repulsion of nuclei, Coulomb's law has been found to hold for radii as small as 10^{-14} m. However, we will seldom be concerned with subatomic or submolecular dimensions in this text. When we speak of charge, we will generally mean an ensemble (i.e., a collection to be treated as a whole) of a great many electrons or protons, in such quantity that the quantum nature of charge is of little consequence. This is commonly referred to as the macroscopic viewpoint, as opposed to the microscopic viewpoint. The macroscopic viewpoint leads to classical electromagnetic theory, and the microscopic to quantum theory.†

† The two theories are related in an interesting way in O. M. Grimes, Electromagnetism and Quantum Theory, *IEEE Spectrum*, May 1966, p. 55.

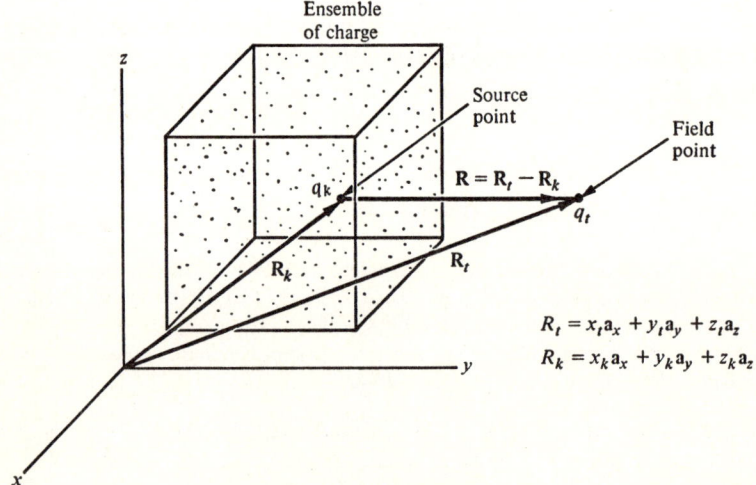

Fig. 1.3.2 Coulomb's law for an ensemble of charge.

FUNDAMENTALS OF FIELDS

If we wish to compute the force on a particular charged particle (say a small test charge q_t) in the vicinity of an ensemble of charge, we can, in theory, compute the coulomb force on q_t due to each charged particle in the ensemble. Forces add in a linear manner, so that the total force on q_t would be the vector sum of the individual forces:

$$\mathbf{F} = q_t \left[\sum_{k=1}^{n} \frac{q_k}{4\pi\varepsilon_0} \frac{\mathbf{R}_t - \mathbf{R}_k}{|\mathbf{R}_t - \mathbf{R}_k|^3} \right] \quad (2)$$

The variables of Eq. (2) are shown in Fig. 1.3.2. Other forces besides the coulomb force will appear when there is relative motion of the charges, and so we assume that a static distribution exists when using Coulomb's law. Also the test charge q_t must be small enough to ensure that its presence does not disturb the ensemble.

CHARGE DENSITY

Suppose we divide the volume containing the ensemble of Fig. 1.3.2 into many small, equal, incremental cubes of volume Δv; for example, by dividing a centimeter cube into 1000 millimeter cubes. The quantity of charge within the incremental cubes may vary from cube to cube, but we can describe an *average* charge density for each cube by dividing the charge enclosed by the corresponding incremental volume. In concept, we could then divide each incremental cube into 1000 smaller cubes, and again compute the average charge density for each of these cubes. Continuing in this fashion, we can eventually arrive at incremental volumes small enough for each Δv to be considered essentially a point in the macroscopic sense. (We are seldom concerned with regions less than 10^{-4} m in diameter from the macroscopic viewpoint.) This leads to a definition of *volume charge density* ρ_v

$$\rho_v = \lim_{\Delta v \to 0} \frac{q \text{ in } \Delta v}{\Delta v} \quad (3)$$

Usually ρ_v will vary from point to point and consequently must be written as a function of the coordinate variables. For example, in a rectangular coordinate system we have $\rho_v(x,y,z)$. If an analytical expression for $\rho_v(x,y,z)$ is known, an integration over the volume will yield the total charge enclosed.

The concept of a density variable, defined in this manner, is not unique to electric charge. In the study of mechanics, mass density and weight density are often similarly defined. The concept is satisfactory as long as the incremental volume remains large compared to molecular diameters.

There are situations where an ensemble of charge is confined to a surface or a sheet, as on the surface of a good conductor. Accordingly, we define the *surface charge density* ρ_s as

$$\rho_s = \lim_{\Delta s \to 0} \frac{q \text{ on } s}{\Delta s} \tag{4}$$

where Δs is an incremental surface area. The conceptual limitations with respect to the macroscopic viewpoint as discussed above apply here. Like ρ_v, ρ_s is a function of the coordinate system variables.

The coulomb force of Eq. (2) can be written in an integral form which is convenient when an expression for charge density is known:

$$\mathbf{F}_t = q_t \left[\int_v \frac{\rho_v(x,y,z)}{4\pi\varepsilon_0} \frac{\mathbf{R}_t - \mathbf{R}_k}{|\mathbf{R}_t - \mathbf{R}_k|^3} \, dv \right] \tag{5}$$

where $dv = dx\,dy\,dz$ and

$$\mathbf{R}_k = x_k \mathbf{a}_x + y_k \mathbf{a}_y + z_k \mathbf{a}_z$$

and the volume integral extends over the region containing the ensemble. The vector $\mathbf{R}_k$ extends from the origin to the point of integration within the charged region, as shown in Fig. 1.3.2. This point is often called the *source point*. The vector $\mathbf{R}_t$ extends from the origin to the test charge at the point where the force is being computed. This point is called the *field point*. The volume integral must extend over all the source points but not necessarily over the field point.

THE E FIELD

When the Coulomb force $\mathbf{F}$ is divided by the test charge in Eqs. (2) or (5), one obtains a vector field of force per unit charge which exists because of the ensemble and which is *independent* of the presence of the test charge.†
This vector field is called the *electric field intensity* $\mathbf{E}$ and is measured in newtons per coulomb or, equivalently, in volts per meter:

$$\mathbf{E} = \frac{\mathbf{F}}{q_t} \tag{6}$$

On comparing Eq. (6) with Eqs. (2) and (5), it appears that the $\mathbf{E}$ field is given by the term in brackets in either of these equations. If we let

$$\mathbf{R} = \mathbf{R}_t - \mathbf{R}_k \tag{7}$$

† This assumes that the test charge q_t is small enough not to disturb the ensemble. Alternately one could define the $\mathbf{E}$ field using the limit process, $\mathbf{E} = \lim_{q_t \to 0} (\mathbf{F}/q_t)$.

FUNDAMENTALS OF FIELDS

then a unit vector† along **R** can be written as

$$\mathbf{a}_R = \frac{\mathbf{R}}{R} \tag{8}$$

and so the equations for the **E** field can be written in the simplified form

$$\mathbf{E} = \sum_{k=1}^{n} \frac{q_k}{4\pi\varepsilon_0} \frac{\mathbf{a}_R}{R^2} \tag{9}$$

$$\mathbf{E} = \int_v \frac{\rho_v(u,v,w)}{4\pi\varepsilon_0} \frac{\mathbf{a}_R}{R^2} \, dv \tag{10}$$

These equations are useful in cases where the geometry of the source points is relatively simple (lines, planes, spheres, etc.), but evaluation becomes very tedious in most cases. Fortunately, simpler analytical techniques have been developed.

The concept of the electric field intensity **E** is one of the most useful concepts of electromagnetic theory. Faraday and Maxwell improved the physical significance of the **E** field by sketching "lines of force" which show the direction of the **E** vector. The lines show the direction of force that would exist on a positively charged particle placed in the field. The density of the lines is often used to indicate the magnitude of the field. One can trace the lines of force, using a small insulated pivoted pointer with a positive charge embedded in the tip, just as one can trace magnetic

† The capital letter **R** is used to denote a vector in a generalized coordinate system whose coordinates are u, v, and w. The corresponding unit vector is denoted $\mathbf{a}_R$, where $\mathbf{a}_R = \mathbf{R}/R$. The variables u, v, and w can be rectangular, cylindrical, or spherical coordinates or any other coordinates convenient for the geometry of the particular problem. The lowercase letter r is used exclusively for the radial variable in cylindrical or spherical coordinates, and $\mathbf{a}_r$ is the corresponding unit vector.

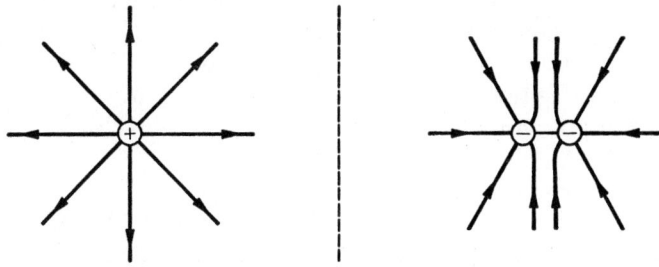

Fig. 1.3.3 E-field maps using lines of force.

lines of force with a compass. Maps for the **E** field for a point charge and a pair of point charges are shown in Fig. 1.3.3.†

1.4 SCALAR AND VECTOR FIELDS

To the engineer or physicist, a field is a set of functions associated with each point in space. *Field theory* is the study of how these functions are affected by the presence of matter or electric charge. The mathematician would say that we are using *vector spaces* to model physical situations. Fortunately, the same mathematical model is used in many areas of physical science, e.g., electrodynamics, fluid dynamics, elasticity, acoustics, and heat conduction, to name a few. Thus when one studies any one of these disciplines, one cannot help gaining some insight into the others. Generally the two types of fields that are useful are the *scalar field* and the *vector field*, of which we now consider some of the most important properties.

SCALAR FIELDS

A field is said to be a scalar field if the function associated with each point in space is a real or complex number. A temperature field is an example of this type. Suppose one put a large piece of ice in the center of a room and used a thermometer to measure the temperature of the air at various points around the room. At a given instant of time, one could associate a real number (temperature) with every point, thereby obtaining a scalar field of temperature due to the presence of the ice, the air, and the walls. (We must take care that the heat of our measuring instruments or of our own body does not distort the field.) The temperature of the walls and of the ice describe *boundary conditions*. The development and solution of differential equations that describe the temperature of all points in space is the study of the *boundary-value problem*. If one could find all those points in the room which were at the same temperature at a given instant of time, they would form an *isothermal surface*. There would be a different isothermal surface for each temperature, and they would be nonintersecting. (A point on an intersection would have two different temperatures at the same instant, a physical impossibility.)

In this text we will be concerned with a *scalar potential field*, caused by the presence of electric charge, whose *equipotential surfaces* are mea-

† Many excellent experimental and numerical methods have been developed for field mapping which will not be presented here. If the reader has difficulty with fields concepts, he will find the following references helpful: W. E. Rogers, "Introduction to Electric Fields," McGraw-Hill Book Company, New York, 1954; Moon and Spencer, *op. cit.*; R. L. Tanner and M. G. Andreasen, Numerical Solution of Electromagnetic Problems, *IEEE Spectrum*, vol. 4, no. 9, September 1967, p. 53.

FUNDAMENTALS OF FIELDS

sured in volts; i.e., each point on the surface has the same voltage with respect to some common reference point. Presently we will develop the relationship between the scalar potential field and the E field of Coulomb and discuss some boundary-value problems.

If we say that a field is invariant, we mean that the field is independent of the coordinate system chosen. This does not necessarily mean invariance with respect to time. For example, a particular point is at a temperature of 50°C whether we describe the point in terms of a rectangular coordinate system or a cylindrical coordinate system. Of course this property seems self-evident in cases where we are guided by strong physical intuition. It is an important property of fields, however, because it enables us to choose any coordinate system (or combination of systems) that seems convenient for the boundary-value problem at hand.

In conclusion, scalar fields are characterized by:

1. A single number associated with each point in space
2. Nonintersecting equipotential surfaces
3. Invariance with respect to the coordinate system

VECTOR FIELDS

In a vector field, every point in space has both a magnitude and a direction associated with it. In general, three components (real or complex) and three unit vectors are associated with each point in a three-dimensional space. The gravitational force field is an example of a vector field. If a small test mass is placed in the gravitational field of the sun, one can find the magnitude and direction of the force at every point in space. In rectangular coordinates, one would express the force at the point (x_1,y_1,z_1) as $F_x(x_1,y_1,z_1)\mathbf{a}_x + F_y(x_1,y_1,z_1)\mathbf{a}_y + F_z(x_1,y_1,z_1)\mathbf{a}_z$. Like scalar fields, vector fields are invariant, and so the force could be expressed in cylindrical, spherical, or other convenient coordinate systems.

The E field is another example of a vector field. The lines of force of Faraday and Maxwell are called *flow lines* in a vector field. Flow lines are continuous lines which are tangent to the field vector at every point and are thus nonintersecting. They are the lines shown in the field mappings of the previous section. In special cases† it is possible to find equipotential surfaces for vector fields. These surfaces are orthogonal to the flow lines and may be shown in the field mapping.

In summary, vector fields are characterized by:

1. A vector (magnitude and direction) associated with each point in space
2. Invariance with respect to the coordinate system

† Equipotential surfaces exist for fields where the curl is zero.

3. Flow lines, or lines of force
4. In certain cases, equipotential surfaces orthogonal to the flow lines

1.5 DELTA FUNCTIONS

In using mathematical models to describe physical systems, a very useful concept is that of a function which is nonzero only at a single point. For example, we think of a point charge as a finite amount of charge located at a point in space. (From the macroscopic viewpoint, we ignore the fact that the point encloses no volume, thus making the charge density infinite.) As another example, in the study of mechanics we can consider a finite weight applied at a point on a beam, causing a deflection of the beam. (We ignore the fact that the force per unit area would be infinite at the point.) The *delta function*, $\delta(x - x_0)$ (unit impulse function, Dirac delta function)† is an ideal mathematical model of such concepts. Formally, we define the notation $\delta(x - x_0)$ for a real variable x by

$$\int_a^b f(x)\delta(x - x_0)\, dx = \begin{cases} 0 & \text{if } x_0 \text{ is not in the interval } (a,b) \\ f(x_0) & \text{if } a < x_0 < b \end{cases} \quad (1)$$

where $f(x)$ is a continuous function at $x = x_0$.

The only properties of $\delta(x)$ that we will need can be obtained easily from the definition, Eq. (1). Suppose we let $f(x) = 1$ and consider the integral of Eq. (1) over the limits $(a,b) = (-\infty, \infty)$. Then one of the properties of the delta function is

$$\int_{-\infty}^{\infty} \delta(x - x_0)\, dx = 1 \quad (2)$$

We can illustrate another property of the delta function by letting $f(x)$ be any continuous function $f(x) \neq 0$ in the interval (a,b). If we choose any interval (a,b) which does not include x_0, then the integral is zero even though $f(x)$ is not zero. Consequently

$$\delta(x - x_0) = 0 \quad \text{for all } x \neq x_0 \quad (3)$$

Thus the delta function is zero except at the point where its argument $x - x_0$ is zero. The graphs for Eqs. (1) and (2) are given in Fig. 1.5.1.

Equation (2) requires unity area under the delta function, but Eq. (3) means that the function has zero width. Hence the delta function is not a function in the usual sense. However, Schwartz's theory of distributions‡

---

† Named in honor of the physicist P. A. M. Dirac, who used the function in his development of quantum mechanics.

‡ Laurent Schwartz, Theory of Distributions, *Actual. Sci. Ind.* 1091, 1122, Hermann & Cie, Paris, 1950, 1951.

FUNDAMENTALS OF FIELDS

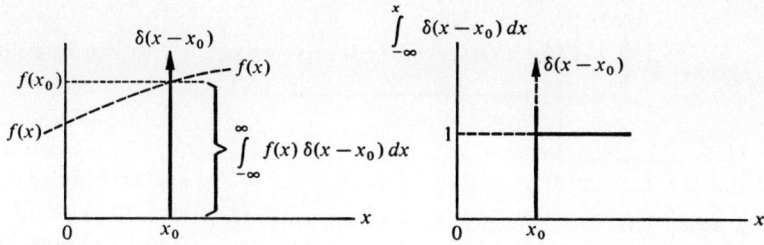

Fig. 1.5.1 Integrals of the delta function.

has made the delta function and other discontinuous functions respectable and has shown that the operations for handling these functions are the same as for normal function theory.

The *unit step function* $U(x - x_0)$ is defined using Eq. (2):

$$U(x - x_0) = \int_{-\infty}^{x} \delta(x - x_0) \, dx = \begin{cases} 0 & \text{if upper limit } x < x_0 \\ 1 & \text{if upper limit } x > x_0 \end{cases} \quad (4)$$

The step function is shown in Fig. 1.5.1. If we differentiate Eq. (4) with respect to x, we obtain

$$\frac{dU(x - x_0)}{dx} = \delta(x - x_0) \quad (5)$$

In our work in Chap. 4 we will find it convenient to use a unit step function to trace a wavefront as the wave moves along the transmission line. To do this we will use a unit step function in which the discontinuity moves along the z axis with a velocity v_p. Thus at any time t, the discontinuity of the unit step function $U(z - v_p t)$ is located at the point z, where its argument $z - v_p t$ is zero. Since time continually increases, the discontinuity moves with a velocity v_p.

The delta function is useful in describing point sources of fields or point sources of charge. Since we often will be concerned with functions in three-dimensional space, we will occasionally need a delta function which is nonzero at the point (u_0, v_0, w_0) when each term of the argument of $\delta(u - u_0, v - v_0, w - w_0)$ is zero. We define such a delta function by

$$\int_v f(u,v,w) \delta(u - u_0, v - v_0, w - w_0) \, dv$$

$$= \begin{cases} 0 & \text{if } (u_0, v_0, w_0) \text{ is not in volume } v \\ f(u_0, v_0, w_0) & \text{if } (u_0, v_0, w_0) \text{ is in volume } v \end{cases} \quad (6)$$

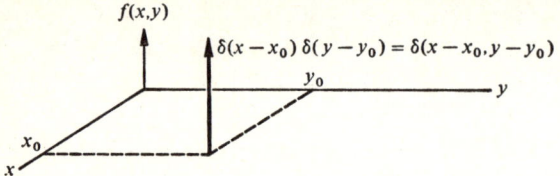

Fig. 1.5.2 The delta function in a two-dimensional space.

The main properties of this unit impulse function located at (u_0, v_0, w_0) are

$$\delta(u - u_0, v - v_0, w - w_0) = 0 \quad \text{for } u \neq u_0$$
$$v \neq v_0$$
$$w \neq w_0 \quad (7)$$

$$\int_v \delta(u - u_0, v - v_0, w - w_0)\, dv = 1 \quad (8)$$

where the integration of Eq. (8) extends over all space. We can sketch the delta function for a two-dimensional space as shown in Fig. 1.5.2. It would require four orthogonal axes to sketch a delta function in three dimensions (an exercise left to the student). However, we can write $\delta(u,v,w)$ as a product of simpler terms:

1. In rectangular coordinates, where $dv = dx\, dy\, dz$,

$$\delta(x,y,z) = \delta(x)\delta(y)\delta(z) \quad (9)$$

2. In cylindrical coordinates, where $dv = r\, dr\, d\phi\, dz$,

$$\delta(r,\phi,z) = \frac{\delta(r)\delta(\phi)\delta(z)}{r} \quad (10)$$

or in polar coordinates

$$\delta(r,\phi) = \frac{\delta(r)\delta(\phi)}{r} \quad (11)$$

3. In spherical coordinates, where $dv = r^2 \sin\theta\, dr\, d\theta\, d\phi$,

$$\delta(r,\theta,\phi) = \frac{\delta(r)\delta(\theta)\delta(\phi)}{r^2 \sin\theta} \quad (12)$$

One can show that Eqs. (9) to (12) are correct by substituting them in Eq. (8). We will find applications of these equations in the sections that follow.

FUNDAMENTALS OF FIELDS

There are many other interesting properties of the delta functions which are not needed in our present work.† Many texts define the delta function as a limit process wherein the area under a curve is required to be unity. For example, a delta function is sometimes defined from a rectangular strip of unit area as the width approaches zero and the height approaches infinity. Such concepts, while intuitively helpful, are not strictly correct, and so the definition given in Eq. (8) is preferred.

1.6 THE GRADIENT AND THE POTENTIAL FIELD

If we let $V(x,y,z)$ denote a scalar-field variable at the point (x,y,z), we can denote the differential change in V by the total differential

$$dV = \frac{\partial V}{\partial x} dx + \frac{\partial V}{\partial y} dy + \frac{\partial V}{\partial z} dz \tag{1}$$

between the points (x,y,z) and $(x + dx, y + dy, z + dz)$. If both points lie on the same equipotential surface $V = V_0$, then $dV = 0$. Therefore the differential equation of the equipotential surfaces of a scalar field is

$$\frac{\partial V}{\partial x} dx + \frac{\partial V}{\partial y} dy + \frac{\partial V}{\partial z} dz = 0 \tag{2}$$

We will show that a charged particle can be moved from point to point on an equipotential surface without requiring any net work from a source.

Suppose we are given an equipotential surface $V = V_0$ of a scalar field, a portion of which is shown in Fig. 1.6.1. Let **R** be a vector from the origin to an arbitrary point $p = (x,y,z)$ on the equipotential surface. Then the differential vector $d\mathbf{R}$ lies in the plane tangent to the surface at the point p, as shown in the figure. Let us construct a vector **N** normal to the tangent plane (and consequently normal to $d\mathbf{R}$) at the point p. The equations for these vectors are

$$\mathbf{R} = x\mathbf{a}_x + y\mathbf{a}_y + z\mathbf{a}_z \tag{3}$$

$$d\mathbf{R} = dx\, \mathbf{a}_x + dy\, \mathbf{a}_y + dz\, \mathbf{a}_z \tag{4}$$

$$\mathbf{N} = f_x\mathbf{a}_x + f_y\mathbf{a}_y + f_z\mathbf{a}_z \tag{5}$$

If we know the equation of the surface, we can use it to find **N**. Conversely, if we know **N**, we can use it to construct the normal surface.

Suppose the components of **N**, namely f_x, f_y, f_z, are unknown at this time. We know that they are functions of the equipotential surface since

† The interested reader can find many of these summarized in G. A. Korn and T. M. Korn, "Mathematical Handbook for Scientists and Engineers," 2d ed., McGraw-Hill Book Company, New York, 1968.

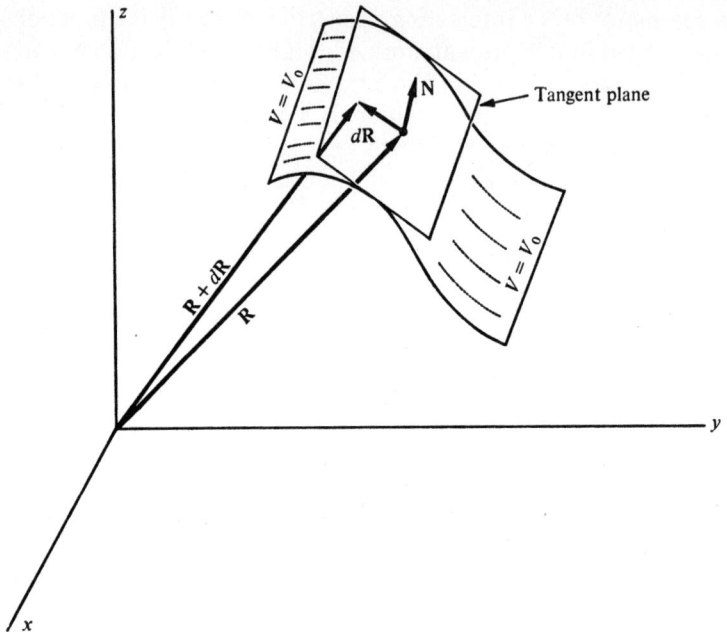

Fig. 1.6.1 An equipotential surface and a tangent plane.

$\mathbf{N}$ depends upon the location of the point p. Furthermore, we require that $\mathbf{N}$ be normal to $d\mathbf{R}$, and so

$$\mathbf{N} \cdot d\mathbf{R} = 0 \tag{6}$$

$$f_x\, dx + f_y\, dy + f_z\, dz = 0 \tag{7}$$

This would be the differential equation of the surface if the components of $\mathbf{N}$ were known. In this case, we know the differential equation of the surface, Eq. (2), and can compare it with Eq. (7) to determine the components of $\mathbf{N}$

$$f_x = \frac{\partial V}{\partial x} \qquad f_y = \frac{\partial V}{\partial y} \qquad f_z = \frac{\partial V}{\partial z} \tag{8}$$

$$\mathbf{N} = \frac{\partial V}{\partial x}\mathbf{a}_x + \frac{\partial V}{\partial y}\mathbf{a}_y + \frac{\partial V}{\partial z}\mathbf{a}_z \tag{9}$$

The vector $\mathbf{N}$ is called the *gradient* of V.

Since $\mathbf{N}$ is normal to the equipotential surface $V = V_0$, it points in the direction of the shortest path to the next equipotential surface $V = V_1$ (suppose $V_1 > V_0$). That is, it points in the direction of the maximum rate of change of V. For example, suppose the scalar field were a two-

FUNDAMENTALS OF FIELDS

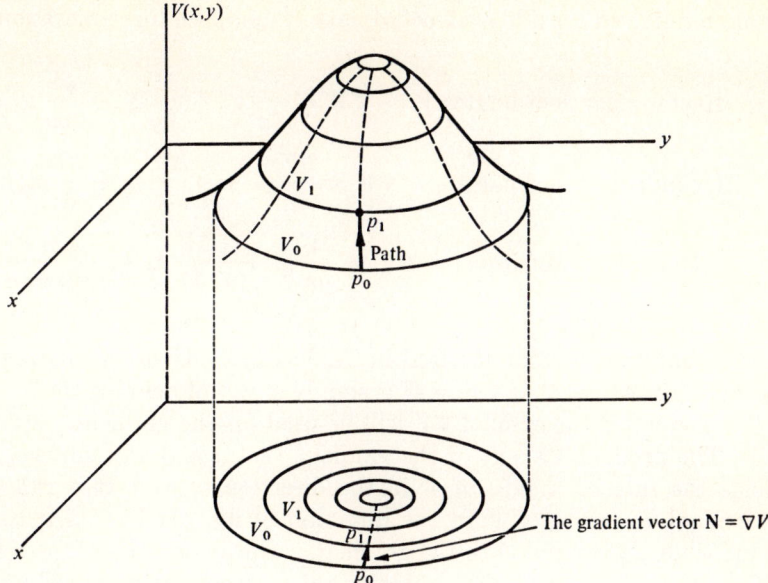

Fig. 1.6.2 The gradient in the xy plane.

dimensional field $V(x,y)$, so that every point of the xy plane corresponds to a particular value of V. The equipotential "surfaces," $V(x,y) =$ constant, then degenerate to equipotential *lines* in the xy plane. A common example is the topographic map. The gradient vector lies in the xy plane, as shown in Fig. 1.6.2, and points in the direction of the steepest uphill path. Furthermore, the greater the magnitude of the gradient $\mathbf{N}$,

$$\mathbf{N} = \left[\left(\frac{\partial V}{\partial x}\right)^2 + \left(\frac{\partial V}{\partial y}\right)^2 + \left(\frac{\partial V}{\partial z}\right)^2\right]^{1/2} \tag{10}$$

the greater the slope of the path, i.e., the steeper the path. Thus both the magnitude and direction of the gradient are important.

One should note carefully that the gradient appears in the two-dimensional plane for a two-dimensional field, and not along the path on the three-dimensional $V(x,y)$ "hill." Usually we will be working with three-dimensional fields, $V(x,y,z)$, and it will be impossible to draw the corresponding four-dimensional hill. However, the gradient will appear in the three-dimensional field, and its magnitude and direction will have the same meaning as in the two-dimensional case.

The gradient vector is used so often in advanced work that it is given special notation

$$\nabla V \equiv \text{grad } V \equiv \text{gradient of scalar } V \tag{11}$$

It has a different form in each coordinate system, the three most common being:

Rectangular coordinates:† $\quad \nabla V = \dfrac{\partial V}{\partial x}\mathbf{a}_x + \dfrac{\partial V}{\partial y}\mathbf{a}_y + \dfrac{\partial V}{\partial z}\mathbf{a}_z \quad (12)$

Cylindrical coordinates: $\quad \nabla V = \dfrac{\partial V}{\partial r}\mathbf{a}_r + \dfrac{1}{r}\dfrac{\partial V}{\partial \phi}\mathbf{a}_\phi + \dfrac{\partial V}{\partial z}\mathbf{a}_z \quad (14)$

Spherical coordinates: $\quad \nabla V = \dfrac{\partial V}{\partial r}\mathbf{a}_r + \dfrac{1}{r}\dfrac{\partial V}{\partial \theta}\mathbf{a}_\theta + \dfrac{1}{r\sin\theta}\dfrac{\partial V}{\partial \phi}\mathbf{a}_\phi$
$$(15)$$

These equations are summarized in Table 1.2.2. Upon comparing Eqs. (9) and (12) we see that $\nabla V = \mathbf{N}$ is simply a way of defining the notation ∇V. Hereafter the symbol ∇V will be used for the gradient.

The product $\nabla V \cdot \mathbf{a}_R$ of the gradient vector and any unit vector $\mathbf{a}_r$ yields the rate of change of V in the direction of $\mathbf{a}_R$. It is called the *directional derivative* of V in the direction of $\mathbf{a}_R$. It has its maximum value when $\mathbf{a}_R$ is parallel to the gradient. However, $\nabla V \cdot d\mathbf{R} = 0$ is the differential equation of the equipotential surfaces, as previously noted.

As an example, consider a scalar potential field given by

$$V(x,y,z) = (x^2 + y^2 + z^2)^{-1/2} \quad (16)$$

For constant values of V, the equipotential surfaces are spherical shells centered about the origin. The values of V decrease with distance from the origin. In spherical coordinates, Eq. (16) is

$$V(r,\theta,\phi) = \frac{1}{r} \quad (17)$$

Intuitively, we expect the gradient to be a radial vector pointing inward toward larger values of V

$$\begin{aligned}\nabla V &= \frac{\partial V}{\partial x}\mathbf{a}_x + \frac{\partial V}{\partial y}\mathbf{a}_y + \frac{\partial V}{\partial z}\mathbf{a}_z \\ &= -x(x^2+y^2+z^2)^{-3/2}\mathbf{a}_x - y(x^2+y^2+z^2)^{-3/2}\mathbf{a}_y \\ &\quad - z(x^2+y^2+z^2)^{-3/2}\mathbf{a}_z \end{aligned}$$

† In rectangular coordinates, the symbol del ∇ may be treated as a vector differential operator, defined as

$$\nabla \equiv \frac{\partial}{\partial x}\mathbf{a}_x + \frac{\partial}{\partial y}\mathbf{a}_y + \frac{\partial}{\partial z}\mathbf{a}_z \quad (13)$$

Then ∇V, $\nabla \cdot \mathbf{E}$, and $\nabla \times \mathbf{E}$ (E is a vector) can be found from the usual rules of differential operators and vector operations. However, no similarly convenient form for ∇ exists in other coordinate systems. Consequently some authors reserve the notation ∇ for rectangular coordinates and use grad V for all other systems.

FUNDAMENTALS OF FIELDS

$$= \frac{-(x\mathbf{a}_x + y\mathbf{a}_y + z\mathbf{a}_z)}{(x^2 + y^2 + z^2)^{3/2}} \quad \text{rectangular coordinates} \quad (18)$$

$$\nabla V = -\frac{\mathbf{r}}{r^3} = -\frac{1}{r^2}\mathbf{a}_r \quad \text{spherical coordinates} \quad (19)$$

From Eq. (19) we see that the gradient is indeed a radial vector pointing toward the origin and furthermore that its magnitude varies inversely with the square of the radius, as shown in Fig. 1.6.3.

The scalar field $V(r,\theta,\phi)$ of this example is said to have a *singularity* at the origin because

$$\lim_{r \to 0} V(r,\theta,\phi) = \lim_{r \to 0} \frac{1}{r} \to \infty \quad (20)$$

The singularities of fields are very important because they determine the functional form of the field. The singularities are analogous to the poles which in circuit analysis determine the functional form of the circuit response.

Since the gradient is a vector at each point in space, *it forms a vector field* whose flow lines are tangent to the gradient vector at each point. It is interesting to note that we can start with a scalar field and derive a related vector field, assuming the field is "smooth" enough for the derivatives of Eq. (8) to exist. It is logical, then, to ask whether we can always find a scalar field if we start with a vector field. That is, can we always find equipotential surfaces for any given vector field? If the two fields always occur in uniquely related pairs, we would prefer to work with the scalar field because of its simplicity. The vector field could easily be obtained by differentiation according to Eqs. (12), (14), or (15). If we

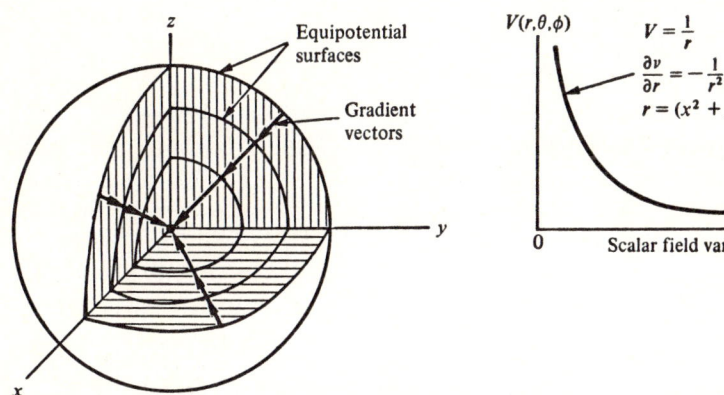

Fig. 1.6.3 A scalar field and its gradient.

are given a vector field, integration is required to find the corresponding scalar field, if one exists, although in some cases we cannot find any meaningful scalar fields or equipotential surfaces.

If an isolated, charged particle with a charge q is located at the origin of a rectangular coordinate system, its **E** field can be obtained from Coulomb's law [Eq. (1) of Sec. 1.3]

$$\mathbf{E} = \frac{\mathbf{F}}{q_t} = \frac{q}{4\pi\varepsilon R^2}\mathbf{a}_R \tag{21}$$

where $\mathbf{R} = x\mathbf{a}_x + y\mathbf{a}_y + z\mathbf{a}_z$. The flow lines of this vector field are shown in Fig. 1.3.3. If Eq. (21) is written in spherical coordinates,

$$\mathbf{E} = \frac{q}{4\pi\varepsilon}\frac{\mathbf{a}_r}{r^2} \tag{22}$$

we see that this field is of the same form as the field of the gradient vector of Eq. (19), except for the constant $q/4\pi\varepsilon$ and the negative sign. Thus there exists a scalar field (this is not always the case) associated with the vector field of Eq. (22), and it must be of the functional form $1/r$, as in Eq. (17). By including the constant of Eq. (22) we let

$$V = \frac{q}{4\pi\varepsilon}\frac{1}{r} \tag{23}$$

for the scalar field. We have assumed the sign of V to be positive. As a first step in justifying this assumption we compute the gradient of V:

$$\nabla V = \nabla\left(\frac{q}{4\pi\varepsilon}\frac{1}{r}\right) = \frac{q}{4\pi\varepsilon}\nabla\frac{1}{r} = -\frac{q}{4\pi\varepsilon}\frac{\mathbf{a}_r}{r^2} \tag{24}$$

Comparing Eqs. (22) and (24), we see that the necessary relationship between the vector field **E** and its scalar field V is

$$\boxed{\mathbf{E} = -\nabla V} \tag{25}$$

When this relationship exists, V is called the *scalar potential field* of **E** (or often simply the potential field if the meaning is clear), and V is measured in volts. The negative sign occurs in Eq. (25) for a simple reason. We have defined the direction of the **E** field as the direction of the force on a positive test charge q_t. It is directed away from the source of the **E** field, for a positive source q in Eq. (22). But the field becomes more intense for points closer to the source. Consequently, the gradient of V is directed inward, as in Fig. 1.6.3, opposite to **E**, and so the negative sign is necessary. This could have been avoided by defining the **E** field in terms of a negative test charge, e.g., an electron, but then the negative sign would arise elsewhere [specifically in Eq. (22)].

FUNDAMENTALS OF FIELDS

Although we developed the relationship between the electric intensity and its scalar potential for a very special case, the relationship given in Eq. (25) is valid for any **E** field which has a corresponding scalar potential field. It is valid for any electrostatic field, but it is not valid when the source charge varies with time. We can show the generality of Eq. (25) by considering the **E** field due to an arbitrary static charge distribution, as given by Eq. (10) of Sec. 1.3. Care must be taken to distinguish the source points (over which we integrate) from the field points (where we find the field), as noted in Sec. 1.3 (see Fig. 1.3.2):

$$\mathbf{E} = \int_v \frac{\rho_v}{4\pi\varepsilon_0} \frac{\mathbf{a}_R}{R^2} dv \tag{26}$$

but

$$\nabla \frac{1}{R} = -\frac{\mathbf{a}_R}{R^2} \tag{27}$$

Therefore the **E** field becomes

$$\mathbf{E} = -\int_v \frac{\rho_v}{4\pi\varepsilon_0} \nabla \frac{1}{R} dv \tag{28}$$

The distribution of the charge density (the source) and the limits of the integral are independent of the position of the field point $E(r,\theta,\phi)$. Consequently they are constants with respect to the differential operator ∇, so that the order of integration and differentiation can be interchanged:

$$\mathbf{E} = -\nabla \left(\int_v \frac{\rho_v}{4\pi\varepsilon_0} \frac{dv}{R} \right) = -\nabla V \tag{29}$$

In conclusion, we find that **E** and its scalar potential field V are related as in Eqs. (25) and (29) for *any* static charge distribution. Thus the scalar potential field is sometimes defined by

$$V = \int_v \frac{\rho_v}{4\pi\varepsilon_0} \frac{dv}{R} \tag{30}$$

although this integral is rarely used to compute V (because we can rarely perform the integration).

We now develop the condition for the existence of a scalar potential field in detail. Suppose we are given an **E** field and its scalar potential field so that an equipotential surface exists; we choose two arbitrary points a and b on the surface. Note that for any line integral on the equipotential surface (such as on path 1 of Fig. 1.6.4)

$$\int_b^a \mathbf{E} \cdot d\mathbf{l} = 0 \tag{31}$$

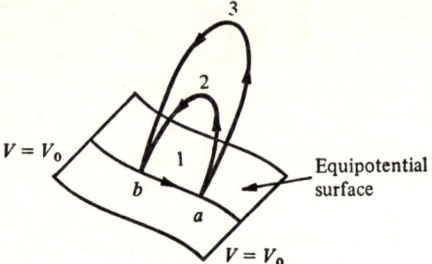

Fig. 1.6.4 Integral paths.

because the E field is normal to the surface. Along any path not on the surface† (such as paths 2 or 3)

$$\int_a^b \mathbf{E} \cdot d\mathbf{l} = -\int_a^b \boldsymbol{\nabla} V \cdot d\mathbf{l}$$

$$= -\int_a^b \left(\frac{\partial V}{\partial x}\mathbf{a}_x + \frac{\partial V}{\partial y}\mathbf{a}_y + \frac{\partial V}{\partial z}\mathbf{a}_z\right)(dx\mathbf{a}_x + dy\mathbf{a}_y + dz\mathbf{a}_z)$$

$$= -\int_a^b \left(\frac{\partial V}{\partial x}dx + \frac{\partial V}{\partial y}dy + \frac{\partial V}{\partial z}dz\right) = -\int_a^b dV$$

$$= V(a) - V(b) = V_{ab} \tag{32}$$

But a and b lie on an equipotential surface, and so $V(a) = V(b) = V_0$. Therefore we conclude that

$$\int_a^b \mathbf{E} \cdot d\mathbf{l} = 0 \tag{33}$$

regardless of which path we choose. If we move points a and b together, or if we use Eq. (31), we obtain the integral for any closed path

$$\oint \mathbf{E} \cdot d\mathbf{l} = 0 \tag{34}$$

Conversely, Eq. (34) implies that a scalar potential field exists. The proof of this will be left as a problem. Therefore Eq. (34) is both a necessary and sufficient condition for the existence of a scalar potential field, given an E field. Presently we will prove that Eq. (34) is equivalent to

$$\boldsymbol{\nabla} \times \mathbf{E} = 0 \tag{35}$$

† The rectangular coordinate system has been chosen for simplicity. One can get the same result in other systems using the corresponding forms for $\boldsymbol{\nabla} V$ and $d\mathbf{l}$:

$d\mathbf{l} = dr\,\mathbf{a}_r + r\,d\phi\,\mathbf{a}_\phi + dz\,\mathbf{a}_z$ cylindrical

$d\mathbf{l} = dr\,\mathbf{a}_r + r\,d\theta\,\mathbf{a}_\theta + r\sin\theta\,d\phi\,\mathbf{a}_\phi$ spherical

The $d\mathbf{l}$ element is always positive, and the direction of integration is accounted for in the limits of the line integral.

FUNDAMENTALS OF FIELDS

Such a vector field is said to be *irrotational*. In an irrotational field any line integral is *independent of the path* between any two arbitrary points.

If one places a test charge q in an irrotational **E** field, there will be a force on the charge given by $\mathbf{F} = q\mathbf{E}$. If the charge moves† through a distance $d\mathbf{l}$, the work done by the **E** field in moving the charge is $\mathbf{F} \cdot d\mathbf{l}$. However, if the test charge is moved against the field by an external source, the work done by that source is $-\mathbf{F} \cdot d\mathbf{l}$. The work done by the external source represents a change in the potential energy of the charged particle. One could allow the particle to return to its original position, and it would do some work in the process. Thus the potential energy of the **E** field is similar to the potential energy of a gravitational field. The work done by an external source in moving a charge q from point a to point b is

$$W_{ba} = -q \int_a^b \mathbf{E} \cdot d\mathbf{l} \tag{36}$$

This work is also the difference in potential energy of the charge at the two points. We can find the work per unit charge by dividing Eq. (36) by q. Comparing this result with Eq. (32), we obtain

$$\frac{W_{ba}}{q} \equiv V_{ba} \equiv -\int_a^b \mathbf{E} \cdot d\mathbf{l} = V(b) - V(a) \tag{37}$$

The quantity V_{ba} is called the *potential difference*‡ between points b and a and is measured in volts (or joules per coulomb). Equation (37) yields a meaningful physical interpretation of the scalar potential field. The potential difference in volts between two equipotential surfaces is the work required to move a unit charge from one surface to the other. This work is independent of the path between the equipotential surfaces if the field is irrotational.

If some point exists where an irrotational **E** field is zero, a charged particle would experience no force at that point and it could not be used to do work. Such a point could be called the *absolute* reference point,

---

† The force on a charged particle moving with a velocity **v** is $\mathbf{F} = q\mathbf{E} + q\mathbf{v} \times \mathbf{B}$. Here we assume that the magnetic field **B** is zero or **v** is negligibly small.

‡ The notation V_{ba} is used, rather than V_{ab}, in accordance with standard practice in circuit theory, where the higher potential point is noted first. For example in **Fig. (A)**, the potential drop (voltage) across the resistor is denoted V_{ba} by standard convention. Similarly in Fig. (B), the potential drop from point b to a is denoted V_{ba} although we integrate $-\int_a^b \mathbf{E} \cdot d\mathbf{l}$ from a to b to find this drop.

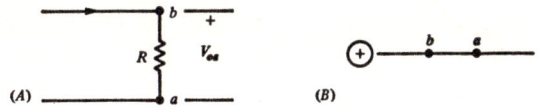

and all potentials could be measured with respect to it. Potentials measured with respect to an absolute reference point are called *absolute potentials*. As a matter of convenience, one usually chooses an *arbitrary* reference point for any given problem and measures all voltages with respect to it. In many practical problems the earth is assumed to be an equipotential surface, and all voltages are measured with respect to ground. In many mathematical problems the **E** field approaches zero at an infinite distance from the origin, and so a point at infinity is used as an absolute reference point. The exact location of the point is immaterial because the work integral is independent of the path. Then we can write

$$V_b = - \int_{\infty}^{b} \mathbf{E} \cdot d\mathbf{l} \tag{38}$$

as absolute potential at point b.

If one moves a charge around any closed path in an irrotational **E** field, work is done when the charge is being moved against the field but is recovered when the charge moves with the field. From

$$\oint \mathbf{E} \cdot d\mathbf{l} = 0 \tag{39}$$

we see that the net energy expended in moving the charge about the closed path must be zero. Thus energy is conserved, and so an irrotational field is also called a *conservative field*.

We have developed a two-way avenue between an irrotational vector field and its scalar potential through the equations

$$\mathbf{E} = -\nabla V \quad \text{and} \quad V_{ba} = - \int_{a}^{b} \mathbf{E} \cdot d\mathbf{l} \tag{40}$$

Given a scalar field V as a function of the generalized coordinates u, v, and w, we can find the electric field intensity as a function of the coordinates from the gradient. However, if we are given the **E** field in generalized coordinates, we can find the potential difference from the line integral between the points a and b. To obtain V as a function of u, v, and w, we choose a to be a fixed reference point, and let b be an arbitrary point (u,v,w).

However, since the operation of differentiation is easier than integration, as a general rule, we usually prefer to derive the scalar potential field independently and find **E** from the gradient.

CURRENT DENSITY

If some region contains a distribution of charge density ρ and an **E** field is generated by some external means, every charged particle q in the region will experience a force $q\mathbf{E}$ which will tend to make the particle move. In free space, the particle will experience a continuous acceleration ($\mathbf{a} = q\mathbf{E}/m$), and its velocity will increase with time. However, in a

FUNDAMENTALS OF FIELDS

conductor, the particle (a valence or free electron, for example) will continuously interact with the crystalline material of the conductor, and a constant average drift velocity will be attained if a constant **E** field is applied. Thus, if $\mathbf{v}_d$ is the average drift velocity,

$$\mathbf{v}_d = \mu_m \mathbf{E} \tag{41}$$

where μ_m, the *mobility constant*, is a function of the particle mass and the structure and temperature of the material.

We can relate the **E** field to a current density **J** in the medium. The differential current dI crossing a differential area ds normally can be written

$$dI = \rho v_d \, ds \tag{42}$$

since ρ is the charge per unit volume and v_d is the average drift velocity of the charge. At a given point we can define a *current density vector* **J** with a magnitude of ρv_d and a direction the same as the direction of the velocity vector $\mathbf{v}_d$:

$$\mathbf{J} = \rho \mathbf{v}_d \tag{43}$$

We could find the net current crossing a given surface by integrating the normal component of **J** over the surface

$$\boxed{I = \int_s \mathbf{J} \cdot d\mathbf{s}} \tag{44}$$

However, $\mathbf{v}_d$ in Eq. (43) can be replaced by $\mu_m \mathbf{E}$ from Eq. (41):

$$\boxed{\mathbf{J} = \rho \mathbf{v}_d = \rho \mu_m \mathbf{E} = \sigma \mathbf{E}} \tag{45}$$

The constant $\sigma = \rho \mu_m$ is called the *conductivity* of the medium, and its units are mhos per meter. Equation (45) is often called *Ohm's law* because of its analogy to Ohm's law of circuit theory. As an exercise, the reader can derive Ohm's law for circuit theory ($I = GV$) from Eq. (45) using Eqs. (32) and (44).

1.7 THE DIVERGENCE

In Sec. 1.6, we found that by mapping points of a *scalar field* to points of a vector field using the gradient operation we obtained the magnitude and direction of the maximum rate of change of the scalar field. Suppose we attempt to use this technique to describe a *vector field* **E** which has both a magnitude and a direction at every point in space. To specify the rate of change, the direction, and the magnitude of **E**, we will find it necessary to map points of **E** to both a scalar field and a vector field. These new mappings are called the *divergence* and the *curl* operations respectively. In this section we study the divergence; the curl will be studied in Sec. 1.8. In physical situations, we will find that both the divergence and the curl of a vector field **E** are associated with energy sources or sinks.

Suppose we are given a vector field **E** and we want to probe, or search, the field to find sources of the field. If we consider a small spherical region of volume Δv with a surface area Δs, we can count the flow lines of the field entering and leaving the spherical region. If more flow lines leave the region than enter it, "something" must be inside the sphere to generate the additional flow lines. We would call the something a *source* of the field. On the other hand, if more flow lines enter the region than leave it, the flow lines terminate in something called a *sink* of the field. However, if there are no sources or sinks in the region, all flow lines that enter also leave the region. We imagine moving the small sphere from point to point in **E** as we search for sources and sinks.

Before we proceed to mathematical models, we should consider two additional ideas: (1) the means by which we probe the field must not disturb or distort the field, or we will not obtain a true picture of the original field; (2) the size of the region we use for probing the field is critical, for if it is too large, it may contain both sinks and sources and we could fail to identify either. Ideally we would allow the sphere to shrink to zero volume so that we could probe point by point.

A measure of the net outflow of flux from a closed surface s is obtained from the integral $\oint_s \mathbf{E} \cdot d\mathbf{s}$, where $d\mathbf{s}$ is an outward normal surface vector, as shown in Fig. 1.7.1. The closed surface s encloses a volume Δv which we will allow to shrink to zero. To put these ideas in a mathematical form, we define the *divergence* of a vector field **E** at a given point as the limit

$$\text{div } \mathbf{E} \equiv \nabla \cdot \mathbf{E} = \lim_{\Delta v \to 0} \frac{\oint_s \mathbf{E} \cdot d\mathbf{s}}{\Delta v} \tag{1}$$

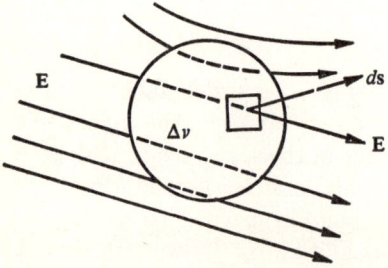

Fig. 1.7.1 Counting flow lines in a region of volume Δv.

FUNDAMENTALS OF FIELDS

Here we restrict s to be any simple closed surface (not necessarily spherical) which encloses the volume Δv. Thus the divergence is a measure of the net outward flow of flux per unit volume over a closed incremental surface.

Example 1 Suppose we use Eq. (1) to compute the divergence of a uniform vector field

$$\mathbf{E} = k\mathbf{a}_x \qquad (2)$$

where k is a real constant. Intuitively we expect the divergence to be zero because the field is uniform. For this example it is simplest to consider a cube oriented as shown in Fig. 1.7.2. Then the dot product $\mathbf{E} \cdot d\mathbf{s}$ is zero over all surfaces except those parallel to the yz plane, where

$$d\mathbf{s} = \pm dy\, dz\, \mathbf{a}_x \qquad (3)$$

The sign of $d\mathbf{s}$ is positive for the surface s_1 and negative for the surface s_2, as shown in the figure. Then the divergence is

$$\begin{aligned}
\mathbf{\nabla} \cdot \mathbf{E} &= \lim_{\Delta v \to 0} \frac{\int_{s_1} \mathbf{E} \cdot d\mathbf{s} + \int_{s_2} \mathbf{E} \cdot d\mathbf{s}}{\Delta v} \\
&= \lim_{\Delta v \to 0} \frac{\int k\, dy\, dz - \int k\, dy\, dz}{\Delta v} \\
&= 0 \qquad (4)
\end{aligned}$$

This result is independent of the location of the incremental volume because the field is uniform. It is also independent of the shape of the simple surface s and of the coordinate system. Thus if we were to choose a spherical surface and a spherical coordinate system, we would still find that $\mathbf{\nabla} \cdot \mathbf{E} = 0$ for the given field.

Example 2 As a second example we can use Eq. (1) to compute the divergence of the $\mathbf{E}$ field of a point charge located at the origin of the coordinate system. In Sec. 1.3 we showed that the $\mathbf{E}$ field in spherical coordinates for a point charge is

$$\mathbf{E} = \frac{q}{4\pi\varepsilon} \frac{\mathbf{a}_r}{r^2} \qquad (5)$$

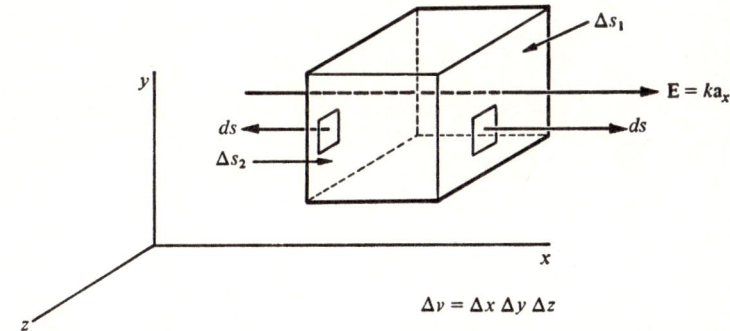

Fig. 1.7.2 The net outflow of flux from the region Δv is zero in a uniform field.

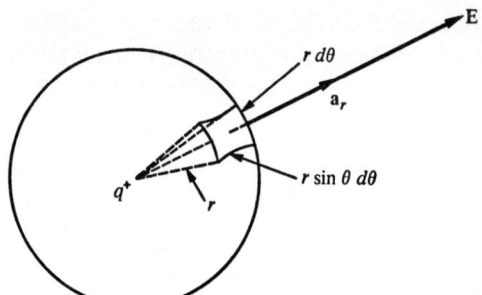

Fig. 1.7.3 A spherical surface for computing the divergence of a point charge.

For this particular example, it is simplest to let Δv be an incremental spherical volume centered at the origin, so that the surface vector is

$$d\mathbf{s} = r^2 \sin\theta \, d\theta \, d\phi \, \mathbf{a}_r \quad (6)$$

as shown in Fig. 1.7.3. If we substitute Eqs. (5) and (6) into Eq. (1), we find the divergence

$$\begin{aligned}
\boldsymbol{\nabla} \cdot \mathbf{E} &= \lim_{r \to 0} \frac{\oint_s \frac{q}{4\pi\varepsilon} \sin\theta \, d\theta \, d\phi}{4\pi r^3/3} \\
&= \lim_{r \to 0} \left(\frac{3q}{16\pi^2 \varepsilon r^3} \int_0^{2\pi} \int_0^{\pi} \sin\theta \, d\theta \, d\phi \right) \\
&= \lim_{r \to 0} \frac{3q}{4\pi\varepsilon r^3} \quad (7)
\end{aligned}$$

This function becomes infinite as we allow the sphere to shrink in volume. However, if we were to move the sphere so that the point charge were not at its origin, we would find that $\boldsymbol{\nabla} \cdot \mathbf{E} = 0$. Thus we find that the divergence is zero everywhere except at the origin, where it is infinite. This suggests the use of a delta function for a mathematical model. We can write the divergence in terms of a delta function†

$$\boldsymbol{\nabla} \cdot \mathbf{E} = \frac{q}{\varepsilon} \delta(r,\theta,\phi) = \frac{q}{\varepsilon} \frac{\delta(r)}{4\pi r^2} \quad (8)$$

† We can show that Eq. (8) is correct by substituting it into the divergence theorem of Sec. 1.8 and showing that an identity is obtained. We assume a spherical neighborhood about the origin with a volume v and a surface s:

$$\int_v \boldsymbol{\nabla} \cdot \mathbf{E} \, dv = \oint_s \mathbf{E} \cdot d\mathbf{s}$$

$$\int_v \frac{q}{\varepsilon} \delta(r,\theta,\phi) \, dv = \oint_s \frac{q}{4\pi\varepsilon} \frac{\mathbf{a}_r}{r^2} \cdot r^2 \sin\theta \, d\theta \, d\phi \, \mathbf{a}_r$$

$$\frac{q}{\varepsilon} \int_v \delta(r,\theta,\phi) \, dv = \frac{q}{4\pi\varepsilon} \int_0^{2\pi} \int_0^{\pi} \sin\theta \, d\theta \, d\phi$$

Since the integral of the delta function is unity, we obtain the identity $1 = 1$.

FUNDAMENTALS OF FIELDS

The term on the right was obtained from Eq. (12) of Sec. 1.5. In conclusion, the divergence of a vector field emanating from a point source is a scalar field which is zero everywhere except at the origin, where it is a delta function.

Since it is often difficult to compute divergence using the basic definition of Eq. (1), equivalent forms for divergence have been developed for each of the common coordinate systems. These equations are given in Table 1.7.1. As an illustration of the method, we will develop the equation for the rectangular coordinate system from the basic definition, Eq. (1), by summing the integrals $\int \mathbf{E} \cdot d\mathbf{s}$ over each face of a cube.

Suppose the $\mathbf{E}$ field at a given point $p_0 = (x_0, y_0, z_0)$ is

$$\mathbf{E}_0 = E_{x0}\mathbf{a}_x + E_{y0}\mathbf{a}_y + E_{z0}\mathbf{a}_z \tag{9}$$

Let p_0, p_1, p_2, p_3, and p_4 be five of the corner points of an incremental rectangular region as shown in Fig. 1.7.4. Assume that the region is small enough for the x component E_x of the $\mathbf{E}$ field to be essentially constant over the face of the cube defined by points p_0, p_2, and p_3, that is, over the "back face." Since p_0 lies on this surface, the x component is essentially E_{x0} over the entire back face. Consequently, the integral of the dot product of the $\mathbf{E}$ field and the outward normal $d\mathbf{s} = -dy\,dz\,\mathbf{a}_x$

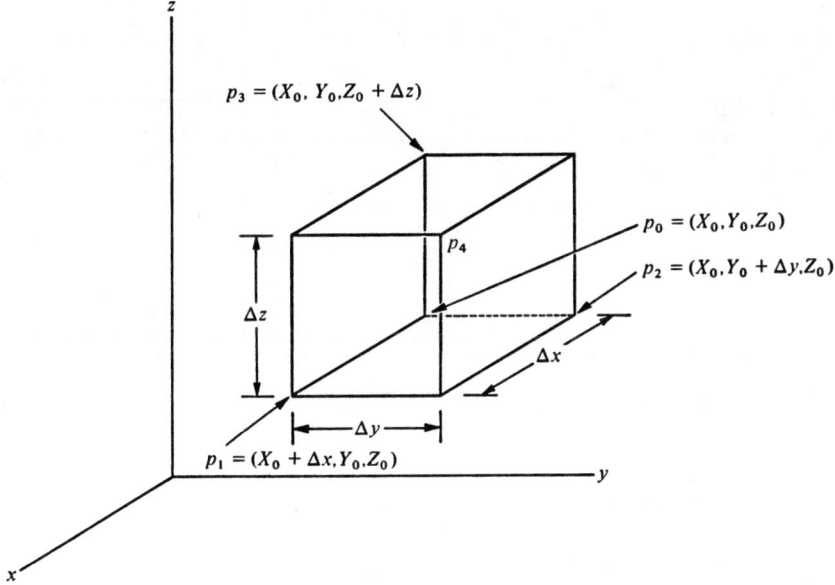

Fig. 1.7.4 An incremental rectangular region for computing divergence.

Table 1.7.1 Gradient, divergence, curl, and scalar laplacian in the common coordinate systems

Vector: $\mathbf{A} = A_u \mathbf{a}_u + A_v \mathbf{a}_v + A_w \mathbf{a}_w$
Scalar: Ψ

Gradient

Rectangular: $\nabla \Psi = \dfrac{\partial \Psi}{\partial x} \mathbf{a}_x + \dfrac{\partial \Psi}{\partial y} \mathbf{a}_y + \dfrac{\partial \Psi}{\partial z} \mathbf{a}_z$ (1)

Cylindrical: $\nabla \Psi = \dfrac{\partial \Psi}{\partial r} \mathbf{a}_r + \dfrac{1}{r}\dfrac{\partial \Psi}{\partial \phi} \mathbf{a}_\phi + \dfrac{\partial \Psi}{\partial z} \mathbf{a}_z$ (2)

Spherical: $\nabla \Psi = \dfrac{\partial \Psi}{\partial r} \mathbf{a}_r + \dfrac{1}{r}\dfrac{\partial \Psi}{\partial \theta} \mathbf{a}_\theta + \dfrac{1}{r \sin \theta}\dfrac{\partial \Psi}{\partial \phi} \mathbf{a}_\phi$ (3)

Divergence

Rectangular: $\nabla \cdot \mathbf{A} = \dfrac{\partial A_x}{\partial x} + \dfrac{\partial A_y}{\partial y} + \dfrac{\partial A_z}{\partial z}$ (4)

Cylindrical: $\nabla \cdot \mathbf{A} = \dfrac{1}{r}\dfrac{\partial r A_r}{\partial r} + \dfrac{1}{r}\dfrac{\partial A_\phi}{\partial \phi} + \dfrac{\partial A_z}{\partial z}$ (5)

Spherical: $\nabla \cdot \mathbf{A} = \dfrac{1}{r^2}\dfrac{\partial r^2 A_r}{\partial r} + \dfrac{1}{r \sin \theta}\left(\dfrac{\partial \sin \theta \, A_\theta}{\partial \theta} + \dfrac{\partial A_\phi}{\partial \phi}\right)$ (6)

Curl

Rectangular: $\nabla \times \mathbf{A} = \left(\dfrac{\partial A_z}{\partial y} - \dfrac{\partial A_y}{\partial z}\right)\mathbf{a}_x + \left(\dfrac{\partial A_x}{\partial z} - \dfrac{\partial A_z}{\partial x}\right)\mathbf{a}_y + \left(\dfrac{\partial A_y}{\partial x} - \dfrac{\partial A_x}{\partial y}\right)\mathbf{a}_z$ (7)

Cylindrical: $\nabla \times \mathbf{A} = \left(\dfrac{1}{r}\dfrac{\partial A_z}{\partial \phi} - \dfrac{\partial A_\phi}{\partial z}\right)\mathbf{a}_r + \left(\dfrac{\partial A_r}{\partial z} - \dfrac{\partial A_z}{\partial r}\right)\mathbf{a}_\phi + \dfrac{1}{r}\left(\dfrac{\partial r A_\phi}{\partial r} - \dfrac{\partial A_r}{\partial \phi}\right)\mathbf{a}_z$ (8)

Spherical: $\nabla \times \mathbf{A} = \dfrac{1}{r \sin \theta}\left(\dfrac{\partial \sin \theta \, A_\phi}{\partial \theta} - \dfrac{\partial A_\theta}{\partial \phi}\right)\mathbf{a}_r + \dfrac{1}{r}\left(\dfrac{1}{\sin \theta}\dfrac{\partial A_r}{\partial \phi} - \dfrac{\partial r A_\phi}{\partial r}\right)\mathbf{a}_\theta$
$\qquad + \dfrac{1}{r}\left(\dfrac{\partial r A_\theta}{\partial r} - \dfrac{\partial A_r}{\partial \theta}\right)\mathbf{a}_\phi$ (9)

Scalar laplacian

Rectangular: $\nabla^2 \Psi = \dfrac{\partial^2 \Psi}{\partial x^2} + \dfrac{\partial^2 \Psi}{\partial y^2} + \dfrac{\partial^2 \Psi}{\partial z^2}$ (10)

Cylindrical: $\nabla^2 \Psi = \dfrac{1}{r}\dfrac{\partial}{\partial r}\left(r \dfrac{\partial \Psi}{\partial r}\right) + \dfrac{1}{r^2}\dfrac{\partial^2 \Psi}{\partial \phi^2} + \dfrac{\partial^2 \Psi}{\partial z^2}$ (11)

Spherical: $\nabla^2 \Psi = \dfrac{1}{r^2}\dfrac{\partial}{\partial r}\left(r^2 \dfrac{\partial \Psi}{\partial r}\right) + \dfrac{1}{r^2 \sin \theta}\dfrac{\partial}{\partial \theta}\left(\sin \theta \dfrac{\partial \Psi}{\partial \theta}\right) + \dfrac{1}{r^2 \sin^2 \theta}\dfrac{\partial^2 \Psi}{\partial \phi^2}$ (12)

FUNDAMENTALS OF FIELDS

over that face is just

$$\int_{\Delta y\, \Delta z} \mathbf{E} \cdot d\mathbf{s} = \int_{\Delta y\, \Delta z} (E_{x0}\mathbf{a}_x + E_{y0}\mathbf{a}_y + E_{z0}\mathbf{a}_z) \cdot (-\mathbf{a}_x\, dy\, dz)$$
$$= -E_{x0}\, \Delta y\, \Delta z \tag{10}$$

With similar reasoning we find that the integral over the surface containing points p_0, p_1, and p_3 is

$$\int_{\Delta x\, \Delta z} \mathbf{E} \cdot d\mathbf{s} = -E_{y0}\, \Delta x\, \Delta z \tag{11}$$

and over the surface containing points p_0, p_1, and p_2 is

$$\int_{\Delta x\, \Delta y} \mathbf{E} \cdot d\mathbf{s} = -E_{z0}\, \Delta x\, \Delta y \tag{12}$$

Thus we have found $\int \mathbf{E} \cdot d\mathbf{s}$ over three of the six surfaces.

To find the value of the integral over the front face, containing points p_1 and p_4, we assume that the change in E_x in moving from point p_0 to point p_1 can be approximated by the first two terms of a Taylor series†

$$E_x = E_{x0} + \frac{\partial E_x}{\partial x} \Delta x \tag{13}$$

In essence, we are assuming that the change in the x component can be described by a linear equation, as shown in Fig. 1.7.5. We see that the error caused by dropping the higher-degree terms of the Taylor series is

† The Taylor series of $f(x)$ in the neighborhood of a point x_0 is

$$f(x) = f(x_0) + f'(x_0)(x - x_0) + \frac{f''(x_0)}{2!}(x - x_0)^2 + \cdots$$
$$+ \frac{f^{(n)}(x_0)}{n!}(x - x_0)^n + \cdots$$

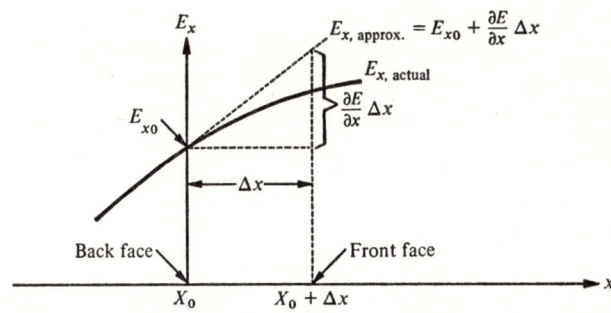

Fig. 1.7.5 Approximation of E_x at the front face of the cube.

negligible if Δx is small enough. We further assume that E_x is essentially constant over the front face. Thus, over the front face

$$\int_{\Delta y\, \Delta z} \mathbf{E} \cdot d\mathbf{s} = \int_{\Delta y\, \Delta z} (E_x \mathbf{a}_x + E_y \mathbf{a}_y + E_z \mathbf{a}_z) \cdot (dy\, dz\, \mathbf{a}_x)$$

$$= E_x\, \Delta y\, \Delta z = \left(E_{x0} + \frac{\partial E_x}{\partial x} \Delta x\right) \Delta y\, \Delta z \tag{14}$$

By similar reasoning, we find the value of the integral over the remaining two faces to be

$$\int_{\Delta x\, \Delta z} \mathbf{E} \cdot d\mathbf{s} = \left(E_{y0} + \frac{\partial E_y}{\partial y} \Delta y\right) \Delta x\, \Delta z \tag{15}$$

for the face containing points p_2 and p_4 and

$$\int_{\Delta x\, \Delta y} \mathbf{E} \cdot d\mathbf{s} = \left(E_{z0} + \frac{\partial E_z}{\partial z} \Delta z\right) \Delta x\, \Delta y \tag{16}$$

for the face containing points p_3 and p_4.

Now we can find the integral over the entire closed surface, $\oint \mathbf{E} \cdot d\mathbf{s}$, by adding the right-hand sides of Eqs. (10), (11), (12), (14), (15), and (16), to obtain

$$\oint \mathbf{E} \cdot d\mathbf{s} = \left(\frac{\partial E_x}{\partial x} + \frac{\partial E_y}{\partial y} + \frac{\partial E_z}{\partial z}\right) \Delta x\, \Delta y\, \Delta z \tag{17}$$

Next the integral is substituted into Eq. (1), the basic definition of the divergence, to obtain the final result

$$\operatorname{div} \mathbf{E} \equiv \nabla \cdot \mathbf{E} = \lim_{\Delta v \to 0} \frac{\oint \mathbf{E} \cdot d\mathbf{s}}{\Delta x\, \Delta y\, \Delta z} = \frac{\partial E_x}{\partial x} + \frac{\partial E_y}{\partial y} + \frac{\partial E_z}{\partial z} \tag{18}$$

This result is valid in the rectangular coordinate system, and generally it is much easier to apply than the basic definition. Simplified forms for the divergence in other coordinate systems are given in Tables 1.2.2 and 1.7.1.

In the first example of this section, we used Eq. (1) to compute the divergence of a uniform field, $\mathbf{E} = k\mathbf{a}_x$ [see Eq. (2)], and we found that $\nabla \cdot \mathbf{E} = 0$. Now this same result can be obtained very easily using Eq. (18):

$$\nabla \cdot \mathbf{E} = \frac{\partial E_x}{\partial x} + \frac{\partial E_y}{\partial y} + \frac{\partial E_z}{\partial z} = \frac{\partial k}{\partial x} = 0 \tag{19}$$

Similarly, in Example 2, with a point charge at the origin of a spherical coordinate system, we found a radial $\mathbf{E}$ field which followed the inverse-square law and which had a divergence $\nabla \cdot \mathbf{E} = q\delta(r,\theta,\phi)/\varepsilon$. It

FUNDAMENTALS OF FIELDS

will be left as an exercise to substitute this **E** field into Eq. (6) of Table 1.7.1, and show that $\nabla \cdot \mathbf{E} = 0$ except at the origin, where it is indeterminate (of the form 0/0).

PROGRAMMED EXERCISE 1.7.1
DIVERGENCE OF THE E FIELD IN A DIODE

Suppose we are given a pair of parallel plates in which separation is small enough to allow us to neglect fringing of the fields or in which we assume plates of infinite extent. The plates are connected across the terminals of a battery of voltage V_0, as shown in Fig. 1.7.6a, and the negative plate is a cathode (an emitter of electrons). The voltage between the plates varies as the $\frac{4}{3}$ power of the distance (see Figure 1.7.6b), as derived from the Child-Langmuir law,† when the current is space-charge-limited. Between the plates is a cloud of electrons, which is very dense near the cathode and which decreases in density in regions near the positive plate. The convection current from the plate to the cathode is constant, so that the current density $\mathbf{i} = \rho \mathbf{v}$ (where ρ is the space-charge density and $\mathbf{v}$ is the electron velocity) is constant.

A series of true statements will now be made about the parallel-plate diode of Fig. 1.7.6. The statements are organized to guide the reader through a certain line of reasoning and to illustrate the relationships of a vector field **E** to its potential field and divergence. The reader should verify each statement.

1. We will assume that the Child-Langmuir law is correct, and so the potential field, measured from the cathode, is

$$V(y) = ky^{4/3} \qquad k = \text{const} \tag{20}$$

   as shown in Fig. 1.7.6b. If there were no charge between the plates (as in a parallel-plate capacitor), the potential field would be a linear function of y; therefore the effect of the space charge is to lower the potential curve. Equipotential surfaces (surfaces of constant voltage) are planes parallel to the plates.

2. Associated with the equipotential surfaces is a gradient (vector) field, and the negative of the gradient is the **E** field (which is proportional to the force a positive particle would experience between the plates). The **E** field is

$$\mathbf{E} = -\nabla V = -\tfrac{4}{3} k y^{1/3} \mathbf{a}_y \tag{21}$$

† Most texts on electron devices derive this law; e.g., K. R. Spangenberg, "Fundamentals of Electron Devices," McGraw-Hill Book Company, New York, 1957.

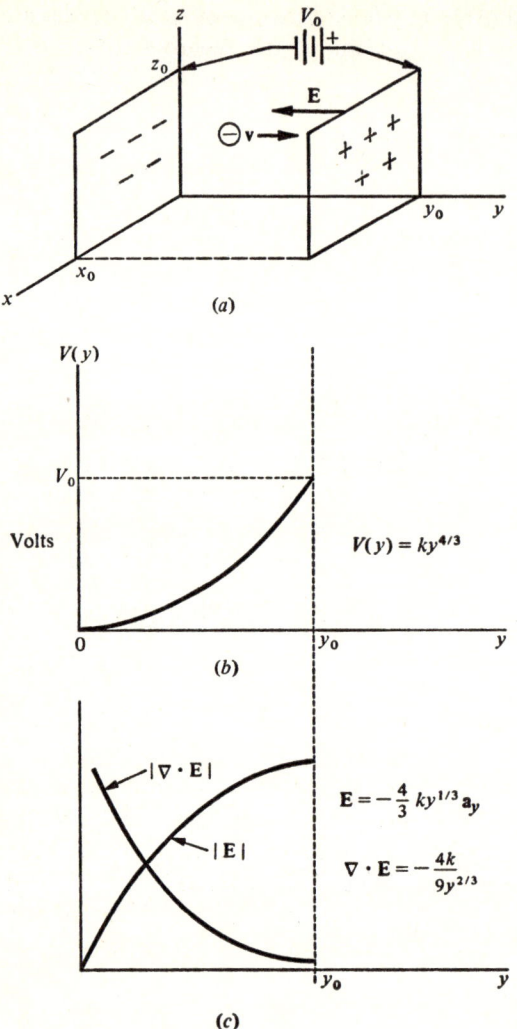

Fig. 1.7.6 Divergence of the **E** field shows the presence of space charge between the plates. (a) A section of the infinite parallel plates. (b) Voltage variation with distance. (c) $|\mathbf{E}|$ and $|\nabla \cdot \mathbf{E}|$ variation with distance.

A sketch of the magnitude of **E** is shown in Fig. 1.7.6c. If there were no charged particles between the plates, **E** would be a constant. Since the **E** field diminishes with smaller values of y, flux lines are terminating on the charged particles.

3. Between the plates, any incremental volume that contains charge will

FUNDAMENTALS OF FIELDS

have more flow lines entering the volume (some of which terminate on charge within the volume) than leaving it. Thus, the divergence of the **E** field is not zero but

$$\nabla \cdot \mathbf{E} = -\frac{4k}{9y^{2/3}} \tag{22}$$

which is sketched in Fig. 1.7.6c. The nonzero, negative divergence indicates the presence of negative charge density.

1.8 GAUSS' LAW, THE DIVERGENCE THEOREM, AND POISSON'S EQUATION

GAUSS' LAW

Coulomb's law (Sec. 1.3) gave us a method of computing the force of a small test charge q_t due to a charge q_1 when the two point charges are R m apart:

$$\mathbf{F} = \frac{q_1 q_t}{4\pi\varepsilon_0 R^2} \mathbf{a}_R \tag{1}$$

We were able to find the electric field intensity **E** by dividing the force by the test charge

$$\mathbf{E} = \frac{\mathbf{F}}{q_t} \tag{2}$$

In these equations it was assumed that the charges are located in free space so that the fields **E** and **F** are inversely proportional to ε_0, the permittivity of free space. If the vector fields are in a medium other than free space, the ε_0 in Eq. (1) must be replaced by the permittivity ε of the medium. The permittivity or dielectric constant is commonly written in terms of the permittivity of free space:

$$\varepsilon = \varepsilon_0 \varepsilon_r \tag{3}$$

where ε_r is called the *relative permittivity* or the *relative dielectric constant*. It varies from unity for free space to about 80 for water to about 1200 for barium titanate. Because of the permittivity, the magnitudes of the vector fields **F** and **E** depend upon the medium for any given geometric charge distribution. It is convenient to define a new vector field, the *flux-density field* **D**, which is independent of the medium in which the charges are embedded. We can remove the effect of the medium by multiplying the **E** field by the permittivity, and thus we define the flux density **D**

$$\mathbf{D} = \varepsilon \mathbf{E} \tag{4}$$

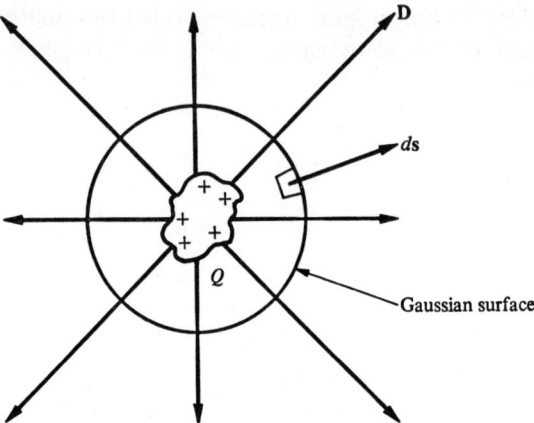

Fig. 1.8.1 The charge can be determined by integrating the flux density over the gaussian surface.

A field mapping for the **D** field is the same as the mapping of the **E** field except for the magnitudes of the vectors.

If we enclose charge inside any simple closed surface, such as the spherical surface shown in Fig. 1.8.1, the number of flux lines that pierce the surface depends upon the amount of charge enclosed. Often a very convenient way to determine the amount of charge enclosed by a simple surface is to integrate the flux-density vector **D** over the entire surface (we assume that D and ρ are single-valued):

$$\boxed{\oint_s \mathbf{D} \cdot d\mathbf{s} = Q = \int_v \rho \, dv} \tag{5}$$

This is called *Gauss' law*. Under the integral on the right-hand side of Gauss' law, the variable ρ represents the charge density, which can vary from point to point inside the region, and v is the volume bounded by the closed surface s, called a *gaussian surface*. According to Gauss' law, one can obtain the total charge Q inside a region either by integrating the charge density ρ over the entire volume or by integrating the flux-density vector **D** over the surface bounding the volume.†

---

† We assume that s is a simple closed surface and that $\oint \mathbf{D} \cdot d\mathbf{s}$ exists over the surface. Discontinuities and singularities of **D** are allowed provided the integral exists and provided the volume v includes the boundary points. As stated, the theorem applies to the case where **D** is time-invariant. When **D** varies with time, propagation time becomes important, as the following example illustrates. Suppose a nuclear explosion at the origin of our coordinate system suddenly creates a **D** field which propagates

Physically, the law has a rather surprising meaning. For example, if a friend gave you a package filled with charged gum wrappers, you could tell how much charge is inside the package without opening it by running a flux detector over the surface. Thus if toys had unique vector fields and children had flux detectors, there would be no surprises at Christmas.

A useful application of Gauss' law is to determine the vector field **D** (and consequently the vector field **E**) when the charge distribution is known. This is particularly convenient for simple, symmetrical charge distributions. Here we assume that the charge Q and its distribution are known and that the variable **D** under the integral sign of Eq. (5) is the unknown. Since Gauss' law is valid for any simple closed surface, we are free to choose any surface that is convenient for the geometry of the problem. In particular, we would like to be able to choose a surface that satisfies the following two conditions:

1. **D** is either normal or tangential to this surface at every point.
2. Where **D** is normal we choose the surface so that the magnitude of the vector is constant over the surface.

When these conditions are satisfied, we can solve for the magnitude of the vector **D**. In this application it is assumed that we know enough about the field mapping to be able to choose a surface which satisfies the above criteria. Often this is the case for simple, symmetrical distributions.

Example 1 An almost trivial example of the above application of Gauss' law will illustrate the technique. Suppose that the moon is uniformly charged throughout its entire volume with a charge density of ρ, so that the total charge is Q (which could be obtained by integrating ρ over the spherical volume of the moon). Because of the symmetrical charge distribution, all the flux lines are radial from the moon, and the vector field **D** is of the form $D_r a_r$, in spherical coordinates, as shown in Fig. 1.8.2. Thus if we choose a concentric spherical gaussian surface which encloses the moon, all the flux lines will be normal to the gaussian surface. At any given radius r, D_r is a constant over the entire spherical shell. Therefore the given spherical surface satisfies the two criteria for a gaussian surface stated above. Assuming that the total charge Q is known, we

---

with a finite velocity. If we choose the gaussian surface s far from the origin, the effect of the explosion will not be felt for a finite time t_0, the time that it takes for the wave to propagate to the surface. Immediately after the explosion, integration over the surface far from the origin will not yield the same result as an integration over a surface very near the origin. Often one can neglect the transient effect by assuming that the change in the field is instantaneously felt over the entire region of interest. Time-varying phenomena will be discussed, starting in Chap. 2.

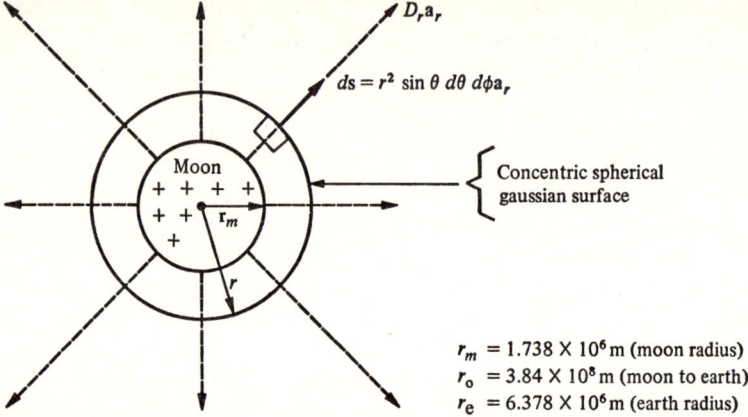

Fig. 1.8.2 Gauss' law can be used to find the **D** field for a known symmetrical charge distribution.

can apply Gauss' law to find the unknown D_r

$$\oint_s \mathbf{D} \cdot d\mathbf{s} = \int_0^{2\pi} \int_0^{\pi} D_r \mathbf{a}_r \cdot r^2 \sin\theta \, d\theta \, d\phi \, \mathbf{a}_r = Q$$

$$= D_r r^2 \int_0^{2\pi} d\phi \int_0^{\pi} \sin\theta \, d\theta = Q$$

$$= D_r r^2 (2\pi)(2) = Q \qquad (6)$$

Thus at any arbitrary radius r we obtain the magnitude of the field **D**

$$D_r = \frac{Q}{4\pi r^2} \qquad (7)$$

Since we know the vector field mapping from the symmetry of the charge distribution, we can write the equations for the vector fields **D** and **E**

$$\mathbf{D} = D_r \mathbf{a}_r \qquad (8)$$

$$\mathbf{E} = \frac{D_r}{\varepsilon_0} \mathbf{a}_r \qquad (9)$$

The resulting **E** field is the same as that obtained using Eq. (10) of Sec. 1.3 (which is based on Coulomb's law), but we find it much simpler to use Gauss' law.

It is not inconceivable that the moon and earth each have some net electrostatic charge and therefore have an electrostatic force of repulsion (or attraction) between them. Let us assume that each body has a uniform charge density of one electron per cubic meter. By using the mean radii given in Fig. 1.8.2 the reader can compute the volume of both the earth and the moon and determine the total charge on each

FUNDAMENTALS OF FIELDS

body, under the above assumption. The reader will find a net charge of $Q_m = -3.52$ C on the moon and a net charge of $Q_e = -173.7$ C on the earth. From Eq. (9) we can compute the magnitude of the **E** field at the surface of the earth due to the moon

$$|\mathbf{E}| = \frac{Q_m}{4\pi\varepsilon_0 r_0^2} = 2.16 \times 10^{-7} \text{ V/m} \tag{10}$$

This field turns out to be surprisingly small because of the large distance between the earth and the moon. If we assume that the **E** field due to the charge on the moon is essentially uniform at the earth (or assume that the charge on the earth is concentrated at one point), we can easily find the electrostatic force of repulsion between the two bodies

$$|\mathbf{F}| = |Q_e \mathbf{E}| = 3.82 \times 10^{-15} \text{ N} \approx 8.53 \times 10^{-16} \text{ lb} \tag{11}$$

It is obvious that the electrostatic forces are negligible compared with the gravitational forces, at least for the charge densities assumed here.

We now derive the differential form (sometimes called the point form) of Gauss' law. The divergence of the flux density field **D** [from Equation (1) of Sec. 1.7]

$$\boldsymbol{\nabla} \cdot \mathbf{D} = \lim_{\Delta v \to 0} \frac{\oint_s \mathbf{D} \cdot d\mathbf{s}}{\Delta v} \tag{12}$$

However, the integral of the flux density over any closed surface s is equal to the charge Q enclosed in the surface

$$\boldsymbol{\nabla} \cdot \mathbf{D} = \lim_{\Delta v \to 0} \frac{Q}{\Delta v} \tag{13}$$

As the volume shrinks to zero, the right-hand side of Eq. (13) reduces to the charge density ρ, and we obtain Gauss' law in point form

$$\boxed{\boldsymbol{\nabla} \cdot \mathbf{D} = \rho} \tag{14}$$

This important vector partial differential equation is one of Maxwell's equations. Theoretically, it is possible to solve this equation for the unknown field vector **D** if the charge distribution ρ is known. Although we will often use Gauss' law in point form, we will not actually solve the partial differential equation in this course. Generally in electrostatics it is simpler to find the scalar potential field V and then find the vector fields **D** and **E** from the gradient of the scalar field rather than solve the vector differential equation above.

THE DIVERGENCE THEOREM

If we substitute Eq. (14) into Eq. (5) for ρ, we obtain the *divergence theorem*

$$\oint_s \mathbf{D} \cdot d\mathbf{s} = \int_v \nabla \cdot \mathbf{D} \, dv \tag{15}$$

This theorem says that the integral of the normal component of **D** over the closed surface s is equal to the integral of the divergence of **D** over the volume v (contained in the closed surface). Note that the integral on the left is a double integral while the integral on the right is a triple integral. K. F. Gauss discovered this famous theorem in 1813 and showed its validity for *any vector function* possessing continuous first partial derivatives throughout the region v. Thus we can apply the theorem to any of the vector fields (**D**, **E**, **B**, and **H**, etc.) that we are concerned with in this course.

Example 2 To gain some familiarity with the divergence theorem, we now compute the total charge in a finite region between the plates of the parallel-plate diode of Fig. 1.7.6. First we compute the charge using the left-hand side of Eq. (15) by integrating over the enclosed surface defined by the rectangular region between the plates. Since the **E** field was found in Eq. (21) of Sec. 1.7, we can now find the **D** field by multiplying by the permittivity of free space. The direction of the **D** field is $-\mathbf{a}_y$, and so the dot product $\mathbf{D} \cdot d\mathbf{s}$ is zero except at the two plates, where

$$d\mathbf{s} = \pm dx \, dz \, \mathbf{a}_y$$

Also the **D** field is zero at $y = 0$, and consequently the left-hand side of the divergence theorem reduces to the following integration over the plate at $y = y_0$:

$$\oint_s \mathbf{D} \cdot d\mathbf{s} = \int_0^{z_0} \int_0^{x_0} \frac{4}{3} \varepsilon_0 k y_0^{1/3} \mathbf{a}_y \cdot (dx \, dz \, \mathbf{a}_y) \tag{16}$$

By Gauss' law, the integral on the left is the charge contained in the volume between the plates, and so

$$Q = -\frac{4}{3} \varepsilon_0 k x_0 y_0^{1/3} z_0 \tag{17}$$

Now we obtain the same result by using the right-hand side of the divergence theorem to compute the total charge. Direct substitution into the integral yields

$$\int_v \nabla \cdot \mathbf{D} \, dv = \int_0^{z_0} \int_0^{y_0} \int_0^{x_0} \frac{-4\varepsilon_0 k}{9 y^{2/3}} \, dx \, dy \, dz$$

$$= -\frac{4}{3} \varepsilon_0 k x_0 y_0^{1/3} z_0 \tag{18}$$

The integration of Eq. (18) was slightly more involved than that of Eq. (16), but the result is the same in either case. The divergence theorem is commonly used to replace a volume integral [such as the one in Eq. (18)] with a simpler surface integral [such as the one in Eq. (16)].

FUNDAMENTALS OF FIELDS

In concluding this example, it is worth recalling that the divergence of the **D** field is equal to the charge density, as given in Eq. (14). Since the divergence of the **D** field and that of the **E** field differ only by the constant ε_0, a plot of the charge density ρ will be proportional to the plot of the divergence of **E**, as given in Fig. 1.7.6c. The equation for the divergence of **D**,

$$\boldsymbol{\nabla} \cdot \mathbf{D} = \frac{-4\varepsilon_0 k}{9y^{2/3}} \tag{19}$$

shows that the charge density approaches infinity as y approaches zero. However, the integral of the charge density is finite, as shown in Eq. (18).

POISSON'S EQUATION

The partial differential equation for the divergence of **D**, as given in Eq. (14), is a differential equation whose dependent variable is a vector. For simplicity, we prefer to work with a differential equation whose dependent variable is a scalar function rather than a vector function. To that end, we now derive Poisson's equation for the scalar potential function V.

The electric field intensity is given by the gradient of the potential field (when $\boldsymbol{\nabla} \times \mathbf{E} = 0$)

$$\mathbf{E} = -\boldsymbol{\nabla} V \tag{20}$$

and so we can obtain a relationship between the **D** field and the potential field by multiplying by the permittivity. We substitute **E** for **D** in Gauss' law in the differential form

$$\boldsymbol{\nabla} \cdot \mathbf{D} = \boldsymbol{\nabla} \cdot (\varepsilon \mathbf{E}) = -\varepsilon \boldsymbol{\nabla} \cdot \boldsymbol{\nabla} V = \rho \tag{21}$$

The divergence of the gradient of a scalar function ($\boldsymbol{\nabla} \cdot \boldsymbol{\nabla} V$), called the *scalar laplacian*, is denoted by $\nabla^2 V$. We obtain *Poisson's equation* from Eq. (21):

$$\boxed{\nabla^2 V = -\frac{\rho}{\varepsilon}} \tag{22}$$

In the event that the charge density is zero in the region of interest, Poisson's equation reduces to *Laplace's equation*

$$\boxed{\nabla^2 V = 0} \tag{23}$$

Equations (22) and (23) are *scalar* partial differential equations for the scalar potential function V. Generally these equations can be solved for V, and then one can obtain the vector fields from the gradient, as shown in Eq. (20).

An equation for the scalar laplacian can be obtained in any coordinate system by performing the gradient and divergence operations as indicated in Eq. (21). Results of these operations for the common coordinate systems are given in Table 1.7.1. For example, in rectangular coordinates Poisson's equation is

$$\frac{\partial^2 V}{\partial x^2} + \frac{\partial^2 V}{\partial y^2} + \frac{\partial^2 V}{\partial z^2} = -\frac{\rho}{\varepsilon} \tag{24}$$

The solution of this equation will depend upon the boundary conditions and the charge distribution ρ. In the literature, such problems are generally called *boundary-value problems*. However, Poisson's equation is valid only where a scalar potential field V exists; i.e., when $\nabla \times \mathbf{E} = 0$. Poisson's equation is useful for finding static fields due to static charge distributions. Since we are largely concerned with dynamic fields, we will not find it necessary to solve Poisson's equation, although a solution integral is given in Sec. 1.11. Throughout most of this text we will be concerned with fields where $\nabla \times \mathbf{E} \neq 0$. It will be shown that electromagnetic fields which have nonzero curl are time-varying and lead to another partial differential equation called the *wave equation*, which will be developed in Chap. 2.

1.9 THE CURL

DESCRIPTION OF THE CURL

Suppose an electric field intensity $\mathbf{E}$ exists in a finite region and is defined by

$$\mathbf{E} = -z\mathbf{a}_y \tag{1}$$

The flow lines are straight lines in the negative $\mathbf{a}_y$ direction, as shown in Fig. 1.9.1. The field increases in magnitude with increasing values of z, but it is uniform in the $\mathbf{a}_x$ and $\mathbf{a}_y$ directions.

We will determine the work required to move a positive test charge q_t around a closed rectangular loop located in the yz plane, as shown in Fig. 1.9.1. The work can be found from the line integral of the force

$$W = -\oint q_t \mathbf{E} \cdot d\mathbf{l} = \left(\int_{P_1}^{P_2} + \int_{P_2}^{P_3} + \int_{P_3}^{P_4} + \int_{P_4}^{P_1} \right) (-q_t \mathbf{E} \cdot d\mathbf{l})$$
$$= W_1 + W_2 + W_3 + W_4 \tag{2}$$

The integrals are (see Fig. 1.9.1)

$$W_1 = q_t \int_{y_1}^{y_2} z_1 \mathbf{a}_y \cdot \mathbf{a}_y \, dy = q_t z_1 (y_2 - y_1) \tag{3}$$

$$W_2 = q_t \int_{z_1}^{z_2} z \mathbf{a}_y \cdot \mathbf{a}_z \, dz = 0 \tag{4}$$

$$W_3 = q_t \int_{y_2}^{y_1} z_2 \mathbf{a}_y \cdot \mathbf{a}_y \, dy = -q_t z_2 (y_2 - y_1) \tag{5}$$

$$W_4 = q_t \int_{z_2}^{z_1} z \mathbf{a}_y \cdot \mathbf{a}_z \, dz = 0 \tag{6}$$

FUNDAMENTALS OF FIELDS

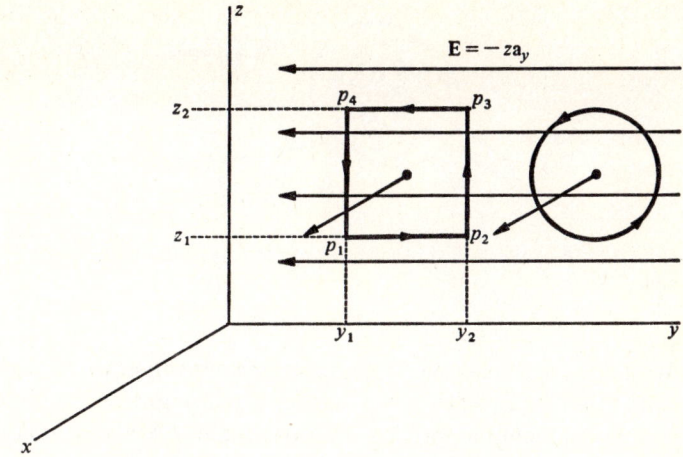

Fig. 1.9.1 Integral paths in a nonconservative field.

The work (from an external source) required to move the test charge around the loop is the sum of the integrals

$$W = q_t(z_1 - z_2)(y_2 - y_1) = -q_t(z_2 - z_1)(y_2 - y_1) \tag{7}$$

We see that if $z_2 > z_1$ and $y_2 > y_1$, the net work required from the source is negative. Thus for the given path of integration, energy from the field is used to move the charge. If we reverse the direction of integration, energy from an external source would be required to move the charge around the path. We can draw the following conclusions from this example:

1. The closed line integral $\oint \mathbf{E} \cdot d\mathbf{l}$ of the field is not zero (as is required for a conservative field). Any field for which this condition exists is called *nonconservative*.
2. The orientation of the plane of the loop, or path, affects the result. For example, if we rotate the loop so that it lies in the xz plane, $W = 0$.
3. $\oint \mathbf{E} \cdot d\mathbf{l}$ is not independent of the path of integration, as shown by the terms in parentheses in Eq. (7).

We would arrive at these same conclusions regardless of the shape of the loop, although particular numerical results will differ. That is, the circular loop of Fig. 1.9.1 would serve our purpose just as well as the rectangular loop.

An important conclusion is that this **E** field is not conservative and could therefore be used to accelerate an electron around a closed path.

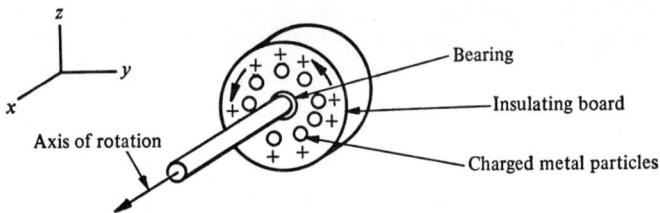

Fig. 1.9.2 A hypothetical testing device for detecting nonconservative **E** fields.

We might ask whether there are other nonconservative fields, in addition to the field of Eq. (1), which can be used to do work and if so, whether we can find a simple way to identify them. We can imagine building a simple testing device to detect such fields by attaching many small equally charged metal particles to the surface of an insulating board. The pieces of metal can be arranged along any path of integration. Ideally, we would mount a bearing at the center of the board so that the axis of rotation would be perpendicular to the board, as shown in Fig. 1.9.2.

If the device is placed in the **E** field of Fig. 1.9.1 (assuming that the device does not distort the field) so the plane of the board is in the yz plane, the charged particles at the top will be in a more intense **E** field than those at the bottom. Since the electrostatic force on any particle is the product of the charge and the field vector, the board will rotate counterclockwise about its x axis. The speed of rotation of the board will increase or decrease until the energy lost due to friction and windage is equal to the net work being done by the field on the particles. Thus the rate of rotation of the device is a measure of how "nonconservative" the field happens to be. However, if we change the orientation of the device, the rate of rotation will change. For example, there would be no rotation if the axis of the device were aligned with the a_y or a_z vectors. If we want a true measure of how much work the field can do, we should search for the orientation where rotation is a maximum.

We can describe the situation mathematically by a single axial vector, as shown in Fig. 1.9.3. The magnitude of the axial vector is proportional to the rate of rotation, and the vector is directed along the axis of rotation in the direction of advance of a right-handed screw. The reader is probably familiar with the use of axial vectors to represent angular velocity, angular acceleration, and torque in the study of mechanics. The line integral $\oint \mathbf{E} \cdot d\mathbf{l}$, called the *circulation* of the field, is a measure of the net tangential forces acting to produce rotation; i.e., in a force field the circulation is the measure of the net torque acting to spin the testing device, or any other symmetrical body, in the field.

FUNDAMENTALS OF FIELDS

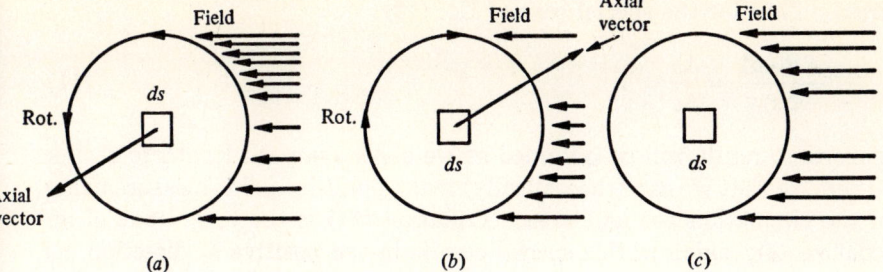

Fig. 1.9.3 Axial vectors for fields with rotation.

Intuitively we can see that a symmetrical body will detect circulation of a field whenever the flow lines are nonsymmetrical about the body, as shown in Fig. 1.9.3. Thus, if circulation is to exist, the field must have spatial variation, and thus its spatial derivatives are not all zero.

DEFINITION OF THE CURL

Our hypothetical testing device will not be a good detector of circulation of the field if the device distorts the field or if the field intensity fluctuates significantly around the circumference of the device, as shown in Fig. 1.9.3c. Ideally our detecting device must be very small so that we can detect variations in the field essentially point by point. We can make the device smaller by decreasing the length of the path of the line integral $\oint \mathbf{E} \cdot d\mathbf{l}$, which corresponds to decreasing the plane surface area bounded by the path. For example, in Fig. 1.9.1 we let the sides of the rectangular path shrink to

$$\Delta y = y_2 - y_1 \quad \text{and} \quad \Delta z = z_2 - z_1 \tag{8}$$

so that the plane surface area bounded by the curve is

$$\Delta s = \Delta y \, \Delta z \tag{9}$$

From Eqs. (2) and (7) the circulation integral is

$$\oint \mathbf{E} \cdot d\mathbf{l} = \frac{-W}{q_t} = (z_2 - z_1)(y_2 - y_1) = \Delta s \tag{10}$$

Now as the path length approaches zero, the circulation integral approaches zero and is not useful if we want point-by-point measurements. However, we can avoid the problem by measuring the circulation *per unit area*. That is, we divide the circulation integral by the area. Thus for the special case of Eq. (10)

$$\frac{\oint \mathbf{E} \cdot d\mathbf{l}}{\Delta s} = \frac{\Delta s}{\Delta s} = 1 \tag{11}$$

This same result will be obtained as we allow the path length to shrink to zero, so that we can (theoretically) make point-by-point measurements of the circulation per unit area. Equation (11) is the magnitude of an axial vector, which in this special case is in the positive $\mathbf{a}_z$ direction, as shown in Fig. 1.9.1. Finally, we can measure the nonconservative property of the field (the ability to do work around a closed path) using the axial vector given by the limit process

$$\left(\lim_{\Delta s \to 0} \frac{\oint \mathbf{E} \cdot d\mathbf{l}}{\Delta s} \right) \mathbf{a}_n = 1 \mathbf{a}_z \tag{12}$$

where $\mathbf{a}_n$ is normal to the plane of the testing device. This result is valid only for the special case where $\mathbf{E} = -z\mathbf{a}_y$ and the testing device is in the yz plane.

In general, for any field $\mathbf{E}(u,v,w)$ in any coordinate system, we define the *curl of the field* at the point (u,v,w)

$$\operatorname{curl} \mathbf{E}(u,v,w) \equiv \nabla \times \mathbf{E} \equiv \left(\lim_{\Delta s \to 0} \frac{\oint_C \mathbf{E} \cdot d\mathbf{l}}{\Delta s} \right)_{\max} \mathbf{a}_n \tag{13}$$

where the area Δs is bounded by the plane curve C and $\mathbf{a}_n$ is a unit vector normal to the surface Δs in the right-hand sense. The plane of the curve C must be oriented so that the magnitude of the vector is a maximum. This definition is independent of the coordinate system but is not in a mathematically convenient form.

SOME PROPERTIES OF THE CURL

Some concepts of the curl, based upon this definition [Eq. (13)], will be stated before we undertake further mathematical work.

1. The curl of a field at any point is an axial (rotational) vector with a magnitude equal to the maximum circulation per unit area.
2. The curl will exist if the flow lines around a symmetrical body tend to cause rotation. Thus in a force field, the circulation per unit area is proportional to the maximum torque on a testing device in the field, which in turn is proportional to the net rate of change of the field with respect to space.
3. The curl is *not* necessarily associated with curvature of the flow lines. For example, the flow lines of Fig. 1.9.1 are straight lines, but the

FUNDAMENTALS OF FIELDS

field has curl. Later we will see an example where the flow lines are curved but the field has no curl at those points. In this sense, the word "curl" is misleading since we are concerned with a rotational vector. In fact, some German authors use the more descriptive word "rotation" and write "rot E" for "curl E."

4. If a field is conservative, it has no curl, since $\oint \mathbf{E} \cdot d\mathbf{l} = 0$. This is the necessary and sufficient condition for the existence of equipotential surfaces. Therefore if a field has a curl, it has no equipotential surfaces.†

Recall that the gradient ∇V of a scalar field V yields a *vector* field. We also found that the divergence $\nabla \cdot \mathbf{E}$ of a *vector* $\mathbf{E}$ field yields a new *scalar* field. Similarly the curl $\nabla \times \mathbf{E}$ of a *vector* field yields a new *vector* field whose flow lines are usually orthogonal to the original flow lines. These three operations (gradient, divergence, and curl) are means of obtaining new vector and scalar fields from given fields. Often these new fields prove useful in understanding the original fields.

THE VECTOR COMPONENTS OF THE CURL

Next we consider the curl written in more useful forms than in Eq. (13). Since the curl is a vector, it can be decomposed into orthogonal components in a generalized coordinates system

$$\nabla \times \mathbf{E} = (\nabla \times \mathbf{E})_u \mathbf{a}_u + (\nabla \times \mathbf{E})_v \mathbf{a}_v + (\nabla \times \mathbf{E})_w \mathbf{a}_w \qquad (14)$$

where the notation for the components is defined by

$$(\nabla \times \mathbf{E})_i \equiv \lim_{\Delta s_i \to 0} \frac{\oint \mathbf{E} \cdot d\mathbf{l}}{\Delta s_i} \qquad i = u, v, w \qquad (15)$$

Thus if we use our hypothetical device to measure curl, we need not search for the orientation of maximum angular velocity. We can determine the curl by measuring its vector components in any three orthogonal directions and computing their sum. Mathematically, the problem of finding the curl reduces to the evaluation of the components of Eq. (15).

We will compute one component of the curl for the rectangular coordinate system to illustrate the method. For simplicity, we translate the origin of the coordinate system to the point $p_0 = (x_0, y_0, z_0)$, as shown in Fig. 1.9.4. To find the $\mathbf{a}_z$ component of the curl, we assume a rectangular path in the yz plane, with sides of differential lengths, so that the point

---

† There are some fields where $\nabla \times \mathbf{E} \neq 0$, but a scalar function f exists such that $\nabla \times f\mathbf{E} = 0$. Then $\mathbf{E} \cdot \nabla \times \mathbf{E} = 0$, and $\mathbf{E} = 1/f \, \nabla V$. The scalar function V is called a *quasipotential* function; see Moon and Spencer, *op. cit.*, p. 54.

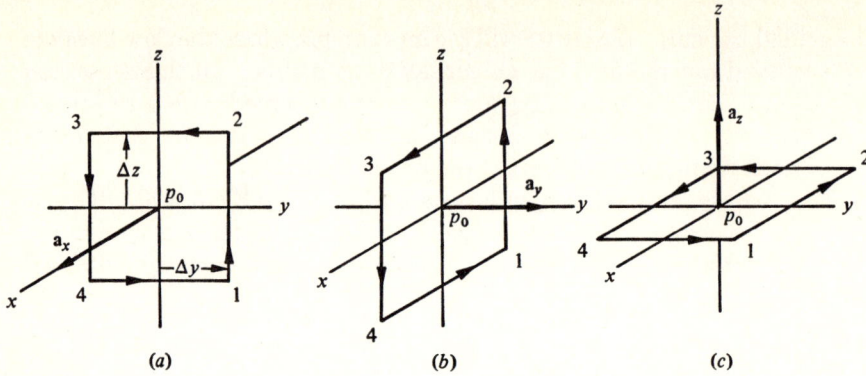

Fig. 1.9.4 The curl is determined by its component vectors in three orthogonal directions. (a) $(\nabla \times \mathbf{E})_x \mathbf{a}_x$. (b) $(\nabla \times \mathbf{E})_y \mathbf{a}_y$. (c) $(\nabla \times \mathbf{E})_z \mathbf{a}_z$.

p_0 is enclosed by the path, as shown in Fig. 1.9.4a. We assume that the rectangle is small enough (we will let its area shrink to zero in the limit) for the **E** field to be constant along any leg of the rectangle. If the **E** field at the point p_0 is $\mathbf{E}_0 = \mathbf{E}_{x0}\mathbf{a}_x + \mathbf{E}_{y0}\mathbf{a}_y + \mathbf{E}_{z0}\mathbf{a}_z$, the field along path 1-2 (from point 1 to 2) is approximately

$$\text{Path 1-2:} \quad \mathbf{E}_0 + \frac{\partial \mathbf{E}}{\partial y}\Delta y \approx \text{const}$$

so that a line integral along this path of length $2\Delta z$ is

$$\left(\mathbf{E}_0 + \frac{\partial \mathbf{E}}{\partial y}\Delta y\right) \cdot dz\, \mathbf{a}_z = \left(E_{z0} + \frac{\partial E_z}{\partial y}\Delta y\right) dz$$

$$= 2\left(E_{z0} + \frac{\partial E_z}{\partial y}\Delta y\right)\Delta z \quad (16)$$

Similarly for the other paths of Fig. 1.9.4a we find:

Path	Field	Line integral	
2-3	$\mathbf{E}_0 + \dfrac{\partial \mathbf{E}}{\partial z}\Delta z$	$-2\left(E_{y0} + \dfrac{\partial E_y}{\partial z}\Delta z\right)\Delta y$	(17)
3-4	$\mathbf{E}_0 - \dfrac{\partial \mathbf{E}}{\partial y}\Delta y$	$-2\left(E_{z0} - \dfrac{\partial E_z}{\partial y}\Delta y\right)\Delta z$	(18)
4-1	$\mathbf{E}_0 - \dfrac{\partial \mathbf{E}}{\partial z}\Delta z$	$2\left(E_{y0} - \dfrac{\partial E_y}{\partial z}\Delta z\right)\Delta y$	(19)

FUNDAMENTALS OF FIELDS

By adding the line integrals we find the circulation integral for the rectangle

$$\oint \mathbf{E} \cdot d\mathbf{l} = 4 \left(\frac{\partial E_z}{\partial y} - \frac{\partial E_y}{\partial z} \right) \Delta y \, \Delta z \tag{20}$$

But the total area of the rectangle is

$$\Delta s_x = 4 \, \Delta y \, \Delta z \tag{21}$$

Now we substitute Eqs. (20) and (21) into (15) to obtain the desired component

$$(\boldsymbol{\nabla} \times \mathbf{E})_x = \lim_{\Delta s_x \to 0} \frac{\oint \mathbf{E} \cdot d\mathbf{l}}{\Delta s_x} = \frac{\partial E_z}{\partial y} - \frac{\partial E_y}{\partial z} \tag{22}$$

By a similar analysis for the $\mathbf{a}_y$ and $\mathbf{a}_z$ orientations shown in Fig. 1.9.4b and c, we can find each of the components of the curl. The result is

$$\boldsymbol{\nabla} \times \mathbf{E} = \left(\frac{\partial E_z}{\partial y} - \frac{\partial E_y}{\partial z} \right) \mathbf{a}_x + \left(\frac{\partial E_x}{\partial z} - \frac{\partial E_z}{\partial x} \right) \mathbf{a}_y \\ + \left(\frac{\partial E_y}{\partial x} - \frac{\partial E_x}{\partial y} \right) \mathbf{a}_z \tag{23}$$

This form of the curl is valid only in rectangular coordinate systems. However, the corresponding forms for other systems are given in Tables 1.2.2 and 1.7.1. For computations, these forms are more useful than Eq. (13), where the curl was defined.

As a memory aid, the curl in *rectangular* coordinates is often written as a determinant

$$\boldsymbol{\nabla} \times \mathbf{E} = \begin{vmatrix} \mathbf{a}_x & \mathbf{a}_y & \mathbf{a}_z \\ \dfrac{\partial}{\partial x} & \dfrac{\partial}{\partial y} & \dfrac{\partial}{\partial z} \\ E_x & E_y & E_z \end{vmatrix} \tag{24}$$

which is the same as Eq. (23).

AN ALTERNATE FORM OF THE CURL

More insight into the physical meaning of the curl can be obtained from the following equation for the curl:

$$\boldsymbol{\nabla} \times \mathbf{E} = \lim_{\Delta v \to 0} \frac{\oint_s \mathbf{a}_n \times \mathbf{E} \, ds}{\Delta v} \tag{25}$$

Here s is a closed surface which bounds the incremental volume Δv.

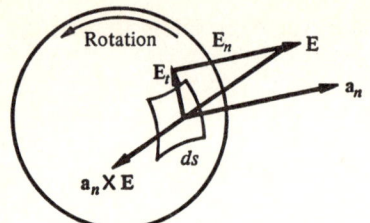

Fig. 1.9.5 The curl is a measure of the tangential components of the **E** field over the closed surface.

Equation (25) could be used instead of Eq. (13) to define the curl.† We will consider a small sphere of volume Δv, as shown in Fig. 1.9.5, although any incremental volume bounded by a simple closed surface will do. The unit vector $\mathbf{a}_n$ is an outward normal vector for the surface element ds. Assume that the small sphere has been placed in an **E** field (without disturbing the field) whose curl we want to measure. We can express the **E** field as the sum of $\mathbf{E}_n$, a component normal to the surface, and $\mathbf{E}_t$, a component tangential to the surface, at every point on the surface of the sphere

$$\mathbf{E} = \mathbf{E}_n + \mathbf{E}_t \qquad (26)$$

Then the cross product $\mathbf{a}_n \times \mathbf{E}$ is a measure of the *tangential* component of **E** at any given point on the surface

$$\mathbf{a}_n \times \mathbf{E} = \mathbf{a}_n \times (\mathbf{E}_n + \mathbf{E}_t) = \mathbf{a}_n \times \mathbf{E}_t \qquad (27)$$

This vector is perpendicular to both $\mathbf{a}_n$ and **E**, as illustrated in Fig. 1.9.5, and its magnitude is $|\mathbf{E}_t|$. The integral of $\mathbf{a}_n \times \mathbf{E}$ over the closed surface yields the *total tangential components* of the field, which would tend to cause any uniformly charged sphere to rotate. As the sphere is allowed to shrink to zero ($\Delta v \to 0$), *the magnitude of the curl at that point is proportional to the rate of rotation of the sphere and the direction of the curl vector is the direction of the axis of rotation, determined in a right-handed sense.*

The fact that a field has curl at a given point means that energy could be extracted from the field. Such a point is called a *source*. If the field is used to absorb energy at that point, the point is called a *sink*. For example, if one made a charged sphere rotate against the field, one could extract energy from the external source and supply it to the field. Thus curl is associated with field sources or sinks. We found that fields with divergence also are associated with sources or sinks. Many practical

---

† It will not be shown here that the two forms are equivalent, but the reader is referred to R. B. McQuistan, "Scalar and Vector Fields," chap. 7, John Wiley & Sons, Inc., New York, 1965.

FUNDAMENTALS OF FIELDS

devices, such as motors, generators, and antennas, utilize electromagnetic fields with nonzero curl to accomplish specific objectives.

EXAMPLES OF THE CURL

Example 1 In the introduction to this section we considered a field of the form

$$\mathbf{E} = -z\mathbf{a}_y \tag{28}$$

whose flow lines are straight lines. Now we can use Eq. (23) to find the curl

$$\nabla \times \mathbf{E} = -\frac{\partial E_y}{\partial z}\mathbf{a}_x = -\frac{\partial(-z)}{\partial z}\mathbf{a}_x = 1\mathbf{a}_x \tag{29}$$

Of course, this is the same result as that obtained in Eq. (12) from the limit process. In this special case, the curl defines a new vector field which is a uniform vector field at every point in space. Note that the flow lines of both $\mathbf{E}$ and $\nabla \times \mathbf{E}$ are straight lines with no "curling" or curvature, even though the curl of $\mathbf{E}$ is not zero.

Example 2 Suppose we are given a vector field

$$\mathbf{R}(r,\phi,z) = \frac{\mathbf{a}_\phi}{r} \tag{30}$$

in cylindrical coordinates. This is a field with circular flow lines which decreases in intensity with distance from the z axis, as shown in Fig. 1.9.6. From Eq. (30), the components are

$$R_r = 0 \qquad R_\phi = \frac{1}{r} \qquad R_z = 0 \tag{31}$$

To find the curl, we must use $\nabla \times \mathbf{R}$ in cylindrical coordinates as given in

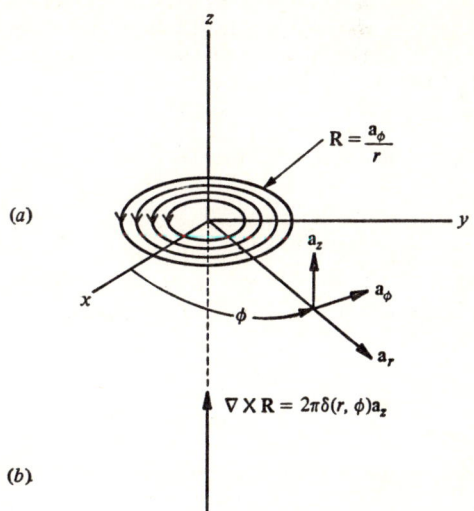

Fig. 1.9.6 A field with zero curl everywhere except at the origin. (*a*) A cross section of the field $\mathbf{R} = \mathbf{a}_\phi/r$. (*b*) The curl of the field $\mathbf{R}$.

Table 1.7.1 [Eqs. (23) and (24) are not valid here; why?]:

$$\nabla \times \mathbf{R} = \left(\frac{1}{r}\frac{\partial R_z}{\partial \phi} - \frac{\partial R_\phi}{\partial z}\right)\mathbf{a}_r + \left(\frac{\partial R_r}{\partial z} - \frac{\partial R_z}{\partial r}\right)\mathbf{a}_\phi$$
$$+ \left(\frac{1}{r}\frac{\partial (rR_\phi)}{\partial r} - \frac{1}{r}\frac{\partial R_r}{\partial \phi}\right)\mathbf{a}_z \quad (32)$$

Upon substitution of Eqs. (31), we find that the $\mathbf{a}_r$ and $\mathbf{a}_\phi$ components are zero and that

$$\nabla \times \mathbf{R} = \frac{1}{r}\frac{\partial (rR_\phi)}{\partial r}\mathbf{a}_z = 0 \quad \text{for } r \neq 0 \quad (33)$$

Thus at any point other than the origin, the curl is zero, even though the flow lines are circular. At the origin $\nabla \times \mathbf{R}$ is indeterminate [of the form 0/0, using Eq. (33)], but by using the circulation integral of Eq. (15) and neglecting the $\mathbf{a}_r$ and $\mathbf{a}_\phi$ components we find that

$$\nabla \times \mathbf{R} = \left[\lim_{s_z \to 0}\frac{\oint \frac{\mathbf{a}_\phi}{r} \cdot (r\, d\phi\, \mathbf{a}_\phi)}{\Delta s_z}\right]\mathbf{a}_z = \left(\lim_{r \to 0}\frac{\int_0^{2\pi} d\phi}{\pi r^2}\right)\mathbf{a}_z = \left(\lim_{r \to 0}\frac{2}{r^2}\right)\mathbf{a}_z \quad (34)$$

This integral was evaluated by considering a circular path of integration, enclosing an area $\Delta s_z = \pi r^2$, in a plane perpendicular to the z axis. Thus the curl is infinite at $r = 0$, that is, at any point on the z axis, but is zero at any other point. When we encounter functions with discontinuities of this type (functions which are zero except at discrete points where they are infinite) we should suspect the presence of delta functions and should investigate to see whether they are appropriate mathematical models. One should recall the difference between an isolated singularity of the types associated with continuous functions (such as poles) and the delta function, which is zero in every neighborhood surrounding the discontinuity. In most cases the delta function cannot be derived directly from the limit process [from Eq. (34) in this case] because the delta function is not a function in the usual sense. We hypothesize that the curl is of the form of a constant times a delta function, in the z direction. From the symmetry of the field, the curl must be independent of the variable z, and so we surmise that

$$\nabla \times \mathbf{R}(r,\phi,z) = k\delta(r,\phi)\mathbf{a}_z \quad (35)$$

where k is a constant. We can verify† that Eq. (35) is correct and that $k = 2\pi$.

† We show that $\nabla \times \mathbf{R}$ of Eq. (35) is the curl of $\mathbf{R}$ by substitution into Stokes' theorem (from Sec. 1.10). We assume a plane surface s defined by the circle C of radius r_0 centered at the origin

$$\int_s \nabla \times \mathbf{R} \cdot d\mathbf{s} = \oint_C \mathbf{R} \cdot d\mathbf{l}$$

$$\int_s \frac{k\delta(r)}{2\pi r}\mathbf{a}_z \cdot (r\, dr\, d\phi\, \mathbf{a}_z) = \oint_C \frac{\mathbf{a}_\phi}{r} \cdot (r\, d\phi\, \mathbf{a}_\phi)$$

$$\frac{k}{2\pi}\int_0^{r_0} \delta(r)\, dr \int_0^{2\pi} d\phi = \int_0^{2\pi} d\phi \quad \text{where } k = 2\pi$$

Thus $\nabla \times \mathbf{R} = 2\pi\delta(r,\phi)\mathbf{a}_z$ is the curl of the field $\mathbf{R} = \mathbf{a}_\phi/r$.

FUNDAMENTALS OF FIELDS

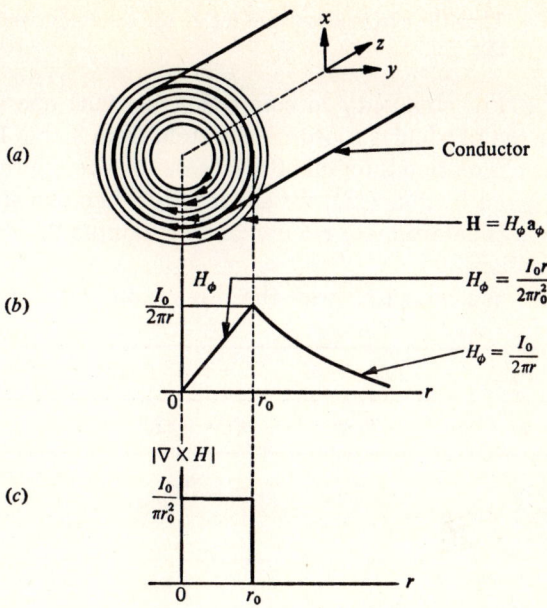

Fig. 1.9.7 The curl of the circular field is zero except inside the conductor. (a) The conductor and its field. (b) The magnitude of the field. (c) The magnitude of the curl of the field.

Equation (11) of Sec. 1.5 can be used to simplify the result:

$$\nabla \times \mathbf{R} = \frac{\delta(r)}{r} \mathbf{a}_z \tag{36}$$

Thus the curl of the field $\mathbf{R} = \mathbf{a}_\phi/r$ is zero at every point except on the z axis, where it is given by Eq. (36), and is shown in Fig. 1.9.6b.

Example 3 Next we consider a vector field $\mathbf{H}$ which is the magnetic field intensity about an infinitely long wire of radius r_0 carrying a constant current I_0. The current is uniformly distributed over the conductor, and the center of the wire lies along the z axis, as shown in Fig. 1.9.7. By using Ampere's law it is usually shown in first courses in field theory that the $\mathbf{H}$ field is

$$\mathbf{H} = H_\phi \mathbf{a}_\phi = \begin{cases} \dfrac{I_0 r}{2\pi r_0^2} \mathbf{a}_\phi & r \leq r_0 \\ \dfrac{I_0}{2\pi r} \mathbf{a}_\phi & r \geq r_0 \end{cases} \tag{37}$$

We can find the curl $\nabla \times \mathbf{H}$ in cylindrical coordinates [see Eq. (8) of Table 1.7.1]

$$\nabla \times \mathbf{H} = \begin{cases} \dfrac{I_0}{\pi r_0^2} \mathbf{a}_z & r < r_0 \\ 0 & r > r_0 \end{cases} \tag{38}$$

Thus the curl is a constant inside the conductor and is zero outside, as shown in Fig. 1.9.7c.

For simplicity in computing the curl one should use the equation for the particular coordinate system from Table 1.2.2 or 1.7.1 rather than the circulation integral of Eq. (13). Although the basic definition of curl is given by Eq. (13), we use it only when the equations from the tables yield indeterminate results, as in Example 2, where we obtained a delta function.

We can summarize the three examples as tabulated. Thus we find

Example	Flow lines	Field	Curl
1	Linear	$\mathbf{E} = -z\mathbf{a}_y$	Constant, $\nabla \times \mathbf{E} = \mathbf{a}_x$
2	Circular	$\mathbf{R} = \dfrac{\mathbf{a}_\phi}{r}$	Zero except along the z axis, where $\nabla \times \mathbf{R} = 2\pi\delta(r,\phi)\mathbf{a}_z$
3	Circular	$\mathbf{H} = \begin{cases} k_1 r \mathbf{a}_\phi & r < r_0 \\ \dfrac{k_2}{r}\mathbf{a}_\phi & r > r_0 \end{cases}$	$\nabla \times \mathbf{H} = 2k_1 \mathbf{a}_z$ $\nabla \times \mathbf{H} = 0$

that curl can be associated with linear or curved flow lines and that curvature in a given region does not necessarily imply curl in that region. One can construct a simple mental image to see which fields have curl and which do not. Imagine the field as an electric field in which we place a small uniformly charged sphere which is free to rotate. If the field forces on the charged particles cause the sphere to rotate, then the field has curl and the direction of the curl is along the axis of rotation in the right-handed sense. If the field does not tend to cause the sphere to rotate, it has no curl.

1.10 STOKES' THEOREM AND PROPERTIES OF VECTOR FIELDS

STOKES' THEOREM

Gauss' theorem and the divergence theorem of Sec. 1.8 are examples of a group of theorems called *integral theorems*. Another integral theorem, called Stokes' theorem, will be developed in this section.

The circulation integral of Sec. 1.9 has an interesting additive property which is useful in the development of Stokes' theorem. Suppose we are given a simple closed curve C (Fig. 1.10.1) which need not be a planar curve. We assume that flux lines from a static vector field $\mathbf{H}$ are encircled by the curve C. We will assume that the vector func-

FUNDAMENTALS OF FIELDS

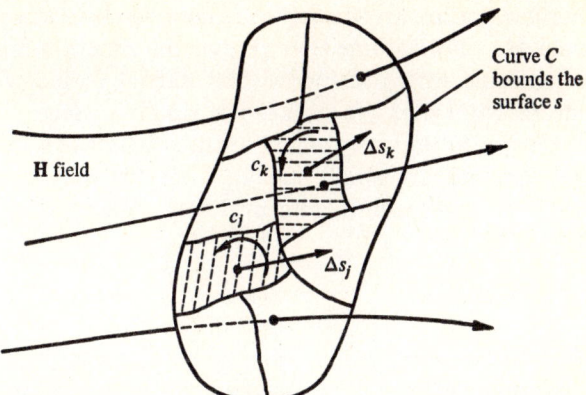

Fig. 1.10.1 The circulation integral can be found either by integrating over the partitions or over the boundary curve C.

tion $\mathbf{H}$ is single-valued and differentiable, with continuous first partial derivatives over the surface of interest. Suppose we choose any simple surface bounded by C and partition it into N arbitrary subdivisions, as shown in Fig. 1.10.1. We can compute the circulation integral $\oint_{C_k} \mathbf{H} \cdot d\mathbf{l}_k$ over the closed boundary curves C_k for every subdivision of the surface. Now if we add the circulation integrals of two adjacent subdivisions, such as the two shaded subdivisions of Fig. 1.10.1, we find that the integrals along the boundary common to both regions will cancel because of the opposite directions of integration. *Thus the sum of the circulation integrals of two adjacent regions is the same as the circulation integral around the curve that bounds both sections.* Inductively, if we extend this reasoning to all the subdivisions of the surface s, we see that the line integrals along all the internal partitions will cancel. The only integrals not canceled will be those along the original curve C, which bounds the entire surface. A mathematical statement of this additive property of the circulation integral is

$$\oint_C \mathbf{H} \cdot d\mathbf{l} = \sum_{k=1}^{N} \oint_{C_k} \mathbf{H} \cdot d\mathbf{l}_k \tag{1}$$

We now use the additive property of the circulation integral to derive Stokes' theorem. In Sec. 1.9 we defined the curl of the vector field as

$$\nabla \times \mathbf{H} \equiv \left(\lim_{\Delta s \to 0} \frac{\oint_C \mathbf{H} \cdot d\mathbf{l}}{\Delta s} \right) \mathbf{a}_n \tag{2}$$

Suppose we are given a surface s bounded by a closed curve C in the vector field $\mathbf{H}$. We assume that the vector function is single-valued and differentiable with continuous first partial derivatives over the surface of interest and that the surface is a simple surface. Next we subdivide the surface into infinitely many small subdivisions, each of area Δs. If the subdivisions are small enough, the curl may be considered essentially constant over each incremental area, i.e.,

$$\nabla \times \mathbf{H} \approx \frac{\oint_{C_k} \mathbf{H} \cdot d\mathbf{l}_k}{\Delta s_k} \mathbf{a}_{n_k} \tag{3}$$

for the kth subdivision. By taking the dot product of each side of the equation with respect to the unit vector $\mathbf{a}_{n_k}$ and multiplying by the incremental area, we arrive at the approximate equation for the circulation integral around the incremental section

$$\nabla \times \mathbf{H} \cdot \Delta s_k \, \mathbf{a}_{n_k} \approx \oint_{C_k} \mathbf{H} \cdot d\mathbf{l}_k \tag{4}$$

Next we add the circulation integrals over the infinite number of incremental subdivisions and use the additive property to obtain the circulation integral around the boundary C

$$\sum_{k=1}^{\infty} \nabla \times \mathbf{H} \cdot \Delta s_k \, \mathbf{a}_{n_k} \approx \sum_{k=1}^{\infty} \oint_{C_k} \mathbf{H} \cdot d\mathbf{l}_k = \oint_C \mathbf{H} \cdot d\mathbf{l} \tag{5}$$

If we consider the summation process on the left in the limit as the incremental areas approach zero, we can replace the summation with an integral and thereby obtain *Stokes' theorem*

$$\boxed{\int_s \nabla \times \mathbf{H} \cdot d\mathbf{s} = \oint_C \mathbf{H} \cdot d\mathbf{l}} \tag{6}$$

Thus, the integral of the normal component of the curl of a vector field over a bounded surface is the same as the integral of the tangential components of that field around the boundary curve C. The vector field and the surface are subject to the constraints previously stated.† Note that

---

† It was assumed that the vector function is single-valued and differentiable with continuous first partial derivatives over the surface of interest and that the surface is a simple surface. However, the continuity condition on the derivatives can be relaxed, and delta functions and some discontinuities are allowed. Also it is assumed that the vector function is independent of time. Suppose we assume that Stokes' theorem is valid for all time-varying fields which satisfy the above conditions. Then suppose a source is suddenly turned on at the origin of our coordinate system and that the field generated by this event, i.e., a transient vector field, begins to propagate through space with a finite velocity. At a given instant of time we can choose a closed curve C to be in the field at the wavefront, so that the line integral around the

FUNDAMENTALS OF FIELDS   71

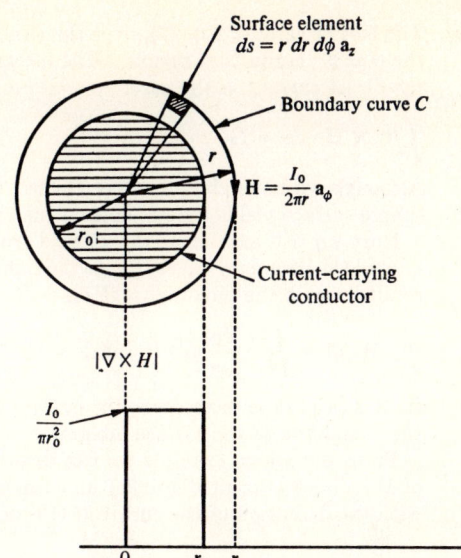

Fig. 1.10.2 Stokes' theorem can be applied to a wire carrying a direct current.

the integral on the left in Eq. (6) is a double integral while the one on the right is a single integral. The theorem is most useful from a theoretical standpoint, but it can also be used to simplify integrations in many cases.

Example 1 To gain some familiarity with Stokes' theorem, we will now apply it to the H field shown in Fig. 1.9.7 of a conductor carrying direct current. Recall that the curl of the H field is zero at all points outside the wire (although the H field itself is not zero) and that the curl is a constant inside the wire, as shown in Fig. 1.10.2. We will compute each of the integrals of Eq. (6), assuming that the boundary curve C is a circle of radius r which encloses the wire, as shown in Fig. 1.10.2. The simplest surface that we can choose is the plane surface (i.e., the drumhead surface) defined by the plane curve C. With the surface defined, we can substitute the value of the curl of H [from Eq. (38) of Sec. 1.9] into the left-hand side of Stokes' theorem:

$$\int_s \nabla \times \mathbf{H} \cdot d\mathbf{s} = \int_0^{2\pi} \int_0^r \nabla \times \mathbf{H} \cdot (r\, dr\, d\phi\, \mathbf{a}_z)$$
$$= \int_0^{2\pi} d\phi \left[\int_0^{r_0} \frac{I_0 r}{\pi r_0^2}\, dr + \int_{r_0}^r (0)\, dr \right] \quad (7)$$

curve is not necessarily zero. At the same instant of time we can choose a surface s, bounded by C, which is beyond the wavefront (where the field over the surface is zero), so that Stokes' theorem is not necessarily valid. However, the theorem is valid if one accounts for the delay time needed for the wavefront to reach the surface. This leads to the topic of retarded potentials and other time-varying phenomena, discussed in the next chapter. Because electromagnetic fields propagate with the velocity of light, one can often neglect the delay time and use Stokes' theorem over a small region of interest.

The last integral of Eq. (7), over the annulus outside the wire, is zero because the curl is zero in that region. The integration yields a constant, I_0, which is the direct current in the wire:

$$\int_s \nabla \times \mathbf{H} \cdot d\mathbf{s} = I_0 \tag{8}$$

Although we have chosen a special case, the integration of the curl over any simple surface yields the net current piercing that surface.

Next we will substitute the $\mathbf{H}$ field from Eq. (37) of Sec. 1.9 (the $\mathbf{H}$ field around the boundary curve C) into the right-hand side of Stokes' theorem. The result is again the direct current I_0:

$$\oint_C \mathbf{H} \cdot d\mathbf{l} = \int_0^{2\pi} \frac{I_0}{2\pi r} \mathbf{a}_\phi \cdot (r\, d\phi\, \mathbf{a}_\phi) = I_0 \tag{9}$$

In this case, as in most cases, the evaluation of the line integral is simpler than the evaluation of the surface integral.

From the above example we can draw a conclusion about the relationship of the direct current density $\mathbf{J}$ and the curl of the magnetic field. Starting with the definition of the curl, Eq. (13) of Sec. 1.9,

$$\nabla \times \mathbf{H} \equiv \left(\lim_{\Delta s \to 0} \frac{\oint_C \mathbf{H} \cdot d\mathbf{l}}{\Delta s} \right) \mathbf{a}_n \tag{10}$$

we replace the circulation integral around the incremental closed curve C (which bounds Δs) by the incremental current ΔI

$$\nabla \times \mathbf{H} = \left(\lim_{\Delta s \to 0} \frac{\Delta I}{\Delta s} \right) \mathbf{a}_n \tag{11}$$

In the limit, the term on the right is just the current density vector $\mathbf{J}$, and so we conclude that

$$\boxed{\nabla \times \mathbf{H} = \mathbf{J}} \tag{12}$$

Equation (12) is a form of Ampere's law and is valid for $\mathbf{H}$ fields generated by direct currents. In Chap. 2, we will modify Ampere's law to account for time-varying fields.

Example 2 We can apply Stokes' theorem to an electric field generated by a point charge, rather than to the magnetic field generated by a direct current. Suppose we choose a closed curve C and place a charged particle at some arbitrary point p in the vicinity of the curve (but not on the curve). We proved in Sec. 1.6 that the line integral $\oint \mathbf{E} \cdot d\mathbf{l}$ is zero for any static electric field around any closed path. From the definition of the curl, Eq. (13) of Sec. 1.9, we find

$$\nabla \times \mathbf{E} = \left(\lim_{\Delta s \to 0} \frac{\oint \mathbf{E} \cdot d\mathbf{l}}{\Delta s} \right) \mathbf{a}_n = 0 \tag{13}$$

for any static $\mathbf{E}$ field. Since $\oint \mathbf{E} \cdot d\mathbf{l} = 0$ and $\nabla \times \mathbf{E} = 0$, Stokes' theorem yields

$$\int_s \nabla \times \mathbf{E} \cdot d\mathbf{s} = \oint_C \mathbf{E} \cdot d\mathbf{l} = 0 \tag{14}$$

FUNDAMENTALS OF FIELDS

Thus Stokes' theorem is valid for static **E** fields although it yields a trivial result. The equation

$$\boxed{\nabla \times \mathbf{E} = 0} \tag{15}$$

is characteristic of all *irrotational fields*, as discussed in Sec. 1.6.

SOME PROPERTIES OF VECTOR FIELDS

We now investigate some of the properties of a vector field obtained from the curl operation. Suppose we let **H** be any vector field which satisfies the criteria for Stokes' theorem. We define the vector field **F** from the curl of **H** as follows:

$$\mathbf{F} \equiv \nabla \times \mathbf{H} \tag{16}$$

and we assume that **F** is not identically zero (as it would be for irrotational fields). Since **F** is a vector field, it can be mapped with flux lines or flow lines which show the direction and magnitude of the field. For example, the flow lines of the field **F** for the current-carrying conductor, as shown in Fig. 1.10.2, is a tube of flow lines of equal magnitude all confined inside the conductor, and the field is zero everywhere outside the conductor. By use of Stokes' theorem we can show that the flow lines for any field **F** given by Eq. (16) must be continuous. Suppose that we choose an arbitrary curve C which encircles some of the flow lines of the field **F**, as shown in Fig. 1.10.3.

If we choose two different surfaces s_1 and s_2, both bounded by the curve C, we can compute the net flow of flux through each of the surfaces. However, Stokes' theorem is true regardless of the surface (bounded by the curve C) that we choose, because the line integral of **H** around the closed curve C is independent of that surface. By applying Stokes'

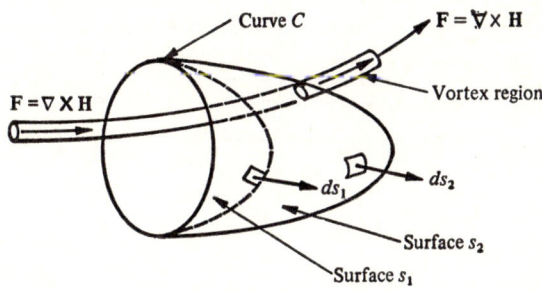

Fig. 1.10.3 The vortex region must pierce all simple surfaces bounded by the curve C.

theorem to each of the surfaces in Fig. 1.10.3 we can write

$$\oint_C \mathbf{H} \cdot d\mathbf{l} = \int_{s_1} \nabla \times \mathbf{H} \cdot d\mathbf{s}_1 = \int_{s_2} \nabla \times \mathbf{H} \cdot d\mathbf{s}_2 \tag{17}$$

However, the surface integrals are just the net flow of flux **F** through each of the two surfaces

$$\int_{s_1} \mathbf{F} \cdot d\mathbf{s}_1 = \int_{s_2} \mathbf{F} \cdot d\mathbf{s}_2 \tag{18}$$

Our conclusion is that the net flow of flux is the same through any simple surface bounded by the curve C. Consequently, *there can be no point sources for the field* **F** *on which flow lines are generated or terminated.* To illustrate this, suppose that there were a point in the volume between surface s_1 and surface s_2 of Fig. 1.10.3 on which a flow line **F** terminated. If such a point exists for the field **F**, more flow lines pierce the surface s_1 than pierce the surface s_2 and consequently the net flow of flux through the two surfaces would not be the same. Thus we would arrive at a contradiction, and we must conclude that no points exist which are either sources or sinks for the field **F** defined by Eq. (16).

This conclusion leads us to another property of the field **F**; namely *either the flow lines of the field* **F** *must be continuous (by forming loops and closing upon themselves), or they must extend from negative infinity to positive infinity.* The flow lines of the vector field **F** (that is, flow lines of a vector field generated from the curl of another vector field) are sometimes called *vorticity lines*, and groups of the flow lines are sometimes called *vorticity tubes*, since they must be continuous.

Another property of the field **F** becomes evident from the above discussion. Since there are no point sources for lines of vorticity, all the flux lines entering any simple closed surface must also leave it. Therefore, there can be no divergence of the field **F**. We could arrive at the same conclusion from the vector identity, which states that the divergence of the curl of any vector is identically equal to zero

$$\boxed{\nabla \cdot \mathbf{F} = \nabla \cdot \nabla \times \mathbf{H} \equiv 0} \tag{19}$$

A vector field **F** whose divergence is identically zero is called a *solenoidal field*.

Physically, if a solenoidal field **F** represents fluid flow, Eq. (19) is equivalent to the statement that the fluid is incompressible; i.e., the amount of fluid entering any given closed surface is equal to the amount leaving at every instant of time. Similarly, if the vector **F** is the current density **J** of electric conduction current, as given in Eq. (12), we can write

$$\boxed{\nabla \cdot \mathbf{J} = 0} \tag{20}$$

This is equivalent to saying that the current must be continuous since the current entering any closed surface must be equal to the current leaving that surface at every instant of time.

Physically, we usually associate static solenoidal fields with vorticity lines generated by external sources. For example, we can think of a battery as a source which generates a current density **J** (a solenoidal field) in a conductor which, in turn, generates the vector field **H**. We can consider the solenoidal field **J** as a *vector source* that generates the vector field **H**. In contrast, the source of static electric fields was shown to be electric charge, which has magnitude only, and we can consider it a *scalar source* of vector fields. Thus we have two types of source (vector and scalar) which generate vector fields of different characteristics. We study some of these differences in more detail in the next section.

1.11 HELMHOLTZ' THEOREM AND A SUMMARY OF STATIC FIELDS
COMBINATIONS OF THE ∇ OPERATORS

We have used the symbol ∇ in connection with three mathematical operations: the gradient, the divergence, and the curl. If we perform two of these operations sequentially, there are nine possible cases to be investigated. Four of the possible nine combinations will not be considered in

	Function		
Operation	Gradient (vector) ∇V	Divergence (scalar) $\nabla \cdot \mathbf{U}$	Curl (vector) $\nabla \times \mathbf{U}$
Gradient ∇	—	$\nabla(\nabla \cdot \mathbf{U})$	—
Divergence ∇·	$\nabla \cdot \nabla V = \nabla^2 V$	—	$\nabla \cdot \nabla \times \mathbf{U} = 0$
Curl ∇×	$\nabla \times \nabla V = 0$	—	$\nabla \times \nabla \times \mathbf{U}$

this course, although they are studied in tensor analysis. For example, when the gradient operation is applied to a scalar V, it yields a vector ∇V. Thus, the gradient of the gradient ($\nabla \nabla V$) has no meaning to us since we have not studied the gradient of a vector. The reader will find it instructive to investigate all four of these cases marked with a blank in the chart above and determine why they are discarded for now.

We are already familiar with three of the remaining five valid cases. To review these, the first familiar entry in the first column of the chart is the divergence of the gradient ($\nabla \cdot \nabla V$), which we have defined as the scalar laplacian and which occurs in Poisson's and Laplace's equations. The curl of the gradient ($\nabla \times \nabla V$), the last entry in column 1, and the divergence of the curl ($\nabla \cdot \nabla \times \mathbf{U}$), in column 3, are familiar vector operations which are identically equal to zero. Thus, the only two entries in the chart which we have not considered are the gradient of the divergence [$\nabla(\nabla \cdot \mathbf{U})$] from column 2, and the curl of the curl ($\nabla \times \nabla \times \mathbf{U}$) from column 3. These two operations on the vector $\mathbf{U}$ are combined to define the *vector laplacian*

$$\nabla^2 \mathbf{U} \equiv \nabla(\nabla \cdot \mathbf{U}) - \nabla \times \nabla \times \mathbf{U} \tag{1}$$

This notation, which defines the meaning of $\nabla^2 \mathbf{U}$, is very similar to that of the scalar laplacian

$$\nabla^2 V = \nabla \cdot \nabla V \tag{2}$$

however, the operations are entirely different. One can distinguish the two operations by noting whether the operation is performed on a vector $\mathbf{U}$ or on a scalar V. The definitions of Eqs. (1) and (2) are valid in any coordinate system; however, some special cases are given in Table 1.7.1. The vector laplacian and the scalar laplacian are not related to each other in any simple manner except in the rectangular coordinate system, as shown in Eq. (9) of Table 1.2.2.

VECTOR POTENTIAL AND POISSON'S EQUATION

We have discussed two types of source of static fields, namely the scalar (point) sources and the vector (vortex) sources. Those fields generated by scalar sources are irrotational (conservative) because the curl of the field is zero. An irrotational field has the following characteristics: (1) a scalar potential V exists, (2) the vector field $\mathbf{U}$ can be obtained from the gradient of the scalar potential at any point, and (3) one can find the scalar potential by solving Poisson's equation (Laplace's equation if there are no sources in the region), subject to the given boundary conditions. As a general rule, it is easier to find the scalar potential and then find the vector field from the gradient than to find the vector field directly.

The vector source is the only other source of static fields, and its continuous flow lines form closed loops. For example, the vector field $\mathbf{J}$ of current density is a solenoidal field, and we think of it as generating the magnetic field $\mathbf{H}$. The essential relationships of $\mathbf{H}$ and $\mathbf{J}$ are

$$\nabla \times \mathbf{H} = \mathbf{J} \neq 0 \qquad \nabla \cdot \mathbf{J} = 0 \tag{3}$$

We assume that the source field **J** is known and we are interested in finding the magnetic field **H**. This problem is analogous to the one in irrotational fields where we know the distribution of the scalar sources ρ and want to find the static field **E** generated by them. The scalar potential proved useful in that case. In an analogous manner it can be shown† that a scalar potential can be developed for magnetic field provided **J** = 0. The scalar potential for magnetic fields is useful in regions where there is no current flow and for magnetic fields generated by permanent magnets. However, a more useful technique is to develop a vector potential **A** associated with the vector source **J**. An equipotential surface of the field **A** is the set of all points where the vector **A** has a given magnitude and direction. Usually these equipotential surfaces appear as shells which encircle the vortex tube of the current density **J**. Unfortunately, the vector potential field **A** does not have a simple physical interpretation like the voltage difference of the scalar potential field V. However, the vector potential **A** proves useful from a mathematical standpoint.

Experimental mapping of the magnetic field about a current-carrying conductor shows that the flux lines of **H** always close upon themselves. Also, we know that there are no point sources for the **H** field, which leads one to speculate that the divergence of the **H** field must be zero. In fact, in Chap. 2 we show, from Maxwell's equations, that the divergence of the **H** field must be zero for all physically realizable cases. Thus, the **H** field must be a solenoidal field like the **J** field, whose characteristics are given in Eq. (3). We speculate that a vector potential field **A** exists because of the **H** field and that the necessary relationships are given by

$$\nabla \times \mathbf{A} = \mu \mathbf{H} \neq 0 \qquad \nabla \cdot \mathbf{H} = 0 \tag{4}$$

[Compare the forms of Eqs. (3) and (4).] From Eq. (4), if we take the divergence of the field $\mu \mathbf{H}$, we find that

$$\nabla \cdot \mu \mathbf{H} = \nabla \cdot (\nabla \times \mathbf{A}) \equiv 0 \tag{5}$$

Thus the relationship between **A** and **H** has been chosen so that it satisfies the requirement that $\nabla \cdot \mathbf{H} = 0$.

The term μ, called the *permeability* of the medium, is a characteristic of the medium in which the field exists, and it is analogous to the permittivity, which we use to describe electric fields. The permeability will be discussed in more detail in Chap. 2, where it will be used to relate the magnetic field intensity **H** to the magnetic flux density **B**

$$\mathbf{B} = \mu \mathbf{H} \tag{6}$$

Our objective is to relate the vector potential **A** to the source **J** by

† W. H. Hayt, Jr., "Engineering Electromagnetics," 2d ed., p. 235, McGraw-Hill Book Company, New York, 1967.

the following derivation, where we assume that μ is a simple scalar constant. First we take the curl of the curl of **A** from Eq. (4),

$$\nabla \times \nabla \times \mathbf{A} = \nabla \times (\mu \mathbf{H}) = \mu(\nabla \times \mathbf{H}) \tag{7}$$

and then replace the curl of **H** by the source **J** from Eq. (3). The result is

$$\boxed{\nabla \times \nabla \times \mathbf{A} = \mu \mathbf{J}} \tag{8}$$

Thus we consider the vector potential field **A** to be generated by the source field **J** as given by the partial differential equation (8). The vector potential **A** can be written in terms of the vector laplacian (just as the scalar potential V was written in terms of the scalar laplacian). To do this, we substitute Eq. (8) into Eq. (1) and obtain

$$\nabla(\nabla \cdot \mathbf{A}) - \nabla^2 \mathbf{A} = \mu \mathbf{J} \tag{9}$$
$$\nabla^2 \mathbf{A} = \nabla(\nabla \cdot \mathbf{A}) - \mu \mathbf{J} \tag{10}$$

Since the scalar potential field **A** is not uniquely specified by Eq. (8), we must specify the divergence of **A**.† The simplest procedure is to choose the divergence of **A** to be zero, which means that **A** will be a solenoidal field, just as **H** and **J** were solenoidal fields. Under the assumption that

$$\nabla \cdot \mathbf{A} = 0 \tag{11}$$

we find that Eq. (10) reduces to

$$\boxed{\nabla^2 \mathbf{A} = -\mu \mathbf{J}} \tag{12}$$

This is called the *vector* Poisson equation, and it is a partial differential equation whose solution depends both upon the boundary conditions and upon the vector source **J**. When this partial differential equation is solved, the **H** field can be obtained from the curl of **A** as given in Eq. (4). Often this is the simplest way to obtain a vector field **H** for a given vector source **J**. For example, in the rectangular coordinate system, the vector Poisson equation can be reduced to three scalar Poisson equations.

SOLUTIONS OF THE SCALAR AND VECTOR POISSON EQUATION

The solution of both the scalar and vector Poisson equation can be written in integral forms, which are often useful. Suppose that at some

---

† It can be shown that if the divergence and the curl of any vector field **A** are specified in some volume v, the field **A** is uniquely determined if the normal component of **A** over the closed surface s which bounds v is given. The proof of this theorem is given in a number of texts, e.g., McQuistan, *op. cit.*, p. 224.

FUNDAMENTALS OF FIELDS

test point (x_t, y_t, z_t) we want to find the scalar potential $V(x_t, y_t, z_t)$ due to an electric charge $\rho(x,y,z)$ distributed over some volume v. The distance from the origin of the coordinate system to the test point is given by the vector $\mathbf{R}_t$, and the distance from the origin to the source points by $\mathbf{R}_s$. The geometry is the same as that shown in Fig. 1.3.2, where

$$\mathbf{R}_t = x_t \mathbf{a}_x + y_t \mathbf{a}_y + z_t \mathbf{a}_z \tag{13}$$

and

$$\mathbf{R}_s = x \mathbf{a}_x + y \mathbf{a}_y + z \mathbf{a}_z \tag{14}$$

The vector from the source points to the test point is then given by

$$\mathbf{R} = \mathbf{R}_t - \mathbf{R}_s \tag{15}$$

Now it can be shown† that Poisson's scalar equation

$$\nabla^2 V = -\frac{\rho}{\varepsilon} \tag{16}$$

has a solution given by the volume integral

$$\boxed{V(x_t, y_t, z_t) = \frac{1}{\varepsilon} \int_v \frac{1}{|R|} \rho(x,y,z) \, dx \, dy \, dz} \tag{17}$$

for unbounded regions. In most cases, this equation is easier to integrate than Eq. (5) of Sec. 1.3, from which we can obtain the **E** field directly. Thus the solution for Poisson's scalar equation can be obtained (provided one can perform the integration) if one knows the charge distribution of the scalar source. Similarly, for the vector Poisson equation,

$$\nabla^2 \mathbf{A} = -\mu \mathbf{J} \tag{18}$$

Morse and Feshbach show that the vector potential $\mathbf{A}(x_t, y_t, z_t)$ at the test point due to the current element $\mathbf{J}(x,y,z)$ is

$$\boxed{\mathbf{A}(x_t, y_t, z_t) = \mu \int_v \frac{1}{|R|} \mathbf{J}(x,y,z) \, dx \, dy \, dz} \tag{19}$$

Equations (17) and (19) are solutions for the potential fields for an unbounded space containing finite steady-state charge and current distributions, under the assumptions that ε and μ remain constant throughout the entire region. When other boundary conditions are imposed, these solutions are not valid.

† These solutions of Poisson's equation are given in P. M. Morse and H. Feshbach, "Methods of Theoretical Physics," vol. 1, pp. 38 and 202, McGraw-Hill Book Company, New York, 1953.

HELMHOLTZ' THEOREM

We have seen that static vector fields are generated by two types of source (scalar and vector sources), and in a linear medium it is logical to expect the fields from each of these sources to add; i.e., superposition can be used in a linear medium. This is the essence of *Helmholtz' theorem*:[†] in any linear medium any static vector field **U** which is finite and continuous and which vanishes at infinity can be expressed as the sum of an irrotational field $\mathbf{U}_1$ and a solenoidal field $\mathbf{U}_2$

$$\mathbf{U} = \mathbf{U}_1 + \mathbf{U}_2 \tag{20}$$

where

$$\nabla \times \mathbf{U}_1 = 0 \quad \text{irrotational field} \tag{21}$$

$$\nabla \cdot \mathbf{U}_2 = 0 \quad \text{solenoidal field} \tag{22}$$

We assume that the irrotational field $\mathbf{U}_1$ is generated by a known scalar source distribution ρ and that the solenoidal field $\mathbf{U}_2$ is generated by a known vector source **J**. We can relate the vector field **U** to the sources by taking the divergence and the curl

$$\nabla \cdot \mathbf{U} = \nabla \cdot (\mathbf{U}_1 + \mathbf{U}_2) = \nabla \cdot \mathbf{U}_1 = \frac{\rho}{\varepsilon} \tag{23}$$

$$\nabla \times \mathbf{U} = \nabla \times (\mathbf{U}_1 + \mathbf{U}_2) = \nabla \times \mathbf{U}_2 = \mathbf{J} \tag{24}$$

Since $\mathbf{U}_1$ is an irrotational field, as given by Eq. (21), there exists a scalar potential V related to $\mathbf{U}_1$ by the gradient

$$\mathbf{U}_1 = -\nabla V \tag{25}$$

When Eq. (25) is substituted into Eq. (23), the divergence of the original field **U** reduces to the scalar Laplace equation

$$\nabla \cdot \mathbf{U} = -\nabla \cdot \nabla V = -\nabla^2 V = \frac{\rho}{\varepsilon} \tag{26}$$

Consequently, V can be found, and then $\mathbf{U}_1$ can be obtained from the gradient.

Similarly, since $\mathbf{U}_2$ is a solenoidal field, a vector potential **A** is related to $\mathbf{U}_2$ by the curl. We assume that the field **A** is also solenoidal, so that its divergence is zero. These relationships are

$$\mathbf{U}_2 = \nabla \times \mathbf{A} \tag{27}$$

$$\nabla \cdot \mathbf{A} = 0 \tag{28}$$

[†] Proof of this theorem is given in *ibid.*, pp. 52-54.

FUNDAMENTALS OF FIELDS

Consequently the curl of the original field **U** reduces to the vector Poisson equation

$$\nabla \times \mathbf{U} = \nabla \times \nabla \times \mathbf{A} = -\nabla^2 \mathbf{A} = \mathbf{J} \tag{29}$$

Thus, once the vector potential **A** is found, the field $\mathbf{U}_2$ can be determined from Eq. (27).

In conclusion, according to Helmholtz' theorem, we can express static fields in terms of the scalar potential field and a vector potential field by substituting Eqs. (25) and (27) into Eq. (20):

$$\boxed{\mathbf{U} = \nabla \times \mathbf{A} - \nabla V} \tag{30}$$

Thus we see that the general static vector field **U** is the superposition of two vector fields, each generated by a different type of source.

SUMMARY OF STATIC FIELDS

We are now in a position to summarize, in very broad terms, the study of static vector fields. We have seen that a field is characterized by its divergence and its curl, and we are especially interested in the situation where either of these operations yields a zero result. Consequently, we can divide the study of static fields into four classes, as shown in the chart. The essential characteristics of each of these classes are summarized in Table 1.11.1.

	$\nabla \cdot \mathbf{U} = 0$	$\nabla \cdot \mathbf{U} \neq 0$
$\nabla \times \mathbf{U} = 0$	Class 1	Class 2
$\nabla \times \mathbf{U} \neq 0$	Class 3	Class 4

SUMMARY OF EQUATIONS FOR STATIC ELECTROMAGNETIC FIELDS

Static electromagnetic fields can be characterized as class 1, 2, or 3, depending upon the nature of the sources of the field. The most important equations for these fields have been developed in this chapter and are summarized below for convenience. The first four equations are called *Maxwell's equations for static fields*. They will be modified to apply to time-varying fields in the next sections.

Gauss' law for electric fields: $\quad \nabla \cdot \mathbf{D} = \rho \quad$ (31)

Gauss' law for magnetic fields: $\quad \nabla \cdot \mathbf{H} = 0 \quad$ (32)

Faraday's law: $\quad \nabla \times \mathbf{E} = 0 \quad$ (33)

Table 1.11.1.

Class	Field name	Field sources	Example	Potential functions	Method of solution
1. $\nabla \cdot \mathbf{U} = 0$, $\nabla \times \mathbf{U} = 0$	Irrotational and solenoidal	None in the region of interest	The E field in a region between the plates of a parallel-plate capacitor or in free space; no charge exists between the plates	Since $\nabla \times \mathbf{U} = 0$, a scalar potential function V exists such that $\mathbf{U} = -\nabla V$	Solve Laplace's scalar equation $\nabla^2 V = 0$ for the given boundary conditions and find $\mathbf{U}$ from $\mathbf{U} = -\nabla V$
2. $\nabla \cdot \mathbf{U} \neq 0$, $\nabla \times \mathbf{U} = 0$	Irrotational (conservative)	Scalar sources appear either as point sources or continuous distributions	The E field generated by a single point charge; also the E field between the plates of a parallel-plate capacitor when charged particles appear fixed between the plates	Since $\nabla \times \mathbf{U} = 0$, a scalar potential function V exists such that $\mathbf{U} = -\nabla V$	Solve Poisson's scalar equation $\nabla^2 V = -\rho/\varepsilon$ for the given boundary conditions and scalar sources ρ and find $\mathbf{U}$ from $\mathbf{U} = -\nabla V$
3. $\nabla \cdot \mathbf{U} = 0$, $\nabla \times \mathbf{U} \neq 0$	Solenoidal	Vector (vortex) sources $\mathbf{J}$ generate the field; flow lines of the vector sources are continuous	The H field generated by direct current of density $\mathbf{J}$ flowing in a conductor; the field inside the conductor (where $\nabla \times \mathbf{H} = \mathbf{J}$) is class 3, and the field outside (where $\nabla \times \mathbf{H} = 0$) is class 1	A vector potential function $\mathbf{A}$ exists such that $\mathbf{U} = \nabla \times \mathbf{A}$; in order to specify $\mathbf{A}$, one generally requires that $\nabla \cdot \mathbf{A} = 0$	Solve the vector Poisson equation $\nabla^2 \mathbf{A} = -\mathbf{J}$ for $\mathbf{A}$, given $\mathbf{J}$, and find $\mathbf{U}$ from $\mathbf{U} = \nabla \times \mathbf{A}$
4. $\nabla \cdot \mathbf{U} \neq 0$, $\nabla \times \mathbf{U} \neq 0$	No specific name; this is the most general static field	Both scalar and vector sources exist, and the field (in a linear medium) is the superposition of the fields generated by each type of source; thus, Helmholtz' theorem applies	The units of the E field and the H field are different, and so no example of superposition exists for static electromagnetic fields	Both a scalar potential function V and a vector potential function $\mathbf{A}$ exist; assume $\nabla \cdot \mathbf{A} = 0$	Helmholtz' theorem applies, and so $\mathbf{U} = \mathbf{U}_1 + \mathbf{U}_2$, where $\mathbf{U}_1$ is irrotational and $\mathbf{U}_2$ is solenoidal; find $\mathbf{U}_1$ as for class 2 and $\mathbf{U}_2$ as for class 3

FUNDAMENTALS OF FIELDS

Ampere's law:	$\nabla \times \mathbf{H} = \mathbf{J}$	(34)
Continuity equation:	$\nabla \cdot \mathbf{J} = 0$	(35)
Equations of the medium:	$\mathbf{J} = \sigma \mathbf{E}$	(36)
	$\mathbf{D} = \varepsilon \mathbf{E}$	(37)
	$\mathbf{B} = \mu \mathbf{H}$	(38)

1.12 THE LORENTZ TRANSFORMATION AND TIME-VARYING FIELDS

INTRODUCTION

Early in the twentieth century, the work of Einstein, Bohr, Born, and many other scientists improved the mathematical model of Maxwell† and showed some of its limitations through its relationships to quantum and relativity theories. Building on Einstein's special theory of relativity, Leigh Page was able to develop all of Maxwell's equations from Coulomb's law in 1912. This approach, which removes the mystery of the origin of the magnetic field and improves our understanding of Maxwell's equations, will be used here to develop a set of mathematical postulates upon which the remainder of our work is based.

While both sound and light are wave motions, the propagation of light differs from that of sound in the following sense. Suppose someone flying in an airplane creates a sound wave by blowing a trumpet. If the plane is flying at subsonic speed, the sound wave moves away from the plane with a relative velocity which is the difference between the velocity of sound and the velocity of the plane. If an observer on the plane could measure the velocity of the sound wave in his direction of motion, he would say that its velocity is less than the velocity determined by an observer on the ground measuring the same sound wave. If the plane could increase its speed sufficiently, it could keep up with the sound wave, or even outrun it in supersonic flight. Consequently, the relative speed of the sound wave with respect to the plane can be a positive number, a negative number, or zero, depending upon the conditions of the experiment.

We can perform similar experiments to determine the relative velocity of light by using the earth as our space ship as it orbits around the sun at approximately 20 mi/s. We can initiate light waves on the earth and measure their velocity in the direction of the travel of the earth. We can also measure the velocity of light as it travels at right angles to the direction of the travel of the earth and opposite to that

† A humorous and authoritative explanation of our heritage from Maxwell can be found in Bern Dibner, James Clerk Maxwell, *IEEE Spectrum*, December 1964, p. 50.

direction. The surprising thing is that in every case the measured relative velocity of the light is a constant and is independent of its direction of travel with respect to the direction of the travel of the earth. This has been confirmed many times to a high degree of accuracy with experiments such as the Michelson-Morley experiment and Frizeau's experiment with moving water. An excellent account of these experiments is given in Elliott's text.†

If one measures the velocity of light from a distant star as the earth orbits around the sun, alternately moving toward and then away from the star, again one can find no measurable difference in the velocity of light. Thus, experimental evidence led Einstein to speculate that *the velocity of light is a constant to any observer, regardless of his velocity*. He assumed that the constant velocity of light is one of the fundamental facts of nature, and consequently he made that fact one of the basic postulates of his special theory of relativity, introduced in 1905. As a second postulate, he assumed the principle of relativity, a much older idea, which states that the fundamental laws of physics are the same everywhere in the universe and are independent of the reference frame, or coordinate system, of the observer. Very readable accounts of the special theory of relativity are given by Elliott (whose work we follow here) and in a book by Steinmetz (based upon his work in 1922).‡ We concentrate on certain aspects of the special theory of relativity which will be of use to us in deriving Maxwell's equations.

DERIVATION OF THE LORENTZ TRANSFORMATION

Suppose two observers are racing through space on parallel paths toward a distant light source, and assume that they are traveling at different, but constant, velocities. Each observer measures the relative velocity of the light coming from the source, and, according to the special theory of relativity, they each arrive at the same number c for the relative velocity of the light. To determine the velocity of light, each of them would have to use some means of measurement of length (distance) and time. We assume that both observers use meters for the units of length and seconds for the units of time.

To be more explicit, we will assume that the axes of the rectangular coordinate systems of the two observers are parallel and that one reference frame is sliding along the z axis of the other reference frame with a constant velocity u, as shown in Fig. 1.12.1. This special orientation of the axes is not necessary in order to obtain the end results, but our work is

---

† *Op. cit.*, chap. 2.
‡ Charles P. Steinmetz, "Four Lectures on Relativity and Space," Dover Publications, Inc., New York, 1967.

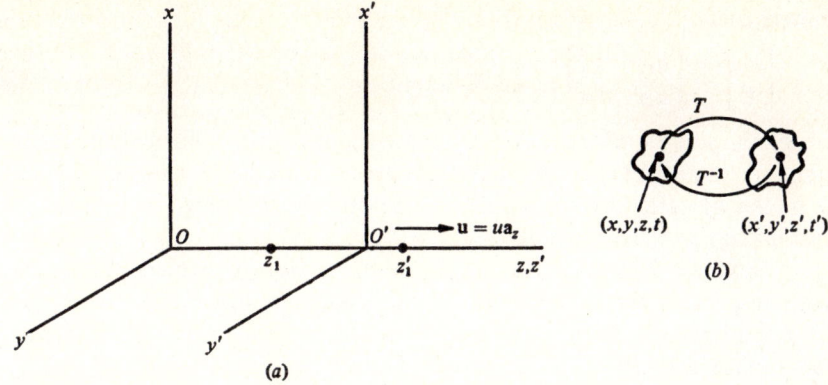

Fig. 1.12.1 The observers O and O' are in fixed rectangular coordinate systems which move with a constant relative velocity **u**. (a) Orientation of the coordinate system. (b) The transformations T and T^{-1}.

greatly simplified by such an assumption. We will assume that there are mathematical transformations which relate the two coordinate systems to each other, as we did in Sec. 1.2, except that we want to include time in the transformation equations. Furthermore, we will assume that each transformation is linear. Thus every point in the coordinate system (x,y,z,t) transforms to a unique point in the coordinate system (x',y',z',t') under the transformation T, as shown in Fig. 1.12.1b. The transformation T must be single-valued, so that at a given instant of time a point z_1 would transform to some point z_1' in a second coordinate system, as shown. We will find that the inverse transform T^{-1}, that is, (x',y',z',t') to (x,y,z,t), has the same properties. Since there is no relative motion along the x or y axes of the two coordinate systems, the corresponding transform equations for either T or T^{-1} are

$$x' = x \tag{1}$$
$$y' = y \tag{2}$$

However, for the z and t axes, we assume a more general linear transformation. We derive the transformation T assuming the linear relationships

$$z' = a_{11}z + a_{12}t \tag{3}$$
$$t' = a_{21}z + a_{22}t \tag{4}$$

In these equations the constants a_{ij} are unknown. We will determine these constants so that the transformation satisfies the principles of the special theory of relativity. The last two equations can be written in matrix form

$$\begin{bmatrix} z' \\ t' \end{bmatrix} = \begin{bmatrix} a_{11} & a_{12} \\ a_{21} & a_{22} \end{bmatrix} \begin{bmatrix} z \\ t \end{bmatrix} \qquad (5)$$

where the square matrix of the unknown constants is a linear transformation. Equations (1) to (4) define the transformation T that an observer O in the system (x,y,z,t) would use to transform known points in this system to the other system.

Next we determine the constants a_{ij} of Eq. (5), which was written with respect to the observer O. Note that points in his system appear to him to remain fixed while points in the system (x',y',z',t') appear to move with a constant velocity $\mathbf{u}$. Thus from Eq. (3) we can find the relationship between the constants a_{11} and a_{12} by considering a fixed point in the system of observer O'. To observer O, it moves a distance dz in time dt:

$$dz' = 0 = a_{11}\, dz + a_{12}\, dt \qquad (6)$$

Observer O sees the point moving with a velocity having magnitude u

$$\frac{dz}{dt} = -\frac{a_{12}}{a_{11}} = u \qquad (7)$$

Consequently,

$$a_{12} = -u a_{11} \qquad (8)$$

and we are left with the three unknowns a_{11}, a_{21}, and a_{22}.

The remaining unknowns can be found from the requirement that the velocity of light be the same in either reference system. Assume that the origin of both coordinate systems are coincident at the time $t = t' = 0$ and that a burst of light is initiated from the origin at that time. The burst of light will travel outward from the origin, and at some later time each observer can measure the distance from the origin of his system to the wavefront of the light burst. The squares of these distances are given by

$$x^2 + y^2 + z^2 = (ct)^2 \qquad (9)$$
$$(x')^2 + (y')^2 + (z')^2 = (ct')^2 \qquad (10)$$

where the constant c for the velocity of light is the same in either system. Therefore the distance given by Eq. (9) must be the same as that of Eq. (10). If we substitute Eqs. (1) to (4) into Eq. (10), we can determine the (x',y',z',t') coordinates in terms of the (x,y,z,t) coordinates. Using Eq. (8) to eliminate the unknown a_{12}, we obtain for the square of the distance

$$x^2 + y^2 + a_{11}^2 (z - ut)^2 = c^2 (a_{21} z + a_{22} t)^2 \qquad (11)$$

FUNDAMENTALS OF FIELDS

By rewriting Eq. (11) we obtain the coefficients for each of the independent variables as follows:

$$x^2 + y^2 + (a_{11}^2 - c^2 a_{21}^2) z^2 = (c^2 a_{22}^2 - u^2 a_{11}^2) t^2 \\ + 2(c^2 a_{21} a_{22} + u a_{11}^2) zt \quad (12)$$

The coefficients of this equation must agree with the coefficients of Eq. (9). By comparing the corresponding coefficients of Eqs. (9) and (12), we obtain three equations in the three unknowns a_{11}, a_{21}, a_{22}:

$$a_{11}^2 - c^2 a_{21}^2 = 1 \quad (13)$$
$$c^2 a_{22}^2 - u^2 a_{11}^2 = c^2 \quad (14)$$
$$c^2 a_{21} a_{22} + u a_{11}^2 = 0 \quad (15)$$

The solution of these equations for the constants is not difficult, but some arbitrary choices do appear with respect to the sign of certain square roots (see Prob. 19). Solving for the unknown constants gives

$$x' = x$$
$$y' = y$$
$$\begin{bmatrix} z' \\ t' \end{bmatrix} = \frac{1}{k} \begin{bmatrix} 1 & -u \\ -\frac{u}{c^2} & 1 \end{bmatrix} \begin{bmatrix} z \\ t \end{bmatrix} \quad (16)$$

which completes our derivation of the transformation T. The constant k in this equation, called the *contraction factor*, is given by

$$\boxed{k = \sqrt{1 - \frac{u^2}{c^2}}} \quad (17)$$

The transformation T of coordinates as given by Eq. (16) is called a *Lorentz transformation*. It is the transformation to be used when one knows coordinates in the system (x,y,z,t).

By similar reasoning one can obtain the transformation T^{-1} to be used when the coordinates (x',y',z',t') are known. In this case, the Lorentz transformation T^{-1}, which is given by Eq. (18), differs from T only in the sign of the relative velocity u:

$$x = x'$$
$$y = y'$$
$$\begin{bmatrix} z \\ t \end{bmatrix} = \frac{1}{k} \begin{bmatrix} 1 & u \\ \frac{u}{c^2} & 1 \end{bmatrix} \begin{bmatrix} z' \\ t' \end{bmatrix} \quad (18)$$

Substituting the matrix product for z and t from Eq. (18) into Eq. (16) shows that the product of the square matrices and the constant $1/k^2$ will yield the identity matrix. This means that the mapping is one to one, so that we can transform coordinates from one system into another and then back again to obtain the original coordinates.

If the relative motion between the two coordinate systems is zero ($u = 0$), then $z = z'$ and $t = t'$, as we can see from Eqs. (16) and (18). However, if u is not zero, z' and t' become functions of both distance z and time t, and vice versa. As the velocity u approaches the speed of light c, the interdependence of time and distance variables becomes more pronounced, and the contraction factor k becomes smaller and smaller, as shown in the table. Thus we see that the contraction factor is essen-

The contraction factor

$$k = \sqrt{1 - \frac{u^2}{c^2}}$$

$\dfrac{u}{c}$	k
0	1.0000
1/10	0.9950
1/5	0.9798
1/4	0.9682
1/3	0.8819
1/2	0.8665
2/3	0.7453
3/4	0.6614
4/5	0.6000
$\sqrt{3}/2$	0.5000
9/10	0.4359
99/100	0.1411
1.0000	0.0000

tially unity unless the relative velocity between the two coordinate systems is very large.

THE DISTORTION OF DISTANCE AND TIME

Suppose the two coordinate systems are at rest so that their origins coincide and that each observer has a meterstick which he uses to mark 1-m intervals along each of his distance axes. While the systems are at rest, the intervals of one coordinate system will be identical to the intervals of the other coordinate system. Furthermore, we assume that each observer has a clock (or some instrument to measure time) and that

FUNDAMENTALS OF FIELDS

their time-interval measurements are identical when the systems are at rest. Next we allow the system (x',y',z',t') to move at the velocity u along the z axis of the system (x,y,z,t), and we assume that both clocks read zero at the instant the origins coincide.

Consider a rod at rest in the (x,y,z,t) system which lies on the z axis and has a length $L = z_2 - z_1$. The moving observer O' measures the length of the rod $L' = z'_2 - z'_1$ in his coordinate system at a given time t'_0 as he moves by the rod. Observer O' uses the Lorentz transformation T^{-1} as given in Eqs. (18) to compare the apparent length L' with the proper length L. From Eqs. (18) we obtain

$$z = \frac{1}{k}(z' + ut') \tag{19}$$

Then from Eq. (19) observer O' can determine z_1 and z_2 from his measurements of z'_1 and z'_2

$$z_1 = \frac{1}{k}(z'_1 + ut'_0) \tag{20}$$

$$z_2 = \frac{1}{k}(z'_2 + ut'_0) \tag{21}$$

The difference between these two equations will be the length L, which is the proper length of the rod. Consequently, under the Lorentz transformation T^{-1}, lengths transform as follows:

$$L = \frac{1}{k}L' \quad \text{from transformation } T^{-1} \tag{22}$$

To reiterate, L' is the length of the rod as measured by the moving observer O', and L is its proper (true) length in the system (x,y,z,t) in which it is at rest. Since the contraction factor k is a number less than unity for any velocity u greater than zero, observer O' will say that the length L' has contracted and is smaller than its proper value L as seen in the reference system (x,y,z,t).

As another example, suppose that a square box which measures 1 m along each edge is placed at rest in the (x,y,z,t) system. If observer O' moves past this box in a direction parallel to one edge at a velocity four-fifths the speed of light, he will see that the length of the box has shrunk to $6/10$ m (as given by the contraction factor) along the edge parallel to his motion. The other edges of the box will appear to remain 1 m in length as long as there is no component of velocity along those edges. Consequently, we see that size is not a fixed property of a body, but is a function of the relative velocity of the observer and the body. We have never observed this effect in our everyday life because we never encounter velocities significant with respect to the velocity of light. However, it can be observed by specially designed experiments. Perhaps a roughly equivalent analogy of everyday life is the pitch of sound which depends upon the conditions under which the sound is observed—specifically, upon the relative velocity of the source and the observer. Similarly, the apparent length of a body depends upon whether an observer is moving rapidly or slowly with respect to that body.

Conversely, if a rod at rest in the (x',y',z',t') system, an observer O in the system (x,y,z,t) can measure the apparent length of the rod $L = z_2 - z_1$ at a given instant of time $t = t_0$ as the rod moves by. Observer O can determine its proper length $L' = z'_2 - z'_1$ in its rest system by using transformation T:

$$L' = \frac{1}{k} L \quad \text{from transformation } T \tag{23}$$

In this case, L is the measured length of the moving rod (as measured by observer O), and L' is its proper length in its rest system. Although the notation is different in Eqs. (22) and (23), the phenomenon is the same in either case. In summary, we see that for any observer, the measurable (apparent) length of a moving object is less than its proper (rest) length, as follows:

$$\text{Proper length} = \frac{1}{k} \text{ (measurable length)} \tag{24}$$

The choice of transformations T or T^{-1} depends only upon the system in which the observer can make measurements, i.e., in his rest system. A graph of Eq. (24) is shown in Fig. 1.12.2a. The straight line of Eq.

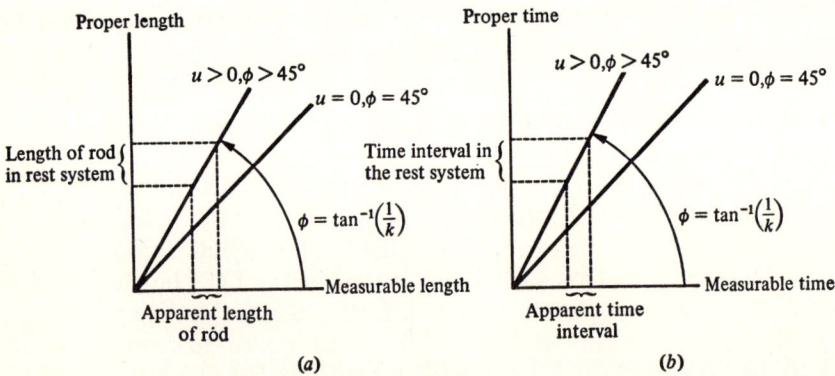

Fig. 1.12.2 For a system in motion, both distance and time are distorted with respect to a system at rest. (a) The apparent length is never greater than the proper length. (b) Clocks in a moving system are always slower than clocks in one's own system.

FUNDAMENTALS OF FIELDS

(24) has a slope of $1/k$, and we can define an angle ϕ as

$$\phi = \tan^{-1}\frac{1}{k} \tag{25}$$

Consequently, when there is no relative velocity between the coordinate systems, the contraction factor k is unity and the angle ϕ is 45°. However, as the relative velocity u increases, the angle ϕ increases from 45° and approaches 90° as u approaches c.

We also find that observers O and O' will not agree on time intervals. Suppose that observer O' is located at a fixed point z'_1 in his reference system and that he measures an interval of time $\Delta t' = t'_2 - t'_1$ (on his clock) that corresponds to a time interval $\Delta t = t_2 - t_1$ on a clock in the system of observer O. By using transformation T^{-1}, observer O' can compare his measured time interval $\Delta t'$ with the proper time interval Δt. The derivation (similar to the one for distance) leads to

$$\Delta t = \frac{1}{k}\Delta t' \quad \text{from transformation } T^{-1} \tag{26}$$

Similarly for transform T, with observer O making the measurements,

$$\Delta t' = \frac{1}{k}\Delta t \quad \text{from transformation } T \tag{27}$$

The results of Eqs. (26) and (27) can be summarized as

$$\text{Proper time} = \frac{1}{k}(\text{measured time}) \tag{28}$$

For example, if the measured (apparent) time interval $\Delta t'$ is 6/10 s to observer O', who is moving with a velocity of four-fifths the speed of light with respect to observer O, the proper interval Δt will be 1 s. Observer O' will say that the clock of observer O is running too slow. The effect of distortion of time is shown in Fig. 1.12.2b.

In conclusion, distance is always a maximum and time is fastest in one's own system. In newtonian mechanics, it was assumed that time and distance were independent variables, and all our equations were based upon a fundamental system of (assumed) independent variables (mks, cgs, etc.). Thus the discovery that time and distance are not independent changed the most fundamental concepts of physics.

THE TRANSFORMATION OF VELOCITY

Since the time and distance variables are interdependent under the Lorentz transformation, we should expect two observers (with a relative velocity u) to disagree about the relative velocity of a commonly observed

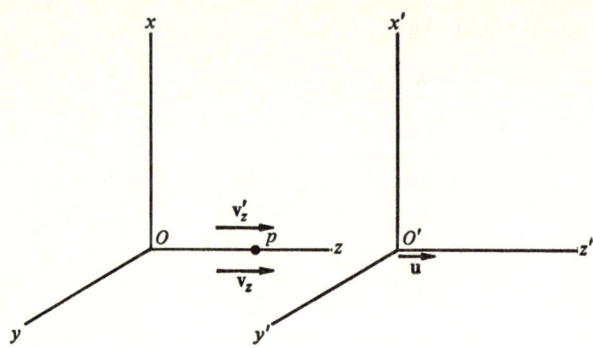

Fig. 1.12.3 To observer O', the velocity v'_z of the particle p is not the vector sum of **u** and **v**$_z$.

particle in motion. For example, suppose the two observers both observe a particle p which is in motion with respect to each of the observers. For simplicity we will assume that the particle is moving along the z axis, as shown in Fig. 1.12.3, with a velocity **v**$_z$ with respect to observer O and velocity $\mathbf{v}'_z$ with respect to observer O'. Assume that observer O measures the velocity $\mathbf{v}_z = dz/dt\, \mathbf{a}_z$ of the particle. Then he can find $\mathbf{v}'_z$ using transformation T of Eq. (16), by taking differentials

$$dz' = \frac{1}{k}(dz - u\,dt) \tag{29}$$

$$dt' = \frac{1}{k}\left(dt - \frac{u}{c^2}dz\right) \tag{30}$$

To determine the magnitude of the velocity $\mathbf{v}'_z$, we take the ratio of dz' to dt'

$$v'_z = \frac{dz'}{dt'} = \frac{dz - u\,dt}{dt - u\,dz/c^2} = \frac{v_z - u}{1 - uv_z/c^2} = (v_z - u)g \tag{31}$$

where

$$g \equiv \frac{1}{1 - uv_z/c^2} \tag{32}$$

Equation (31) is the Lorentz transformation for the z component of velocity for the reference systems given in Fig. 1.12.3. If the velocity of the particle is not collinear with any of the axes, there will be velocity components in all three of the distance coordinates. The transformations for the x and y axes can be derived from Eqs. (1), (2), and (30).

FUNDAMENTALS OF FIELDS

The derivations are similar to the one above, and the results are

$$v'_x = \frac{dx'}{dt'} = \frac{dx}{(1/k)(dt - u\,dz/c^2)} = kgv_x \tag{33}$$

$$v'_y = \frac{dy'}{dt'} = \frac{dy}{(1/k)(dt - u\,dz/c^2)} = kgv_y \tag{34}$$

Consequently, Eqs. (31), (33), and (34) constitute the Lorentz transformation for the velocity of the particle p from **v** to **v'**.

PROGRAMMED EXERCISE 1.12.1

The effects of the Lorentz transformation of velocity will now be explored by a series of examples. In all the examples which follow we assume the configuration given in Fig. 1.12.1 unless specifically stated otherwise. The reader should study each of the following cases in detail and verify the results using Eq. (31).

Case 1 $u = 0\mathbf{a}_z$, $\mathbf{v}_z = -\frac{4}{5}c\mathbf{a}_z$. To observer O the particle moves with a velocity $-\frac{4}{5}c\mathbf{a}_z$. To observer O' the particle moves with a velocity $-\frac{4}{5}c\mathbf{a}_z$. Therefore, the observers measure the same velocity of p when there is no relative motion between them.

Case 2 $u = \frac{4}{5}c\mathbf{a}_z$, $\mathbf{v}_z = -\frac{4}{5}c\mathbf{a}_z$. To observer O the particle moves with a velocity $-\frac{4}{5}c\mathbf{a}_z$. To observer O' the particle moves with a velocity $-0.976c\mathbf{a}_z$. By newtonian mechanics, observer O' would say p moves with a velocity $(v_z - u)\mathbf{a}_z = -\frac{8}{5}c\mathbf{a}_z$, or faster than the speed of light. By relativistic mechanics, p approaches the speed of light as u or v_z increase, but it will not exceed it.

Case 3 $u = c\mathbf{a}_z$, $\mathbf{v}_z = -c\mathbf{a}_z$. To observer O the particle moves with a velocity $-c\mathbf{a}_z$. To observer O' the particle moves with a velocity $-c\mathbf{a}_z$. Here we assume that observer O' is "riding a light wave" and he passes a light wave p moving in the opposite direction. Both observers find that p has a velocity c.

Case 4 Suppose that the velocity of the particle p is in the positive z direction (reversed from the above cases). If p moves with the velocity of light, then observer O' finds $v'_z = c$ regardless of his velocity u. This is also true in the limit as $u \to c$.

Case 5 Suppose observer O' has a light source (such as a flashlight) which he uses to initiate a light wave in the positive z direction. Also suppose he is moving in the positive z direction at a velocity **u** when he initiates the light wave. Will the light wave travel with a velocity $v_z\mathbf{a}_z = (u + c)\mathbf{a}_z$ with respect to a "stationary" observer O? We can use Eq. (18) to see what observer O observes. We see that

$$dz = \frac{1}{k}(dz' + u\, dt')$$
$$dt = \frac{1}{k}\left(dt' + \frac{u}{c^2} dz'\right)$$
(35)

Therefore observer O finds a velocity v_z of

$$v_z = \frac{v_z' + u}{1 + uv_z'/c^2} = (v_z' + u)g' \tag{36}$$

where

$$g' \equiv \frac{1}{1 + uv_z'/c^2} \tag{37}$$

We can use this equation to answer the above question.

In conclusion, observers in motion with respect to one another will disagree about velocities, except for the velocity of light, which is always the same for every observer. Observed velocities will be less than the velocities predicted by newtonian mechanics because of the factor $1 \pm uv/c^2$ in the denominator of the equations. The velocities are smaller than expected due to the distortions of time and distance under the Lorentz transformation.

MASS AND ENERGY

As an observer O observes a particle in motion, it appears that the size of the particle decreases in the direction of motion. Consequently, the density of the particle must change unless possibly mass decreases in such a way that it compensates for the decrease in size. However, this would be in violation of the newtonian law of conservation of mass. In postulating the special theory of relativity, Einstein found that the law of conservation of mass and the law of conservation of momentum could not both be valid and one or the other had to be discarded. Experimental evidence indicated that the law of conservation of momentum was correct, and Einstein went on to investigate the variation of mass.†
We omit the derivations here because they lead us too far from the objectives of this text, but it was found that mass of a particle increases with

---

† The reasons for this decision are discussed in Elliott, *op. cit.*, chap. 2.

respect to an external observer as the relative velocity of the particle increases, according to

$$m = \frac{m_0}{k} \tag{38}$$

where m_0 is the rest mass of the particle. The rest mass is the minimal mass of the particle, and it occurs when k has its maximum value of unity. Since the contraction factor k approaches zero as the velocity approaches that of light, the mass of a particle increases without limit as the particle approaches the speed of light. This means that an external observer must supply more and more energy to a particle if he wants to increase its velocity and that it would require an infinite amount of energy to cause the particle to move at the speed of light. Because the particle size is also decreasing, the density of the particle is not constant but must increase also. All these are effects that appear to an external observer. An observer moving with the particle would detect no change in its size, mass, or density.

The relationship between mass and energy can be demonstrated by the following derivation. If we expand the factor $1/k$ of Eq. (38), using the binomial expansion,

$$\frac{1}{k} = \left(1 - \frac{v^2}{c^2}\right)^{-\frac{1}{2}} = 1 + \frac{1}{2}\left(\frac{v}{c}\right)^2 + \frac{3}{8}\left(\frac{v}{c}\right)^4 + \cdots \tag{39}$$

and substitute the series into Eq. (38), we obtain the following equation for mass:

$$m = m_0 + \frac{m_0}{2}\left(\frac{v}{c}\right)^2 + \frac{3m_0}{8}\left(\frac{v}{c}\right)^4 + \cdots \tag{40}$$

According to newtonian mechanics, the kinetic energy of a body in motion is $m_0 v^2/2$. In comparing this with the second term of the series of Eq. (40), it is apparent that the second term represents the kinetic energy divided by the square of the velocity of light. Consequently, if we multiply every term of Eq. (40) by the square of the velocity of light, the second term will have units of energy and so must every other term of the series. Thus we are led to conclude that the total energy E associated with the particle in motion is

$$E = mc^2 = m_0 c^2 + \frac{m_0}{2} v^2 + \frac{3 m_0 v^4}{8 c^2} + \cdots \tag{41}$$

The second term of the series represents the conventional kinetic energy as given by newtonian mechanics. The higher-degree terms of the series were never discovered experimentally in pre-Einstein days because for $v \ll c$ these terms were negligible with respect to the second term. However, the most interesting term is the first, m_0c^2. According to this term, there is a tremendous amount of energy E_0 associated with the mass m_0, even if the velocity v is zero. This led to the famous equation

$$E_0 = m_0c^2 \tag{42}$$

which has been demonstrated in a most dramatic way by the detonation of nuclear explosives. Because the term c^2 is such a large number, there is a tremendous amount of energy associated even with a small amount of mass.

Steinmetz† has a descriptive comparison between the energy associated with the rest mass and the kinetic energy of the same mass moving at a high velocity; i.e., he has compared the first and second terms of the series of Eq. (41) in the following way. The earth is moving around the sun at about 20 mi/s, which is approximately 25 times as fast as the fastest rifle bullet. The kinetic energy associated with 1 kg of matter traveling at that velocity (the energy given by the second term of the series) is about 150 kWh, which is about 15 times as much energy as would be given up by the heat of combustion for an equivalent amount of coal. However, if the rest mass of 1 kg were converted into energy according to Eq. (42), it would yield 250×10^8 kWh.

THE TRANSFORMATION OF MASS AND FORCE

Suppose that a particle moves with a velocity $v_z\mathbf{a}_z$ with respect to a stationary observer O. To the observer, the mass of the particle appears to increase as its velocity increases according to

$$m = \frac{m_0}{\sqrt{1 - v^2_z/c^2}} \tag{43}$$

[from Eqs. (17) and (38) with v_z substituted for u].

Suppose that a second observer O', who is moving with a velocity $\mathbf{u}$ with respect to observer O, also observes the particle in motion. To him, the velocity of the particle is $v'_z\mathbf{a}_z$, and the mass of the particle appears to be

$$m' = \frac{m_0}{\sqrt{1 - (v'_z/c)^2}} \tag{44}$$

The rest mass m_0 is the same in Eqs. (43) and (44) because both observers will measure the same rest mass if they move with the particle. We can determine the transformations between m and m' by eliminating m_0

† *Op. cit.*

between Eqs. (43) and (44). The result is

$$m' = m \left[\frac{1 - (v_z/c)^2}{1 - (v_z'/c)^2} \right]^{1/2} \tag{45}$$

If we eliminate the v_z' from Eq. (45) by a substitution from Eq. (31), we obtain the way in which the mass transforms from m (as seen by observer O) to m' (as seen by observer O')

$$m' = \frac{m}{k}\left(1 - \frac{uv_z}{c^2}\right) = \frac{m}{kg} \tag{46}$$

The masses m and m' are functions of both time and distance coordinates and must be treated accordingly if they are involved in differentiation or integration formulas. The number kg in the denominator in Eq. (46) can be either greater than unity or less than unity, depending upon the velocity v_z of the particle with respect to the relative velocity u of the observers. If $v_z = 0$ so that the particle is not in motion with respect to an observer O, then $kg = k$ and consequently observer O' measures m' as being larger than m_0. Moreover, if the particle moves with a velocity u so that it is fixed with respect to observer O', then $kg = 1/k$ and observer O will see an increase in the mass of the particle. It will be left as a problem to show that the number kg is bounded as follows:

$$k \leq kg < \sqrt{\frac{c+u}{c-u}} \tag{47}$$

as the velocity v_z of the particle increases from zero to the velocity of light.

One of the laws of physics, valid in any reference frame, is that the total force on a mass is equal to the time derivative of the momentum of that mass. That is, in either of our reference frames, the force on a particle is

$$\mathbf{F} = \frac{d}{dt} m\mathbf{v} \tag{48}$$

$$\mathbf{F}' = \frac{d}{dt'} m'\mathbf{v}' \tag{49}$$

It is of immediate interest to determine how the forces $\mathbf{F}$ and $\mathbf{F}'$ are related under the Lorentz transformation. For simplicity, we will continue to use rectangular coordinate systems and express forces and velocities in the usual manner in terms of components and unit vectors. For example,

$$\mathbf{F}' = F_x' \mathbf{a}_x + F_y' \mathbf{a}_y + F_z' \mathbf{a}_z \tag{50}$$

$$\mathbf{v}' = v_x' \mathbf{a}_x + v_y' \mathbf{a}_y + v_z' \mathbf{a}_z \tag{51}$$

The actual derivation of the relationship between **F** and **F'** is given by Elliott and will not be repeated here, since we are primarily concerned with certain applications of the Lorentz transformations and not with their derivations. To observer O the Lorentz transformation of the components of **F** to the components of **F'** is given by

$$F'_x = kgF_x \tag{52}$$

$$F'_y = kgF_y \tag{53}$$

$$F'_z = F_z - \frac{gu}{c^2}(v_x F_x + v_y F_y) \tag{54}$$

We will also be interested in the inverse transformation of the force components. To an observer O' the relationships between the components of **F** and **F'** are given by

$$F_x = \frac{1}{kg}F'_x \tag{55}$$

$$F_y = \frac{1}{kg}F'_y \tag{56}$$

$$F_z = F'_z + \frac{u}{kc^2}(v_x F'_x + v_y F'_y) \tag{57}$$

In these transformations, we have assumed the same orientation of the coordinate axes and the same relative velocity u along the z axis as previously. Note that the components of **v** are given in the system (x,y,z,t).

Suppose that a particle p is moving in the z direction, as shown in Fig. 1.12.4, and that a known force **F'** exists on the particle. We want

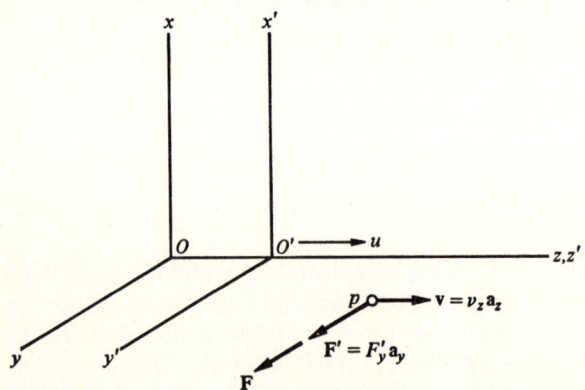

Fig. 1.12.4 The observers measure different forces on the particle p.

FUNDAMENTALS OF FIELDS

to determine the force **F** as seen by an observer O; that is, we want the Lorentz transformation of **F'** to **F**. We will assume that the force **F** is of the form

$$\mathbf{F} = F_x \mathbf{a}_x + F_y \mathbf{a}_y + F_z \mathbf{a}_z \tag{58}$$

This force can be found in terms of the known components of **F'** using Eqs. (55) to (57). By substituting these equations into Eq. (58) and rearranging terms it is possible to obtain

$$\mathbf{F} = \left(\frac{1}{k} F'_x \mathbf{a}_x + \frac{1}{k} F'_y \mathbf{a}_y + F'_z \mathbf{a}_z \right) + \left[\frac{u}{kc^2} \mathbf{v} \times (\mathbf{a}_z \times \mathbf{F'}) \right] \tag{59}$$

However, it is easier to show that Eq. (59) is equivalent to Eq. (58) by expanding the vector triple product and then using Eqs. (55) to (57). The advantage of Eq. (59) is that the velocity **v** of the particle appears only once in the equation, and often this proves to be convenient. In Fig. 1.12.4 we have assumed that the particle p has only an $\mathbf{a}_z$ component of velocity and that the force **F'** has only an $\mathbf{a}_y$ component F'_y. If we substitute these vectors **v** and **F'** into Eq. (59), we find the force **F** on the particle with respect to observer O

$$\mathbf{F} = \frac{1}{k} F'_y \mathbf{a}_y + \frac{u}{kc^2} v_z \mathbf{a}_z \times (\mathbf{a}_z \times F'_y \mathbf{a}_y)$$

$$= \frac{F'_y}{k} \left(1 - \frac{uv_z}{c^2} \right) \mathbf{a}_y = \frac{F'_y}{kg} \mathbf{a}_y \tag{60}$$

Consequently, the force **F** as seen by observer O is in the $\mathbf{a}_y$ direction, and its magnitude can be greater or less than F'_y, depending upon the value of kg. In this particular example, the result could have been obtained most easily by computation of individual components from Eqs. (55) to (57). It is suggested that the reader use these equations to obtain Eq. (60).

1.13 THE DEVELOPMENT OF MAXWELL'S EQUATIONS

THE ELECTRIC AND MAGNETIC FIELDS

Suppose the coordinate axes of two observers O and O' are aligned as shown in Fig. 1.13.1 so that the coordinate system of observer O' slides along the z axis of the system of observer O. It is assumed that the relative velocity between the two coordinate systems is **u** and that an electric charge q is located at the point z'_1 and is fixed in the coordinate system (x',y',z',t'). Since the static charge is fixed with respect to observer O', he sees a static electric field **E'** throughout the entire region. If a test charge q_t exists in the region, it will experience a force **F'** because

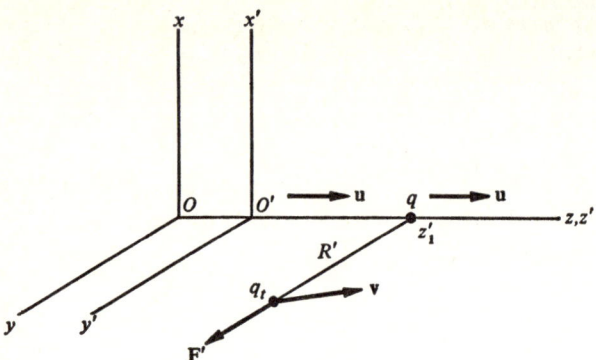

Fig. 1.13.1 Charge q moving along the z axis exerts a force on the test charge q_t.

of the charge q. Because the charge q is fixed with respect to observer O', the only force that observer O' will be able to detect upon the test charge will be the coulomb force as given by Coulomb's law, even though q_t might be in motion with respect to q. We will assume that Coulomb's law

$$\mathbf{F}' = \frac{qq_t}{4\pi\varepsilon_0} \frac{\mathbf{R}'}{(R')^3} = q_t \mathbf{E}' \qquad (1)$$

is valid, so that we can derive the electrostatic field $\mathbf{E}'$ as detected by observer O'. We will assume that the force field is given in rectangular coordinates, so that if the observer O' knows the force $\mathbf{F}'$ and its components,

$$\mathbf{F}' = F'_x \mathbf{a}_x + F'_y \mathbf{a}_y + F'_z \mathbf{a}_z \qquad (2)$$

he can determine the components of the electric field $\mathbf{E}'$

$$\mathbf{E}' = \frac{F'_x}{q_t} \mathbf{a}_x + \frac{F'_y}{q_t} \mathbf{a}_y + \frac{F'_z}{q_t} \mathbf{a}_z \qquad (3)$$

To observer O', both the force field and the electric field are static-fields, and the theory developed earlier in Chap. 1 is adequate for his use.

To observer O, the charge q is in motion along the z axis, and consequently the force fields and electric fields will appear to vary with time as q moves. Assuming that we can determine the force field from Coulomb's law for the static case, as seen by observer O', we can use the Lorentz transformation of the previous section to determine the dynamic (time-varying) force field as seen by observer O. We will assume that *charge is invariant* under the Lorentz transformation. By starting with

FUNDAMENTALS OF FIELDS

a single experimental law (Coulomb's law) and using the Lorentz transformation we can develop Maxwell's equations for time-varying fields. One of the advantages of this approach is to show how time-varying electric and magnetic fields are related.

Assuming that we know the force **F'** and its components, as shown in Fig. 1.13.1, we can find the force that observer O would detect on the particle q_t from the force equation [Eq. (59) of Sec. 1.12, repeated below]

$$\mathbf{F} = \left(\frac{1}{k} F'_x \mathbf{a}_x + \frac{1}{k} F'_y \mathbf{a}_y + F'_z \mathbf{a}_z\right) + \frac{u}{kc^2} [\mathbf{v} \times (\mathbf{a}_z \times \mathbf{F'})] \tag{4}$$

In this equation we assume that the velocity **v** is the velocity of the particle q_t with respect to observer O. Now suppose that the velocity **v** is zero in Eq. (4), so that observer O detects a force field and an electric field of

$$\mathbf{F} = \frac{1}{k} F'_x \mathbf{a}_x + \frac{1}{k} F'_y \mathbf{a}_y + F'_z \mathbf{a}_z \tag{5}$$

$$\mathbf{E} = \frac{\mathbf{F}}{q_t} = \frac{1}{k} E'_x \mathbf{a}_x + \frac{1}{k} E'_y \mathbf{a}_y + E'_z \mathbf{a}_z \tag{6}$$

In effect, this is the Lorentz transformation of the electric field as seen by observer O. The transformations of the individual components are

$$\boxed{E_x = \frac{E'_x}{k} \qquad E_y = \frac{E'_y}{k} \qquad E_z = E'_z} \tag{7}$$

Note that if the velocity **u** with respect to the two reference frames is also zero, the contraction factor k is unity and the electric fields in the two reference frames are identical, as one would expect. The force **F** measured by observer O on the fixed test charge q_t (that is, for the case where **v** = 0) is called the electric field force or the coulomb force

$$\mathbf{F} = q_t \mathbf{E} = \mathbf{F}_{\text{coul}} \qquad \mathbf{v} = 0 \tag{8}$$

As long as observer O does not move his test charge, the coulomb force is the only force he can detect, even though the source charge q is in motion. We draw this conclusion from Eq. (4). If the source charge q is in motion, the coulomb force on q_t will vary with time.

Next we consider the more general case of Eq. (4), where the velocity **v** is nonzero. We can rewrite Eq. (4) in terms of the electric fields **E** and **E'**,

$$\mathbf{F} = q_t \mathbf{E} + \frac{u}{kc^2} [\mathbf{v} \times (\mathbf{a}_z \times q_t \mathbf{E'})] \tag{9}$$

and then rearrange the constants in the bracketed term to obtain the convenient form

$$\mathbf{F} = q_t\mathbf{E} + q_t\left[\mathbf{v} \times \left(\frac{u}{kc^2}\mathbf{a}_z \times \mathbf{E}'\right)\right] \tag{10}$$

It is obvious that the total force on the charged particle is the sum of the coulomb force and a force which depends upon the term in the brackets. This term has units of the field **E** (by comparison with the first term on the right), but it is the cross product of the velocity **v** and a new field which depends upon u and $\mathbf{E}'$. This new field is called a *magnetic field* **B** and is defined by

$$\boxed{\mathbf{B} \equiv \frac{u}{kc^2}\mathbf{a}_z \times \mathbf{E}' \qquad \text{Wb/m}^2} \tag{11}$$

The units of **B** are webers per square meter, also called *teslas*. This is the magnetic field for the special orientation of axes and the charge distribution which we have chosen. We can see that the magnetic field is independent of the test charge or its velocity, but it does depend upon the $\mathbf{E}'$ field of the charge q and upon the velocity **u** of that charge. We see that it is not caused by some new phenomenon but rather by the electric field of a charge moving with respect to an observer.

Next we rewrite the equation for force in terms of our newly defined **B** field

$$\mathbf{F} = q_t\mathbf{E} + q_t\mathbf{v} \times \mathbf{B} \tag{12}$$

Then by factoring the charge q_t we obtain

$$\boxed{\mathbf{F} = q_t(\mathbf{E} + \mathbf{v} \times \mathbf{B})} \tag{13}$$

which is called the *Lorentz force equation*. The force given by Eq. (13) is sometimes called the *Lorentz force*, and we see that it is the sum of a coulomb force and a force due to the magnetic field:

$$\mathbf{F} = \mathbf{F}_{\text{coul}} + \mathbf{F}_{\text{mag}} \tag{14}$$

One of the advantages of the derivation of the force fields, using relativity, is that the Lorentz force can be derived from the basic postulates of the special theory of relativity. Thus we need not treat the Lorentz force as an experimental law.

FUNDAMENTALS OF FIELDS

The Lorentz transformation for the individual components of the **B** field can be obtained from Eq. (11) by expanding the curl

$$\mathbf{B} = \frac{u}{kc^2} \begin{vmatrix} \mathbf{a}_x & \mathbf{a}_y & \mathbf{a}_z \\ 0 & 0 & 1 \\ E'_x & E'_y & E'_z \end{vmatrix} \tag{15}$$

$$\mathbf{B} = \frac{u}{kc^2}(-E'_y \mathbf{a}_x + E'_x \mathbf{a}_y + 0 \mathbf{a}_z) \tag{16}$$

The Lorentz transformation is

$$\boxed{B_x = -\frac{u}{kc^2} E'_y \qquad B_y = \frac{u}{kc^2} E'_x \qquad B_z = 0} \tag{17}$$

Thus according to the theory, we can use Coulomb's law to determine the components of the static electric field $\mathbf{E}'$ as seen by observer O' and transform them, according to Eq. (17), to determine the magnetic field **B** as seen by the observer O. The component B_z is zero because of the particular orientation of our coordinate axes. (Under what conditions will $B_z \neq 0$?)

It is of interest to use Eqs. (7) and (17) to compute the ratios of the electric and magnetic field components as seen by observer O:

$$\boxed{\frac{E_x}{B_y} = \frac{c^2}{u} = -\frac{E_y}{B_x}} \tag{18}$$

The constant (c^2/u) has units of velocity, but when multiplied by the permeability constant of free space, it has units of impedance. If the velocity u is equal to the velocity of light, the impedance is called the *intrinsic impedance*, a constant which will be important in our study of transmission lines and waveguides.

We have determined the Lorentz transformations [Eqs. (7) and (17)] for the electric and magnetic fields for an observer O, assuming the fields of the observer O' are known. It is also convenient to have the inverse transformation to determine the fields seen by observer O', given the fields of observer O. From the force equations [Eqs. (52) to (54) of Sec. 1.12] we can show that this inverse transformation is

$$\boxed{E'_x = \frac{1}{k}(E_x - uB_y) \qquad E'_y = \frac{1}{k}(E_y + uB_x) \qquad E'_z = E_z} \tag{19}$$

The interested reader will find that this derivation is essentially the same as the one just completed.

TRANSFORMATION OF STATIC SOURCES

Earlier in Chap. 1 we found that static fields are due to scalar sources q and vector sources **J**. It is interesting to determine how these sources are affected by the Lorentz transformation.

Since charge is invariant, a given amount of charge dq will be constant in either reference system; therefore

$$dq = \rho \, dv = \rho' \, dv' = dq' \tag{20}$$

However, the differential volume dv' will be distorted when in motion because of the distortion of the distance axes. For our particular coordinate systems, the volume transforms as follows:

$$dv' = dx' \, dy' \, dz' = dx \, dy \, \frac{dz}{k} = \frac{dv}{k} \tag{21}$$

We can substitute Eq. (21) into Eq. (20) to determine how charge densities are distorted under transformation:

$$\boxed{\rho = \frac{\rho'}{k}} \tag{22}$$

Note that the charge distribution ρ' is a static distribution of charge in the system (x',y',z',t') but the charge density ρ is a function of time because the contraction factor k contains the velocity u.

To obtain the transformation of a vector source **J**, we start with the definition of current as the time rate of change of charge, which in turn can be written in terms of the charge density

$$dI_z = \frac{dq}{dt} = \frac{\rho \, dv}{dt} = \frac{\rho'}{k} \frac{dv}{dt} \tag{23}$$

For the geometry of interest, shown in Fig. 1.13.2, we can determine the current density over a differential surface by taking the ratio of the differ-

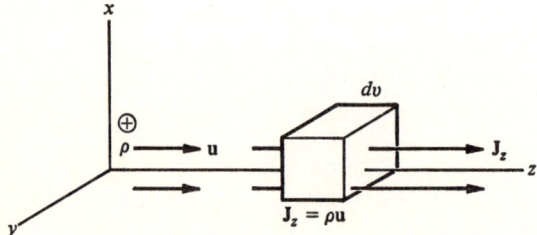

Fig. 1.13.2 The current density J_z is determined by the charge density ρ and its velocity.

FUNDAMENTALS OF FIELDS

ential current to the differential surface, where the current and the surface are at right angles. By dividing Eq. (23) by the differential surface $dx\,dy$

$$\frac{dI_z}{dx\,dy} = \frac{\rho'}{k}\frac{dz}{dt} = \frac{\rho'}{k}u \tag{24}$$

we can obtain the current density $\mathbf{J}_z$ and show how it is related to the charge densities ρ and ρ':

$$\mathbf{J}_z = \frac{\rho'}{k}\mathbf{u} = \rho\mathbf{u} \tag{25}$$

The direction of the current-density vector is determined by the direction of the velocity vector $\mathbf{u}$, but its magnitude is the product of the charge density and the velocity magnitude u. Because the charge density ρ is time-variant, the current density $\mathbf{J}$ also varies with time.

MAXWELL'S EQUATIONS

Suppose that a charge distribution ρ' is fixed in the reference frame of observer O', as shown in Fig. 1.13.3. The electric field that observer O' sees is an irrotational static field whose characteristics are given by the curl and divergence of Eqs. (26) and (27), as discussed earlier:

$$\nabla' \times \mathbf{E}' = 0 \tag{26}$$

$$\nabla' \cdot \mathbf{E}' = \frac{\rho'}{\varepsilon} \tag{27}$$

We will develop Maxwell's equations for the dynamic electromagnetic field, seen by observer O, from Eqs. (26) and (27). From Eq. (26), we

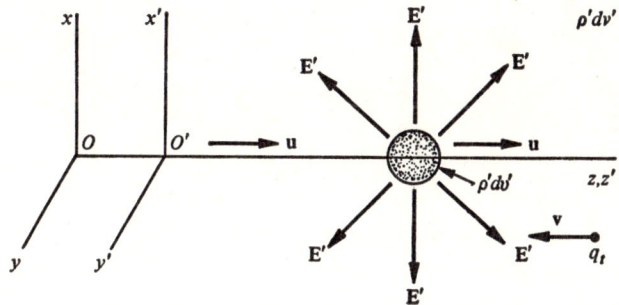

Fig. 1.13.3 The static electric field seen by observer O' is a dynamic electromagnetic field to observer O.

expand the curl of **E'** to obtain each of its components

$$\nabla' \times \mathbf{E}' = \left(\frac{\partial E'_z}{\partial y'} - \frac{\partial E'_y}{\partial z'}\right) \mathbf{a}_x + \left(\frac{\partial E'_x}{\partial z'} - \frac{\partial E'_z}{\partial x'}\right) \mathbf{a}_y$$
$$+ \left(\frac{\partial E'_y}{\partial x'} - \frac{\partial E'_x}{\partial y'}\right) \mathbf{a}_z = 0 \quad (28)$$

Each component in parentheses must be zero because the curl of **E'** is zero. Consequently, we can work with each component individually, and first we equate the z component to zero

$$\frac{\partial E'_y}{\partial x'} - \frac{\partial E'_x}{\partial y'} = 0 \quad (29)$$

We can use the Lorentz transformation of the fields as given by Eq. (19) to obtain this component of the curl as seen by observer O. Also, the partial derivatives must be transformed according to the usual chain rule.† Using the Lorentz transformation and the chain rule on Eq. (29), we obtain

$$\frac{1}{k}\left(\frac{\partial E_y}{\partial x} - \frac{\partial E_x}{\partial y}\right) + \frac{u}{k}\left(\frac{\partial B_x}{\partial x} + \frac{\partial B_y}{\partial y}\right) = 0 \quad (30)$$

The second set of parentheses in Eq. (30) encloses the divergence of **B** (in this particular case, $B_z = 0$). If we had chosen a more general orientation for our axes, all the terms of the divergence would have been obtained. Rewriting Eq. (30) in terms of the divergence of **B**, we obtain

$$\left(\frac{\partial E_y}{\partial x} - \frac{\partial E_x}{\partial y}\right) + u(\nabla \cdot \mathbf{B}) = 0 \quad (31)$$

To prove that the divergence of **B** is identically zero, we use the trans-

---

† Given $x = x'$, $y = y'$, $z = (z' + ut')/k$, $t = (t' + uz'/c^2)/k$, and some function $f = f(x,y,z,t)$, then

$$\frac{\partial f}{\partial x'} = \frac{\partial f}{\partial x}, \quad \frac{\partial f}{\partial y'} = \frac{\partial f}{\partial y} \quad (A)$$

$$\frac{\partial f}{\partial z'} = \frac{\partial f}{\partial z}\frac{\partial z}{\partial z'} + \frac{\partial f}{\partial t}\frac{\partial t}{\partial z'} = \frac{1}{k}\frac{\partial f}{\partial z} + \frac{u}{kc^2}\frac{\partial f}{\partial t} \quad (B)$$

$$\frac{\partial f}{\partial t'} = \frac{\partial f}{\partial z}\frac{\partial z}{\partial t'} + \frac{\partial f}{\partial t}\frac{\partial t}{\partial t'} = \frac{u}{k}\frac{\partial f}{\partial z} + \frac{1}{k}\frac{\partial f}{\partial t} \quad (C)$$

From (C), if $\partial f/\partial t' = 0$, then

$$\frac{\partial f}{\partial t} = -u\frac{\partial f}{\partial z} \quad (D)$$

FUNDAMENTALS OF FIELDS

formations of Eq. (17) to write the divergence in the form

$$\nabla \cdot \mathbf{B} = \frac{\partial B_x}{\partial x} + \frac{\partial B_y}{\partial y} + \frac{\partial B_z}{\partial z} = \frac{u}{kc^2}\left(\frac{\partial E'_x}{\partial y} - \frac{\partial E'_y}{\partial x}\right)$$
$$= \frac{u}{kc^2}\left(\frac{\partial E'_x}{\partial y'} - \frac{\partial E'_y}{\partial x'}\right) \quad (32)$$

where the term in the last parentheses is the $\mathbf{a}_z$ component of the curl $\nabla' \times \mathbf{E}'$, as seen by observer O', and is identically zero. Consequently, we conclude that

$$\boxed{\nabla \cdot \mathbf{B} = 0} \quad (33)$$

This is one of Maxwell's equations for time-varying fields and is the same as in static fields. In conclusion, we see that the $\mathbf{B}$ field (either static or dynamic) is a solenoidal field.

Next we work with the $\mathbf{a}_x$ component of the curl from Eq. (28)

$$\frac{\partial E'_z}{\partial y'} - \frac{\partial E'_y}{\partial z'} = 0 \quad (34)$$

We substitute Eqs. (18), (19), and the chain rule into Eq. (34) to find

$$\frac{\partial E'_z}{\partial y} - \left(\frac{1}{k}\frac{\partial E'_y}{\partial z} + \frac{u}{kc^2}\frac{\partial E'_y}{\partial t}\right) = 0 \quad (35)$$

$$\left(\frac{\partial E_z}{\partial y} - \frac{\partial E_y}{\partial z}\right) + \frac{\partial B_x}{\partial t} = 0 \quad (36)$$

Equation (36) is the $\mathbf{a}_x$ component of the curl $\nabla' \times \mathbf{E}'$ as seen by observer O.

The same process can be used to obtain the $\mathbf{a}_y$ component of the curl $\nabla' \times \mathbf{E}'$ as seen by observer O. Starting with the $\mathbf{a}_y$ component from Eq. (28),

$$\frac{\partial E'_x}{\partial z'} - \frac{\partial E'_z}{\partial x'} = 0 \quad (37)$$

and by using Eqs. (18) and (19) and the chain rule, we obtain

$$\left(\frac{\partial E_x}{\partial z} - \frac{\partial E_z}{\partial x}\right) + \frac{\partial B_y}{\partial t} = 0 \quad (38)$$

We have transformed each of the components of the curl of Eq. (28) to the reference frame of observer O, and so now we can determine how the curl $\nabla' \times \mathbf{E}'$ appears to him. Direct substitution of each of these components from Eqs. (31), (33), (36), and (38) into Eq. (28) accomplishes this result:

$$\nabla' \times \mathbf{E}' = \nabla \times \mathbf{E} + \frac{\partial \mathbf{B}}{\partial t} = 0 \tag{39}$$

Consequently, we obtain Maxwell's equation for the curl of the **E** field as seen by the observer O.

$$\boxed{\nabla \times \mathbf{E} = -\frac{\partial \mathbf{B}}{\partial t}} \tag{40}$$

In conclusion, the **E** field is not irrotational with respect to observer O (**E**' was irrotational with respect to observer O') unless the time variation of the **E** field is zero. Recall that in the case of the static magnetic field (where $\nabla \times \mathbf{H} = \mathbf{J}$), we considered **J** as a vector source of the field **H**. By comparing this with Eq. (40) we see that we can consider the vector of the time-varying **B** field as a source of the **E** field. Consequently, the concept of what constitutes a source is not as simple and clear-cut for dynamic fields as for static fields. It is probably better to think of the time-varying electric and magnetic fields as being interdependent than to think of one field generating the other. Also, we can reinforce our conclusion that the divergence of **B** is zero, given in Eq. (33), by taking the divergence of each side of Eq. (40). Since the divergence of the curl is identical to zero,

$$\nabla \cdot \nabla \times \mathbf{E} = -\nabla \cdot \frac{\partial \mathbf{B}}{\partial t} = \frac{-\partial}{\partial t} \nabla \cdot \mathbf{B} = 0 \tag{41}$$

We have interchanged the order of differentiation with respect to the space and time derivatives in Eq. (41), assuming that the function **B** is well-behaved. Also, the variables x, y, z, and t are independent in any given reference system. This completes our work with the curl of the **E**' field as derived from Eq. (26).

Next we can show that the divergence operation in one reference frame is related to the divergence operation in the other reference frame by the contraction factor k. We write the divergence for the observer O and use the transformation of Eq. (7) and the chain rule to write

$$\begin{aligned}
\nabla \cdot \mathbf{E} &= \frac{\partial E_x}{\partial x} + \frac{\partial E_y}{\partial y} + \frac{\partial E_z}{\partial z} \\
&= \frac{1}{k}\frac{\partial E'_x}{\partial x} + \frac{1}{k}\frac{\partial E'_y}{\partial y} + \frac{\partial E'_z}{\partial z} \\
&= \frac{1}{k}\left(\frac{\partial E'_x}{\partial x'} + \frac{\partial E'_y}{\partial y'} + \frac{\partial E'_z}{\partial z'}\right)
\end{aligned} \tag{42}$$

Consequently,

$$\nabla \cdot \mathbf{E} = \frac{1}{k} \nabla' \cdot \mathbf{E}' \tag{43}$$

We can write Maxwell's equation for the divergence of the **E** field as seen by observer O in terms of the time-varying charge distribution ρ by substituting Eqs. (22) and (43) into Eq. (27). [Recall that Eq. (27) is Maxwell's equation for static fields.] Thus Maxwell's equation for time-varying fields is

$$\boxed{\nabla \cdot \mathbf{E} = \frac{\rho}{\varepsilon}} \tag{44}$$

The relationship between the electric flux density and the electric field intensity is the same as for static fields,

$$\mathbf{D} = \varepsilon \mathbf{E} \tag{45}$$

Therefore, Eq. (44) can be written in the alternate form

$$\boxed{\nabla \cdot \mathbf{D} = \rho} \tag{46}$$

Thus we see that the **D** field and the **E** field are not solenoidal, for either the static case or the time-varying case, unless the charge density is zero (as in a charge-free region). Whenever the divergence is not zero for the **D** and **E** fields, we can expect to find point sources for those fields. However, no point sources will exist for the **B** field or the **H** field because the divergence for those fields is zero.

In the above derivation, we did not actually transform the components of $\nabla' \cdot \mathbf{E}'$ of Eq. (27) to show the relationships between the components in one coordinate system and the components in the other coordinate system. Rewriting Eq. (27),

$$\nabla' \cdot \mathbf{E}' = \frac{\partial E'_x}{\partial x'} + \frac{\partial E'_y}{\partial y'} + \frac{\partial E'_z}{\partial z'} = \frac{\rho'}{\varepsilon} \tag{47}$$

we transform each of the components of the divergence using the transformations of Eq. (19) and the chain rule. Also, we transform the source charge density using Eq. (22). After a fair amount of algebra, we obtain

$$\nabla' \cdot \mathbf{E}' = \frac{1}{k} \nabla \cdot \mathbf{E} + \frac{u}{k}\left(\frac{\partial B_x}{\partial y} - \frac{\partial B_y}{\partial x}\right) + \frac{u}{kc^2}\frac{\partial E_z}{\partial t} = \frac{k\rho}{\varepsilon} \tag{48}$$

Next we replace the divergence of **E** with the ratio ρ/ε,

$$\frac{1}{k}\frac{\rho}{\varepsilon} + \frac{u}{k}\left(\frac{\partial B_x}{\partial y} - \frac{\partial B_y}{\partial x}\right) + \frac{u}{kc^2}\frac{\partial E_z}{\partial t} = \frac{k\rho}{\varepsilon} \tag{49}$$

and then, after a minor algebraic rearrangement, we obtain

$$\left(\frac{\partial B_y}{\partial x} - \frac{\partial B_x}{\partial y}\right) = \frac{u\rho}{c^2\varepsilon} + \frac{1}{c^2}\frac{\partial E_z}{\partial t} \tag{50}$$

One may recognize this as the $\mathbf{a}_z$ component of the curl of **B**, but the right-hand side can be written in a more recognizable form by the definition of the *permeability* μ. This new constant is defined as

$$\boxed{\mu \equiv \frac{1}{c^2\varepsilon}} \tag{51}$$

Since the permittivity ε is a constant determined by the medium, the permeability μ is also.

Next we substitute Eq. (25) into Eq. (50) to replace the moving charge density ρu with the current density J_z. As a result, the $\mathbf{a}_z$ component of the curl of **B** is

$$(\nabla \times \mathbf{B})_z = \mu J_z + \frac{1}{c^2}\frac{\partial E_z}{\partial t} \tag{52}$$

Here we can use the equations of the medium which relate the flux-density fields to the intensity fields to simplify Eq. (52):

$$\mathbf{B} = \mu \mathbf{H} \qquad \mathbf{D} = \varepsilon \mathbf{E} \tag{53}$$

The results of the substitutions are

$$(\nabla \times \mathbf{H})_z = J_z + \frac{\partial D_z}{\partial t} \tag{54}$$

In Eq. (54), we obtained only the $\mathbf{a}_z$ component of the curl of **H** because we have chosen a very special orientation of the coordinate axes of observer O with respect to those of observer O'. If we were to choose a more general orientation, we would complicate the mathematical derivations considerably, but we could obtain a complete vector expression for the curl of **H**. We avoid the long derivations and simply state that components similar to the one in Eq. (54) will be obtained for each component of the curl. As a result, we can write Maxwell's equation for the curl of **H**

$$\boxed{\nabla \times \mathbf{H} = \mathbf{J} + \frac{\partial \mathbf{D}}{\partial t}} \tag{55}$$

It is interesting to note that the time-varying **D** field on the right-hand side of Eq. (55) must have units of current density. The time variation of the **D** field is called *displacement-current density*, a name which originated with Maxwell. The addition of the displacement-current-density vector to the right-hand side of Eq. (55) has not changed the fact that the **H** field is solenoidal [see Eq. (33)], but it does show that the time-varying **H** and **D** fields are interdependent. Again, one can think of the displacement-current-density vector as a vector source of the solenoidal **H** field in the same way that we consider **J** a vector source of **H**.

Since the special theory of relativity was not known in Maxwell's day, it is interesting to investigate how he discovered the displacement-current-density term of Eq. (55). At that time, only the static form of Eq. (55) was known. According to an unverified story, he got up early one sunny morning and after breakfast and a second cup of tea decided that it was a fine day to take the divergence of the curl. When he took the divergence of the curl of **E**, as we did in Eq. (41), he obtained results which were consistent with the known facts. However, when he took the divergence of the curl of **H** for the known static equation $\nabla \times \mathbf{H} = \mathbf{J}$, he found that the divergence of the current-density vector **J** was zero. He knew that $\nabla \cdot \mathbf{J}$ was not zero in the general case but was dependent upon the time rate of change of charge. By differentiating Eq. (46) with respect to time he arrived at an expression which led him to the displacement-current density. Consequently he wrote the divergence of the curl as

$$\nabla \cdot \nabla \times \mathbf{H} = \nabla \cdot \mathbf{J} + \nabla \cdot \frac{\partial \mathbf{D}}{\partial t} = 0 \tag{56}$$

He was satisfied with this equation because it led to consistent results, and consequently he added to the equation of the curl of **H** the time-varying term $\partial \mathbf{D}/\partial t$. Also, the equation for the continuity of current (which was known to Maxwell) can be obtained directly from Eq. (56) by interchanging the order of the time- and space-differentiation operations. The result is

$$\boxed{\nabla \cdot \mathbf{J} = -\frac{\partial \nabla \cdot \mathbf{D}}{\partial t} = -\frac{\partial \rho}{\partial t}} \tag{57}$$

This is known as the *continuity equation*, and its physical significance will be explored in the next chapter.

In conclusion, we have developed Maxwell's equations for time-varying fields by applying the Lorentz transformation to Coulomb's law and assuming that charge is conserved. Maxwell's equations will be used as the basis of our work in the remainder of the text.

PROBLEMS

1. Derive the matrix in Eq. (34) of Sec. 1.2 for transforming a vector from rectangular to spherical coordinates.
2. Derive the expression for arc length given in Eq. (47) of Sec. 1.2 in the generalized orthogonal coordinate system.
3. Three equal, positive point charges Q are located at the corners of an equilateral triangle which has sides of length L. Find the force on one of the charges. What can you say about $\mathbf{E}$, both inside and outside the triangle?
4. A positive charge q_1 and a negative charge q_2 are located at $z = +a/2$ and $z = -a/2$, respectively, to form an electric dipole. Find the $\mathbf{E}$ field and the scalar potential field at points where the distance to either charge is much greater than a.
5. Show that the delta function can be expressed in the various coordinate systems as given in Eqs. (9) to (12) in Sec. 1.5.
6. If $\oint \mathbf{E} \cdot d\mathbf{l} = 0$ for any closed path [see Eq. (34) in Sec. 1.6], show that a scalar potential field exists.
7. The capacitor is charged to K V, and plate spacing is z m. Find $\oint \mathbf{E} \cdot d\mathbf{l}$ for the path $abcda$ shown. What assumptions are involved?

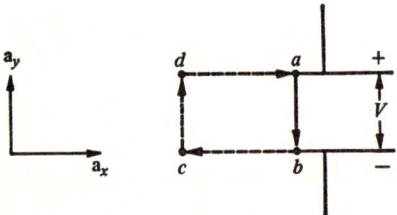

Fig. P.7 The charged capacitor for Prob. 7.

8. Three infinite charged lines are placed equidistant from each other, as shown in the figure. The linear charge density on line A is $\lambda_A = 5 \sin \omega t$ C/m; on line B is $\lambda_B =$

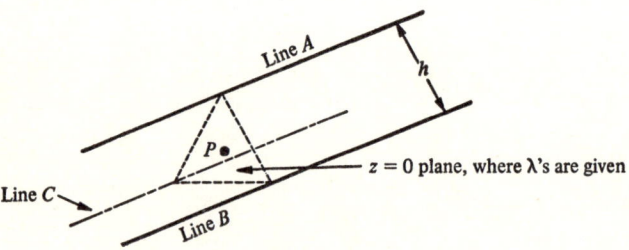

Fig. P.8 The charged lines for Prob. 8.

5 sin $(\omega t + 120°)$; and on line C is $\lambda_C = 5 \sin(\omega t + 240°)$. Find the electric field intensity at point P (at the centroid of the triangle).

9. A linear charge density of λ C/m occupies the sides of a square with sides of length S, as shown. Find E at the point P, directly above the center of the square.

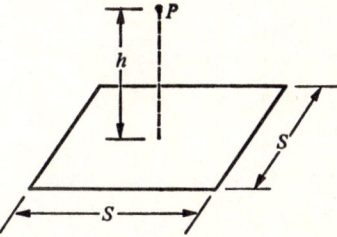

Fig. P.9 Linear charge density around a square for Prob. 9.

10. Suppose that $\mathbf{H} = z\mathbf{a}_x + x\mathbf{a}_y + y\mathbf{a}_z$ and $\mathbf{F} = \nabla \times \mathbf{H}$. Show that the vector field $\mathbf{F}$ is not everywhere perpendicular to the vector field $\mathbf{H}$, although many vector fields are perpendicular to the field generated by taking the curl. Can you describe the conditions which must be met by a vector field $\mathbf{H}$ so that $\mathbf{F} = \nabla \times \mathbf{H}$ is everywhere perpendicular to $\mathbf{H}$?

11. As an illustration of the work required to move a charge in an electric field, let $\mathbf{E} = (-y + 2)\mathbf{a}_x + (x - 1)\mathbf{a}_y + \mathbf{a}_z$, and find the work required to move a charge of 100 μC along an incremental path of 0.1 mm, beginning at (1,1,1) and directed toward (a) (2,1,1) and (b) (0,1,1).

12. Find the work required to move a charge of 4 C in the field $\mathbf{E} = x\mathbf{a}_x + y\mathbf{a}_y + z\mathbf{a}_z$ from (0,0,0) to (1,1,0) (a) along the straight line $y = x$ and (b) on the parabola $y = x^2$.

13. If $\mathbf{E} = (z + 3)\mathbf{a}_x + (x - 2)\mathbf{a}_y + (y + 1)\mathbf{a}_z$, find the potential difference between the points (a) (0,0,0) and (1,2,3), (b) (1,1,1) and (1,2,3), and (c) (1,1,1) and (0,0,0).

14. The square shown is filled with a surface charge density of γ C/m². Find the electric field intensity at point P.

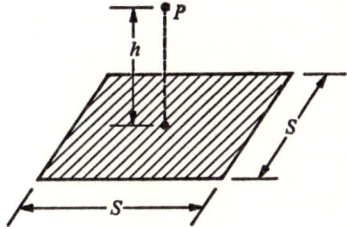

Fig. P.14 Surface charge density on a square for Prob. 14.

15. If E is a vector field and there exist scalar fields V_1 and V_2 such that $V_1\mathbf{E} = \nabla V_2$, show that E and $\nabla \times \mathbf{E}$ are orthogonal; i.e., show that $\mathbf{E} \cdot \nabla \times \mathbf{E} = 0$.

16. A dipole field is due to a positive charge $+Q$ on the z axis at $d/2$ m from the origin and a negative charge $-Q$ at $z = -d/2$ m. At any arbitrary point r m from the

origin, where $r \gg d$, the electric field in spherical coordinates is

$$\mathbf{E} = \frac{Qd}{4\pi\varepsilon_0 r^3} (2 \cos\theta \, \mathbf{a}_r + \sin\theta \, \mathbf{a}_\theta)$$

Find the divergence and the curl of the field.

17. Given a scalar field

$$V = \frac{1}{\sqrt{x^2 + y^2 + z^2}}$$

verify that $\nabla \times \nabla V = 0$.

18. Given the vector field

$$\mathbf{H} = 2x\mathbf{a}_x + y\mathbf{a}_y - 3z\mathbf{a}_z$$

find $\nabla \cdot \mathbf{H}$, $\nabla \times \mathbf{H}$, $\nabla \cdot \nabla \times \mathbf{H}$ and all vectors $\mathbf{A}$ such that $\nabla \times \mathbf{A} = \mathbf{H}$.

19. From Eqs. (13) to (15) in Sec. 1.12, derive the constants used in the Lorentz transformation, given in Eq. (16).

20. A meterstick fixed in the xz plane makes an angle θ with the x axis. What angle θ' does the stick appear to make with the x' axis to an observer O'_b in $x'y'z'$?

21. A jet airliner 150 ft long is cruising at a ground speed of 600 mi/h. By how much does the plane appear shortened to a ground observer? How long need the pilot fly at this speed before his clock appears to have lost 1 s to a ground observer?

22. Show that the constant kg is bounded as given in Eq. (47) of Sec. 1.12, and prove that kg is less than unity when $v_z < v_0$. From Eq. (46) describe the physical significance of kg being more or less than unity.

23. A particle moves in a circle of unit radius, centered at the origin, in the xz plane. To an observer moving along the $\mathbf{a}_x$ axis with a velocity u the circle appears as an ellipse whose major axis is 2 units and minor axis is 1 unit long. How fast is the observer moving?

24. Using the transformations for field components given in Eq. (19) of Sec. 1.13, and the transformation for charge density given in Eq. (22), derive the transformation for the divergence of the electric field, as expressed in Eq. (48) of Sec. 1.13.

2
Time-varying Fields

2.1 POSTULATES OF ELECTROMAGNETIC THEORY AND THEIR PHYSICAL INTERPRETATIONS

A SET OF MATHEMATICAL POSTULATES

We now formulate a set of mathematical postulates (rules or equations) which can be used as a basis for our future work in electromagnetic theory. Our set of postulates has been selected largely on the basis of convenience for solving field-theory problems, rather than attempting to select the most basic set of laws. (We have shown that we could start with Coulomb's law and the special theory of relativity, but this is not a very convenient starting point for most practical problems.) These equations and their variables are the everyday working tools for the engineer involved in electromagnetic theory. Consequently, this section serves a dual purpose: it serves as a convenient place to summarize Maxwell's equations, and it can serve as a starting point for the work that follows.

Since the postulates are rules that define how variables are interrelated, our first step is to list all the variables with which we are concerned. They have been divided somewhat arbitrarily into two groups, basic variables and defined variables. Given a charged particle, we assume that one has a means of determining the amount of charge, the velocity of the particle, and the force on the particle. We have called these the basic variables from which all the other variables can be defined.

Basic variables

q = free charge (scalar), C
$\mathbf{v}$ = velocity of free charge (vector), m/s
$\mathbf{F}$ = Lorentz force on free charge (vector), N

The following is a list of field variables which we have been able to define from the basic variables above. These variables are interrelated by the set of mathematical postulates.

Defined variables

ρ = free charge density (scalar), C/m^3
$\mathbf{J}$ = current density (vector), A/m^2
$\mathbf{E}$ = electric field intensity (vector), V/m
$\mathbf{D}$ = electric flux density (vector), C/m^2
$\mathbf{H}$ = magnetic field intensity (vector), A/m
$\mathbf{B}$ = magnetic flux density (vector), W/m^2
$\mathbf{A}$ = vector magnetic potential (vector), W/m
V = scalar electric potential (scalar), V
I = electric current (scalar), A

Certain constants are properties of media in which electromagnetic fields exist. These constants, listed below, will receive further attention in Sec. 2.6.

Constants of isotropic, homogeneous media

ε = permittivity (scalar), F/m
$\varepsilon_0 = 8.854 \times 10^{-12}$ F/m for free space
μ = permeability (scalar), H/m
$\mu_0 = 4\pi \times 10^{-7}$ H/m for free space
σ = conductivity (scalar), mhos/m

The mathematical postulates which relate the variables and constants are given in Table 2.1.1. The differential forms are familiar from prior work, and the integral forms will be explained in this section.

There is an analogy between the study of electromagnetic field theory and the game of chess. Given the chess board, the various pieces,

TIME-VARYING FIELDS

Table 2.1.1 Mathematical postulates of electromagnetic theory

Name (law)	Differential form	Integral form
Ohm	$\mathbf{J} = \sigma \mathbf{E}$	
Permittivity	$\mathbf{D} = \varepsilon \mathbf{E}$	
Permeability	$\mathbf{B} = \mu \mathbf{H}$	
Lorentz force	$\mathbf{F} = q(\mathbf{E} + \mathbf{v} \times \mathbf{B})$	
Continuity equation (conservation of charge)	$\nabla \cdot \mathbf{J} = -\dfrac{\partial \rho}{\partial t}$	$I = \oint_s \mathbf{J} \cdot d\mathbf{s} = -\dfrac{dq}{dt}$
Gauss (Maxwell)	$\nabla \cdot \mathbf{D} = \rho$	$q = \oint_s \mathbf{D} \cdot d\mathbf{s} = \int_v \rho \, dv$
Gauss (Maxwell)	$\nabla \cdot \mathbf{B} = 0$	$\oint_s \mathbf{B} \cdot d\mathbf{s} = 0$
Faraday (Maxwell)	$\nabla \times \mathbf{E} = \dfrac{-\partial \mathbf{B}}{\partial t}$	$-\oint_c \mathbf{E} \cdot d\mathbf{l} = \int_s \dfrac{\partial \mathbf{B}}{\partial t} \cdot d\mathbf{s} = \dfrac{d\phi}{dt}$
Ampere (Maxwell)	$\nabla \times \mathbf{H} = \mathbf{J} + \dfrac{\partial \mathbf{D}}{\partial t}$	$\oint_c \mathbf{H} \cdot d\mathbf{l} = I + \int_s \dfrac{\partial \mathbf{D}}{\partial t} \cdot d\mathbf{s}$

and the postulates or rules of the game, one proceeds in some logical manner to investigate the significance of the various moves, as governed by the rules. The objective is to maneuver the pieces so as to accomplish some desired result, i.e., to win the game. In electromagnetic theory, the dependent variables of the fields are analogous to the chess pieces, in that they can be maneuvered to accomplish some desired result. These maneuvers are governed by the set of mathematical postulates (the rules of the game). The postulates are consistent in the sense that one does not lead to the contradiction of another, in the same way that a given rule in chess does not violate some other rule. If our objective were simply to study the mathematical structure of electromagnetic field theory, the analogy with the chess game would be complete. However, we are interested in using the mathematical model to represent physical phenomena, and for this reason the study of field theory is more than just a mathematical game. In order to investigate the physical significance of some of the postulates, we will develop the integral equations given in Table 2.1.1. Generally the integral forms are easier to explain physically, but often the differential forms are more convenient in our mathematical work.

PHYSICAL INTERPRETATIONS

The equation for the continuity of current will be considered first. In the differential form, it is

$$\nabla \cdot \mathbf{J} = \frac{-\partial \rho}{\partial t} \tag{1}$$

According to this equation the divergence of the current density at a particular point equals the rate at which charge density is changing at that point. Instead of considering a single point, suppose that we consider all the points enclosed in a rectangular box of volume v, as shown in Fig. 2.1.1. Next we integrate both sides of Eq. (1) over the volume

$$\int_v \nabla \cdot \mathbf{J} \, dv = -\int_v \frac{\partial \rho}{\partial t} \, dv \tag{2}$$

The divergence theorem can now be applied to the integral on the left-hand side of Eq. (2). On the right-hand side of Eq. (2) we interchange the operations of integration and differentiation in order to integrate the charge density over the entire volume v, thereby obtaining the charge q contained in the volume. The result is

$$I = \oint_s \mathbf{J} \cdot d\mathbf{s} = -\frac{\partial}{\partial t} \int_v \rho \, dv = -\frac{\partial q}{\partial t} \tag{3}$$

The integral on the left expresses the net outflow of current through the surface of the box. The equation says that the net outflow of current from the box is equal to the rate at which the charge inside the box is decreasing with time. In essence, it is a statement of the law of conservation of charge. In ordinary circuit theory, the corresponding continuity equation is

$$I = -\frac{dq}{dt} \tag{4}$$

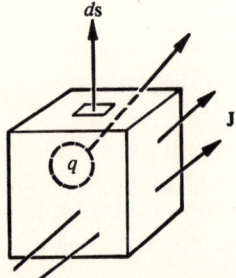

Fig. 2.1.1 The net outflow of current depends upon the rate of change of charge.

TIME-VARYING FIELDS

and often the negative sign is omitted. In conclusion, Eqs. (1), (3), and (4) all express the same basic idea.

The first of Maxwell's equations to be considered is Gauss' law for electric fields

$$\nabla \cdot \mathbf{D} = \rho \tag{5}$$

We studied Gauss' law for static fields in Chap. 1. The same ideas are valid here, but now $\mathbf{D}$ and ρ can be time-varying. To review, Gauss' law says that divergence of flux density from a particular point in space is equal to the charge density at that particular point. Instead of considering charge at a single point, suppose we consider all the charge contained in a box, as shown in Fig. 2.1.2. First we integrate both sides of Eq. (5) over the volume and then apply the divergence theorem to the integral on the left, as in the previous example. The result is

$$\oint_s \mathbf{D} \cdot d\mathbf{s} = \int_v \rho \, dv = q \tag{6}$$

The result does not depend upon a particular shape of the volume (a box, a sphere, a cylinder, etc., would be satisfactory), but the closed surface s must enclose the volume v and only that volume. Therefore the surface s is defined by the volume (and vice versa). For transient phenomena, a delay time will be encountered between the time of occurrence of an event inside the volume and the time the effect of that event reaches the surface. Because electromagnetic waves travel at the velocity of light, the delay time is often ignored for small volumes.

The procedure used in the previous two paragraphs can be applied

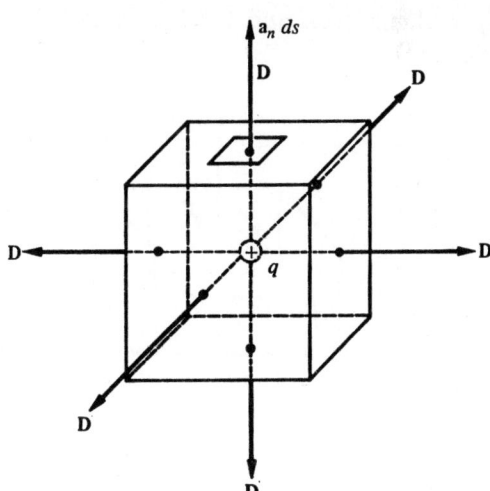

Fig. 2.1.2 The charge enclosed can be determined from the net flux density piercing the surface.

to Gauss' law for magnetic fields to obtain an integral form from the differential form

$$\nabla \cdot \mathbf{B} = 0 \tag{7}$$

This law states that the divergence of the magnetic flux density is zero at every point in space. If we integrate over the volume of a box, as shown in Fig. 2.1.3, and apply the divergence theorem, the result is

$$\oint_s \mathbf{B} \cdot d\mathbf{s} = 0 \tag{8}$$

The law in this form says that the net amount of flux piercing the surface must be zero, or the same amount of flux that enters the box must leave the box. This means that there can be no isolated magnetic charges or sources comparable to those for electric fields, as shown in Fig. 2.1.2. If one places a permanent magnet in the box in Fig. 2.1.3 so that both the north and the south poles are enclosed, all the flux lines leaving the north pole enter at the south pole. Consequently, there will be no divergence of the magnetic field. It would appear that one could place one end of the bar magnet in the box and obtain an isolated source for a magnetic field. However, this is not true because the hypothetical surface of the box must at some point cut through the bar magnet unless the entire magnet is enclosed inside the box. Then all the flux coming through the bar magnet pierces the surface of the box. Usually we will be concerned with magnetic fields caused by some time-varying source and not from permanent magnets. An example of this type is the electromagnetic field in the vicinity of a radiating antenna. However, regardless of the nature of the source, magnetic flux lines always close upon themselves, and the **B** and **H** fields are solenoidal (assuming μ is constant).

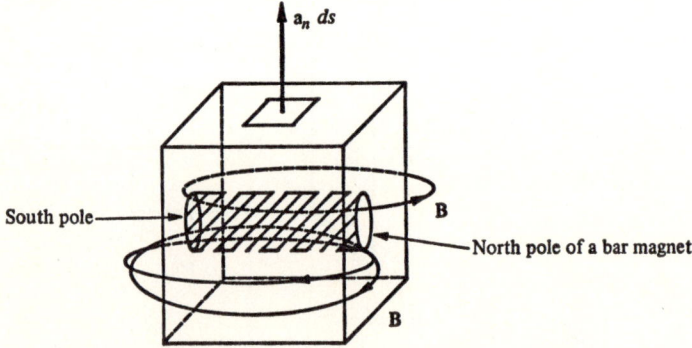

Fig. 2.1.3 The magnetic field is solenoidal, and no point sources exist.

TIME-VARYING FIELDS

We now consider Faraday's law, which in its differential form is

$$\nabla \times \mathbf{E} = -\frac{\partial \mathbf{B}}{\partial t} \tag{9}$$

In static fields, $\nabla \times \mathbf{E} = 0$, and the field is irrotational. From Eq. (9), we see that both $\mathbf{E}$ and $\mathbf{B}$ can be time-varying. If an $\mathbf{E}$ field exists in a given region, a force will be exerted on any test charge in that region, and consequently if we move the charge from one point to another against the force, we must do work. Since the voltage difference is the work per unit charge in moving the charge along some path from one point to another, we can expect a voltage difference to exist in regions where an $\mathbf{E}$ field exists. To demonstrate this, suppose we choose some simple surface over which a time-varying $\mathbf{B}$ field exists, as shown in Fig. 2.1.4. Because of the time-varying $\mathbf{B}$ field, there will also be a time-varying $\mathbf{E}$ field over the surface. If we integrate both sides of Eq. (9) over the surface,

$$\int_s \nabla \times \mathbf{E} \cdot d\mathbf{s} = -\int_s \frac{\partial \mathbf{B}}{\partial t} \cdot d\mathbf{s} \tag{10}$$

we can apply Stokes' theorem to the left-hand side of the equation by integrating around the closed curve that bounds the surface s. On the right-hand side of Eq. (10) we can interchange the operations of differentiation with respect to time and integration with respect to the space variables over the surface. The result is

$$\oint_C \mathbf{E} \cdot d\mathbf{l} = -\frac{\partial}{\partial t}\int_s \mathbf{B} \cdot d\mathbf{s} \tag{11}$$

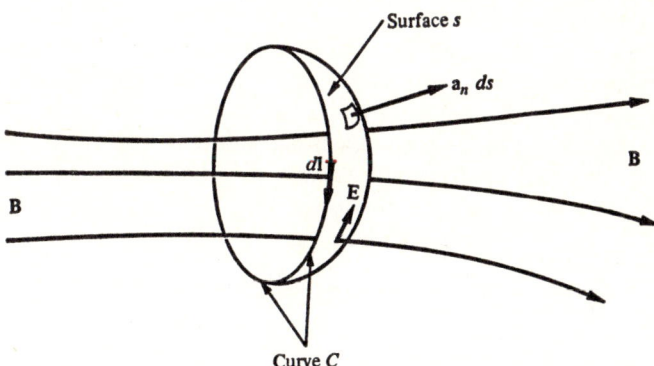

Fig. 2.1.4 The voltage induced in the loop depends upon the rate of change of the magnetic flux through the loop.

The integral on the left is the voltage difference we would measure if we were to move a charged particle around the curve C. Since this voltage varies instant by instant, we assume the integration takes place instantaneously. This voltage is commonly called the *electromotive force* (emf). Since the integral on the right-hand side of Eq. (11) is the integration of the flux-density vector over a surface, the integral is simply the magnetic flux Φ which cuts the surface s. Thus Eq. (11) can be written

$$\text{emf} = -\frac{\partial \Phi}{\partial t} \tag{12}$$

Consequently, we see that the electromotive force or voltage induced in the loop is equal to the negative of the time rate of change of the magnetic flux which links the loop. If one were to integrate around the loop N times, the induced voltage would be N times greater than the voltage given in Eq. (12). In circuit theory, this equation is called *Lenz's law*, normally written

$$\text{emf} = -N\frac{d\Phi}{dt} \tag{13}$$

Note that if the **B** field is a static field, the voltage induced around the loop is zero. Recall that in static fields, the line integral around any closed curve is always zero, and the voltage between any two points is independent of the path we take between those points. However, the value of the line integral in Eq. (11) is not independent of the path of integration. For example, if we choose a loop of very small diameter, the magnetic flux cutting it would be less than the flux cutting one of a larger diameter and consequently the voltage in the two loops would be different. Therefore we assume that the loop is stationary with respect to our reference frame. Also, since the fields are varying from instant to instant at every point, we assume that time is stopped while we perform the integration or that the integration over both the curve and the surface takes place instantaneously. If we were to perform the integration over and over again, thereby obtaining different values for the emf instant by instant, we could plot the voltage as a function of time.

If we use a piece of copper wire for our loop, as shown in Fig. 2.1.4, we can attach a cathode-ray oscilloscope to the ends of the wire and observe the voltage variations instant by instant. In this case, the wires leading from our loop to the oscilloscope become part of the loop, and voltage will also be induced in them. Thus if we want to measure the voltage induced in the wire loop itself, we must either shield the leads to the oscilloscope or orient them so that no flux cuts the loop formed by the oscilloscope leads.

TIME-VARYING FIELDS

The last of Maxwell's equations to be considered is Ampere's law

$$\nabla \times \mathbf{H} = \mathbf{J} + \frac{\partial \mathbf{D}}{\partial t} \qquad (14)$$

Ampere's law of static fields is of the same form except that the displacement-current density $\partial \mathbf{D}/\partial t$ is zero. From our study of the special theory of relativity, we know that the displacement-current term arises from the divergence of an electric field of a charge in motion. In fact, without displacement current, electromagnetic waves could not exist and propagate through free space. Maxwell was the first to hypothesize the existence of the displacement current, and that was long before the special theory of relativity had been postulated. The integral form of Ampere's law is helpful in understanding the physical significance of the law. Suppose we have a vortex region in which a current density $\mathbf{J}$ is flowing, as in a conductor. The current density $\mathbf{J}$ is due to the motion of charged particles in the conductor, and we will assume that $\mathbf{J}$ is time-varying. As a result of the presence of the moving charge, there will be both an electric and a magnetic field in the region surrounding the vortex. If we choose some closed curve C surrounding the vortex region, as shown in Fig. 2.1.5, we can choose any simple surface bounded by that curve and integrate both sides of Ampere's law over the chosen surface. By integrating the curl of $\mathbf{H}$ from Eq. (14) over the surface,

$$\int_s \nabla \times \mathbf{H} \cdot d\mathbf{s} = \int_s \left(\mathbf{J} + \frac{\partial \mathbf{D}}{\partial t} \right) \cdot d\mathbf{s} \qquad (15)$$

we can apply Stokes' theorem to the integral on the left to obtain a line integral around the closed curve C. The integral on the right can be written as a sum

$$\oint_C \mathbf{H} \cdot d\mathbf{l} = \int_s \mathbf{J} \cdot d\mathbf{s} + \int_s \frac{\partial \mathbf{D}}{\partial t} \cdot d\mathbf{s} \qquad (16)$$

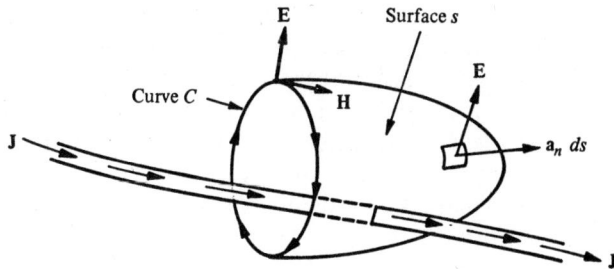

Fig. 2.1.5 The integration of $\mathbf{H}$ around the closed curve equals the sum of all current passing through the surface.

The first integral on the right represents both the conduction and convection currents passing through the surface, and the second term represents the displacement current (due to the time-varying **D** field) passing through the surface. Consequently, we can say that the line integral of **H** around the closed curve is equal to the total current passing through that curve:

$$\oint_C \mathbf{H} \cdot d\mathbf{l} = I_c + I_d = I_t \tag{17}$$

where I_c denotes the convection and conduction current and I_d the displacement current.

As described by the continuity equation, whenever current flows, it must be continuous. If one thinks of current flow only in terms of moving, charged particles, a dilemma arises if a simple series circuit contains a capacitor. Since no charged particles move between the plates of the capacitor, it might appear that the continuity equation could not be valid. However, the time-varying charge on the plate of the capacitor induces a time-varying **D** field in the region between the capacitor plates. Therefore displacement-current density exists between the plates, and because of this, energy is transferred through the fields at the speed of light, and the total current I_t is continuous.

2.2 THE POTENTIAL SOLUTIONS OF MAXWELL'S EQUATIONS

In Chap. 1 we found that the scalar potential V of the **E** field and the vector potential **A** of the **H** field could be used to simplify electromagnetic-field problems. By using the potentials, the static-field problems reduce to finding solutions of Poisson's scalar or vector equations for a given set of boundary conditions. In most cases the solution of Poisson's equation is simpler than the solution of the original field problem. Without introducing any new potentials, we can write the solutions of Maxwell's equations in terms of the potentials **A** and V. In essence, what we do is trade the problem of solving Maxwell's equations in their original form for the problem of solving two other partial differential equations for the potentials **A** and V, in the hope that these solutions will be obtained more easily than the original ones. This is one of those instances where we can often violate the law of conservation of difficulty and actually reduce a hard problem to an easier one. Consequently, potential functions are an important part of electromagnetic theory.

As in Chap. 1, we define the vector potential **A** by the following relationship to the **B** field:

$$\nabla \times \mathbf{A} = \mathbf{B} \tag{1}$$

Because of the way we have defined **A**, Maxwell's equation for the diver-

TIME-VARYING FIELDS

gence of **B** is automatically satisfied

$$\nabla \cdot \mathbf{B} = \nabla \cdot \nabla \times \mathbf{A} \equiv 0 \tag{2}$$

However, we must show that the vector potential as defined will also satisfy the rest of Maxwell's equations. If we substitute Eq. (1) into Faraday's law and interchange the time and space derivatives, we obtain

$$\nabla \times \mathbf{E} = -\frac{\partial \mathbf{B}}{\partial t} = -\frac{\partial (\nabla \times \mathbf{A})}{\partial t}$$

$$= \nabla \times \left(-\frac{\partial \mathbf{A}}{\partial t}\right) \tag{3}$$

By comparing the left- and right-hand sides of Eq. (3), we see that we can write the electric intensity field **E** as the negative of the time rate of change of the vector potential field **A**, to within an arbitrary constant

$$\mathbf{E} = -\frac{\partial \mathbf{A}}{\partial t} \tag{4}$$

However, we know that the electric intensity field **E** can be written in terms of a scalar potential field V

$$\mathbf{E} = -\nabla V \tag{5}$$

It would seem that the **E** field is ambiguously defined by examining Eqs. (4) and (5). There is no ambiguity, however, since the two fields are generated by different sources, and both fields can exist simultaneously. The total electric intensity field is simply the superposition of these two:

$$\boxed{\mathbf{E} = -\nabla V - \frac{\partial \mathbf{A}}{\partial t}} \tag{6}$$

We must show that this equation, as well as Eq. (1), will satisfy all of Maxwell's equations. By direct substitution we can show that Eqs. (1) and (6) satisfy Faraday's law

$$\nabla \times \mathbf{E} = -\nabla \times \nabla V - \nabla \times \frac{\partial \mathbf{A}}{\partial t} = -\frac{\partial \mathbf{B}}{\partial t} \tag{7}$$

In the above equation, recall that the curl of the gradient is identically zero and, using Eq. (1), Eq. (7) reduces to an identity.

Next we substitute Eq. (6) into Gauss' law for electric fields and obtain a partial differential equation which the potential fields **A** and V must satisfy. We find that

$$\nabla \cdot \mathbf{E} = -\nabla \cdot \nabla V - \nabla \cdot \frac{\partial \mathbf{A}}{\partial t} = \frac{\rho}{\varepsilon} \tag{8}$$

and from Eq. (8) we obtain

$$\nabla^2 V + \frac{\partial (\nabla \cdot \mathbf{A})}{\partial t} = -\frac{\rho}{\varepsilon} \qquad (9)$$

The potentials $\mathbf{A}$ and V must satisfy this partial differential equation for both a given charge distribution ρ and a given set of boundary conditions. When the values of $\mathbf{A}$ and V are found to satisfy Eq. (9), the electric field will satisfy Gauss' law, given by Eq. (8).

Next we must show that Eqs. (1) and (6) for the $\mathbf{B}$ and $\mathbf{E}$ fields will satisfy Ampere's law,

$$\nabla \times \mathbf{H} = \mathbf{J} + \frac{\partial \mathbf{D}}{\partial t} \qquad (10)$$

which can be written

$$\nabla \times \mathbf{B} = \mu \mathbf{J} + \mu \varepsilon \frac{\partial \mathbf{E}}{\partial t} \qquad (11)$$

We will assume that the current density $\mathbf{J}$ is due to a known independent source. By substituting Eqs. (1) and (6) into Eq. (11) we obtain a partial differential equation relating the potentials V and $\mathbf{A}$ to the source $\mathbf{J}$

$$\nabla \times \nabla \times \mathbf{A} = \mu \mathbf{J} - \mu \varepsilon \frac{\partial}{\partial t} \left(\nabla V + \frac{\partial \mathbf{A}}{\partial t} \right) \qquad (12)$$

Next we use a vector identity to write the curl of the curl of $\mathbf{A}$ in terms of the divergence and the vector laplacian, and we also interchange the time and space derivatives

$$\nabla (\nabla \cdot \mathbf{A}) - \nabla^2 \mathbf{A} = \mu \mathbf{J} - \mu \varepsilon \nabla \frac{\partial V}{\partial t} - \mu \varepsilon \frac{\partial^2 \mathbf{A}}{\partial t^2} \qquad (13)$$

or

$$\nabla^2 \mathbf{A} - \mu \varepsilon \frac{\partial^2 \mathbf{A}}{\partial t^2} = -\mu \mathbf{J} + \mu \varepsilon \nabla \frac{\partial V}{\partial t} + \nabla (\nabla \cdot \mathbf{A}) \qquad (14)$$

Although Eq. (14) does not look like a simplification of Maxwell's equations, if we can find the scalar and vector potentials $\mathbf{A}$ and V that satisfy Eq. (14) for a given set of boundary conditions, Ampere's law will be satisfied. However, the potential fields must also be chosen so as to satisfy Eq. (9). When this is accomplished, Eqs. (1) and (6) for the $\mathbf{B}$ and $\mathbf{E}$ fields are solutions of Maxwell's equations.

In the above development, we did not have to specify the divergence of $\mathbf{A}$ in order to obtain the $\mathbf{B}$ and $\mathbf{E}$ fields that satisfy Maxwell's equations. Of course, in a given physical situation, the measurable $\mathbf{B}$ and $\mathbf{E}$ fields satisfy the given source and boundary conditions and are not influenced in the least by our attempts to describe them mathematically. The

TIME-VARYING FIELDS

potential fields **A** and V are introduced largely for our convenience in problem solving, and, as it turns out, the **B** and **E** fields are independent of our choice of the divergence of **A**. Since we are free to choose the divergence of **A**, we should choose it in such a way as to simplify Eqs. (9) and (14), since those are the equations we must solve for the potential fields. Our choice for the divergence of **A** is called a *gauge* or a *gauge transformation*, and the two most common gauges are the *Coulomb gauge* and the *Lorentz gauge*:

$$\nabla \cdot \mathbf{A} = 0 \qquad \text{Coulomb gauge} \qquad (15)$$

$$\nabla \cdot \mathbf{A} = -\mu\varepsilon \frac{\partial V}{\partial t} \qquad \text{Lorentz gauge} \qquad (16)$$

These two particular gauges have been chosen because of the way in which they simplify Eqs. (9) and (14) and thereby simplify our problem of solving for the potential fields.

Because we are attempting to simplify Eqs. (9) and (14), the Coulomb gauge of Eq. (15) is certainly a logical choice. It is the best gauge to use with static fields, because then Eqs. (9) and (14) reduce to Poisson's equations, as demonstrated in Chap. 1. However, if we use the Coulomb gauge for time-varying fields, Eq. (9) still reduces to Poisson's scalar equation

$$\nabla^2 V = -\frac{\rho}{\varepsilon} \qquad (17)$$

We will assume that this equation can be solved for the scalar potential V. Using the Coulomb gauge, Eq. (14) reduces to

$$\nabla^2 \mathbf{A} - \mu\varepsilon \frac{\partial^2 \mathbf{A}}{\partial t^2} = -\mu \mathbf{J} + \mu\varepsilon \nabla \frac{\partial V}{\partial t} \qquad (18)$$

All the terms on the right-hand side of Eq. (18) are known or can be determined, and they become a forcing function for the differential equation in the unknown vector potential **A**. This equation is called the *inhomogeneous wave equation*. If the forcing function is zero, the wave equation is said to be homogeneous. Equation (18) is said to be a *vector wave equation* because the unknown dependent variable **A** is a vector. However, in rectangular coordinates, the vector wave equation can be reduced to three scalar wave equations, one in each of the rectangular components of the vector **A**. (By comparison, recall that the vector Poisson equation can be reduced to scalar Poisson equations in each of the rectangular coordinates.) In conclusion, we can use the Coulomb gauge to reduce the problem of solving Maxwell's vector equations to the problem of solving a scalar Poisson equation for V and three scalar wave equations for **A**. Once **A** and V have been determined, we can find the **B** and **E** fields using Eqs. (1) and (6).

The other commonly used gauge transformation, the Lorentz gauge, is given in Eq. (16). If we substitute Eq. (16) into Eqs. (9) and (14), we obtain the inhomogeneous wave equations

$$\nabla^2 V - \mu\varepsilon \frac{\partial^2 V}{\partial t^2} = -\frac{\rho}{\varepsilon} \tag{19}$$

and

$$\nabla^2 \mathbf{A} - \mu\varepsilon \frac{\partial^2 \mathbf{A}}{\partial t^2} = -\mu \mathbf{J} \tag{20}$$

If the time variations are sinusoidal, these equations can be reduced to the *Helmholtz equations* (Chap. 7). Equation (19) is an inhomogeneous *scalar* wave equation of the dependent variable V, and Eq. (20) is an inhomogeneous *vector* wave equation for the vector potential **A**. The advantage of the Lorentz gauge over the Coulomb gauge is that Eqs. (19) and (20) each have only one dependent variable (the equations are not coupled), whereas Eq. (18) has two dependent variables, **A** and V. For the sake of comparison, the Coulomb gauge and the Lorentz gauge and the pertinent equations are tabulated.

Comparison of the gauge transformations

Term or equation	Coulomb gauge	Lorentz gauge
$\nabla \cdot \mathbf{A}$	$\nabla \cdot \mathbf{A} = 0$	$\nabla \cdot \mathbf{A} = -\mu\varepsilon \frac{\partial V}{\partial t}$
B	$\mathbf{B} = \nabla \times \mathbf{A}$	$\mathbf{B} = \nabla \times \mathbf{A}$
E	$\mathbf{E} = -\nabla V - \frac{\partial \mathbf{A}}{\partial t}$	$\mathbf{E} = -\nabla V - \frac{\partial \mathbf{A}}{\partial t}$
Eq. (9)	$\nabla^2 V = -\frac{\rho}{\varepsilon}$	$\nabla^2 V - \mu\varepsilon \frac{\partial^2 V}{\partial t^2} = -\frac{\rho}{\varepsilon}$
Eq. (14)	$\nabla^2 \mathbf{A} - \mu\varepsilon \frac{\partial^2 \mathbf{A}}{\partial t^2} = \mu \mathbf{J} + \mu\varepsilon \nabla \frac{\partial V}{\partial t}$	$\nabla^2 \mathbf{A} - \mu\varepsilon \frac{\partial^2 \mathbf{A}}{\partial t^2} = -\mu \mathbf{J}$

2.3 THE WAVE EQUATION

Imagine a boy on the bank of a quiet pond of water dropping pebbles in the water in an effort to overturn bugs waterskiing on the surface of the pond. As each stone strikes the water, a surface wave travels across the pond in a circle concentric with the point of impact. The falling stone has kinetic energy, and it loses most of this energy at the time of impact. By the conservation of energy, the energy lost by the stone is transferred to the water, and most of it appears in the form of the wave motion of the water.

A bug in the immediate vicinity of the point of impact may be overturned by the kinetic energy of the water. However, it can be argued that the wave contains energy even though it may be some distance from the point of impact. For example, a bug several yards from the point of impact of the stone will be raised a perceptible amount as it rides over the crest of the wave. When the bug is on the crest of the wave, it has a greater potential energy than it had a moment before. Some energy of the wave has been used to raise the mass of the bug to a new height. This energy is only temporarily stored in the body of the bug, for a moment after the passage of the wave the bug has its original potential energy again. The wave passes on, to dissipate its energy on the banks of the pond. This example illustrates that energy can be stored in the form of wave motion.

A somewhat more obvious point is that the energy travels with a finite velocity. For example, a nearsighted bug several yards from the point of impact may not even be aware of the disturbance until the wave reaches him. These phenomena of energy storage and travel at a finite velocity are characteristic of all types of wave motion.

A similar example of wave motion familiar to astronomers is that of the nova, or new star. Probably because of internal nuclear reactions a star can suddenly burst into much greater brilliance than normal for a period of time. During this period it dissipates a tremendous amount of energy, energy that may not reach the earth until thousands or millions of years after the initial explosion. Here one imagines the wavefront of light as a great spherical shell, concentric to the nova, traveling outward with the velocity of light, in much the same manner as the surface wave on the pond traveled in a concentric circle from the impact of the stone. Again, the two characteristics of wave phenomena are evident: energy storage and travel at a finite velocity.

As another example, imagine two astronauts, one in the vicinity of the earth and the other in the vicinity of Mars, and suppose that their separation distance is 100 million mi. Suppose that they are trying to communicate with each other via laser beams by pulsing the lasers on and off according to International Morse Code. In this case the bursts of energy are constrained by the laser to follow very narrow beams, and of course the energy travels with the velocity of light. With the given separation distance, an astronaut would be transmitting for about $8\frac{1}{2}$

min before the other astronaut would even receive the first letter of the message. It is obvious that each wave must contain a detectable amount of energy at the receiver for satisfactory communications. Thus, even though this wave is constrained to travel in a very narrow beam, it has much the same characteristics as the electromagnetic wave from the nova or the wave on the surface of the pond.

Although wave motion prevails throughout many fields of engineering and physics, we immediately connect it with waves in water, such as the waves on the ocean. However, sound waves or acoustic waves are equally common but are not visible to the eye. Mechanical vibrations, such as the vibrations of a steel beam or a stretched rubber band, are also examples of wave motion. All these forms of wave motion we can sense by sight, hearing, or feeling, i.e., by one of the human senses. However, there are many other forms of wave phenomena which go undetected by the human senses. One example of this is electromagnetic waves at frequencies other than visible light. Fortunately, wave motion is described by the same mathematical equations, regardless of the medium in which the wave phenomenon occurs. We will now derive the wave equation for an electric intensity field in a source-free medium.

It will be assumed that some type of disturbance has caused an electromagnetic field to exist. The cause of the disturbance could be man-made, as by a radio transmitter, or a natural phenomenon such as a lightning discharge; however, the cause of the disturbance is not of concern here. What matters is that the electromagnetic field exists, and the object is to find the differential equation describing the motion of this field in time and space. We will assume that the wave travels in a medium in which the permittivity ε, permeability μ, and conductivity σ are constants.

The derivation begins with Maxwell's equation for the curl of **E**.

$$\nabla \times \mathbf{E} = -\frac{\mu \, \partial \mathbf{H}}{\partial t} \tag{1}$$

The immediate objective is to eliminate the variable **H** in the equation and thereby obtain a differential equation involving only **E**. First, we take the curl of each side of Eq. (1)

$$\nabla \times \nabla \times \mathbf{E} = -\mu \nabla \times \frac{\partial \mathbf{H}}{\partial t} \tag{2}$$

On the right-hand side of Eq. (2) we interchange the order of differentiation and the curl of **H**

$$\nabla \times \frac{\partial \mathbf{H}}{\partial t} = \frac{\partial}{\partial t} (\nabla \times \mathbf{H}) = \frac{\partial}{\partial t} \left(\sigma \mathbf{E} + \varepsilon \frac{\partial \mathbf{E}}{\partial t} \right) \tag{3}$$

When Eq. (3) is substituted into Eq. (2), a differential equation of the **E** field is obtained. However, the left-hand side of Eq. (2) can be reduced to a more recognizable form by use of the vector laplacian

$$\nabla \times \nabla \times \mathbf{E} = -\nabla^2 \mathbf{E} + \nabla(\nabla \cdot \mathbf{E}) = -\mu\sigma \frac{\partial \mathbf{E}}{\partial t} - \mu\varepsilon \frac{\partial^2 \mathbf{E}}{\partial t^2} \tag{4}$$

It will be assumed that there is no free charge in the medium, and so the divergence of **E** is zero. This is simply a special statement of one of Maxwell's equations

$$\nabla \cdot \mathbf{E} = \frac{\rho}{\varepsilon} = 0 \tag{5}$$

When Eq. (5) is substituted into Eq. (4), the desired differential equation involving only one dependent variable (the electric field **E**) is obtained. This differential equation will be called the *damped-wave equation*

$$\boxed{\nabla^2 \mathbf{E} = \mu\sigma \frac{\partial \mathbf{E}}{\partial t} + \mu\varepsilon \frac{\partial^2 \mathbf{E}}{\partial t^2}} \tag{6}$$

In this equation, note that the space derivatives are all on the left-hand side and are all of second order. On the other hand, only time derivatives are involved on the right-hand side of Eq. (6). This is a linear, second-order, vector partial differential equation which can be solved by advanced techniques which will not be considered here. However, in the next chapter a functional form of the solution of this equation will be obtained.

If a lossless medium is assumed, the conductivity σ will be zero and the first term on the right-hand side of Eq. (6) will be zero. ($\sigma\mathbf{E} = \mathbf{J}$ represents a conduction current through the medium, and is assumed to be zero.) Using this assumption of a lossless medium, we obtain the homogeneous vector wave equation

$$\boxed{\nabla^2 \mathbf{E} = \mu\varepsilon \frac{\partial^2 \mathbf{E}}{\partial t^2} \qquad \sigma = 0} \tag{7}$$

The constant $\mu\varepsilon$ appears in the wave equations, and we know that

$$c^2 = \frac{1}{\mu\varepsilon} \tag{8}$$

where c is the velocity of light. It will be shown in Chap. 3 that $\sqrt{\mu\varepsilon}$ is the reciprocal of the velocity of propagation of the wave.

The student should now repeat this derivation starting with Maxwell's equation for the curl of the magnetic field vector **H** rather than the electric field vector **E**. By following essentially the same derivation one can show that the differential equations for the magnetic field are

$$\nabla^2 \mathbf{H} = \mu\sigma \frac{\partial \mathbf{H}}{\partial t} + \mu\varepsilon \frac{\partial^2 \mathbf{H}}{\partial t^2} \tag{9}$$

$$\nabla^2 \mathbf{H} = \mu\varepsilon \frac{\partial^2 \mathbf{H}}{\partial t^2} \qquad \sigma = 0 \tag{10}$$

Equation (10) is the wave equation in a lossless medium. Note that the constant $\mu\varepsilon$ appears here, as it did in Eq. (7). This means that the electric and magnetic fields travel together with the same velocity.

Throughout most of the remainder of this text we will be concerned either directly or indirectly with the wave equations [Eqs. (7) and (10)] and their solutions. To solve these differential equations, a set of initial conditions and boundary conditions must be given for the particular physical problem under consideration. To distinguish between these two types of conditions, recall that the initial conditions describe the state of the magnetic and electric fields in the medium at time $t = 0$. Boundary conditions, on the other hand, usually describe the limitations placed on the electromagnetic fields at the boundaries of the medium. For example, the free space inside a rectangular waveguide is bounded by the conducting walls of the waveguide. At these conducting walls certain conditions on the normal and tangential fields must be satisfied. These conditions are called the boundary conditions, and normally they remain invariant with time. Initial conditions, on the other hand, exist only at time $t = 0$. For a given set of initial conditions and boundary conditions, the solution of the wave equation is unique. This, of course, is a very fortunate situation which makes a wave equation a very powerful mathematical model and one which applies to a wide variety of physical phenomena. For example, using the Lorentz gauge in a lossless, source-free region, the following functions all satisfy homogeneous wave equations: scalar potential V, vector potential **A**, electric field vector **E**, and magnetic field vector **H**. Furthermore all these wave equations have the same constant coefficient $\mu\varepsilon$, and so wave propagation for any of these variables occurs at the speed of light.

2.4 VOLTAGE, ELECTROMOTIVE FORCE, AND POTENTIAL DIFFERENCE

If a charged particle is moved from point a to point b in space, work is sometimes required from an external source to move the charge, because

TIME-VARYING FIELDS

of force on the particle due to electromagnetic fields. The force due to the electromagnetic fields is described by the Lorentz force equation

$$\mathbf{F} = q_t(\mathbf{E} + \mathbf{v} \times \mathbf{B}) \tag{1}$$

We define the *voltage drop* between points a and b along some given curve C to be the work per unit charge supplied by an external source in moving a positively charged particle along the path from point a to point b. Thus, the voltage drop V_{ba} is defined by the equation

$$\boxed{V_{ba} = -\int_a^b \frac{\mathbf{F}}{q_t} \cdot d\mathbf{l}} \tag{2}$$

For example, in a static field, if we assume that the point b is at a higher potential than the point a, the force $\mathbf{F}$ due to the field would tend to move a positively charged particle from b to a. An external source would be required to move the particle against this force [hence the minus sign in Eq. (2)], and in the process work would be supplied by the source to the particle. If we substitute the Lorentz force equation into Eq. (2), we obtain an expression for the voltage drop between the two points a and b along some curve C

$$V_{ba} = -\int_C \mathbf{E} \cdot d\mathbf{l} - \int_C (\mathbf{v} \times \mathbf{B}) \cdot d\mathbf{l} \tag{3}$$

If the curve C changes with time, i.e., is in motion, and the particle moves along the curve, the velocity v is the sum, or the total velocity of the particle, and V_{ba} becomes a function of time. In this equation, $\mathbf{E}$ and $\mathbf{B}$ can be time-varying or static fields, or both. We can write $\mathbf{E}$ in terms of its potential functions V and $\mathbf{A}$,

$$\mathbf{E} = -\nabla V - \frac{\partial \mathbf{A}}{\partial t} \tag{4}$$

substitute Eq. (4) into Eq. (3), and then obtain the voltage drop

$$\boxed{V_{ba} = \int_C \nabla V \cdot d\mathbf{l} + \int_C \frac{\partial \mathbf{A}}{\partial t} \cdot d\mathbf{l} - \int_C (\mathbf{v} \times \mathbf{B}) \cdot d\mathbf{l}} \tag{5}$$

The first term on the right is the potential difference between the two points due to any $\mathbf{E}$ field that may exist. We will show that the second term is related to the emf or induced voltage. The last term on the right in Eq. (5) is the voltage that would be induced in a conductor in motion with respect to the $\mathbf{B}$ field. When it is observed in semiconductors and conductors, this voltage is often called the *Hall voltage*, after Harvard professor Edmund Hall, who first observed the effect in a conductor in

1879. The Hall voltage is important in the study of electromechanical generators and motors. Thus we have assigned a name to each term on the right-hand side of Eq. (5), and consequently we can write the equation for the voltage drop as

$$V_{ba} = \text{potential difference} - \text{emf} + \text{Hall voltage} \tag{6}$$

We now consider each term of Eq. (6) in more detail. Suppose that the path C is stationary (independent of time) and that the **E** field is a static field. Then the emf and the Hall voltage are both zero, and the voltage drop V_{ba} is simply the potential difference between the two points

$$V_{ba} = \int_a^b \nabla V \cdot d\mathbf{l} = V_b - V_a \tag{7}$$

The potential difference was studied in Sec. 1.6, where it was shown that the voltage drop is independent of the path between the points a and b. Thus for static **E** fields, it makes no difference whether the path of integration is in motion (time-dependent) or not as long as the end points remain fixed.

However, we will show that the voltage drop due to the second term on the right-hand side of Eq. (5) depends both upon the end points and the path of integration, and thus the voltage between the points a and b is not unique (there are an infinite number of voltages). For a given curve, the voltage is a unique function of time for that curve

$$\text{emf} = -\int_C \frac{\partial \mathbf{A}}{\partial t} \cdot d\mathbf{l} \tag{8}$$

If we choose a closed path for the curve, the potential difference [the first term of Eq. (5)] is zero and the second term is the emf described in Lenz's law. To show this, we assume a *closed* curve C, and we apply Stokes' theorem to a surface bounded by the closed curve C

$$\text{emf} = -\oint_C \frac{\partial \mathbf{A}}{\partial t} \cdot d\mathbf{l} = -\frac{\partial}{\partial t}\int_s \nabla \times \mathbf{A} \cdot d\mathbf{s} \tag{9}$$

Replacing $\nabla \times \mathbf{A}$ with **B**, we obtain Lenz's law,

$$\text{emf} = -\frac{\partial}{\partial t}\int_s \mathbf{B} \cdot d\mathbf{s} = \frac{-\partial \Phi}{\partial t} \tag{10}$$

which we can think of as the voltage generated by the time-varying magnetic field. If a conductor which does not form a closed loop is placed in a time-varying field, a voltage is induced in it by the field even if the conductor is not in motion. In this case, we must use Eq. (8) rather than Eq. (10). An example of this is the voltage induced on a television antenna due to an impinging electromagnetic field.

TIME-VARYING FIELDS

If a conductor is in motion in a static **B** field, a voltage is induced in the conductor because the Lorentz force on the free electrons within the conductor causes a redistribution of the charge. Consequently, when the rotor conductors of a generator are turned in a magnetic field (generated by the field windings), a voltage appears at the terminals of the generator.

If a current is flowing through the rotor conductors of a dc motor, the free electrons in the conductor move in a magnetic field between the pole faces and a Lorentz force is exerted on them. This force is transferred to the conductor, and consequently the rotor turns. Similarly, if a current is flowing through a semiconductor material in a magnetic field, the polarity of the voltage induced at the side surfaces of the semiconductor depends upon whether the conductors are electrons or holes. This is called the *Hall effect*, and it can be used to determine the type of material and the extent of the doping. Conversely, if a piece of semiconductor material of known characteristics is moved at a known velocity through a magnetic field, the Hall effect can be used to measure the magnetic field. We define the Hall voltage as

$$\text{Hall voltage} = -\int_C \mathbf{v} \times \mathbf{B} \cdot d\mathbf{l} \tag{11}$$

In conclusion, we see that a voltage drop between two points depends upon the three terms given on the right-hand side of Eq. (5). The voltage drop depends upon the end points, the curve of integration, the velocity of the curve with respect to the **B** field, the static electric field, and, finally, the time-varying magnetic field.

2.5 POWER, ENERGY, AND POYNTING'S THEOREM

In our derivation of the Lorentz force equation, we found that a charged particle dq has a force exerted on it by an electric field, and also by a magnetic field if the particle is in motion. Because of this force, an electromagnetic field can do work on a charged particle. In moving through a differential (vector) distance $d\mathbf{l}$, the work done on the charged particle is

$$dW = \mathbf{F} \cdot d\mathbf{l} \tag{1}$$

Next we can use the Lorentz force equation to write

$$dW = dq(\mathbf{E} + \mathbf{v} \times \mathbf{B}) \cdot d\mathbf{l} \tag{2}$$

Here the vector **v** represents the velocity of the particle as it moves along its trajectory, and the vector $d\mathbf{l}$ represents a differential distance along that trajectory; consequently these two vectors are parallel to each other. However, the cross product of **v** and **B** is a vector which is perpendicular

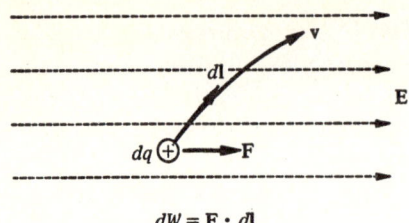

Fig. 2.5.1 The electric field does work on the particle dq as it moves through the field.

to both **v** and **B**, so that the component of the force due to the magnetic field is perpendicular to the direction of motion. Thus the dot product of the force due to the magnetic field and the vector $d\mathbf{l}$ is zero, and the magnetic field does no work on the charged particle.

Now consider the force due to the electric field, as shown in Fig. 2.5.1. Unlike the magnetic field, the electric field will do work on the charged particle as it moves through the field. Suppose that instead of a single charged particle in the field we have a charge density ρ distributed over the volume dv, so that the differential work is

$$dW = \rho \, dv \, \mathbf{E} \cdot d\mathbf{l} \tag{3}$$

From this equation we can obtain the differential power supplied by the field to the particle

$$dP = \frac{dW}{dt} = \mathbf{E} \cdot \left(\rho \frac{d\mathbf{l}}{dt}\right) dv = \mathbf{E} \cdot \rho \mathbf{v} \, dv \tag{4}$$

Since the motion of the charged particles represents the current density **J**,

$$\mathbf{J} = \rho \mathbf{v} \tag{5}$$

we can write the differential power per unit volume as

$$\boxed{\frac{dP}{dv} = \mathbf{E} \cdot \mathbf{J}} \tag{6}$$

Equation (6) is based upon the law of conservation of energy, and from it we will derive equations for the storage of energy in the electric and magnetic fields and for the power flow out of the given volume.

We begin the derivation by substituting Ampere's law for the current density **J** into the right-hand side of Eq. (6):

$$\mathbf{E} \cdot \mathbf{J} = \mathbf{E} \cdot \left(\nabla \times \mathbf{H} - \frac{\partial \mathbf{D}}{\partial t}\right) \tag{7}$$

The first term on the right-hand side of Eq. (7) can be recognized as one

TIME-VARYING FIELDS

term in the vector identity

$$\nabla \cdot (\mathbf{H} \times \mathbf{E}) = \mathbf{E} \cdot (\nabla \times \mathbf{H}) - \mathbf{H} \cdot (\nabla \times \mathbf{E}) \tag{8}$$

By using this vector identity in Eq. (7), we obtain the following form for the power density per unit volume:

$$\mathbf{E} \cdot \mathbf{J} = \nabla \cdot (\mathbf{H} \times \mathbf{E}) + \mathbf{H} \cdot (\nabla \times \mathbf{E}) - \mathbf{E} \cdot \frac{\partial \mathbf{D}}{\partial t} \tag{9}$$

The curl of $\mathbf{E}$ can be replaced by the time variation of the $\mathbf{B}$ field according to Faraday's law, and with some minor algebraic rearrangement we can write

$$\mathbf{E} \cdot \mathbf{J} + \nabla \cdot (\mathbf{E} \times \mathbf{H}) + \mathbf{H} \cdot \frac{\partial \mathbf{B}}{\partial t} + \mathbf{E} \cdot \frac{\partial \mathbf{D}}{\partial t} = 0 \tag{10}$$

Finally, to obtain the power over a given volume, we integrate throughout that volume,

$$\int_v \left[\mathbf{E} \cdot \mathbf{J} + \nabla \cdot (\mathbf{E} \times \mathbf{H}) + \mathbf{H} \cdot \frac{\partial \mathbf{B}}{\partial t} + \mathbf{E} \cdot \frac{\partial \mathbf{D}}{\partial t} \right] dv = 0 \tag{11}$$

and then apply the divergence theorem to the second integral, thereby converting it to a surface integral

$$\int_v \mathbf{E} \cdot \mathbf{J} \, dv + \oint_s (\mathbf{E} \times \mathbf{H}) \cdot d\mathbf{s} + \frac{\partial}{\partial t} \int_v \mathbf{H} \cdot \mathbf{B} \, dv + \frac{\partial}{\partial t} \int_v \mathbf{E} \cdot \mathbf{D} \, dv = 0 \tag{12}$$

This equation is again a statement of the law of conservation of energy, and since the terms must sum to zero, some of them must be negative. We now analyze each integral term of Eq. (12).

The first integral in Eq. (12) represents the power applied to the charged particles by the field. If that integral is negative, the charged particles are being forced against the field by an external source which must supply power. However, if the first integral is positive, the field is doing work on the charged particles. The second integral in Eq. (12) represents the net outflow of a vector $\mathbf{E} \times \mathbf{H}$ over the closed surface defined by the volume. If this integral is positive, it represents a power outflow through the surface s; if it is negative, it represents the net flow of power into the volume. The third term in Eq. (12) represents the time rate of change of the energy stored in the volume in the magnetic field. If this term is positive, an external source is supplying energy to the field and the magnitude of the field will be increasing. Conversely, if the term is negative, the field is decaying and energy is being extracted from it. The fourth term in Eq. (12) is the time rate of change of energy stored in the electric field, and its interpretation is similar to that of the third term. Equation (12) can be summarized in words as

$$\begin{pmatrix} \text{Power} \\ \text{from} \\ \text{source} \end{pmatrix} + \begin{pmatrix} \text{Power} \\ \text{out of} \\ \text{surface} \end{pmatrix} + \begin{pmatrix} \text{Rate of energy storage} \\ \text{in magnetic field and} \\ \text{in electric field} \end{pmatrix} = 0 \quad (13)$$

In the second term of Eq. (12), the volume v is bounded by the closed surface s, and the power flowing out of that surface is

$$P_{\text{out}} = \oint_s (\mathbf{E} \times \mathbf{H}) \cdot d\mathbf{s} \quad (14)$$

The vector $\mathbf{E} \times \mathbf{H}$ is called the *Poynting vector*† and is denoted by

$$\boxed{\mathbf{S} = \mathbf{E} \times \mathbf{H} = \text{Poynting vector, W/m}^2} \quad (15)$$

The Poynting vector $\mathbf{S}$ is a power-density vector which shows the direction and magnitude of power flow through a surface at any point where $\mathbf{E}$ and $\mathbf{H}$ are known. In connection with wave propagation, the Poynting vector is often useful in determining the direction of power flow.

Next we will investigate the energy-storage terms of Eq. (12). The differential power per unit volume dP_s can be obtained from either Eq. (10) or (12):

$$\frac{dP_s}{dv} = \mathbf{E} \cdot \frac{\partial \mathbf{D}}{\partial t} + \mathbf{H} \cdot \frac{\partial \mathbf{B}}{\partial t} \quad (16)$$

However, since the flux-density fields are related to the intensity fields by the equations

$$\mathbf{D} = \varepsilon \mathbf{E} \qquad \mathbf{B} = \mu \mathbf{H} \quad (17)$$

we can simplify the right-hand side of Eq. (16) by writing the terms as derivatives of dot products. For example, the time derivative of the dot product of the electric field intensity and the flux-density field is

$$\frac{\partial (\mathbf{E} \cdot \mathbf{D})}{\partial t} = \mathbf{D} \cdot \frac{\partial \mathbf{E}}{\partial t} + \mathbf{E} \cdot \frac{\partial \mathbf{D}}{\partial t} = 2\mathbf{E} \cdot \frac{\partial \mathbf{D}}{\partial t} \quad (18)$$

for media in which ε is considered to be a scalar constant. A similar form holds for the magnetic fields (if μ is a scalar constant), and so Eq. (16) becomes

$$\frac{dP_s}{dv} = \frac{\partial}{\partial t} \left(\frac{\mathbf{E} \cdot \mathbf{D}}{2} + \frac{\mathbf{H} \cdot \mathbf{B}}{2} \right) \quad (19)$$

† See Sec. 3.4 for further discussion of the complex Poynting vector. In that section $\mathbf{E}$ and $\mathbf{H}$ are functions of a complex variable, whereas they are considered as real functions above.

TIME-VARYING FIELDS

When both sides of Eq. (19) are integrated with respect to time, we obtain the differential energy stored per unit volume (the energy density). The energy is the sum of the energy density U_E due to the electric field and the energy density U_H of the magnetic field

$$\frac{dW_e}{dv} = \tfrac{1}{2}\mathbf{E}\cdot\mathbf{D} + \tfrac{1}{2}\mathbf{H}\cdot\mathbf{B} = U_E + U_H \tag{20}$$

If we assume the rectangular form for the vectors **E** and **H**,

$$\mathbf{E} = E_x\mathbf{a}_x + E_y\mathbf{a}_y + E_z\mathbf{a}_z$$
$$\mathbf{H} = H_x\mathbf{a}_x + H_y\mathbf{a}_y + H_z\mathbf{a}_z \tag{21}$$

and then use Eqs. (17), we can evaluate the dot products $\mathbf{E}\cdot\mathbf{D}$ and $\mathbf{H}\cdot\mathbf{B}$. Therefore the energy densities can be written in the following useful forms:

$$\boxed{U_E = \frac{\varepsilon}{2}(E_x{}^2 + E_y{}^2 + E_z{}^2)} \tag{22}$$

$$\boxed{U_H = \frac{\mu}{2}(H_x{}^2 + H_y{}^2 + H_z{}^2)} \tag{23}$$

As an example of the use of Eq. (22), we compute the energy stored in the electric field between the plates of a parallel capacitor, neglecting any fringing of the fields. The assumed geometry is shown in Fig. 2.5.2.

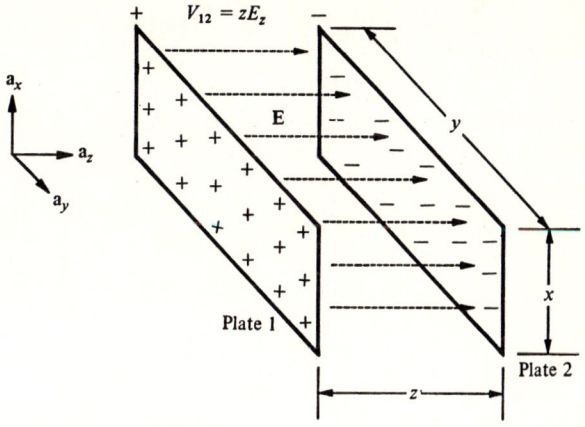

Fig. 2.5.2 Energy is stored in electric fields as shown for the parallel-plate capacitor.

We assume a uniform electric field of the form

$$\mathbf{E} = E_z \mathbf{a}_z \tag{24}$$

so that by integrating the electric field along a line from plate 2 to plate 1 we obtain the potential difference V_{12} between the plates. Using Eq. (24) and integrating the energy density of Eq. (22) over the volume between the plates, we obtain the energy stored in the capacitor

$$W_E = \int_v U_E \, dv = \frac{\varepsilon}{2} \int_v E_z^2 \, dx \, dy \, dz \tag{25}$$

Since the electric field is constant over the region, the integration yields the volume xyz. With some algebraic rearrangement, we can write the equation for the stored energy in the familiar form

$$W_E = \frac{\varepsilon}{2}(xyz)E_z^2 = \frac{\varepsilon A V^2}{2z} = \tfrac{1}{2} C V^2 \tag{26}$$

where $A = xy$ is the surface area of one of the plates and C is the capacitance of the capacitor. By a similar derivation we could show that the energy stored in the magnetic field in a coaxial cable carrying a direct current I is given by

$$W_H = \tfrac{1}{2}(L_i + L_e)I^2 \tag{27}$$

where L_i is the internal inductance per unit meter and L_e is the external inductance per unit meter. However, this derivation will be left for a problem.

2.6 RADIATION

As an application of Maxwell's equations and the vector potential, we will consider an example of electromagnetic radiation from an elementary antenna. In a previous section we showed that the Lorentz gauge

$$\nabla \cdot \mathbf{A} = -\mu\varepsilon \frac{\partial V}{\partial t} \tag{1}$$

allows us to reduce Maxwell's equations to two inhomogeneous wave equations of the potential functions $\mathbf{A}$ and V

$$\nabla^2 V - \mu\varepsilon \frac{\partial^2 V}{\partial t^2} = -\frac{\rho}{\varepsilon} \tag{2}$$

$$\nabla^2 \mathbf{A} - \mu\varepsilon \frac{\partial^2 \mathbf{A}}{\partial t^2} = -\mu \mathbf{J} \tag{3}$$

TIME-VARYING FIELDS

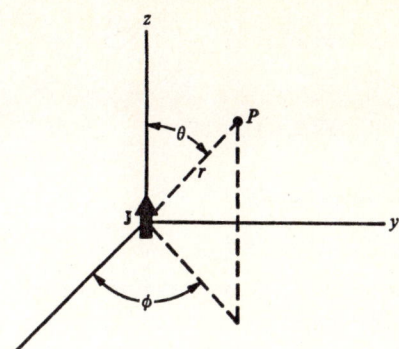

Fig. 2.6.1 The elemental current **J** produces fields at point P.

If we can find **A** for a given current distribution **J**, we can find V from Eq. (1) and consequently the electric and magnetic fields

$$\mathbf{B} = \nabla \times \mathbf{A} \tag{4}$$

$$\mathbf{E} = -\nabla V - \frac{\partial \mathbf{A}}{\partial t} \tag{5}$$

Therefore we are most interested in finding **A**.

For some antennas of simple geometry (such as the loop antenna and the dipole) it is possible to obtain excellent analytical results, especially if one is interested in the *far field* (usually taken to mean at least 10 wavelengths removed from the source) under sinusoidal time variations. Although it is beyond the scope of this text, antenna theory has advanced to the point where very complicated structures and arrays of structures can be predictably designed. We will consider only the elementary case where a point source of current

$$\mathbf{J} = J_0 e^{j\omega t} \delta(x) \delta(y) \delta(z) \mathbf{a}_z \tag{6}$$

located at the origin flows in the $\mathbf{a}_z$ direction and produces an alternating electromagnetic field in the surrounding space. In particular, we will determine the fields at some point P in space, as shown in Fig. 2.6.1. We further assume that the current flow has existed long enough in the homogeneous isotropic medium for any transient phenomena (associated with turning the current on and establishing the fields) to have passed, so that steady-state sinusoidal oscillations exist at the frequency ω.

To simplify the analysis we align the z axis with the direction of the **J** vector so that the vector wave equation (3) can be reduced to three scalar wave equations:

$$\nabla^2 A_z(t) - \mu\varepsilon \frac{\partial^2 A_z(t)}{\partial t^2} = -\mu J_0 e^{j\omega t} \delta(x)\delta(y)\delta(z) \tag{7}$$

$$\nabla^2 A_x(t) - \mu\varepsilon \frac{\partial^2 A_x(t)}{\partial t^2} = 0 \qquad (8)$$

$$\nabla^2 A_y(t) - \mu\varepsilon \frac{\partial^2 A_y(t)}{\partial t^2} = 0 \qquad (9)$$

Consequently,

$$A_x(t) = 0 = A_y(t) \qquad (10)$$

since the forcing function is zero for Eqs. (8) and (9). Furthermore, the time-variation term $e^{j\omega t}$ can be canceled from both sides of Eq. (7):

$$\nabla^2 A_z e^{j\omega t} + \omega^2 \mu\varepsilon A_z e^{j\omega t} = \mu J_0 e^{j\omega t} \delta(x)\delta(y)\delta(z) \qquad (11)$$

Therefore

$$\nabla^2 A_z + \omega^2 \mu\varepsilon A_z = \mu J_0 \delta(x)\delta(y)\delta(z) \qquad (12)$$

where A_z is a function of x, y, and z but not of t. Since the right-hand side of Eq. (12) is zero everywhere except at a single point, the origin, we expect to find spherical symmetry for the function A_z. Consequently, for this ideal point source, A_z is only a function of the radial distance r from the source, and it is simplest to work with spherical coordinates.

Elsewhere in this text we consider methods of solving the wave equation. For now, it is sufficient to show that the solution for Eq. (12) is

$$A_z = \frac{\mu J_0 e^{-\gamma r}}{4\pi r} \qquad (13)$$

where

$$\gamma = j\omega \sqrt{\mu\varepsilon} \qquad (14)$$

We can accomplish this through a sequence of easy steps in a programmed exercise, which enables the student to take an active part.

PROGRAMMED EXERCISE 2.6.1

RADIATION FROM A POINT SOURCE

1. To show that A_z is a solution everywhere except possibly at the origin, substitute Eq. (13) into

$$\nabla^2 A_z - \gamma^2 A_z = 0 \qquad (15)$$

Next, if we assume a radial solution

$$\mathbf{A} = \frac{\mu J_0 e^{-\gamma r}}{4\pi r} \mathbf{a}_r$$

TIME-VARYING FIELDS

instead of the form given in Eq. (13), will **A** satisfy the vector wave equation [from Eq. (3)]

$$\nabla^2 \mathbf{A} - \gamma^2 \mathbf{A} = 0$$

in spherical coordinates? Note the differences between the scalar and vector laplacian in these two substitutions.

2. To show that A_z satisfies Eq. (12) at the origin, substitute Eq. (13) into (12) and integrate over a small spherical volume, using $dv = r^2 \sin\theta \, dr \, d\theta \, d\phi$. This results in

$$\int (\nabla^2 A_z - \gamma^2 A_z) \, dv = \int \mu J_0 \delta(x) \delta(y) \delta(z) \, dv \tag{16}$$

Use Eqs. (9) and (12) of Sec. 1.5

$$\delta(x)\delta(y)\delta(z) = \delta(r,\theta,\phi) = \frac{\delta(r)\delta(\theta)\delta(\phi)}{r^2 \sin\theta} \tag{17}$$

to show that the integral on the right-hand side equals μJ_0. Of the integrals on the left-hand side, use Gauss' theorem on the first to obtain a surface integral which tends to μJ_0 as r approaches zero, and show that the second approaches zero as r approaches zero. Consequently, we show that A_z as given by Eq. (13) is a solution of Eq. (12) at the origin as well as elsewhere.

3. Find the magnetic field **H** from

$$\mathbf{H} = \frac{1}{\mu} \nabla \times \mathbf{A} \tag{18}$$

where $\mathbf{A} = A_z \mathbf{a}_z$, as given in Eq. (13). First transform **A** to spherical coordinates, finding that

$$\mathbf{A} = \frac{\mu J_0 e^{-\gamma r}}{4\pi r} (\cos\theta \, \mathbf{a}_r - \sin\theta \, \mathbf{a}_\theta) \tag{19}$$

$$\mathbf{H} = \frac{J_0 e^{-\gamma r}}{4\pi} \left(\frac{\gamma}{r} + \frac{1}{r^2} \right) \sin\theta \, \mathbf{a}_\phi \tag{20}$$

4. Find the electric field **E** from

$$\mathbf{E} = -\nabla V - \frac{\partial \mathbf{A}}{\partial t} = \nabla \frac{\nabla \cdot \mathbf{A}}{j\omega\mu\varepsilon} - j\omega \mathbf{A} \tag{21}$$

In spherical coordinates, the components are

$$E_r = \frac{J_0 e^{-\gamma r}}{4\pi} \left(\sqrt{\frac{\mu}{\varepsilon}} \frac{2}{r^2} + \frac{2}{j\omega\varepsilon r^3} \right) \cos\theta \tag{22}$$

$$E_\theta = \frac{J_0 e^{-\gamma r}}{4\pi} \left(\frac{j\omega\mu}{r} + \sqrt{\frac{\mu}{\varepsilon}} \frac{1}{r^2} + \frac{1}{j\omega\varepsilon r^3} \right) \sin\theta \tag{23}$$

and

$$E_\phi = 0 \tag{24}$$

Equations (20), (22), (23), and (24) describe the **E** and **H** fields produced by a point source at the origin, and our analysis is complete.

5. In many cases we are interested in the fields at some point far removed from the source, e.g., in radio transmission or radar applications. In such cases those terms which vary as r^{-2} or r^{-3} (called the *near-field* terms) are negligible compared with the r^{-1} terms (called the *far-field* or *radiation* terms). Significant simplifications can be made when we are interested in the far fields. Show that the far fields are

$$\mathbf{A} = A_z \mathbf{a}_z = \frac{\mu J_0 e^{-\gamma r}}{4\pi r} \mathbf{a}_z \tag{25}$$

$$\mathbf{H} = j\omega \sqrt{\frac{\varepsilon}{\mu}} A_z \sin\theta \, \mathbf{a}_\phi \tag{26}$$

$$\mathbf{E} = j\omega A_z \sin\theta \, \mathbf{a}_\theta \tag{27}$$

6. Sketch the magnitude of **E** as a function of θ for values of θ between 0 and 360° for any plane ϕ = constant. Assume that J_0, r, and ω are constants. The sketch is called the *radiation pattern* of the elemental antenna. Because of the symmetry with respect to ϕ, the three-dimensional radiation pattern is shaped like a doughnut.

7. Find the ratio of the magnitudes of **E** to **H** in the far field. Compute the numerical value of this ratio, assuming free-space constants. This ratio, called the *intrinsic impedance*, will be discussed in more detail in Chap. 3. Because the ratio is a constant, it is not necessary to solve for both the **E** and **H** fields.

8. To obtain the time-varying fields, multiply **H** and **E** of Eqs. (26) and (27) by $e^{j\omega t}$ and find the real parts. Before finding the real parts of these complex equations, however, first replace γ with $j\omega\sqrt{\mu\varepsilon}$. Then it is a simple algebraic step to show that

$$\mathbf{E}(r,\theta,t) = \frac{-\omega\mu J_0}{4\pi r} \sin\theta \sin(\omega t - \omega\sqrt{\mu\varepsilon}\,r)\mathbf{a}_\theta = E_\theta \mathbf{a}_\theta \tag{28}$$

$$\mathbf{H}(r,\theta,t) = \sqrt{\frac{\varepsilon}{\mu}} E_\theta \mathbf{a}_\phi \tag{29}$$

Consequently, we can find the far field for any point in space at any time.

9. Find the Poynting vector from Eqs. (28) and (29) and determine the direction of power flow. The total power radiated from the elemental antenna can be found by integrating the Poynting vector over the

TIME-VARYING FIELDS

surface of a sphere centered at the origin with a radius of r_0. The total power available for detection at some distant receiving antenna can be found by integrating the Poynting vector over the effective area of the antenna.

With this last paragraph, you have essentially derived the *radiation resistance* of the antenna. The essential idea is that the input impedance to the antenna (as seen by a generator feeding the antenna) must have a real part (a resistance) that absorbs power from the generator at the same rate as power is radiated from a spherical surface surrounding the antenna. Thus the antenna could be replaced by an equivalent resistance and some reactance, as far as the generator is concerned. The reactance depends upon the frequency and the geometry of the antenna. For the small antenna of length Δz m, the radiation resistance in ohms is

$$R_{ra} = 80\pi^2 \left(\frac{\Delta z}{\lambda}\right)^2 \tag{30}$$

In addition to the radiation resistance, the input impedance will also contain a real term which accounts for the ohmic losses of current flowing on the metallic surfaces. The ratio of the radiation resistance to the total resistance is a measure of the efficiency of the antenna. These are topics typically considered in the study of linear antennas.[†] Although we have considered only the most elementary antenna, the student now has the background and mathematical maturity to do some independent studies.

Perhaps the elemental antenna considered in this section seems too simple to be useful, but more realistic antennas can be analyzed as a superposition of such elemental antennas. An equally important accomplishment in this section was to see one example of how an electromagnetic field is associated with a source current. Throughout most of this text we assume the existence of the fields and bypass consideration of how they are generated.

2.7 BOUNDARY CONDITIONS AND PROPERTIES OF MATERIALS
THE DIELECTRIC TENSOR

In our study of electromagnetic theory we have assumed that the flux-density fields and the intensity fields are related by simple constants, the permeability and the permittivity. For example, in free space

$$\mathbf{D} = \varepsilon_0 \mathbf{E}$$
$$\mathbf{B} = \mu_0 \mathbf{H} \tag{1}$$

[†] For example, see W. L. Weeks, "Antenna Engineering," McGraw-Hill Book Company, New York, 1968.

However, many materials (in particular crystals) do not exhibit such simple relationships between the field vectors. It is more general to assume that each component of the **D** vector is a linear function of each of the components of the **E** vector

$$D_x = \varepsilon_{11}E_x + \varepsilon_{12}E_y + \varepsilon_{13}E_z$$
$$D_y = \varepsilon_{21}E_x + \varepsilon_{22}E_y + \varepsilon_{23}E_z \qquad (2)$$
$$D_z = \varepsilon_{31}E_x + \varepsilon_{32}E_y + \varepsilon_{33}E_z$$

These equations express a linear relationship between the **D** and **E** fields; a medium which displays these characteristics is said to be a *linear medium*. In a linear medium superposition is valid because the constants ε_{ij} do not depend upon the magnitude of the electric intensity components. However, dielectric materials generally break down under very high field intensity, and then the linear relationships of Eqs. (2) no longer apply. We will not consider any nonlinear media in this book. By assuming a linear medium, as described by Eqs. (2), we can write the permittivity constants in a matrix form which is useful in describing other properties of the medium

$$[\varepsilon_{ij}] = \begin{bmatrix} \varepsilon_{11} & \varepsilon_{12} & \varepsilon_{13} \\ \varepsilon_{21} & \varepsilon_{22} & \varepsilon_{23} \\ \varepsilon_{31} & \varepsilon_{32} & \varepsilon_{33} \end{bmatrix} \qquad (3)$$

The matrix of Eq. (3) is called the *dielectric tensor* or the *permittivity tensor* of the linear medium.

Suppose that we know the nine elements of the matrix of Eq. (3) at some given point in space. Now suppose that at some other point in the medium we measure these nine constants, by some means, and find that they are not the same as at the original point. We would then say that the properties of the material are a function of the position, and such a medium is said to be *inhomogeneous*. Conversely, if the constants in Eq. (3) do not change with position throughout the region of interest, the medium is said to be *homogeneous*. We will be concerned only with homogeneous media in this text.

The parameters of Eq. (3) can also vary with time, frequency, and temperature. For example, when a cloud drifts into the field of a radar beam, it changes the characteristics of the path of propagation. Good dielectrics used in transmission lines and waveguides are relatively insensitive to temperature, but often they are sensitive to frequency. In this case the parameters of Eq. (3) at a given point in space will be different at different frequencies. Such characteristics of the medium are often considered in more advanced courses in electromagnetic theory or solid-state theory, but they will not be considered here.

The dielectric properties of a medium can also be described by certain characteristics of the matrix of Eq. (3). We will describe three cases.

Case 1 Free space

$$\varepsilon_{ij} = \begin{cases} 0 & \text{if } i \neq j \\ \varepsilon_0 & \text{if } i = j \end{cases} \tag{4}$$

In this case all the nondiagonal elements of the matrix are zero, and the diagonal elements are identical and equal to the permittivity of free space. The fields are then related as in Eqs. (1).

Case 2 Isotropic medium

$$\varepsilon_{ij} = \begin{cases} 0 & \text{if } i \neq j \\ \varepsilon_r \varepsilon_0 & \text{if } i = j \end{cases} \tag{5}$$

Again the nondiagonal elements of the matrix are zero, and the diagonal elements are identical and can be written in terms of the *relative dielectric constant* ε_r. In an isotropic medium, **D** and **E** vectors are parallel, and the behavior of the medium is independent of the direction of the field vectors. We will study the propagation of electromagnetic waves in isotropic media in the remainder of this book.

Case 3 Anisotropic medium

$$[\mathbf{D}] = [(\varepsilon_{ij})] [\mathbf{E}] \tag{6}$$

In an *anisotropic* medium, the nondiagonal elements of the matrix of Eq. (3) are not necessarily zero, and the **D** and **E** vectors are not necessarily parallel. Many crystalline substances are anisotropic, and the behavior of such a medium depends upon the direction of the field vectors. However, it is shown in more advanced courses that the matrix of Eq. (3) is symmetric. We will not become involved with anisotropic media in this text.†

The relationship of the magnetic field vectors **B** and **H** can also be described by a permeability matrix, similar to the permittivity matrix in Eq. (3). However, magnetic properties of a medium are more complicated than the dielectric properties in many cases. If we consider a single atom, the electrons in motion about the nucleus constitute moving charge, which in turn generates a small magnetic field. Furthermore, both the electron spin and the nuclear spin have an influence on the small magnetic field. When an external magnetic field is applied to the medium, the atom attempts to reorient itself so that its magnetic field aligns itself

† The reader seeking more information is referred to Max Born and Emil Wolf, "Principles of Optics," The Macmillan Company, New York, 1964.

with the external field. Hence a torque is applied to the atom by the external field, and if the atom returns to its original position when the external field is removed, the energy can be stored, in much the same way that energy can be stored in a compressed spring. The material is said to be *polarized* by the magnetic field, and the atoms act as small dipoles with small magnetic moments. The total moment for any atom is a combination of the orbital moment, electron-spin moment, and the nuclear-spin moment. Because there are many ways in which these three moments can combine, a very wide range of properties is possible. When the atomic dipoles are such that the individual atoms have no magnetic moments, the material is said to be *diamagnetic*. If the alignment of the atomic dipoles is such as to produce a magnetic moment, the materials are classified as paramagnetic, ferromagnetic, antiferromagnetic, ferrimagnetic, or superparamagnetic, depending upon how the atomic dipoles are aligned. In our work we will assume that the medium is linear, diamagnetic, homogeneous, and isotropic.†

BOUNDARY CONDITIONS

By now the reader has probably concluded, correctly, that we are taking the simplest possible cases of the properties of materials in our study of electromagnetic fields. Assuming even the simplest properties of materials, however, we must still consider conditions where there is a boundary or sharp discontinuity from one medium to another. As an example, consider the situation shown in Fig. 2.7.1, where there is a sharp discontinuity between medium 1 and medium 2. Assume that we know the constants of both media and that we know the flux-density vector in medium 1. We consider a small rectangular volume which encloses an incremental section of the boundary, as shown in Fig. 2.7.1, and we let the dimension d approach zero, so that in the limit the box shrinks to an incrementally thin plane with zero volume, at the surface. We can apply Gauss' law over the closed surface of the volume

$$\oint \mathbf{D} \cdot \mathbf{a}_n \, ds = \rho_s S = Q \tag{7}$$

Since there may be charge on either side of the boundary or there may be charge on the surface at the boundary, we assume that some charge is enclosed in the incremental volume, as shown in the right-hand side of Eq. (7), where ρ_s is the surface charge density and S is the total surface area. Consider S_1 to be the surface area in medium 1 and S_2 to be the surface area in medium 2, as shown in the figure. In integrating over the volume we can neglect the integral over the sides of the rectangular region

† The interested reader can find more information in *ibid.* or in books on solid-state electronics, such as Albert Van Der Ziel, "Solid State Physical Electronics," Prentice-Hall, Inc., Engelwood Cliffs, N.J., 1957.

TIME-VARYING FIELDS

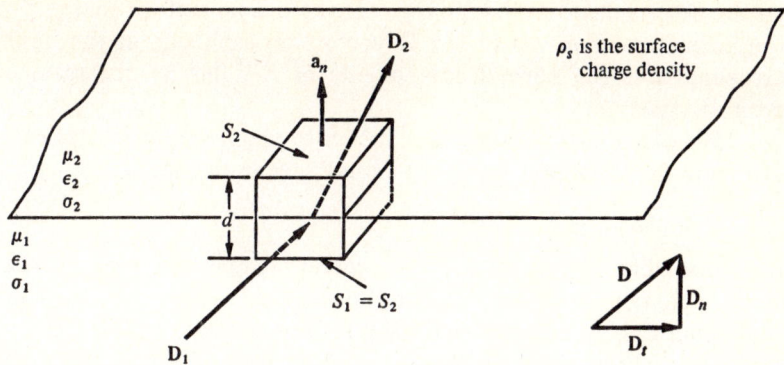

Fig. 2.7.1 Boundary conditions for the flux-density vector **D**.

because that dimension is approaching zero

$$\int_{S_1} \mathbf{D}_1 \cdot \mathbf{a}_n \, ds + \int_{S_2} \mathbf{D}_2 \cdot \mathbf{a}_n \, ds + \lim_{d \to 0} \overset{0}{\int_{\text{sides}}} = \rho_s S \tag{8}$$

For the normal vector $\mathbf{a}_n$ in the second medium, the dot products of Eq. (8) can be evaluated in terms of the normal components of the flux-density vectors

$$\mathbf{D}_1 \cdot \mathbf{a}_n = -D_{1n} = \text{normal component} \tag{9}$$

$$\mathbf{D}_2 \cdot \mathbf{a}_n = D_{2n} = \text{normal component} \tag{10}$$

When these equations are substituted into Eq. (8), Gauss' law reduces to the simple form

$$\oint \mathbf{D} \cdot \mathbf{a}_n \, ds = -D_{1n} \int_{S_1} ds + D_{2n} \int_{S_2} ds$$
$$= (D_{2n} - D_{1n})S = \rho_s S \tag{11}$$

which shows the relationship between the normal components of the flux-density vectors. The results of the derivation are

$$\boxed{D_{2n} - D_{1n} = \rho_s \quad \text{or} \quad \mathbf{a}_n \cdot (\mathbf{D}_2 - \mathbf{D}_1) = \rho_s} \tag{12}$$

and therefore

$$\boxed{\varepsilon_2 E_{2n} - \varepsilon_1 E_{1n} = \rho_s} \tag{13}$$

From these equations we can find the normal component of either the flux-density vector **D** or the electric intensity vector **E** on one side of the boundary if we know the surface charge density and the electric fields on the other side of the boundary. If there is no surface charge density, the normal components of the **D** vectors on each side of the boundary are equal but the normal components of **E** differ by the ratio of the permittivities.

We can apply the same derivation to magnetic fields that can exist at a boundary. Again we apply Gauss' law at the boundary,

$$\oint \mathbf{B} \cdot \mathbf{a}_n \, ds = 0 \tag{14}$$

but since there are no isolated magnetic charges, there is no magnetic surface charge density and the right-hand side of the equation is zero. Following the same derivation used for the electric flux density, we are led to the following conclusions:

$$\boxed{B_{2n} - B_{1n} = 0} \tag{15}$$

or

$$\boxed{\mathbf{a}_n \cdot (\mathbf{B}_2 - \mathbf{B}_1) = 0} \tag{16}$$

therefore

$$\boxed{\mu_2 H_{2n} - \mu_1 H_{1n} = 0} \tag{17}$$

Consequently the normal components of the magnetic flux densities are equal on each side of the boundary, but the normal components of the magnetic intensities **H** differ by the ratio of the permeabilities.

In the above derivation we are not able to draw any conclusions about the tangential components because their dot product with the normal vector $\mathbf{a}_n$ is zero. However, we can obtain the boundary conditions on the tangential components by using Ampere's law and Faraday's law around an enclosed loop at the boundary, as shown in Fig. 2.7.2. We will assume that the charge at the surface is in motion so that a surface current density $\mathbf{J}_s$ exists at the boundary. By considering a rectangular loop of height d and length l, as shown in the figure, we let the dimension d go to zero; thus we integrate the vector **H** around the loop right at the surface. Using Ampere's law

$$\oint_C \mathbf{H} \cdot d\mathbf{l} = \int_s \mathbf{J} \cdot \mathbf{a}_n \, ds + \int_s \frac{\partial \mathbf{D}}{\partial t} \cdot \mathbf{a}_n \, ds \tag{18}$$

we see that the displacement current is zero because the surface area approaches zero as d approaches zero

$$\lim_{d \to 0} \left(\int_s \frac{\partial \mathbf{D}}{\partial t} \cdot \mathbf{a}_n \, ds \right) = 0 \tag{19}$$

TIME-VARYING FIELDS

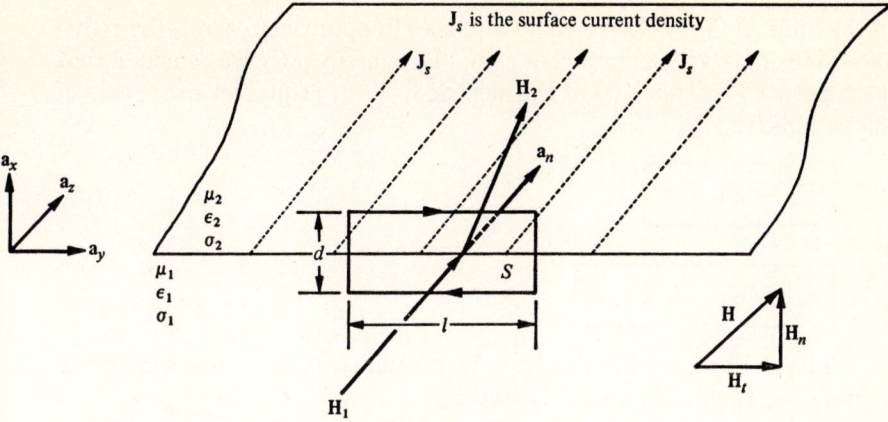

Fig. 2.7.2 Boundary conditions for **H**.

However, the surface current density appears as an impulse function, and when integrated over the surface, this term yields the surface current

$$\int_s \mathbf{J}_s \cdot \mathbf{a}_n \, ds = I_s \tag{20}$$

Thus the right-hand side of Eq. (18) reduces to the surface current I_s, which pierces the loop shown in Fig. 2.7.2. Since the dimension d is approaching zero, the left-hand side of Eq. (18) reduces to the form

$$\oint \mathbf{H} \cdot d\mathbf{l} = \int \mathbf{H}_1 \cdot (-\mathbf{a}_y) \, dl + \int \mathbf{H}_2 \cdot \mathbf{a}_y \, dl = (-H_{1t} + H_{2t})l \tag{21}$$

Now we can equate the right-hand sides of Eqs. (20) and (21) and divide both sides of that equation by the length l. However, the ratio of the surface current I_s to the length l is just the magnitude of the current density $\mathbf{J}_s$. Therefore the derivation leads us to the following conclusion:

$$\boxed{H_{2t} - H_{1t} = J_s} \tag{22}$$

or

$$\boxed{\mathbf{a}_n \times (\mathbf{H}_2 - \mathbf{H}_1) = \mathbf{J}_s} \tag{23}$$

Note that if the surface current is zero, the tangential components of the **H** vectors are equal on each side of the boundary.

We can repeat the above derivation for the **E** field, using Faraday's law instead of Ampere's law

$$\oint_C \mathbf{E} \cdot d\mathbf{l} = -\int_s \frac{\partial \mathbf{B}}{\partial t} \cdot \mathbf{a}_n \, ds \tag{24}$$

In the limit as the surface area goes to zero (d approaches zero), the right-hand side of Eq. (24) approaches zero, and consequently we conclude that the tangential components of the electric fields are equal on either side of the boundary:

$$\boxed{E_{2t} - E_{1t} = 0} \tag{25}$$

or

$$\boxed{\mathbf{a}_n \times (\mathbf{E}_2 - \mathbf{E}_1) = 0} \tag{26}$$

The derivations for the boundary conditions of media with finite values of μ, ε, and σ are summarized by

$$\boxed{\begin{array}{ll} D_{2n} - D_{1n} = \rho_s & B_{2n} - B_{1n} = 0 \\ E_{2t} - E_{1t} = 0 & H_{2t} - H_{1t} = J_s \end{array}} \tag{27}$$

One other case we must consider is the boundary condition on a perfect conductor, where the conductivity is infinite. The conclusions we draw from this case are often sufficiently accurate for good conductors such as copper or silver. Consider the situation shown in Fig. 2.7.3 and assume that a **D** field with both a normal and tangential component exists at the surface of the conductor, so that a surface charge density ρ_s exists at the surface. Let $\mathbf{a}_n$ be a unit vector normal to the surface and pointing into the conductor. Consider a point at the surface just inside the conductor and write Ampere's law at that point

$$\nabla \times \mathbf{H} = \mathbf{J} + \frac{\partial \mathbf{D}}{\partial t} = \sigma \mathbf{E} + \varepsilon \frac{\partial \mathbf{E}}{\partial t} \tag{28}$$

We now divide both sides of the equation by the conductivity σ, which we let approach infinity,

$$\lim_{\sigma \to \infty} \left[\frac{1}{\sigma} (\nabla \times \mathbf{H}) - \frac{\varepsilon}{\sigma} \frac{\partial \mathbf{E}}{\partial t} \right] = \mathbf{E} \to 0 \tag{29}$$

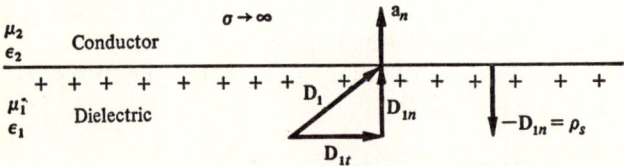

Fig. 2.7.3 Boundary conditions at a perfect conductor.

TIME-VARYING FIELDS

Since **E** and **H** and the derivatives indicated in Eq. (29) must remain finite, we conclude that the electric intensity vector **E** approaches zero inside the perfect conductor as the conductivity σ approaches infinity. Therefore no **E** fields exist inside a perfect conductor. Since the **E** fields are zero, the **D** fields are also zero inside a perfect conductor.

Now consider a point at the surface of the conductor just inside the dielectric material. From Eq. (12),

$$\mathbf{a}_n \cdot (\mathbf{D}_2 - \mathbf{D}_1) = \rho_s \tag{30}$$

where the vector $\mathbf{D}_2$ must be zero since it exists inside the conductor. Consequently, we see that the magnitude of the normal component $-\mathbf{D}_1$ of the **D** vector in the dielectric is equal to the surface charge density ρ_s.

$$|\mathbf{D}| = \rho_s \tag{31}$$

Similarly, if we consider Eq. (25) right at the boundary, where E_{2t} is zero inside the perfect conductor, we conclude that there can be no tangential component of electric field at the surface of a perfect conductor

$$E_{1t} = 0 \tag{32}$$

Since **E** = 0 inside the conductor, $\nabla \times \mathbf{E} = 0$, and from Faraday's law we find that

$$\nabla \times \mathbf{E} = -\frac{\partial \mathbf{B}}{\partial t} = -\mu \frac{\partial \mathbf{H}}{\partial t} = 0 \tag{33}$$

This implies either that **B** = 0 and **H** = 0 inside a perfect conductor or that **B** and **H** are constants. The latter means that **B** and **H** have always existed, with no possibility of change. Since this is not physically observed, we choose

$$\mathbf{B} = 0 = \mathbf{H} \tag{34}$$

Then, noting that $\mathbf{a}_n$ is a normal vector into the conductor, we write

$$\mathbf{a}_n \times (\mathbf{H}_2 - \mathbf{H}_1) = \mathbf{J}_s \Rightarrow \mathbf{J}_s = -\mathbf{a}_n \times \mathbf{H}_{1t} \tag{35}$$

In conclusion, we find that no electromagnetic fields exist inside the perfect conductor, that the electric field in the dielectric at the surface of the conductor is perpendicular to the surface and is equal in magnitude to the surface charge density, and that the **H** field is tangential to the surface and is equal in magnitude to the surface current.

For convenience and future reference, the most important equations of Chap. 2 are summarized in Table 2.7.1.

Table 2.7.1 Summary of the elementary field equations

Basic equations

Equation	Point form	Integral form	
Lorentz force equation	$\mathbf{F} = q\mathbf{E} + q(\mathbf{v} \times \mathbf{B})$		(1)
Conservation of charge	$\nabla \cdot \mathbf{J} + \dfrac{\partial \rho}{\partial t} = 0$	$I = \oint_s \mathbf{J} \cdot \mathbf{a}_n \, ds = -\dfrac{\partial q}{\partial t}$	(2)
Gauss' law	$\nabla \cdot \mathbf{B} = 0$	$\oint_s \mathbf{B} \cdot \mathbf{a}_n \, ds = 0$	(3)
Gauss' law	$\nabla \cdot \mathbf{D} = \rho$	$\oint_s \mathbf{D} \cdot \mathbf{a}_n \, ds = q = \int_v \rho \, dv$	(4)
Faraday's law	$\nabla \times \mathbf{E} = -\dfrac{\partial \mathbf{B}}{\partial t}$	$\oint_C \mathbf{E} \cdot d\mathbf{l} = -\int_s \dfrac{\partial \mathbf{B}}{\partial t} \cdot \mathbf{a}_n \, ds = -\dfrac{d\phi}{dt}$	(5)
Ampere's law	$\nabla \times \mathbf{H} = \mathbf{J} + \dfrac{\partial \mathbf{D}}{\partial t}$	$\oint_C \mathbf{H} \cdot d\mathbf{l} = I + \int_s \dfrac{\partial \mathbf{D}}{\partial t} \cdot \mathbf{a}_n \, ds$	(6)

Equations of a homogeneous isotropic medium

$\mathbf{J} = \sigma \mathbf{E}$	$\sigma = 0$ in free space	(7)
$\mathbf{D} = \varepsilon \mathbf{E}$	$\varepsilon_0 = 8.854 \times 10^{-12} \approx \dfrac{10^{-9}}{36\pi}$ F/m in free space	(8)
$\mathbf{B} = \mu \mathbf{H}$	$\mu_0 = 4\pi \times 10^{-7}$ H/m in free space	(9)

Poynting's theorem (for conservation of energy)

$$P_{in} = P_d + \dfrac{\partial}{\partial t}(W_E + W_M) + \int_s \mathbf{E} \times \mathbf{H} \cdot \mathbf{a}_n \, ds \tag{10}$$

Boundary conditions

$$D_{2n} - D_{1n} = \rho_s \quad B_{2n} - B_{1n} = 0 \quad E_{2t} - E_{1t} = 0 \quad H_{2t} - H_{1t} = J_s \tag{11}$$

The wave equations

$$\nabla^2 \mathbf{A} = \mu\sigma \dfrac{\partial \mathbf{A}}{\partial t} + \mu\varepsilon \dfrac{\partial^2 \mathbf{A}}{\partial t^2} \tag{12}$$

In free space:

$$\sigma = 0 \quad c = 3 \times 10^8 \text{ m/s} \quad \nabla^2 \mathbf{A} = \dfrac{1}{c^2} \dfrac{\partial^2 \mathbf{A}}{\partial t^2} \tag{13}$$

TIME-VARYING FIELDS

PROBLEMS

1. Show that the following fields do *not* satisfy Maxwell's equations for static conditions:
(a) $\mathbf{E} = 10\mathbf{a}_y$; (b) $\mathbf{H} = 5r\mathbf{a}_r$.

2. A parallel-plate capacitor with a plate surface of S is connected across a battery. An electron of mass m and charge e is placed in the free space between the plates. The electron begins to fall due to the gravitational acceleration g. What voltage must be applied to just stop its downward motion?

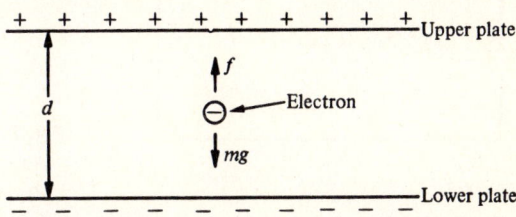

Fig. P.2 Sketch for Prob. 2.

3. You are assigned the job of designing a variable capacitor with two movable and three fixed plates, as shown. The fixed plates are all connected to one terminal and the movable ones to another. Assume that the capacitance is due to the parallel plates of the shaded areas only. Assume a constant plate separation d for all plates. (a) Find the total capacitance as a function of θ and d for $0 \leq \theta \leq \pi/2$, and sketch C versus θ. (b) If $C_{max} = 100$ pF and the plate separation $d = 1$ mm, what value of radius r is required?

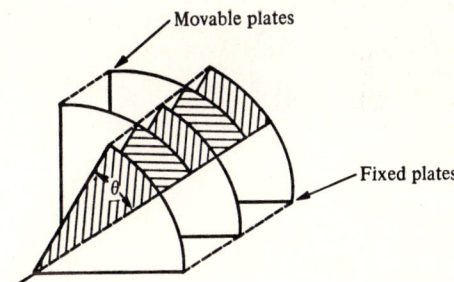

Fig. P.3 Variable capacitor for Prob. 3.

4. Given a potential field $V = x$, find the charge enclosed in a vacuous cube 2 cm on an edge centered at (0,0,0).

5. A fly falls in a bowl of coulombs and swallows one before he can get out. He crawls out at point (0,1,0) and flys to point (3,7,3) along the path $y = 2x + 1$, $z = x^2/3$, through an electric field of $\mathbf{E} = \mathbf{a}_x + 2xy\mathbf{a}_y + z\mathbf{a}_z$. At (3,7,3) he lands on a window screen which is at the same potential as the point (0,1,0). Calculate the potential difference between the fly and the screen as a function of his position.

6. If there is distributed charge in the region of the fly's flight (Prob. 5), what is the charge density at the point $(1,3,\tfrac{1}{3})$?

7. The switch is closed at $t = 0$, and $V_C = 0$ at $t = 0$. Show that current is continuous in the circuit by finding $\partial D/\partial t$. The capacitor has area A, plate separation z, and permittivity ε.

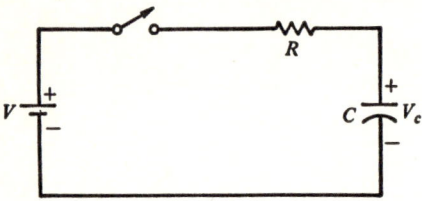

Fig. P.7 The RC circuit for Prob. 7.

8. For a coaxial conductor, with a cross section as shown in the figure, derive expressions for the internal and external inductance per meter, referred to as L_i and L_e respectively in Eq. (27) of Sec. 2.5. The corresponding terms of Eq. (27) yield the energy stored inside the conductors and between the conductors respectively. *Hint:* Use cylindrical coordinates, assume a current I in each conductor, and evaluate $\mathbf{H} \cdot \mathbf{B}$ in Eq. (20) to obtain U_H, and then integrate U_H over the volume of interest.

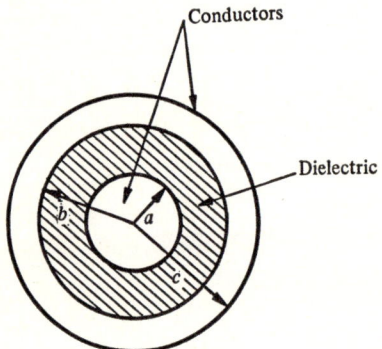

Fig. P.8 Coaxial cable for Prob. 8.

9. A battery of voltage $V_0 = 100$ V is connected to a parallel-plate capacitor (area of one plate $= 0.9$ m²). The plates are mechanically pumped so that the separation distance d of the plates varies as

$$d = 2 \times 10^{-3}(2 + \cos 3600\pi t)$$

Find an equation for the current that flows from the battery.

10. Attempt to derive $W_H = \tfrac{1}{2}LI^2$ for a long straight wire carrying a current I from the magnetic fields using the expression for energy density given in Eq. (20) of Sec. 2.5. What physical significance can be associated with the result?

TIME-VARYING FIELDS

11. The sketch shows a conductor moving through a magnetic field of $\mathbf{B} = 60\mathbf{a}_y$ Wb/m^2 with a velocity of $\mathbf{u} = 5\mathbf{a}_x$ m/s. (a) Indicate the polarity of the voltage at voltmeter V. (b) Calculate the magnitude of the voltage.

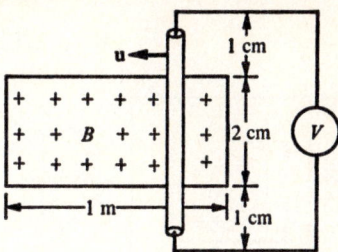

Fig. P.11 Sketch for Prob. 11.

12. Two equal point charges of $Q = \frac{1}{9}$ C are located at $(0,2,0)$ and $(0,-2,0)$ m in a vacuum. (a) Find the electric intensity at any point along the x axis. (b) Evaluate E at $x = 0$ and $x = 1$ m on the x axis. (c) Find the force on a charge of $\frac{1}{10}$ C located at $(1,0,0)$.

13. An aluminum disk of radius r is uniformly charged and has a surface charge density of k C/m^2. Find the E field at any point above the center of the disk.

14. Ivan smuggles a mason jar containing 10^{27} protons out of the missile range under the hood of his red Volkswagen. He drives north parallel to the earth's magnetic field B and turns east at Highway 66 with a velocity U_0. What is the acceleration on the Volkswagen as it moves east? Sketch its path of motion.
Given (all units in mks):

Earth's field $= \mathbf{B} = -5.8 \times 10^{-5}\mathbf{a}_x$ Wb/m^2

Proton charge $= Q = 1.6 \times 10^{-19}$ C

$U_0 = 100$ km/hr $= 62$ mi/hr $= 27.8$ m/s

Car weight $= 4000$ N ≈ 900 lb

Car mass $= 408$ kg

Number of protons $= 10^{27}$

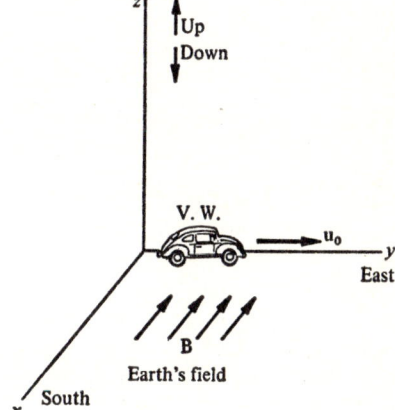

Fig. P.14 Sketch for Prob. 14.

15. Ivan (Prob. 14) discovers his error after a short time and places the jar in a box made of Permalloy 2-81 (2% Mo, 81% Ni) whose relative permeability is 130. Calculate and then sketch the **B** field within the walls of the box and within the interior of the box.

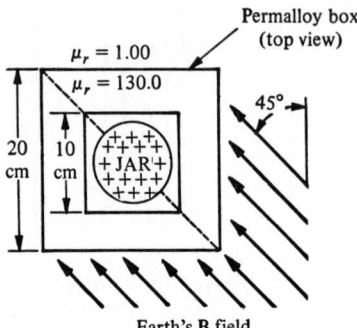

Fig. P.15 Earth's **B** field affects the box.

16. Find the far-field equations $\mathbf{A}(r,\theta,\phi)$, $\mathbf{E}(r,\theta,\phi)$, $\mathbf{H}(r,\theta,\phi)$ at an arbitrary point p for a current density of the form $\mathbf{J} = J_0 e^{i\omega t}\delta(x)\delta(y)\mathbf{a}_z$. In this case, the current is confined to a line along the z axis.

17. Find the total power radiated from the point-source antenna by integration of the Poynting vector for the far fields over a spherical surface, as suggested in step 9 of Programmed Exercise 2.6.1.

18. Derive the radiation resistance given in Eq. (30) of Sec. 2.6.

19. A plumber accidentally connects a water line to a waveguide and partially fills the waveguide. If the equation for **D** just above the water is $\mathbf{D} = \mathbf{a}_x + 10\mathbf{a}_z$ at a given instant, find **D** just below the water at that time.

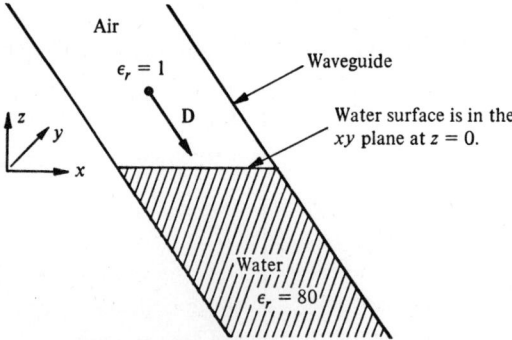

Fig. P.19 Sketch for Prob. 19.

3
Plane-wave Propagation

3.1 SOLUTION OF THE WAVE EQUATION

An important example of the application of Maxwell's equations is wave propagation, and in this chapter we consider the properties of plane waves in a linear, homogeneous, isotropic, time-invariant medium. A plane wave is one in which the electric field intensities, and also the magnetic field intensities, are in time phase at all points in a plane normal to the direction of propagation. For example, a spherical wavefront emanating from a point source appears to be a plane wave in a small region a long distance from the source. Initially we consider waves propagating in an unbounded region, but later in the chapter we consider the reflection of plane waves from a perfect conductor as well as from a dielectric material. We will also consider polarization of plane waves and plane waves in dissipative media.

From Maxwell's equations the equations of a plane wave will be derived, assuming the medium described above, in which μ and ε are scalar constants. For the present, we assume that the medium is nonconducting, although this restriction is removed in a later section. We will use rectangular coordinates and assume that there are no field variations in the x or y direction. Maxwell's curl equations, for the medium assumed, are given by Eqs. (1) and (5). Equating components of each curl equation, we obtain the three equations written below each curl equation:

$$\nabla \times \mathbf{E} = -\mu \frac{\partial \mathbf{H}}{\partial t} \tag{1}$$

$$\frac{\partial E_y}{\partial z} = \mu \frac{\partial H_x}{\partial t} \tag{2}$$

$$\frac{\partial E_x}{\partial z} = -\mu \frac{\partial H_y}{\partial t} \tag{3}$$

$$0 = \mu \frac{\partial H_z}{\partial t} \tag{4}$$

$$\nabla \times \mathbf{H} = \varepsilon \frac{\partial \mathbf{E}}{\partial t} \tag{5}$$

$$\frac{\partial H_y}{\partial z} = -\varepsilon \frac{\partial E_x}{\partial t} \tag{6}$$

$$\frac{\partial H_x}{\partial z} = \varepsilon \frac{\partial E_y}{\partial t} \tag{7}$$

$$0 = \varepsilon \frac{\partial E_z}{\partial t} \tag{8}$$

From Eqs. (4) and (8) it can be seen that E_z and H_z must be either zero or constants. The constant, or static, parts of the solutions are not pertinent to the present discussion of wave propagation and will be neglected. Thus the electric and magnetic fields which vary with time lie in the xy plane. It will be shown that these fields propagate in the z direction as a wave.

We now derive the wave equation for the wave described above. If Eq. (3) is differentiated with respect to z and Eq. (6) is differentiated with respect to t, we get

$$\frac{\partial^2 E_x}{\partial z^2} = -\mu \frac{\partial^2 H_y}{\partial z\, \partial t} \tag{9}$$

$$\frac{\partial^2 H_y}{\partial t\, \partial z} = -\varepsilon \frac{\partial^2 E_x}{\partial t^2} \tag{10}$$

Interchanging the order of the derivatives† of H_y in Eq. (9) and substitut-

† The derivatives can be interchanged if H_y is a uniformly convergent function of z and t.

PLANE-WAVE PROPAGATION

ing into (10) gives the one-dimensional form of the wave equation for E_x

$$\frac{\partial^2 E_x}{\partial z^2} = \mu\varepsilon \frac{\partial^2 E_x}{\partial t^2} \tag{11}$$

In similar fashion, we could show that the fields E_y, H_x, and H_y also satisfy Eq. (11) under the assumptions made. These four field components are known as the *transverse components* because they lie in the plane normal to the direction of propagation. Since this wave has no electric or magnetic field components in the *longitudinal* direction (the direction of propagation), it is called a *transverse electromagnetic wave* (TEM wave). In Chap. 7 we study both *transverse magnetic* (TM) waves and *transverse electric* (TE) waves.

A TM wave has the following characteristics:

1. The magnetic field lies entirely in the transverse plane.
2. The electric field has a longitudinal component as well as a transverse component.

A TE wave has the following characteristics:

1. The electric field lies entirely in the transverse plane.
2. The magnetic field has a longitudinal component as well as a transverse component.

For the present, however, we will continue our development of TEM waves, which have only transverse **E** and **H**, by solving Eq. (11).

The wave equation for E_x can be written

$$\frac{\partial^2 E_x}{\partial z^2} - \mu\varepsilon \frac{\partial^2 E_x}{\partial t^2} = 0 \tag{12}$$

If $\partial/\partial z$ and $\partial/\partial t$ are regarded as operators, Eq. (12) can be written†

$$\left(\frac{\partial}{\partial z} - \sqrt{\mu\varepsilon}\,\frac{\partial}{\partial t}\right)\left(\frac{\partial}{\partial z} + \sqrt{\mu\varepsilon}\,\frac{\partial}{\partial t}\right) E_x = 0 \tag{13}$$

From this equation it can be seen that if E_x satisfies the conditions imposed by Eq. (14) or (15), then E_x is also a solution of the wave equation

$$\frac{\partial E_x}{\partial z} - \sqrt{\mu\varepsilon}\,\frac{\partial E_x}{\partial t} = 0 \tag{14}$$

$$\frac{\partial E_x}{\partial z} + \sqrt{\mu\varepsilon}\,\frac{\partial E_x}{\partial t} = 0 \tag{15}$$

† This operator method is valid for linear constant-coefficient differential equations.

By applying the chain rule for derivatives it can be shown that *any* analytic function of the form $F(t - \sqrt{\mu\varepsilon}\, z)$ is a solution to Eq. (15) and *any* analytic function of the form $G(t + \sqrt{\mu\varepsilon}\, z)$ is a solution of Eq. (14). Any constant E_0 is also a solution to the wave equation, since all derivatives vanish. This solution can be thought of as a static field in space. Since the wave equation is linear, superposition applies and a solution can be written in the form

$$E_x = F(t - \sqrt{\mu\varepsilon}\, z) + G(t + \sqrt{\mu\varepsilon}\, z) + E_0 \qquad (16)$$

This solution can be shown to be the most general solution of the one-dimensional wave equation; it consists of two arbitrary functions and an arbitrary constant, which is the expected result for a second-order linear partial differential equation.

The solution of the wave equation for the other transverse field quantities E_y, H_x, and H_y can be derived in exactly the same manner. The result, of course, must be in the same form as Eq. (16), the only differences in the various solutions being due to the boundary conditions of these other field quantities.

The particular TEM wave discussed in this section is called a *uniform* plane wave because **E** (and **H**) has the same magnitude throughout any transverse plane, defined by $z = $ constant. Since a uniform plane wave must extend to infinity in at least two directions, it cannot physically exist. However, the definition of a uniform plane wave is sometimes modified to describe a wave that is uniform only in a finite region of any transverse plane. For example, the field of a radio or radar transmitting antenna is essentially a uniform plane wave in the vicinity of a distant receiving antenna or target.

An example of a *nonuniform* plane wave is a TEM wave traveling down a coaxial transmission line. It will be shown in Sec. 4.2 that both **E** and **H** lie in the transverse $r\phi$ plane (assuming perfect conductors) and that the time phase of **E** (and of **H**) is constant in a transverse plane. Thus the wave is a plane wave. Since fields are zero outside the outer coaxial conductor, the field magnitudes in a transverse plane clearly are dependent on the coordinates; since the fields exist in only a part of the transverse plane, the wave is nonuniform. Even if we restrict our attention to the region between conductors, where fields exist, we will again find that the wave is nonuniform, by the following reasoning. Both **E** and **H** are proportional to $1/r$ in the region between the coaxial conductors, and so the magnitude of the wave in a transverse plane does not have the same (uniform) value throughout the region where fields exist. The proportionality of **E** and **H** to $1/r$ will be shown in Sec.

PLANE-WAVE PROPAGATION

4.2. Thus the wave is nonuniform, even in the region between coaxial conductors.†

3.2 TRAVELING WAVES

The immediate objective is to obtain as much information as possible about the solution of the wave equation

$$E_x = F(t - \sqrt{\mu\varepsilon}\, z) + G(t + \sqrt{\mu\varepsilon}\, z) \tag{1}$$

without assuming a specialized waveform. Since we are interested in wave propagation, we will neglect the constant (static) term. It is obvious that E_x is a linear combination of two functions, F and G. We will find that F represents a wave traveling in the positive z direction while G represents a wave traveling in the negative z direction.

Suppose that F is represented by an arbitrary waveform, as shown in Fig. 3.2.1, and at some particular time t_1 the wave is located at position z_1. At some later time t_2 the wave will have traveled to a new position z_2. To an observer riding the peak of the wave, the height of the wave always appears constant, as shown in Fig. 3.2.1, assuming the wave is not distorted:

$$F(t - \sqrt{\mu\varepsilon}\, z) = F(M) = \text{const} \tag{2}$$

In order for F to appear constant to such an observer, its argument must be a constant

$$t - \sqrt{\mu\varepsilon}\, z = M \tag{3}$$

Equation (3) is the equation of a straight line; that is, t and z are linearly related. We now solve Eq. (3) for z and differentiate with respect to t to

† For a more complete discussion of uniform and nonuniform plane waves, see R. B. Adler, L. J. Chu, and R. M. Fano, "Electromagnetic Energy Transmission and Radiation," chap. 7, John Wiley & Sons, Inc., New York, 1960.

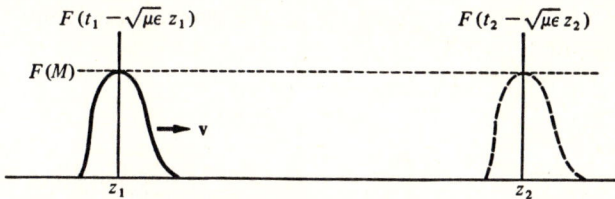

Fig. 3.2.1 A traveling wave.

obtain the *velocity of propagation* v_p, also called the *phase velocity*,

$$\frac{dz}{dt} = \frac{1}{\sqrt{\mu\varepsilon}} = v_p \tag{4}$$

Equation (4) states that the rate of change of distance with respect to time must be a constant of the medium in order for the argument in Eq. (2) to be a constant. Therefore the derivative in Eq. (4) represents the velocity with which the wave must travel in order to satisfy Eq. (2). The wave F is called a *traveling wave*. If the same analysis is applied to G, the velocity of propagation of this traveling wave is found to be

$$\frac{dz}{dt} = \frac{-1}{\sqrt{\mu\varepsilon}} = -v_p \tag{5}$$

where the negative sign simply means that the wave is traveling in the negative z direction. Thus the waves F and G travel with the same velocity but in opposite directions. The positive-going wave for the x component of **E** will be denoted by E_x^+ and the negative-going wave by E_x^-. The mathematical forms of these definitions are

$$F(t - \sqrt{\mu\varepsilon}\, z) = E_x^+\left(t - \frac{z}{v_p}\right) = E_x^+ \tag{6}$$

$$G(t + \sqrt{\mu\varepsilon}\, z) = E_x^-\left(t + \frac{z}{v_p}\right) = E_x^- \tag{7}$$

In conclusion, the solution of the wave equation is the superposition of waves traveling in opposite directions but each having a velocity v_p:

$$E_x(z,t) = E_x^+\left(t - \frac{z}{v_p}\right) + E_x^-\left(t + \frac{z}{v_p}\right) \tag{8}$$

It has been shown that the velocity of propagation is a constant if the medium is distortionless; conversely, it will now be shown that waveforms are not distorted if the velocity of propagation is a constant.† Suppose that the negative-going wave is zero ($E_x^- = 0$), so that we have only a single positive-going wave, as shown in Fig. 3.2.1. At a particular time t_1 and a given position z_1, the wave has a functional form given by

$$E_x(z_1,t_1) = E_x^+\left(t_1 - \frac{z_1}{v_p}\right) \tag{9}$$

† If the function F is decomposed into its Fourier components, every component must travel with the same velocity to avoid distortion of the waveform. In many practical applications μ and ε are essentially independent of frequency over a wide range. In these cases the velocity of propagation is essentially constant.

PLANE-WAVE PROPAGATION

At some later time t_2 the wave will have traveled to a new position z_2, as shown in Fig. 3.2.1. The functional form of the wave at this new time and position is given by

$$E_x(z_2,t_2) = E_x^+\left(t_2 - \frac{z_2}{v_p}\right) \tag{10}$$

Now express t_2 and z_2 in terms of t_1 and z_1 as follows

$$t_2 = t_1 + \Delta t \tag{11}$$
$$z_2 = z_1 + \Delta z \tag{12}$$

and make the appropriate substitutions in Eq. (10):

$$E_x(z_2,t_2) = E_x^+\left(t_1 + \Delta t - \frac{z_1 + \Delta z}{v_p}\right)$$
$$= E_x^+\left(t_1 - \frac{z_1}{v_p} + \Delta t - \frac{\Delta z}{\Delta z/\Delta t}\right)$$
$$= E_x^+\left(t_1 - \frac{z_1}{v_p}\right) \tag{13}$$

The result of the substitutions is

$$E_x(z_2,t_2) = E_x(z_1,t_1) \tag{14}$$

Equation (14) states that the functional form of the wave at position z_2 is the same as the functional form of the wave at position z_1. In conclusion, the wave does not change its form as it travels, and the medium is said to be *distortionless*. Although the functional form of E_x as given in Eq. (1) was derived for a TEM wave, it can be shown that components of TE and TM waves (which can exist in waveguides) have a similar functional form. The phase velocity is different from that of the TEM wave in most of these cases, but the characteristics of the traveling wave remain the same.

We now summarize the notation to be used for traveling-wave vectors and their components for the transverse fields of the plane waves illustrated in Fig. 3.2.2. The *magnitude* of the **E** field in a given direction is

$$E_x = E_x^+ + E_x^-$$
$$E_y = E_y^+ + E_y^- \tag{15}$$

where all the components are functions of z and t. The instantaneous electric field *vector* at any given point is the vector sum

$$\mathbf{E} = E_x \mathbf{a}_x + E_y \mathbf{a}_y \tag{16}$$
$$\mathbf{E} = \mathbf{E}_x + \mathbf{E}_y \tag{17}$$

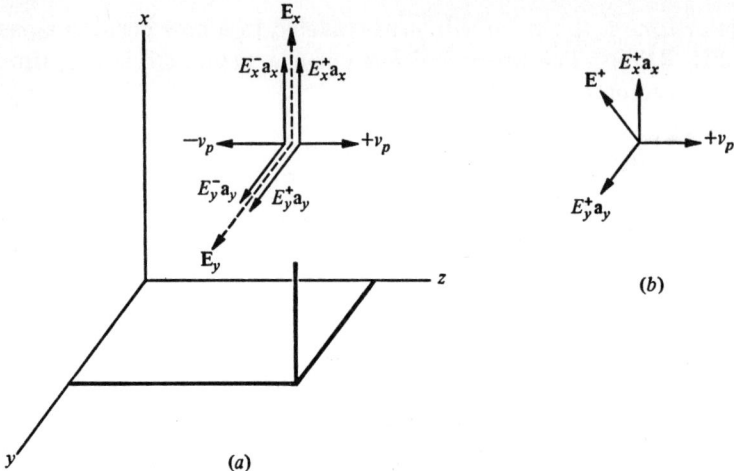

Fig. 3.2.2 The vector components of traveling waves.

From Eqs. (15) and (16) we see that the vector can be written as the vector sum of the positive- and negative-going waves

$$\mathbf{E} = (E_x^+ \mathbf{a}_x + E_y^+ \mathbf{a}_y) + (E_x^- \mathbf{a}_x + E_y^- \mathbf{a}_y) \tag{18}$$

$$\mathbf{E} = \mathbf{E}^+ + \mathbf{E}^- \tag{19}$$

In the work that follows one must carefully distinguish between the vector components of Eq. (15), the directional field vectors $\mathbf{E}_x$ and $\mathbf{E}_y$ of Eq. (17), and the traveling-wave vectors $\mathbf{E}^+$ and $\mathbf{E}^-$ of Eq. (19). Similar notation will be used to describe field vectors and components of $\mathbf{E}$ and $\mathbf{H}$ fields for the more general case where components in all three directions exist.

3.3 CHARACTERISTICS OF A UNIFORM PLANE WAVE

To determine some of the characteristics of a uniform plane wave, we now consider a positive-going wave. First we derive relationships between the transverse field components E_x^+, E_y^+, H_x^+, and H_y^+. From Eq. (1) of Sec. 3.2 we see that E_x^+ is of the form

$$E_x^+ = F\left(t - \frac{z}{v_p}\right) \tag{1}$$

From Eq. (3) of Sec. 3.1, using the chain rule for differentiation, it follows that

$$\frac{\partial H_y^+}{\partial t} = -\frac{1}{\mu} \frac{\partial E_x^+}{\partial z} = \frac{1}{\mu v_p} F'\left(t - \frac{z}{v_p}\right) = \sqrt{\frac{\varepsilon}{\mu}} F'\left(t - \frac{z}{v_p}\right) \tag{2}$$

PLANE-WAVE PROPAGATION

where

$$F'\left(t - \frac{z}{v_p}\right) = \frac{\partial F(t - z/v_p)}{\partial(t - z/v_p)}$$

Since H_y^+ is of the same functional form, $H_y^+(t - z/v_p)$, it follows that

$$\frac{\partial H_y^+}{\partial t} = \frac{\partial H_y^+}{\partial(t - z/v_p)}$$

and we can integrate both sides of Eq. (2). Neglecting the constants of integration, we have

$$H_y^+ = \sqrt{\frac{\varepsilon}{\mu}} F\left(t - \frac{z}{v_p}\right) = \frac{E_x^+}{\eta} \tag{3}$$

where

$$\eta = \sqrt{\frac{\mu}{\varepsilon}} \tag{4}$$

The quantity η, called the *intrinsic impedance*, is a constant of the medium and has dimensions of ohms.

Similarly, we could use Eq. (2) of Sec. 3.1 to find that $H_x^+ = -E_y^+/\eta$. Thus we find that the four components of the positive-going wave are related as follows:

$$\frac{E_x^+}{H_y^+} = -\frac{E_y^+}{H_x^+} = \eta \tag{5}$$

For a negative-going wave, in similar fashion, we find that

$$\frac{E_x^-}{H_y^-} = -\frac{E_y^-}{H_x^-} = -\eta \tag{6}$$

The negative sign occurs with η because the velocity of propagation is negative (recall that $\eta = \mu v_p$).

The total electric field for the positive-going wave is

$$\mathbf{E}^+ = E_x^+ \mathbf{a}_x + E_y^+ \mathbf{a}_y \tag{7}$$

and from Eq. (5) it is seen that

$$\eta \mathbf{H}^+ = -E_y^+ \mathbf{a}_x + E_x^+ \mathbf{a}_y \tag{8}$$

Forming the dot product $\mathbf{E}^+ \cdot \eta \mathbf{H}^+$ always yields zero, which indicates that the electric and magnetic fields in a positive-going wave are perpendicular, as shown in Fig. 3.3.1. Similar analysis from Eq. (6) would show that the electric and magnetic fields in a negative-going wave are also perpendicular.

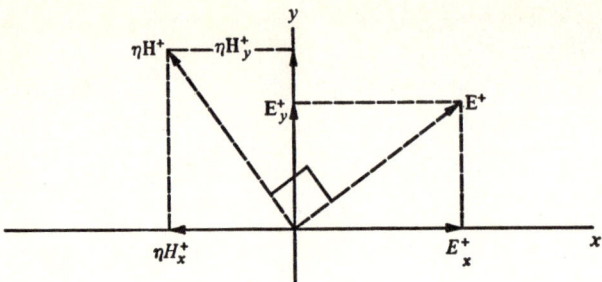

Fig. 3.3.1 Electric and magnetic fields in a positive-going wave are perpendicular.

Equations (5) and (6) also require that the instantaneous magnitude of the electric field be η times the instantaneous magnitude of the magnetic field in each of the traveling waves. Therefore, if we know one of the field intensities of a traveling wave, we can easily find the other. The relationship is analogous to Ohm's law in circuits, where the voltage and current are related by a constant called *impedance*. Thus the simple relationships of Eqs. (5) and (6) are very useful. In addition, we see that $\mathbf{E}^+ \times \mathbf{H}^+$ yields a power-density wave (with units of watts per square meter) in the positive z direction and $\mathbf{E}^- \times \mathbf{H}^-$ yields a power-density wave in the negative z direction.

From Sec. 2.5, the energy stored in the electric field per unit volume is

$$U_E = \frac{\varepsilon E^2}{2} = \frac{\varepsilon}{2}(E_x^2 + E_y^2) \tag{9}$$

and the energy stored in the magnetic field per unit volume is

$$U_H = \frac{\mu H^2}{2} = \frac{\mu}{2}(H_x^2 + H_y^2) \tag{10}$$

In Eqs. (9) and (10), field components such as H_y must represent the sum of the corresponding field components in both the positive- and negative-going wave, as discussed in Sec. 3.2. For example,

$$H_y = H_y^+ + H_y^-$$

By Eq. (5) or (6) one can show that U_E and U_H are equal for a single traveling wave, and so the energy density per unit volume at any point in a positive-going (or negative-going) wave is always divided equally between the electric and magnetic fields.

In general, however, the total energy density in the $\mathbf{E}$ field differs from the total energy density in the $\mathbf{H}$ field at a point where both positive-

PLANE-WAVE PROPAGATION

and negative-going traveling waves exist. Thus superposition can be used only in special cases to find the total energy density.

The Poynting vector for the positive-going wave is

$$S_z{}^+ \mathbf{a}_z = (E_x{}^+ H_y{}^+ - E_y{}^+ H_x{}^+)\mathbf{a}_z = \frac{1}{\eta}(E_x{}^{+2} + E_y{}^{+2})\mathbf{a}_z \qquad (11)$$

and for the negative-going wave is

$$S_z{}^- \mathbf{a}_z = (E_x{}^- H_y{}^- - E_y{}^- H_x{}^-)\mathbf{a}_z = \frac{-1}{\eta}[(E_x{}^-)^2 + (E_y{}^-)^2]\mathbf{a}_z \qquad (12)$$

Since no energy can be dissipated in the lossless medium, the time-average value of $S_z{}^+$ (or of $S_z{}^-$) must be the same for any plane described by $z =$ constant. The instantaneous magnitudes, however, may be different at different planes. Thus the power density in a single traveling wave remains constant in a lossless medium.

Many practical problems involve sinusoidal time variation, and it will be convenient to use phasor notation. This will be developed in the next section, using the damped-wave equation in E_x obtained from Maxwell's equations for a lossy medium.

3.4 PLANE WAVES IN DISSIPATIVE MEDIA

In this section, first, we remove the time variation from the one-dimensional wave equation for lossy media by assuming sinusoidal waveforms and using phasor techniques. The phasor notation can, of course, be similarly applied to sinusoidal waves in a lossless medium simply by setting conductivity equal to zero. Once the time variation has been removed from the wave equation, we can solve the remaining second-order differential equation. We then discuss some of the characteristics of plane waves in a lossy medium, including the skin effect. It will be assumed throughout this section that all losses of the medium are incorporated in the value for conductivity σ and that the medium is isotropic. The actual loss mechanisms in a dissipative medium can be rather complicated, and will not be discussed here.†

The damped-wave equation for the $E_x(z,t)$ component of a plane wave with no variations in the x or y direction is

$$\frac{\partial^2 E_x(z,t)}{\partial z^2} = \sigma\mu \frac{\partial E_x(z,t)}{\partial t} + \varepsilon\mu \frac{\partial^2 E_x(z,t)}{\partial t^2} \qquad (1)$$

† More complete discussions of such losses will be found in S. Ramo, J. R. Whinnery, and T. Van Duzer, "Fields and Waves in Communication Electronics," John Wiley & Sons, Inc., New York, 1965, and W. L. Weeks, "Electromagnetic Theory for Engineering Applications," John Wiley & Sons, Inc., New York, 1964.

which can be derived in the same fashion as the wave equation for a lossless medium but assuming $\sigma \neq 0$.

It will be assumed that the function $E_x(z,t)$ has a sinusoidal time variation. A closely related function which is helpful in removing the time dependence can be written as the product of a distance variation $E_x(z)$ and a time variation $e^{j\omega t}$, as shown in Eq. (2). This function is a complex variable, and can be separated into real and imaginary parts as shown:

$$E_x(z)e^{j\omega t} = \text{Re}\,[E_x(z)e^{j\omega t}] + j\,\text{Im}\,[E_x(z)e^{j\omega t}] \tag{2}$$

The function $e^{j\omega t} = \cos\omega t + j\sin\omega t$ denotes the complex sum of two monochromatic (single-frequency) sinusoids, where $\omega = 2\pi f$. Since Maxwell's equations in simple media are linear, we need only consider the monochromatic case, in which all fields variables are simply periodic in time. Then, by Fourier analysis, any linear field of arbitrary time dependence can be synthesized from a solution of the monochromatic field.

Any of the functions—$E_x(z)e^{j\omega t}$, its real part, or its imaginary part—can be substituted for $E_x(z,t)$ in Eq. (1). Each of these substitutions would reduce the damped-wave equation to an ordinary differential equation in the variable z (independent of time). Since $E_x(z,t)$ is a measurable electric field, it must exist physically, and we must choose one of the real functions Re $[E_x(z)e^{j\omega t}]$ or Im $[E_x(z)e^{j\omega t}]$ to represent $E_x(z,t)$. The choice is arbitrary, and we will hereafter use

$$E_x(z,t) = \text{Re}\,[E_x(z)e^{j\omega t}] \tag{3}$$

We first remove the time dependence in Eq. (1) by means of the substitution given in Eq. (3) and then solve the resulting ordinary differential equation. In this way we justify our assumption of the functional form of $E_x(z,t)$ given in Eq. (3).

The following derivatives which occur in Eq. (1) are obtained by substituting Re $[E_x(z)e^{j\omega t}]$ for $E_x(z,t)$:

$$\frac{\partial E_x(z,t)}{\partial t} = \text{Re}\,[j\omega E_x(z)e^{j\omega t}] \tag{4}$$

$$\frac{\partial^2 E_x(z,t)}{\partial t^2} = \text{Re}\,[-\omega^2 E_x(z)e^{j\omega t}] \tag{5}$$

$$\frac{\partial^2 E_x(z,t)}{\partial z^2} = \text{Re}\left[e^{j\omega t}\frac{\partial^2 E_x(z)}{\partial z^2}\right] \tag{6}$$

When these equations are substituted into the damped-wave equation, assuming that σ, μ, and ε are independent of frequency, we get

$$\text{Re}\left\{e^{j\omega t}\frac{\partial^2 E_x(z)}{\partial z^2}\right\} = \text{Re}\,\{e^{j\omega t}[j\omega\mu(\sigma + j\omega\varepsilon)E_x(z)]\} \tag{7}$$

PLANE-WAVE PROPAGATION

From complex-variable theory it can be shown, for the functions of interest, that if the real parts of two functions are always equal, then the two functions themselves are equal.† Equating the functions within the braces from Eq. (7), we find that the term $e^{j\omega t}$ cancels out of each side of the equation, leaving the differential equation

$$\frac{\partial^2 E_x(z)}{\partial z^2} = j\omega\mu(\sigma + j\omega\varepsilon)E_x(z) \tag{8}$$

Since the time variation has been removed, the partial derivative can be written as a total derivative, yielding the ordinary second-order constant-coefficient differential equation

$$\frac{d^2 E_x(z)}{dz^2} = \gamma^2 E_x(z) \tag{9}$$

When $E_x(z)$ is a solution of Eq. (9), Re $[E_x(z)e^{j\omega t}]$ is a solution of the damped-wave equation.

In Eq. (9), γ is called the *propagation constant*, and is a complex number

$$\gamma = (-\omega^2\mu\varepsilon + j\omega\mu\sigma)^{1/2} = \text{propagation constant} \tag{10}$$

The propagation constant can be written as the sum of its real and imaginary parts, each of which is given a special name, as follows:

$$\gamma = \alpha + j\beta$$
$$\alpha = \textit{attenuation constant}, \text{Np/m (neper/meter)}$$
$$\beta = \textit{phase constant}, \text{rad/m} \tag{11}$$

It is an elementary problem to solve the second-order differential equation given by Eq. (9), and the reader can easily show that the solution is of the form

$$E_x(z) = C_1 e^{-\gamma z} + C_2 e^{+\gamma z} \tag{12}$$

The complex function $E_x(z)$ will be called a *phasor*;‡ C_1 and C_2 are arbitrary constants determined by the boundary conditions and are generally complex numbers.

It has just been shown, by using $E_x(z)$ as in Eq. (12), that Re $[E_x(z)e^{j\omega t}]$ is a solution of the damped-wave equation, Eq. (1). Similarly it can be shown that Im $[E_x(z)e^{j\omega t}]$ and $E_x(z)e^{j\omega t}$ are solutions of the damped-wave

---

† The functions in braces in Eq. (7) must satisfy the Cauchy-Riemann conditions.

‡ A phasor is defined to be any complex number. Thus $E_x(z)$ is an electric field phasor, and $e^{j\omega t}$ is a phasor which rotates in the complex plane as a function of time [as does the product $E_x(z)e^{j\omega t}$]. However, $E_x(z,t)$ is always real and therefore is not a phasor.

equation if $E_x(z)$ has the same form as in Eq. (12). However, values of C_1 and C_2 may differ in the three cases, depending upon the boundary conditions.

To *regain* the time dependence, we multiply both sides of Eq. (12) by $e^{j\omega t}$ and then take the real part:

$$E_x(z,t) = \text{Re}\,[E_x(z)e^{j\omega t}] = \text{Re}\,[(C_1 e^{-\gamma z} + C_2 e^{+\gamma z})e^{j\omega t}] \tag{13}$$

This expression for $E_x(z,t)$ can be rewritten

$$E_x(z,t) = \text{Re}\,(C_1 e^{-\alpha z}e^{j(\omega t-\beta z)} + C_2 e^{+\alpha z}e^{j(\omega t+\beta z)}) \tag{14}$$

The first term on the right-hand side of Eq. (14) will be shown to be a positive-going damped sinusoidal waveform. The factor C_1 is a constant, as explained above. The next factor $e^{-\alpha z}$ is an exponential which causes the amplitude of that term to approach zero as z becomes very large. This exponential factor causes an attenuation (damping) of the electric field with respect to distance, and consequently α is called the *attenuation constant*. If $\alpha z = -1$, there is 1 Np of attenuation. The next factor, $e^{j(\omega t-\beta z)}$, is a sinusoid which is a function of both time and distance. By holding the phase angle constant ($\omega t - \beta z =$ constant) one can find the velocity of propagation of the sinusoid by differentiation

$$\frac{dz}{dt} = v_p = \frac{\omega}{\beta} \quad \text{m/s} \tag{15}$$

Equation (15) shows that the velocity of propagation is positive for the first term on the right-hand side of Eq. (14). Consequently, the first term represents a positive-going damped sinusoidal waveform. A similar analysis on the last term of Eq. (14) shows it to be a negative-going damped sinusoidal waveform, with amplitude decreasing as z decreases. The velocity of propagation can also be expressed as

$$v_p = f\lambda \tag{16}$$

where

$$f = \text{frequency of sinusoid, Hz} \tag{17}$$

and

$$\lambda = \text{wavelength, m} \tag{18}$$

Equating (15) and (16), we obtain another expression for the phase constant β

$$\beta = \frac{2\pi}{\lambda} \tag{19}$$

Therefore, for a change in distance z of 1 wavelength (at a fixed time), βz in Eq. (14) changes 2π rad, which is one complete cycle of the sinusoid.

PLANE-WAVE PROPAGATION

A solution similar to that given for $E_x(z,t)$ in Eq. (14) would be obtained for the other transverse field components $E_y(z,t)$, $H_x(z,t)$, and $H_y(z,t)$ of this plane wave in a lossy medium. It will be left as a problem to show that the following relations hold between the phasors of the various transverse field components of this plane wave:

$$\frac{E_x^+(z)}{H_y^+(z)} = -\frac{E_y^+(z)}{H_x^+(z)} = \left(\frac{j\omega\mu}{\sigma + j\omega\varepsilon}\right)^{1/2} = \eta \qquad (20)$$

$$\frac{E_x^-(z)}{H_y^-(z)} = -\frac{E_y^-(z)}{H_x^-(z)} = -\left(\frac{j\omega\mu}{\sigma + j\omega\varepsilon}\right)^{1/2} = -\eta \qquad (21)$$

In these equations, η reduces to the value previously obtained for the intrinsic impedance of a lossless medium if σ is zero.

SKIN EFFECT

As an example of wave propagation in a dissipative medium, we now consider the penetration of a sinusoidal plane wave into a good conductor. The plane wave propagates in free space, in a direction normal to the plane surface of a good conductor filling the half space in which $z \geq 0$. This is illustrated in Fig. 3.4.1.

The wave induces conduction and displacement currents in the conductor, as given by Maxwell's equation

$$\nabla \times \mathbf{H} = (\sigma + j\omega\varepsilon)\mathbf{E} \qquad (22)$$

We now show that $\sigma \gg \omega\varepsilon$, even in a poor metallic conductor. Nichrome, a metallic compound sometimes used for winding precision resistors, has a conductivity of approximately 10^6 mhos/m. At a frequency of 1 GHz (10^9 c/s),

$$\frac{\sigma}{\omega\varepsilon} \approx 1.8 \times 10^7 \gg 1$$

so that we can neglect the displacement current in comparison with the conduction current in Eq. (22), even for a poor metallic conductor.

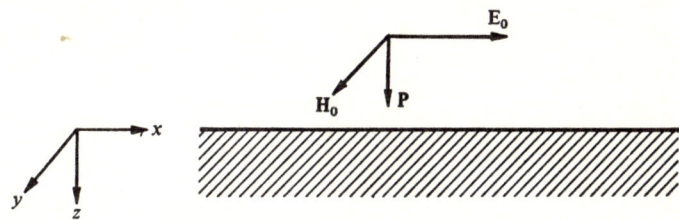

Fig. 3.4.1 A plane wave entering a good conductor.

When we use the inequality $\sigma \gg \omega\varepsilon$ in Eq. (10), γ becomes $\sqrt{j\omega\mu\sigma}$. The equation satisfied by the phasors of the transverse components (E_x and H_y, we will assume) of the plane wave in the conductor is

$$\frac{d^2 E_x(z)}{dz^2} = j\omega\mu\sigma E_x(z) = \gamma^2 E_x(z) \tag{23}$$

where

$$\gamma = \sqrt{j\omega\mu\sigma} = (1+j)\sqrt{\pi f \mu \sigma} = \frac{1+j}{\delta} \tag{24}$$

and

$$\delta = \frac{1}{\sqrt{\pi f \mu \sigma}} \quad \text{m} \tag{25}$$

In terms of γ, the solution of Eq. (23) is given by

$$E_x(z) = C_1 e^{-\gamma z} + C_2 e^{+\gamma z} \tag{26}$$

which is a special case of Eq. (12). Or, in terms of the quantity δ defined by Eq. (25),

$$E_x(z) = C_1 \exp \frac{-(1+j)z}{\delta} + C_2 \exp \frac{+(1+j)z}{\delta} \tag{27}$$

Since δ is a positive constant, C_2 must be zero; otherwise, the electric field intensity would increase exponentially to the impossible value of infinity as z approached infinity (deep in the conductor). If we let E_0 be the phasor of the electric field intensity at the surface of the conductor ($z = 0$), then

$$E_x(z) = E_0 e^{-z/\delta} e^{-jz/\delta} \tag{28}$$

Similarly,

$$H_y(z) = H_0 e^{-z/\delta} e^{-jz/\delta} \tag{29}$$

and since the current density is $\mathbf{i} = \sigma \mathbf{E}$,

$$i_x(z) = \sigma E_0 e^{-z/\delta} e^{-jz/\delta} = i_0 e^{-z/\delta} e^{-jz/\delta} \tag{30}$$

In Eqs. (29) and (30), H_0 and i_0 are the magnitudes of the magnetic field and current-density phasors at the surface.

From Eqs. (28) to (30) we see that $E_x(z)$, $H_y(z)$, and $i_x(z)$ decrease exponentially with penetration into the conductor and δ is the depth at which they have decreased to e^{-1} (about 36.9 percent) of their values at

PLANE-WAVE PROPAGATION

the surface. Thus δ is called the *skin depth* or *depth of penetration*. For a (nonexistent) perfect conductor with $\sigma = \infty$, $\delta = 0$ and all fields and currents reside only on the surface. From Eq. (25) we see that this condition would also occur if either $f = \infty$ or $\mu = \infty$. Otherwise, the fields and currents *decay exponentially with depth*.

This derivation for the skin depth of a semi-infinite plane conductor is strictly valid only for this geometry in the rectangular coordinate system. However, the resulting equations are close approximations to many practical situations. Typically, these are cases where the frequency is high enough for δ to be much less than either the conductor thickness or the surface curvature.

For example, copper has a conductivity of about 5.8×10^7 mhos/m; and if $f = 100$ MHz, δ is not quite 0.007 mm. Thus a round copper wire 1 mm in diameter would appear essentially as a semi-infinite plane conductor to a plane wave impinging on its surface. This is true whether we consider the source of the fields to be a distant source of radiation or a current in the wire. If the source of the fields is a current in the wire, the resulting cylindrical fields do not form a plane wave. However, if we consider a rectangular area on the surface having sides small with respect to the diameter, the tangential **E** field and the (circumferential) **H** field are approximately a plane wave in this small region. Therefore approximately 63.1 percent of the current flow (for a source frequency of 100 MHz) is in the outer 0.007 mm thickness of the wire.

THE COMPLEX POYNTING VECTOR AND POWER

In circuit theory, for sinusoidal time variations, complex power is a complex number obtained from the product of a phasor voltage and the conjugate of a phasor current,

$$\text{Complex power} = VI^* = \text{watts} + j(\text{vars}) \tag{31}$$

where the real part is the time-average of power, the reactive (imaginary) part is the rate of change of stored energy, and the asterisk denotes the complex conjugate of the current phasor. In the preceding discussion of skin depth we worked with phasor components of electromagnetic fields, and so it is logical to ask whether a complex-power phasor, analogous to Eq. (31), exists for fields. We will find that there is a direct analogy between the complex power of circuit theory and complex-power density for fields.

Consider an electric field vector of the form

$$\mathbf{E}(x,y,z) = E_x \mathbf{a}_x + E_y \mathbf{a}_y + E_z \mathbf{a}_z \tag{32}$$

where the components E_x, E_y, and E_z are phasors (and hence are complex numbers) which vary with distance but not with time.† Similarly, the magnetic field vector **H** is considered to be a vector whose directional components are phasors, and **H*** denotes a vector whose components are complex conjugates of the components of **H**. We define the *complex Poynting vector* ⟨**S**⟩ as

$$\langle \mathbf{S} \rangle \equiv \langle \mathbf{E}(x,y,z,t) \times \mathbf{H}(x,y,z,t) \rangle = \tfrac{1}{2} \operatorname{Re} [\mathbf{E}(x,y,z) \times \mathbf{H}^*(x,y,z)] \quad (33)$$

if $\mathbf{E}(x,y,z)$ and $\mathbf{H}(x,y,z)$ are *peak* values of the phasors. If the phasors are given in terms of *rms* values, the factor of $\tfrac{1}{2}$ is removed. We will find that ⟨**S**⟩ has complex components,

$$\langle S_k \rangle = \langle P_k \rangle + jQ_k \qquad k = x, y, \text{ or } z \quad (34)$$

where S_k is a complex-power density of the kth components analogous to the complex power of Eq. (31). The angle-bracket notation is used to denote the time average taken over one cycle, which distinguishes the complex Poynting vector from the instantaneous Poynting vector discussed in Sec. 2.5. (Note that Q_k is not a time average.)

We now study the complex Poynting vector for the special case of a positive-going wave, using the positive-going phasor

$$\mathbf{E}_x(z) = C_1 e^{-\gamma z} \mathbf{a}_x \quad (35)$$

from Eq. (12). Generally C_1 is a complex constant, independent of frequency, which we write as

$$C_1 = E_{xr} + jE_{xi} = |E_x| e^{j\theta_x} \quad (36)$$

However, the propagation constant $\gamma = \alpha + j\beta$ is also a complex constant, so that we can rewrite $\mathbf{E}_x$ as

$$\mathbf{E}_x = |E_x| e^{-\alpha z} e^{j(\theta_x - \beta z)} \mathbf{a}_x \quad (37)$$

† One is tempted to write **E** as a "complex vector" $\mathbf{E} = \mathbf{E}_r + j\mathbf{E}_i$ by writing Eq. (32) as $\mathbf{E} = (E_{xr} + jE_{xi})\mathbf{a}_x \oplus (E_{yr} + E_{yi})\mathbf{a}_y \oplus (E_{zr} + jE_{zi})\mathbf{a}_z = (E_{xr}\mathbf{a}_x \oplus E_{yr}\mathbf{a}_y \oplus E_{zr}\mathbf{a}_z) + j(E_{xi}\mathbf{a}_x \oplus E_{yi}\mathbf{a}_y \oplus E_{zi}\mathbf{a}_z)$, where $\oplus$ denotes vector addition and $+$ denotes addition in the complex-number field. However, the operations of addition and multiplication in the complex-number field are defined for complex numbers $c = a + jb$, where a and b are *real numbers*, not vectors. Consequently, one should not be surprised to encounter trouble with "complex vectors." For example, if $\mathbf{E} = \mathbf{E}_r \oplus j\mathbf{E}_i$, how does one define a polar form of the vector? Or how would one define the cross (vector) product of complex vectors? We can properly avoid such problems by considering a "complex vector" **E** to be a vector with *complex components* $E_x = E_{xr} + jE_{xi}$, $E_y = E_{yr} + jE_{yi}$, and $E_z = E_{zr} + jE_{zi}$ (where all double-subscripted terms are real numbers) and by forbidding commutativity (or interchange) of the operations of vector addition and addition of complex numbers. Although these operations were distinguished by $\oplus$ and $+$, respectively, for clarity in this footnote, it is not customary to make this distinction; hereafter, we will use $+$ to denote both types of addition.

PLANE-WAVE PROPAGATION

Similarly, we let

$$\mathbf{E}_y = |E_y|e^{-\alpha z}e^{j(\theta_y-\beta z)}\mathbf{a}_y \tag{38}$$

The intrinsic impedance η is often complex

$$\eta = |\eta|e^{j\zeta} \tag{39}$$

and can be used, with Eq. (20), to find the magnetic field vectors

$$\mathbf{H}_x = \frac{-\mathbf{E}_y}{\eta} = \frac{-|E_y|}{|\eta|}e^{-\alpha z}e^{j(\theta_y-\beta z-\zeta)}\mathbf{a}_x \tag{40}$$

$$\mathbf{H}_y = \frac{\mathbf{E}_x}{\eta} = \frac{|E_x|}{|\eta|}e^{-\alpha z}e^{j(\theta_x-\beta z-\zeta)}\mathbf{a}_y \tag{41}$$

For this plane wave, $H_z = 0$ and $E_z = 0$, and so the complex vectors needed to find $\langle\mathbf{S}\rangle$ from Eq. (33) are formed from Eqs. (37), (38), (40), and (41):

$$\mathbf{E} = \mathbf{E}_x + \mathbf{E}_y = e^{-\alpha z}e^{-j\beta z}(|E_x|e^{j\theta_x}\mathbf{a}_x + |E_y|e^{j\theta_y}\mathbf{a}_y) \tag{42}$$

and

$$\mathbf{H}^* = \mathbf{H}_x^* + \mathbf{H}_y^* = \frac{e^{-\alpha z}e^{j(\beta z+\zeta)}}{|\eta|}(-|E_y|e^{-j\theta_y}\mathbf{a}_x + |E_x|e^{-j\theta_x}\mathbf{a}_y) \tag{43}$$

We can now compute the complex Poynting vector, and we will assume that $|E_x|$ and $|E_y|$ are peak values rather than rms:

$$\langle\mathbf{S}\rangle = \tfrac{1}{2}(\mathbf{E}\times\mathbf{H}^*) = \frac{e^{-2\alpha z+j\zeta}}{2|\eta|}\begin{vmatrix} |E_x|e^{j\theta_x} & |E_y|e^{j\theta_y} & 0 \\ -|E_y|e^{-j\theta_y} & |E_x|e^{-j\theta_x} & 0 \\ \mathbf{a}_x & \mathbf{a}_y & \mathbf{a}_z \end{vmatrix} \tag{44}$$

Expanding the determinant and using Euler's identity, we obtain

$$\langle\mathbf{S}\rangle = \langle\mathbf{S}_z\rangle = e^{-2\alpha z}\frac{|E_x|^2 + |E_y|^2}{2|\eta|}(\cos\zeta + j\sin\zeta)\mathbf{a}_z \tag{45}$$

We can draw the following conclusions from Eq. (45):

1. The vector $\langle\mathbf{S}\rangle$ has only an $\mathbf{a}_z$ component for this plane wave, which coincides with the direction of propagation of the wave.
2. The vector attenuates as a function of distance because of the $e^{-2\alpha z}$ term, but it is independent of time.
3. If the intrinsic impedance is real, then $\zeta = 0$ and $\langle\mathbf{S}\rangle$ is real even though $\mathbf{E}$ and $\mathbf{H}$ have complex components.
4. If the intrinsic impedance is complex, then $\langle\mathbf{S}\rangle$ is complex

$$\langle\mathbf{S}\rangle = (\langle P_z\rangle + jQ_z)\mathbf{a}_z \tag{46}$$

and ζ, the power-factor angle, is analogous to the power-factor angle of circuit theory.

It was shown in Sec. 2.5 that the integral of the Poynting vector over a surface yields the power passing through that surface. There the vector $\mathbf{S}(z,t)$, a function of both distance and time, was interpreted as a power-density vector pointing in the direction of power flow. For sinusoidal waves, the complex Poynting vector has a similar interpretation, in terms of a complex-power density rather than real power. For the plane wave, we will show that the real part of the complex-power density $\langle \mathbf{S} \rangle$, as defined in Eq. (33), is the *time average* of the instantaneous real-power density. This derivation need not be restricted to the special case of the plane wave, but this simple example will suffice to illustrate the principle.

The instantaneous real-power-density vector can be found from the ordinary Poynting vector, defined and discussed in Sec. 2.5

$$\mathbf{S}(x,y,z,t) = \{\text{Re } [\mathbf{E}(x,y,z)e^{j\omega t}]\} \times \{\text{Re } [\mathbf{H}(x,y,z)e^{j\omega t}]\} \tag{47}$$

As usual, the operation Re means "take the real part of the complex component," and this must be done *after* multiplying each directional component or phasor, given by Eqs. (37), (38), (40), and (41), by $e^{j\omega t}$. Then we evaluate the vector product† and obtain the following result:

$$\mathbf{S}(z,t) = \frac{e^{-2\alpha z}}{|\eta|} [|E_x|^2 \cos(\omega t + \theta_x - \beta z) \cos(\omega t + \theta_x - \beta z - \zeta) \\ + |E_y|^2 \cos(\omega t + \theta_y - \beta z) \cos(\omega t + \theta_y - \beta z - \zeta)] \mathbf{a}_z \tag{48}$$

For this example, $\mathbf{E}$ and $\mathbf{H}$ are functions only of z and t, so that

$$\mathbf{S}(x,y,z,t) = \mathbf{S}(z,t)$$

A trigonometric identity

$$\cos u \cos(u - \zeta) = \tfrac{1}{2}[\cos \zeta + \cos(2u + \zeta)] \tag{49}$$

can be used to show that $\mathbf{S}(z,t)$ is composed of a time-independent term (which we will show is $\langle P_z \rangle$),

$$\langle P_z \rangle = e^{-2\alpha z} \frac{|E_x|^2 + |E_y|^2}{2|\eta|} \cos \zeta \tag{50}$$

and a time-dependent term containing the sinusoids

$$\cos(2\omega t + 2\theta_x - 2\beta z + \zeta) \quad \text{and} \quad \cos(2\omega t + 2\theta_y - 2\beta z + \zeta)$$

These sinusoids have a frequency of 2ω, which is twice that of the $\mathbf{E}$ or $\mathbf{H}$ fields. (Again, this is a direct analogy to circuit theory.) To obtain the

---

† We cannot evaluate the vector products and then take the real part since Re $(A_1 + jB_1)$ Re $(A_2 + jB_2) \neq$ Re $[(A_1 + jB_1)(A_2 + jB_2)]$.

time-average value of $\mathbf{S}(z,t)$ we integrate over one period $T = 2\pi/\omega$ and divide by that period

$$\text{Time average of } \mathbf{S}(z,t) = \frac{1}{T}\int_0^T \mathbf{S}(z,t)\, dt = \langle P \rangle \tag{51}$$

However, the average of the sinusoidal terms over one period is zero, and the average of the time-independent (constant) term is just the constant, which in our special case is given by Eq. (50). Thus the real part of the complex Poynting vector $\langle \mathbf{S}_z \rangle$, from Eq. (45), is equal to the time-average value of the power density $\langle P_z \rangle$, which was calculated using instantaneous values of $\mathbf{E}$ and $\mathbf{H}$.

In the study of guided as well as unguided waves, we will find that both the instantaneous and the complex forms of the Poynting vector are useful.†

3.5 POLARIZATION OF PLANE WAVES

Thus far, we have studied the propagation of a single plane wave in a homogeneous medium. The wave equation is linear, and therefore any plane-wave solution to it can be constructed as the sum of other plane-wave solutions. In this way, many complex electromagnetic waves can be considered as the superposition of a number of plane waves having different magnitudes, phases, and directions of propagation. This viewpoint is of value primarily as a concept, since other methods presented later are more convenient for actual analysis in most cases of interest.

There is an important practical case, however, where we consider the superposition of sinusoidal plane waves of the same frequency, all propagating in the same direction. The orientation of the field vectors in such waves is usually described by the *polarization* of the wave.

UNPOLARIZED WAVES

A number of sinusoidal plane waves, each propagating in the same direction, with arbitrary orientation of the field vectors, random phases, and random magnitudes, constitutes an *unpolarized* wave.

PLANE-POLARIZED WAVES

If the total electric field vector of the wave always lies in a given plane parallel to the direction of propagation, the wave is said to be *plane-polarized* or *linearly polarized*. The total electric field vector of the wave is the sum of all the electric field vectors for each wave, with all waves propagating in the same direction and with the same frequency. For example, this type of polarization exists when all the superposed waves

† For a complete derivation of the relationship between these two forms, see Adler, Chu, and Fano, *op. cit.*, chap. 1.

have electric fields in the same direction. The magnitudes of all of the
E and H vectors can change with time, but they cannot change direction.
Linearly polarized waves also exist when all the superposed waves have
the same time phase. In this case, the component vectors have different
directions, but they sum to a total vector which has a fixed direction and
has the same time variation as any of the component vectors. Figure
3.5.1*a* and *b* illustrates these two cases for the superposition of two waves.
A third way in which linear polarization can occur will be discussed later.
If either or both of these conditions (coincident direction or time phase)
are present for more than two waves, the total wave is also plane-polarized,
by superposition. Care must be taken not to confuse the time phase,
drawn in a phasor diagram, with the spatial vectors illustrated below.

For radio waves, it is customary to describe the polarization of a
plane-polarized wave according to the direction of the total electric
field vector. Because of the polarizing properties of most earth-based
antennas, many radio waves can be described as being either *vertically
polarized* (vertical **E** field) or *horizontally polarized* (horizontal **E** field). In
optics, however, the term "vertically polarized wave" would describe a
plane wave with a vertical *magnetic* field, rather than with a vertical
electric field. Thus it is best to avoid ambiguity by specifying which field
quantity is vertical or horizontal.

ELLIPTICALLY POLARIZED WAVES

If there are two or more plane waves of the same frequency but of different
phases, magnitudes, and orientations of the field vectors, the superposition
of these waves is an *elliptically polarized* wave. This is the most general

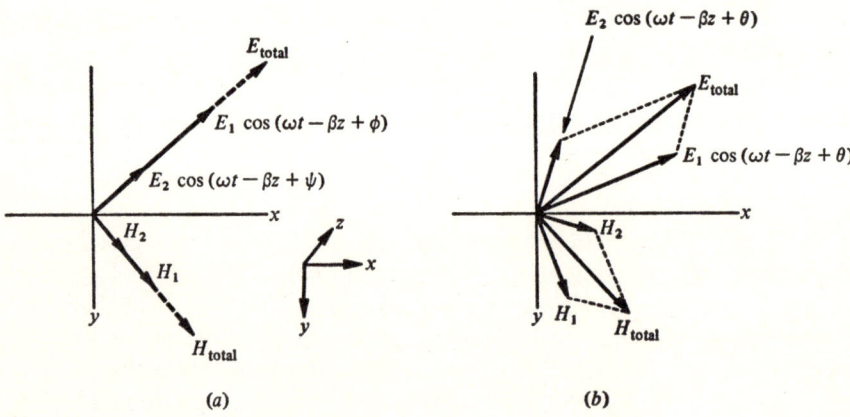

Fig. 3.5.1 (*a*) Electric fields in the same direction. (*b*) Electric fields with the same time phase.

PLANE-WAVE PROPAGATION

type of wave polarization, since the other types are degenerate cases of it. In an elliptically polarized wave, the end point of the electric field vector (and the magnetic field vector) traces out an ellipse as a function of time, in any given plane normal to the direction of propagation. This is shown in the following derivation. The sum of the x components of the electric field vectors will be our time reference, and will be written

$$E_x = E_1 \cos(\omega t - \beta z) \tag{1}$$

The sum of the y components of all the waves will then be

$$E_y = E_2 \cos(\omega t - \beta z + \theta) \tag{2}$$

where θ is the angle by which E_y leads (or lags, if θ is negative) E_x. In a given plane normal to the direction of propagation, such as the plane defined by $z = 0$, Eqs. (1) and (2) reduce to the parametric equations

$$E_x = E_1 \cos \omega t \tag{3}$$

and

$$E_y = E_2 \cos(\omega t + \theta) \tag{4}$$

respectively. These equations are the parametric equations of an ellipse, as can be seen by plotting $\mathbf{E} = E_x \mathbf{a}_x + E_y \mathbf{a}_y$ for given values of E_1, E_2, and θ or by eliminating the time t from Eqs. (3) and (4).

In Eq. (4), if $\theta = \pm 90°$, the ellipse is oriented with its major and minor axes superimposed on the x and y coordinate axes; if $E_1 > E_2$, the major axis is $2E_1$ and the minor axis is $2E_2$. Figure 3.5.2 illustrates the wave described by the following equations in the $z = 0$ plane, where we have arbitrarily assumed $E_2 = 2E_1$.

$$E_x = E_1 \cos \omega t \tag{5}$$
$$E_y = 2E_1 \cos(\omega t + 90°) \tag{6}$$

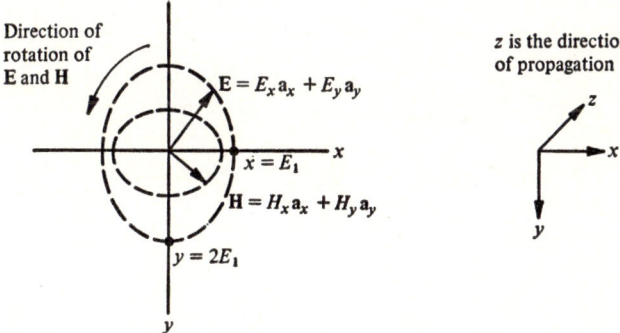

Fig. 3.5.2 An elliptically polarized wave.

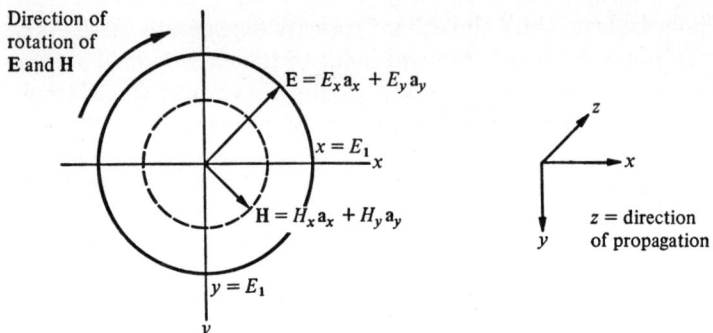

Fig. 3.5.3 A circularly polarized wave.

If $\theta \neq \pm 90°$, the major and minor axes of the ellipse do not coincide with the coordinate axes and the ellipse appears tilted. If θ is positive (negative), the electric field vector rotates counterclockwise (clockwise) in the $z = 0$ plane, looking in the direction of propagation.

CIRCULARLY POLARIZED WAVES

If there are two or more plane waves which combine so that the magnitudes of the x and y components E_1 and E_2 of the electric field are equal and one of the components leads the other by $\theta = 90°$, the wave is *circularly polarized*. Figure 3.5.3 illustrates the circularly polarized wave described by

$$E_x = E_1 \cos \omega t \tag{7}$$

and

$$E_y = E_1 \cos (\omega t - 90°) \tag{8}$$

in the $z = 0$ plane.

Because of the relationships

$$\frac{E_x}{H_y} = -\frac{E_y}{H_x} = \eta \tag{9}$$

derived earlier, it is easy to show that the magnetic field vector rotates in the same direction and at the same rate as the electric field vector. The end point of the magnetic field vector transcribes (1) an ellipse if the **E** field is elliptical, or (2) a circle if the **E** field is circular, or (3) a straight line if **E** is linearly polarized. The latter two are special cases of elliptical polarization.

3.6 VELOCITIES OF PROPAGATION

Until now, we have considered only the velocity of propagation, or phase velocity, of a single plane wave. If one traveled with the wave at the phase velocity in a lossy medium, the field strength would diminish in proportion to $e^{-\alpha z}$, where z is the direction of propagation and $\alpha = \text{Re } \gamma$, as in Sec. 3.4. From Eqs. (10) and (15) of Sec. 3.4, repeated here for convenience, it is seen that α and β are, in general, functions of frequency:

$$\gamma = \sqrt{-\omega^2\mu\varepsilon + j\omega\mu\sigma} = \alpha + j\beta \tag{1}$$

and

$$v_p = \frac{\omega}{\beta} \tag{2}$$

In the lossless case, where $\sigma = 0$,

$$\gamma = j\beta = j\omega\sqrt{\mu\varepsilon} \tag{3}$$

and

$$v_p = \frac{\omega}{\beta} = \frac{1}{\sqrt{\mu\varepsilon}} \tag{4}$$

Thus, in a lossless medium where μ and ε are independent of frequency, the velocity of propagation is the same whether we are considering a transient wave or a sinusoidal wave. That is, the Fourier components of a wave with a complicated shape all travel at the same velocity in this distortionless medium, thus preserving the original waveshape.

In dissipative media, however, $\alpha \neq 0$, and α and β are functions of frequency. In addition, μ, ε, and σ generally vary slightly with frequency, which is most noticeable when a wide frequency range is being considered. For many cases, the result is that β does not increase linearly with frequency, so that $v_p = \omega/\beta$ is also a function of frequency. Thus the Fourier components of a transient (nonsinusoidal) wave would all travel at different velocities, and the waveshape of the total electric (or magnetic) field would be continually changing. Because of the varying phase velocity, such a medium is said to be *dispersive*. A signal propagating in a dispersive medium thus arrives at its destination somewhat distorted.

Theoretically, at least, if we could plot α versus ω and β versus ω for a homogeneous medium, we could reconstruct the signal that was sent from the signal that was actually received, thereby eliminating the distortion. Practically, however, the medium is often inhomogeneous (like the atmosphere), and noise in any transmitter or receiver makes accurate regeneration of the original signal difficult.

GROUP VELOCITY

The *group velocity* v_g is often convenient for describing some characteristics of a dispersive medium. We now discuss the propagation of a wave which will lead to the definition of group velocity.

Let us consider the traveling plane wave containing two slightly different frequencies, $\omega_0 - d\omega$ and $\omega_0 + d\omega$, in which

$$E_x = E_0\{\cos[(\omega_0 - d\omega)t - (\beta_0 - d\beta)z] \\ + \cos[(\omega_0 + d\omega)t - (\beta_0 + d\beta)z]\} \quad (5)$$

This is a wave traveling in the positive z direction, in which β is a function of frequency. Equation (5) can be rewritten

$$E_x = E_0\{\cos[(\omega_0 t - \beta_0 z) - (d\omega\, t - d\beta\, z)] \\ + \cos[(\omega_0 t - \beta_0 z) + (d\omega\, t - d\beta\, z)]\} \quad (6)$$

A trigonometric identity reduces Eq. (6) to

$$E_x = 2E_0 \cos(d\omega\, t - d\beta\, z) \cos(\omega_0 t - \beta_0 z) \quad (7)$$

Equation (7) shows that the electric field intensity of this wave (and similarly the magnetic field) can be interpreted as a high-frequency wave whose amplitude varies at the low-frequency rate of $d\omega$. At any particular plane determined by $z = $ constant, Eq. (7) is simply a wave of frequency ω_0, with amplitude modulation at a lower frequency $d\omega$. The envelope of the wave is then

$$2E_0 \cos(d\omega\, t - d\beta\, z) \quad (8)$$

Similar envelopes are used to represent the information, or signal, being sent by means of amplitude modulation. This envelope is a sinusoidal traveling wave, and to ascertain the velocity with which any part of the envelope travels, we set

$$d\omega\, t - d\beta\, z = \text{const}$$

Thus

$$v_g = \frac{dz}{dt} = \frac{d\omega}{d\beta} = \text{group velocity} \quad (9)$$

which is the velocity of an observer who stays on the same point of the envelope. Thus v_g is the velocity of the modulation of the wave. Since the phase velocity $v_p = \omega/\beta$, it will be left for a problem to show that the group velocity can be written

$$v_g = \frac{v_p}{1 - (\omega/v_p)(dv_p/d\omega)} \quad (10)$$

For a more complex signal, made up of many sinusoidal components, the group velocity closely approximates the velocity of the composite envelope, providing there is relatively small dispersion over the frequency band required to describe the signal. For these cases, the group velocity

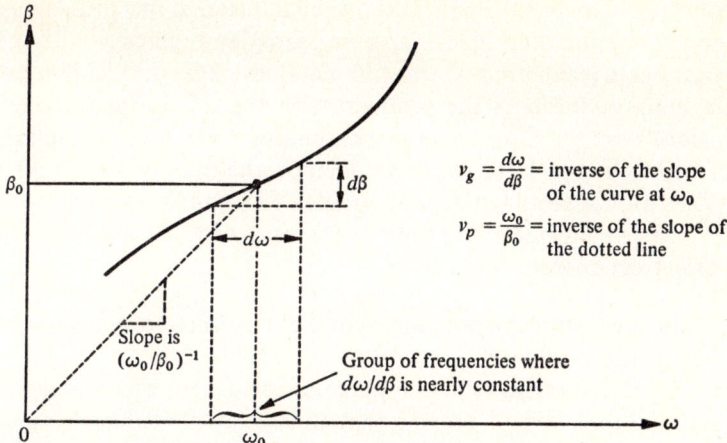

Fig. 3.6.1 Phase and group velocities from a plot of β versus ω.

can be used as the signal velocity, which is the velocity at which the information travels.

If dispersion is large over the frequency range of interest, however, the composite waveshape is constantly being distorted as it travels. Because of this, a "signal velocity" is difficult, if not impossible, to define. The meaning of group velocity, as given in Eq. (9) or (10), is ambiguous in these cases, since $d\omega/d\beta$ may not be a constant throughout the frequency range of interest. If β is plotted as a function of ω for a particular medium, both the phase velocity v_p and the group velocity v_g can be found at a particular frequency ω_0, as shown in Fig. 3.6.1.

For a wave in a nondispersive medium, the phase velocity does not vary with frequency, and we see from Eq. (10) that the phase and group velocities are equal.†

3.7 A PLANE WAVE INCIDENT ON A PERFECT CONDUCTOR

We now consider a single-frequency uniform plane wave propagating in a lossless dielectric filling the space $z < 0$ and incident on a perfect conductor located at $z = 0$. If the conductor is perfect, the fields will not penetrate to the interior and the conductor can be arbitrarily thin. Since the fields and the conductor must be infinite in at least two directions, it may appear that this is not a practical problem. However, there are numerous cases

---

† For a more complete discussion of the velocities of propagation, the interested reader should refer to J. A. Stratton, "Electromagnetic Theory," McGraw-Hill Book Company, New York, 1941, or L. Brillouin, "Wave Propagation and Group Velocity," Academic Press Inc., New York, 1960.

which are closely approximated by the solution to this problem, and it also serves as a building block for more complex problems. For example, a radar beam incident on a large flat metallic conductor yields approximately the same solution as the problem with the infinite perfect conductor in regions near the finite metallic conductor. The more complex problems, to be studied in later chapters, include the field distributions in coaxial cables and in rectangular and circular waveguides and cavities.

It is convenient to study three cases of a plane wave incident on a perfect conductor:

1. The incident wave propagates in the $+z$ direction, normal to the surface of the conductor.
2. The electric field is in the plane of incidence, which is determined by the direction of propagation and a normal to the surface of the conductor. This is the first case of oblique incidence to be considered.
3. The electric field is perpendicular to the plane of incidence, the second case of oblique incidence to be considered.

Oblique incidence is divided into the two cases because the solutions are easier to obtain. This does not restrict the problems which can be solved, however, since any plane wave obliquely incident on a perfect conductor can be separated into the vector sum of two plane waves, one having its electric field in the plane of incidence and the other having its electric field normal to the plane of incidence. Since the wave equation is linear, we can find the fields due to each component wave separately and then add the results to obtain the total field distribution. In this manner, we can find the field distribution due to a uniform plane wave incident at any angle on a perfect conductor.

NORMAL INCIDENCE

First, however, we consider case 1, in which the electric and magnetic fields are in a plane parallel to the conductor, so that the direction of propagation is normal to the surface of the conductor. We will assume that the incident plane wave has electric field E_x^+ in the x direction, and magnetic field H_y^+ in the y direction, as shown in Fig. 3.7.1, and that both are single-frequency sinusoids. Furthermore, we will assume that the electric field in the incident wave is given by the phasor

$$E_x^+ = E_0 e^{-j\beta z} \tag{1}$$

so that

$$H_y^+ = \frac{E_0}{\eta} e^{-j\beta z} \tag{2}$$

Thus the incident wave is a uniform plane wave.

PLANE-WAVE PROPAGATION

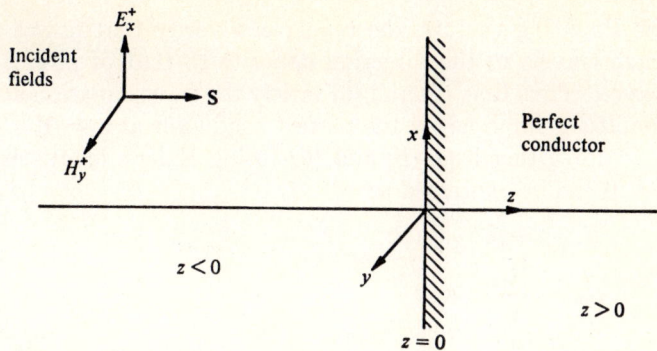

Fig. 3.7.1 A plane wave normally incident on a perfect conductor.

The plane wave defined by Eqs. (1) and (2) does not satisfy the boundary conditions at the surface of the perfect conductor, however, because the electric field is not zero (in these equations) for all time at $z = 0$. The solution of the wave equation consists of the sum of a positive-going wave and a negative-going wave. The positive-going **E** field and **H** field are the incident waves given in Eqs. (1) and (2). The negative-going waves are reflections of the incident fields and are necessary to satisfy the boundary conditions at the perfect conductor. These boundary conditions require that the electric field be zero at $z = 0$ for all time, so that

$$\mathbf{E}_{tot} = \mathbf{E}_x = \mathbf{E}_x{}^+ + \mathbf{E}_x{}^- = 0 \qquad \text{at } z = 0 \tag{3}$$

where $\mathbf{E}_x{}^-$ denotes the reflected electric field. Inserting Eq. (1) into Eq. (3), we conclude that the magnitudes of $\mathbf{E}_x{}^+$ and $\mathbf{E}_x{}^-$ are equal but their directions are opposite. Therefore

$$\mathbf{E}_x{}^- = -E_0 e^{+j\beta z} \mathbf{a}_x \tag{4}$$

The reflected **H** field is found from the relation derived in Sec. 3.3 for negative-going waves,

$$\frac{E_x{}^-}{H_y{}^-} = -\eta$$

Thus the total fields in the region $z < 0$ are given by

$$E_x = E_x{}^+ + E_x{}^- = E_0 e^{-j\beta z} - E_0 e^{+j\beta z} \tag{5}$$

$$H_y = H_y{}^+ + H_y{}^- = \frac{E_0}{\eta} e^{-j\beta z} + \frac{E_0}{\eta} e^{+j\beta z} \tag{6}$$

The existence of the reflected wave is also verified from the viewpoint of energy. Since no energy can enter the perfect conductor, the energy carried by the incident wave must be returned by a reflected wave. From the Poynting vectors, the reader can verify that power in the reflected wave travels in the direction opposite to that of power in the incident wave. Thus Eqs. (5) and (6) satisfy the wave equation and the boundary conditions imposed by the perfect conductor at $z = 0$.

Rewriting Eqs. (5) and (6) (using Euler's identities), the fields for $z < 0$ can be expressed as

$$E_x = -j2E_0 \sin \beta z \tag{7}$$

$$H_y = \frac{2E_0}{\eta} \cos \beta z \tag{8}$$

These equations are phasors of the field quantities and are functions of distance only. To regain the time dependence (as discussed in Sec. 3.4), we multiply by $e^{j\omega t}$ and take the real part

$$E_x(z,t) = \text{Re } (-j2E_0 \sin \beta z \, e^{j\omega t}) \tag{9}$$

$$H_y(z,t) = \text{Re } \left(\frac{2E_0}{\eta} \cos \beta z \, e^{j\omega t}\right) \tag{10}$$

The phasors of Eqs. (7) and (8) are *standing waves*, which are said to exist if two (or more) sinusoidal traveling waves of the same frequency are traveling in opposite directions. The phasor of the standing wave is the sum of the phasors of the traveling waves, which is evident from Eqs. (5) and (6) for this example. Therefore the standing-wave phasor is a function of distance only, as were the phasors of the individual traveling waves of which the standing wave is composed.

Equations (7) and (8) reveal three types of quadrature:

1. E_x and H_y are perpendicular, as discussed in Sec. 3.3.
2. E_x and H_y have a spatial difference of 90°, or $\lambda/4$, between peaks and nulls, as illustrated in the standing-wave patterns shown in Fig. 3.7.2.

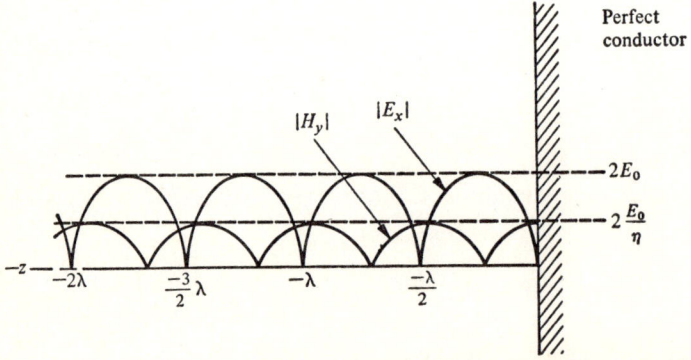

Fig. 3.7.2 Standing-wave patterns showing $|E_x|$ and $|H_y|$ for $z < 0$.

PLANE-WAVE PROPAGATION

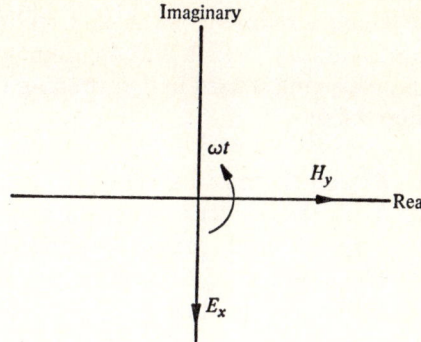

Fig. 3.7.3 A phasor diagram showing the time relationship of E_x and H_y.

3. E_x lags H_y by 90° in time, in any plane z = constant, as indicated by the phasor diagram in Fig. 3.7.3.

OBLIQUE INCIDENCE

E field in the plane of incidence We now consider a plane wave obliquely incident on a perfect conductor, polarized with the electric field in the plane of incidence, as illustrated in Fig. 3.7.4. The plane of incidence is defined by the direction of propagation and a normal to the surface and coincides with the plane of the paper in Fig. 3.7.4.

The direction of propagation of the incident wave, denoted the ζ direction, makes an angle θ with the normal to the surface (or with the z direction). Since energy cannot pass into the perfect conductor, there

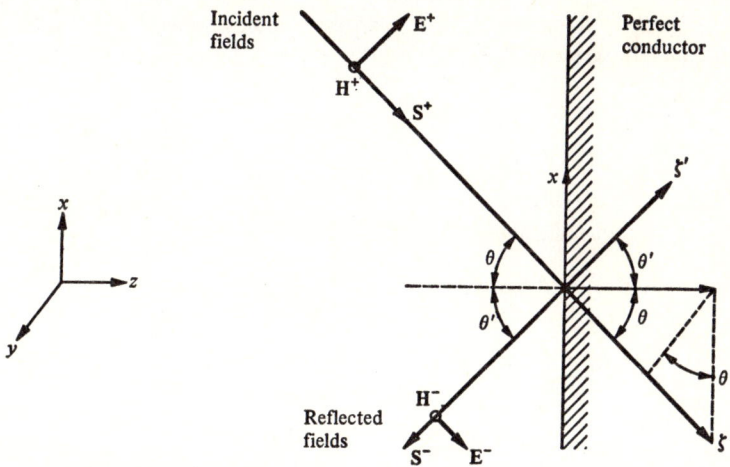

Fig. 3.7.4 Oblique incidence on a perfect conductor, with the E field in the plane of incidence.

must be a reflected wave, and we label the negative of its direction of propagation ζ', which is at an unknown angle θ' in the figure. The following reasoning is used in determining the directions of $\mathbf{E}^-$ and $\mathbf{H}^-$ shown in Fig. 3.7.4:

1. $\mathbf{E}^+$ and $\mathbf{E}^-$ must cancel at the surface of the perfect conductor.
2. As in the incident wave, the reflected fields $\mathbf{E}^-$ and $\mathbf{H}^-$ must be perpendicular to each other and to the direction in which they propagate.
3. The direction of $\mathbf{E}^- \times \mathbf{H}^-$ must coincide with the $-\zeta'$ direction, away from the conductor.

Furthermore, with field directions as shown in Fig. 3.7.4,

$$\frac{\mathbf{E}^+}{\mathbf{H}^+} = \frac{\mathbf{E}^-}{\mathbf{H}^-} = \eta \qquad (11)$$

The phasor for the total electric field at any point in the region $z < 0$ can be written

$$\mathbf{E}(x,z) = \mathbf{E}^+ e^{-j\beta\zeta} + \mathbf{E}^- e^{+j\beta\zeta'} \qquad (12)$$

Here, $\mathbf{E}^+$ and $\mathbf{E}^-$ are reference values of incident and reflected waves respectively, at $\zeta = 0 = \zeta'$, which coincides with the y axis of our rectangular coordinate system (with x, y, and z axes).

It is usually desirable to express all fields equations in terms of the coordinate system most naturally fitting the boundaries, rather than a coordinate system oriented with one axis in the direction of propagation. For this problem, we can write equations in terms of x, y, and z by means of the following transformations, obtained from Fig. 3.7.4:

$$\zeta = -x \sin \theta + z \cos \theta \qquad (13)$$
$$\zeta' = x \sin \theta' + z \cos \theta' \qquad (14)$$

We will now rewrite Eq. (12), separating the wave into its x and z components and substituting from Eqs. (13) and (14) for ζ and ζ',

$$E_x(x,z) = E^+ \cos \theta \exp[-j\beta(-x \sin \theta + z \cos \theta)] \\ - E^- \cos \theta' \exp[+j\beta(x \sin \theta' + z \cos \theta')] \qquad (15)$$

$$E_z(x,z) = E^+ \sin \theta \exp[-j\beta(-x \sin \theta + z \cos \theta)] \\ + E^- \sin \theta' \exp[+j\beta(x \sin \theta' + z \cos \theta')] \qquad (16)$$

The total magnetic field is

$$H_y(x,z) = H^+ \exp[-j\beta(-x \sin \theta + z \cos \theta)] \\ + H^- \exp[+j\beta(x \sin \theta' + z \cos \theta')] \qquad (17)$$

PLANE-WAVE PROPAGATION

We can now apply the boundary condition at the perfect conductor

$$E_x(x,z) = 0 \quad \text{at } z = 0 \text{ for all } x \tag{18}$$

or

$$E_x(x,0) = E^+ \cos\theta \exp(+j\beta x \sin\theta) \\ - E^- \cos\theta' \exp(+j\beta x \sin\theta') = 0 \tag{19}$$

For this equation to be satisfied for all values of x, the exponential terms must factor out, requiring that

$$\sin\theta = \sin\theta' \tag{20}$$

or

$$\theta = \theta'$$

Thus the angle of reflection is equal to the angle of incidence. With this result substituted into Eq. (19), we obtain

$$E^+ = E^- \tag{21}$$

so that the reflected amplitude is equal to the incident amplitude of the E field. Applying Euler's identities and Eqs. (20) and (21) to Eqs. (15), (16), and (17), respectively, we obtain the field components

$$E_x(x,z) = -j2E^+ \cos\theta \sin(\beta z \cos\theta) \exp(+j\beta x \sin\theta) \tag{22}$$

$$E_z(x,z) = 2E^+ \sin\theta \cos(\beta z \cos\theta) \exp(+j\beta x \sin\theta) \tag{23}$$

$$H_y(x,z) = \frac{2E^+}{\eta} \cos(\beta z \cos\theta) \exp(+j\beta x \sin\theta) \tag{24}$$

This field distribution is a traveling wave with respect to the x direction, since the phase velocity in this direction is

$$v_{px} = \frac{-\omega}{\beta \sin\theta} \tag{25}$$

In the z direction, however, the field is a standing wave, since E_x is zero for all time at the surface of the conductor and at regularly spaced planes normal to the z axis, where z satisfies

$$\sin(\beta z \cos\theta) = 0 \tag{26}$$

or

$$z = \frac{-m\lambda}{2 \cos\theta} = -md \quad m = 0, 1, 2, 3, \ldots$$

The magnitude of the E_z phasor is a maximum in planes which are odd multiples of $d/2$ from the conductor. E_z and H_y are zero where E_x is maximum, are maximum where E_x is zero, and lead E_x by 90° in time phase at any point. It is interesting to note that the distance d between

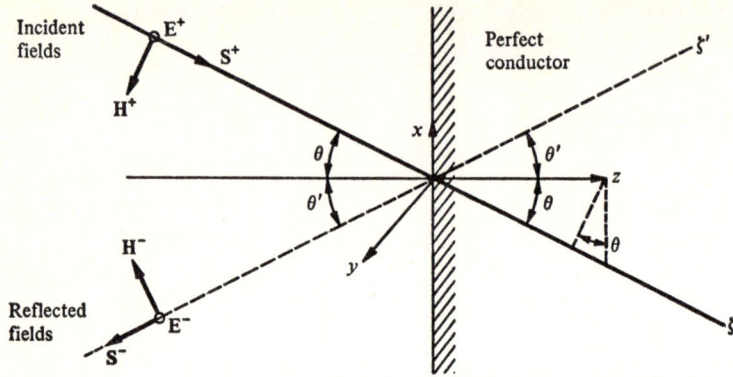

Fig. 3.7.5 Oblique incidence on a perfect conductor, with the **H** field in the plane of incidence.

successive minima, measured along the z axis, becomes greater as the angle of incidence becomes more and more oblique (θ increases). If θ is zero, however, Eqs. (22) to (24) reduce to the equations obtained previously for a normally incident wave.

E field normal to the plane of incidence For this type of polarization, **E**$^+$ and **E**$^-$ are normal to the plane of the paper (as shown in Fig. 3.7.5), and we will assume that **H**$^+$ and **H**$^-$ are in the directions illustrated in the figure. With field directions as shown, $E^+/H^+ = \eta = E^-/H^-$, and we will show that **E**$^-$ is actually in the opposite direction to **E**$^+$. Proceeding as before, we write the field components as functions of the x and z coordinates

$$E_y(x,z) = E^+ \exp\left[-j\beta(-x\sin\theta + z\cos\theta)\right] \\ + E^- \exp\left[+j\beta(x\sin\theta' + z\cos\theta')\right] \quad (27)$$

$$\eta H_x(x,z) = -E^+ \cos\theta \exp\left[-j\beta(-x\sin\theta + z\cos\theta)\right] \\ + E^- \cos\theta' \exp\left[+j\beta(x\sin\theta' + z\cos\theta')\right] \quad (28)$$

$$\eta H_z(x,z) = -E^+ \sin\theta \exp\left[-j\beta(-x\sin\theta + z\cos\theta)\right] \\ - E^- \sin\theta' \exp\left[+j\beta(x\sin\theta' + z\cos\theta')\right] \quad (29)$$

Applying the boundary condition at the perfectly conducting plane, we see that E_y must be zero at $z = 0$ for all x. As before, the exponents must be equal so that they can be factored out, and this requires that $\theta = \theta'$. In turn, this condition requires that $E^+ = -E^-$. Using Euler's identities once more, the field components of Eqs. (27) to (29) can be rewritten

$$E_y = -j2E^+ \sin(\beta z \cos\theta) \exp(+j\beta x \sin\theta) \quad (30)$$

PLANE-WAVE PROPAGATION

$$H_x = \frac{-2E^+}{\eta} \cos\theta \cos(\beta z \cos\theta) \exp(+j\beta x \sin\theta) \tag{31}$$

$$H_z = \frac{j2E^+}{\eta} \sin\theta \sin(\beta z \cos\theta) \exp(+j\beta x \sin\theta) \tag{32}$$

These equations have the combined characteristics of a wave traveling in the $-x$ direction and a standing wave in the z direction (for $z < 0$), as obtained previously for polarization with the E field in the plane of incidence. Here, E_y and H_z have zeros and H_x has maxima at the conducting plane and at equally spaced parallel planes md meters apart, with d as given in Eq. (26).

We now conclude our discussion of reflection from an infinite, perfectly conducting plane, since we can use superposition to express any obliquely incident wave as the sum of the oblique waves discussed in this section.

3.8 A PLANE WAVE INCIDENT ON A PERFECT DIELECTRIC

In this section, we consider a single-frequency uniform plane wave propagating in a lossless medium (a perfect dielectric) which fills the space $z < 0$ and which has an intrinsic impedance $\eta_1 = \sqrt{\mu_1/\varepsilon_1}$. The space $z > 0$ is filled with a perfect dielectric which has an intrinsic impedance $\eta_2 = \sqrt{\mu_2/\varepsilon_2}$, and we will solve the following three cases:

1. The plane wave is normally incident upon the boundary at $z = 0$.
2. The plane wave is obliquely incident upon the boundary, with E in the plane of incidence.
3. The plane wave is obliquely incident upon the boundary, with E normal to the plane of incidence.

These three cases are sufficient to completely characterize the problem of a uniform plane wave incident on a plane dielectric boundary, since we can use superposition to solve the following additional problems:

1. A plane wave propagating in the $+z$ direction, with the following parallel-plane dielectric boundaries:

   For $z < 0$: $\quad \eta = \eta_1 = \sqrt{\dfrac{\mu_1}{\varepsilon_1}}$

   For $0 < z < z_1$: $\quad \eta = \eta_2 = \sqrt{\dfrac{\mu_2}{\varepsilon_2}}$

   For $z_1 < z$: $\quad \eta = \eta_3 = \sqrt{\dfrac{\mu_3}{\varepsilon_3}}$

2. A plane wave obliquely incident upon a dielectric boundary, with **E** neither in the plane of incidence nor normal to the plane of incidence.
3. The plane wave obliquely incident (as in 2) upon the parallel-plane dielectric boundaries described in 1.

We could, of course, similarly solve problems with four (or more) different dielectric media if the three (or more) boundary planes separating different media are parallel.

NORMAL INCIDENCE

We will assume sinusoidal time variations, and we will also assume that the incident wave has the following field vectors, given in phasor form:

$$\mathbf{E}_i = E_i e^{-j\beta_1 z} \mathbf{a}_x \tag{1}$$

$$\mathbf{H}_i = H_i e^{-j\beta_1 z} \mathbf{a}_y = \frac{E_i}{\eta_1} e^{-j\beta_1 z} \mathbf{a}_y \tag{2}$$

The problem is to find the phasors of the reflected fields $\mathbf{E}_R$ and $\mathbf{H}_R$ and the field components $\mathbf{E}_T$ and $\mathbf{H}_T$ which are transmitted into the second medium, as illustrated in Fig. 3.8.1. Since there are now *two* positive-going waves, we use subscripts i, R, and T to denote the incident, reflected, and transmitted waves, respectively.

Since the tangential electric and magnetic fields must be continuous at the boundary $z = 0$, we can write the transmitted fields as the sum of the incident and reflected fields

$$\mathbf{E}_T = \mathbf{E}_i + \mathbf{E}_R \tag{3}$$

$$\mathbf{H}_T = \mathbf{H}_i + \mathbf{H}_R \tag{4}$$

Although these are vector equations, it can be shown that $\mathbf{E}_R$ and $\mathbf{E}_T$ have components only in the x direction (as does $\mathbf{E}_i$) and $\mathbf{H}_R$ and $\mathbf{H}_T$ have components only in the y direction (as does $\mathbf{H}_i$). To prove this, we can assume that $\mathbf{E}_R$ or $\mathbf{E}_T$ has a nonzero y component. We then arrive at the contradictory result that either (1) $\mathbf{E}_R$ is not perpendicular to $\mathbf{H}_R$ or (2) $\mathbf{E}_T$ is not perpendicular to $\mathbf{H}_T$. Thus the directions of the field components as shown in Fig. 3.8.1 are correct. The details of this derivation will be left for a problem.

As a result of the preceding derivation, the reflected and transmitted field vectors can be written

$$\mathbf{E}_R = E_R e^{+j\beta_1 z} \mathbf{a}_x \tag{5}$$

$$\mathbf{H}_R = H_R e^{+j\beta_1 z} \mathbf{a}_y = \frac{-E_R}{\eta_1} e^{+j\beta_1 z} \mathbf{a}_y \tag{6}$$

$$\mathbf{E}_T = E_T e^{-j\beta_2 z} \mathbf{a}_x \tag{7}$$

$$\mathbf{H}_T = H_T e^{-j\beta_2 z} \mathbf{a}_y = \frac{E_T}{\eta_2} e^{-j\beta_2 z} \mathbf{a}_y \tag{8}$$

PLANE-WAVE PROPAGATION

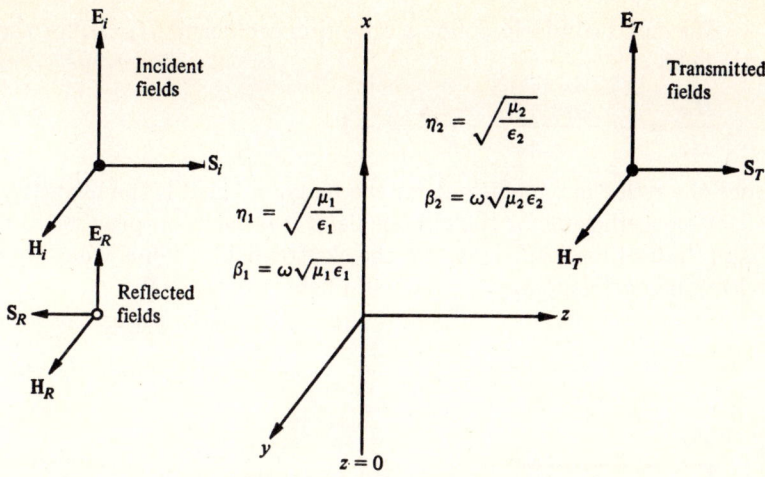

Fig. 3.8.1 Reflection and transmission at a dielectric boundary.

It should be noted from these equations, and from Fig. 3.8.1, that β differs in the two media. The boundary conditions at the dielectric interface $z = 0$, given in Eqs. (3) and (4), simplify to scalar equations, since all **E** field components are in the x direction and all **H** field components are in the y direction

$$E_T = E_i + E_R \tag{9}$$
$$H_T = H_i + H_R \tag{10}$$

We now find E_R and E_T in terms of E_i by means of the following derivation. Just to the right of the dielectric boundary at $z = 0$, and with field directions as shown in Fig. 3.8.1,

$$\eta_2 = \frac{E_T}{H_T} = \frac{E_i + E_R}{H_i + H_R} = \frac{E_i + E_R}{E_i/\eta_1 - E_R/\eta_1} \tag{11}$$

Dividing numerator and denominator by E_i, we obtain

$$\eta_2 = \frac{\eta_1(1 + E_R/E_i)}{1 - E_R/E_i} \tag{12}$$

Solving for the ratio E_R/E_i,

$$\boxed{\frac{E_R}{E_i} = \frac{\eta_2 - \eta_1}{\eta_2 + \eta_1} = \Gamma} \tag{13}$$

This ratio is called the *reflection coefficient*, and is denoted by Γ.

We could similarly define a reflection coefficient H_R/H_i for the magnetic field.

$$\frac{H_R}{H_i} = -\frac{\eta_2 - \eta_1}{\eta_2 + \eta_1} = \frac{-E_R}{E_i} = -\Gamma \tag{14}$$

Since the reflection coefficient for the magnetic field is the negative of the reflection coefficient for the electric field, it is common practice to reserve Γ and "reflection coefficient" for the electric field. Thus when we need a reflection coefficient for the magnetic field, we use $-\Gamma$.

It is convenient to define a *transmission coefficient* as the ratio

$$\tau = \frac{E_T}{E_i} = \frac{E_i + E_R}{E_i} = 1 + \frac{E_R}{E_i} \tag{15}$$

so that

$$\boxed{\tau = \frac{E_T}{E_i} = 1 + \Gamma = \frac{2\eta_2}{\eta_2 + \eta_1}} \tag{16}$$

We can now write the reflected and transmitted fields in terms of the incident fields.

$$\mathbf{E}_R = \Gamma E_i e^{+j\beta_1 z} \mathbf{a}_x \tag{17}$$

$$\mathbf{H}_R = -\Gamma H_i e^{+j\beta_1 z} \mathbf{a}_y = \frac{-\Gamma E_i}{\eta_1} e^{+j\beta_1 z} \mathbf{a}_y \tag{18}$$

$$\mathbf{E}_T = \tau E_i e^{-j\beta_2 z} \mathbf{a}_x \tag{19}$$

$$\mathbf{H}_T = \frac{\tau E_i}{\eta_2} e^{-j\beta_2 z} \mathbf{a}_y \tag{20}$$

Here Γ and τ are constants determined by the two dielectric media, as given in Eqs. (13) and (16). Thus we have solved the problem of a plane wave normally incident on a dielectric interface.

OBLIQUE INCIDENCE

E field in the plane of incidence We now consider a plane wave obliquely incident on a boundary between two lossless dielectrics. As we found for oblique incidence on a perfect conductor, this polarization is one of two cases which completely characterize oblique incidence on a dielectric boundary. The other type of polarization, which we will consider soon, is with the **E** field normal to the plane of incidence. Then, to solve a problem in which an arbitrarily polarized plane wave is obliquely incident on a dielectric boundary, we can express the plane wave as the sum of two plane waves. One of these two plane waves is polarized with the **E** field in the plane of incidence, and the other is polarized with the **E** field normal to the plane of incidence. Thus the vector sum of these two waves yields

PLANE-WAVE PROPAGATION

the original arbitrarily polarized plane wave. We now consider in detail the case in which the **E** field is in the plane of incidence, as shown in Fig. 3.8.2.

We will assume that the fields in the incident wave, which propagates in the ζ direction (Fig. 3.8.2), are known and are given by

$$\mathbf{E}_i = E_i e^{-j\beta_1 \zeta}(\cos\theta\, \mathbf{a}_x + \sin\theta\, \mathbf{a}_z) \tag{21}$$

$$\mathbf{H}_i = H_i e^{-j\beta_1 \zeta} \mathbf{a}_y = \frac{E_i}{\eta_1} e^{-j\beta_1 \zeta} \mathbf{a}_y \tag{22}$$

The problem is to find the fields in the reflected and transmitted waves in terms of the incident fields. From Fig. 3.8.2, the reflected and transmitted fields are

$$\mathbf{E}_R = E_R e^{+j\beta_1 \zeta'}(-\cos\theta'\, \mathbf{a}_x + \sin\theta'\, \mathbf{a}_z) \tag{23}$$

$$\mathbf{H}_R = H_R e^{+j\beta_1 \zeta'} \mathbf{a}_y = \frac{E_R}{\eta_1} e^{+j\beta_1 \zeta'} \mathbf{a}_y \tag{24}$$

$$\mathbf{E}_T = E_T e^{-j\beta_2 \zeta''}(\cos\theta''\, \mathbf{a}_x + \sin\theta''\, \mathbf{a}_z) \tag{25}$$

$$\mathbf{H}_T = H_T e^{-j\beta_2 \zeta''} \mathbf{a}_y = \frac{E_T}{\eta_2} e^{-j\beta_2 \zeta''} \mathbf{a}_y \tag{26}$$

In these equations, θ' and θ'' must be expressed in terms of the angle of incidence θ and the constants of the media, and E_R and E_T must be expressed in terms of E_i and the constants of the media. With field directions as shown in Fig. 3.8.2, we note that $E_R = +\eta_1 H_R$, as given in Eq. (24).

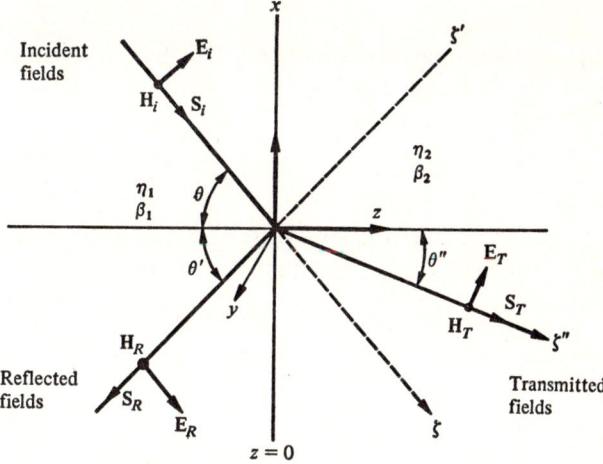

Fig. 3.8.2 Oblique incidence on a dielectric boundary, with the **E** field in the plane of incidence.

From Fig. 3.8.2 we see that ζ, ζ', and ζ'' are related to x and z as follows:

$$\zeta = -x \sin \theta + z \cos \theta \tag{27}$$

$$\zeta' = x \sin \theta' + z \cos \theta' \tag{28}$$

$$\zeta'' = -x \sin \theta'' + z \cos \theta'' \tag{29}$$

Inserting these values in Eqs. (21) to (26), we now satisfy the boundary condition which requires that the tangential **E** field be continuous at a boundary. We therefore equate the x component of $\mathbf{E}_i + \mathbf{E}_R$ evaluated at $z = 0$ to the x component of $\mathbf{E}_T$ evaluated at $z = 0$

$$E_i \cos \theta \exp[-j\beta_1(-x \sin \theta)] - E_R \cos \theta' \exp[+j\beta_1(x \sin \theta')]$$
$$= E_T \cos \theta'' \exp[-j\beta_2(-x \sin \theta'')] \tag{30}$$

If Eq. (30) is to be valid for all values of x, the exponential terms must be equal (and therefore cancel) for all values of x. Thus

$$\beta_1 \sin \theta = \beta_1 \sin \theta' = \beta_2 \sin \theta'' \tag{31}$$

The first equality states that

$$\theta = \theta' \tag{32}$$

so that the angle of reflection is equal to the angle of incidence. The second equality is Snell's law, which is probably more familiar written in the form (using $\beta_j = \omega/v_j$ and canceling ω)

$$\frac{\sin \theta}{v_1} = \frac{\sin \theta''}{v_2} \tag{33}$$

We can use Eqs. (32) and (33) to simplify Eq. (30) (which equates tangential components of the electric field at the boundary) and to write a similar equation expressing the continuity of the magnetic field at the boundary. Rewriting Eq. (30),

$$E_i \cos \theta - E_R \cos \theta = E_T \cos \theta'' = E_T \left(1 - \frac{v_2^2}{v_1^2} \sin^2 \theta\right)^{1/2} \tag{34}$$

Equating tangential components of the magnetic field at $z = 0$ and canceling exponentials as in Eq. (30), we obtain

$$H_i + H_R = H_T$$

which can be written (from Fig. 3.8.2)

$$\frac{E_i}{\eta_1} + \frac{E_R}{\eta_1} = \frac{E_T}{\eta_2} \tag{35}$$

From Eqs. (34) and (35) we can find both E_R and E_T as functions of either E_i or H_i. It is customary to use the incident field component which

PLANE-WAVE PROPAGATION

is parallel to the dielectric boundary as the reference (H_i in this case), so that

$$E_R = \eta_1 H_i \left(\frac{\eta_1 \cos\theta - \eta_2 \cos\theta''}{\eta_1 \cos\theta + \eta_2 \cos\theta''} \right) = \eta_1 H_i \Gamma_m \tag{36}$$

$$E_T = \eta_2 H_i \left(\frac{2\eta_1 \cos\theta}{\eta_1 \cos\theta + \eta_2 \cos\theta''} \right) = \eta_2 H_i \tau_m \tag{37}$$

The term in parentheses in Eq. (36) can be thought of as the *reflection coefficient for a magnetic field with incident angle θ*, which we will abbreviate Γ_m. The term in parentheses in Eq. (37) can be thought of as the *transmission coefficient for a magnetic field with incident angle θ*, abbreviated τ_m. These expressions are valid only for **E** in the plane of incidence. From Eq. (33), if $\theta = 0$, then $\theta'' = 0$. For these values, Γ_m and τ_m in Eqs. (36) and (37) reduce to expressions for the reflection coefficient and transmission coefficient, respectively, for a magnetic field normally incident on a dielectric boundary. Using Eqs. (36) and (37), Eqs. (23) to (26) can be rewritten in terms of H_i:

$$\mathbf{E}_R = \eta_1 H_i \Gamma_m \exp\left[j\beta_1(x\sin\theta + z\cos\theta)\right](-\cos\theta\, \mathbf{a}_x + \sin\theta\, \mathbf{a}_z) \tag{38}$$

$$\mathbf{H}_R = H_i \Gamma_m \exp\left[j\beta_1(x\sin\theta + z\cos\theta)\right]\mathbf{a}_y \tag{39}$$

$$\mathbf{E}_T = \eta_2 H_i \tau_m \exp\left[-j\beta_2(-x\sin\theta'' + z\cos\theta'')\right](\cos\theta''\, \mathbf{a}_x + \sin\theta''\, \mathbf{a}_z) \tag{40}$$

$$\mathbf{H}_T = H_i \tau_m \exp\left[-j\beta_2(-x\sin\theta'' + z\cos\theta'')\right]\mathbf{a}_y \tag{41}$$

where

$$\sin\theta'' = \frac{v_2}{v_1}\sin\theta = \frac{\beta_1}{\beta_2}\sin\theta \tag{42}$$

We have thus succeeded in describing the reflected and transmitted traveling waves in terms of the incident wave.

We can demonstrate the existence of standing waves in the region $z < 0$ by writing the total **E** and **H** fields as the sum of the respective incident and reflected fields. The total fields in this region have the characteristics of standing waves in the z direction and traveling waves in the x direction, namely

$$\mathbf{E} = \eta_1 H_i \exp(+j\beta_1 x \sin\theta)\{[\exp(-j\beta_1 z \cos\theta) \\ - \Gamma_m \exp(+j\beta_1 z \cos\theta)]\cos\theta\, \mathbf{a}_x + [\exp(-j\beta_1 z \cos\theta) \\ + \Gamma_m \exp(+j\beta_1 z \cos\theta)]\sin\theta\, \mathbf{a}_z\} \tag{43}$$

$$\mathbf{H} = H_i \exp(+j\beta_1 x \sin\theta)[\exp(-j\beta_1 z \cos\theta) \\ + \Gamma_m \exp(+j\beta_1 z \cos\theta)]\mathbf{a}_y \tag{44}$$

The total fields in the region $z > 0$, described by Eqs. (40) and (41), are traveling waves, since waves traveling in opposite directions are required to form a standing wave.

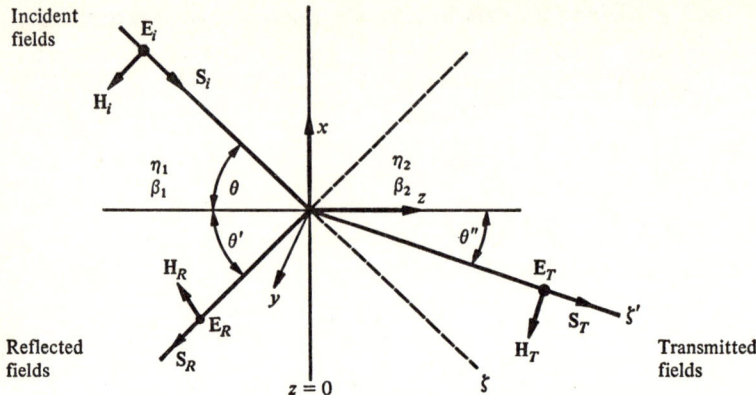

Fig. 3.8.3 Oblique incidence on a dielectric boundary, with the **E** field normal to the plane of incidence.

E field normal to the plane of incidence For this type of polarization, the fields $\mathbf{E}_i$, $\mathbf{E}_R$, and $\mathbf{E}_T$ are normal to the plane of the paper (as shown in Fig. 3.8.3), and we will assume that the corresponding magnetic fields are in the directions illustrated in the figure. We will find the reflected and transmitted fields in terms of the incident electric field. Proceeding as we did in the previous case, we write the field components as functions of the x and z coordinates, using the field directions shown in Fig. 3.8.3,

$$\mathbf{E}_i = E_i \exp\left[-j\beta_1(-x\sin\theta + z\cos\theta)\right]\mathbf{a}_y \tag{45}$$

$$\mathbf{H}_i = \frac{E_i}{\eta_1} \exp\left[-j\beta_1(-x\sin\theta + z\cos\theta)\right](-\cos\theta\,\mathbf{a}_x - \sin\theta\,\mathbf{a}_z) \tag{46}$$

$$\mathbf{E}_R = E_R \exp\left[+j\beta_1(x\sin\theta' + z\cos\theta')\right]\mathbf{a}_y \tag{47}$$

$$\mathbf{H}_R = \frac{E_R}{\eta_1} \exp\left[+j\beta_1(x\sin\theta' + z\cos\theta')\right](\cos\theta'\,\mathbf{a}_x - \sin\theta'\,\mathbf{a}_z) \tag{48}$$

$$\mathbf{E}_T = E_T \exp\left[-j\beta_2(-x\sin\theta'' + z\cos\theta'')\right]\mathbf{a}_y \tag{49}$$

$$\mathbf{H}_T = \frac{E_T}{\eta_2} \exp\left[-j\beta_2(-x\sin\theta'' + z\cos\theta'')\right](-\cos\theta''\,\mathbf{a}_x - \sin\theta''\,\mathbf{a}_z) \tag{50}$$

We now apply the boundary condition which requires the tangential components of the total electric field to be continuous at the boundary $z = 0$

$$E_i \exp\left[-j\beta_1(-x\sin\theta)\right] + E_R \exp\left[+j\beta_1(x\sin\theta')\right] \\ = E_T \exp\left[-j\beta_2(-x\sin\theta'')\right] \tag{51}$$

PLANE-WAVE PROPAGATION

As in the previous case of oblique incidence, the exponentials must be equal if Eq. (51) is to be true for all values of x. Therefore

$$\theta = \theta' \quad \text{and} \quad \beta_1 \sin \theta = \beta_2 \sin \theta'' \tag{52}$$

as before. Thus Eq. (51) reduces to

$$E_i + E_R = E_T \tag{53}$$

which can be written

$$\eta_1 H_i + \eta_1 H_R = \eta_2 H_T \tag{54}$$

The second boundary condition requires that the tangential magnetic fields be continuous at $z = 0$. Equating the x components of the total **H** field on each side of the boundary, setting $z = 0$, and canceling exponents, we obtain

$$-H_i \cos \theta + H_R \cos \theta = -H_T \cos \theta''$$

$$= -H_T \left(1 - \frac{v_2^2}{v_1^2} \sin^2 \theta \right)^{\frac{1}{2}} \tag{55}$$

From Eqs. (53) to (55) we can express E_R and E_T in terms of E_i, which is parallel to the dielectric boundary

$$E_R = \left(\frac{\eta_2 \cos \theta - \eta_1 \cos \theta''}{\eta_2 \cos \theta + \eta_1 \cos \theta''} \right) E_i = \Gamma_e E_i \tag{56}$$

$$E_T = \left(\frac{2\eta_2 \cos \theta}{\eta_2 \cos \theta + \eta_1 \cos \theta''} \right) E_i = \tau_e E_i \tag{57}$$

The term in parentheses in Eq. (56) can be thought of as the *reflection coefficient for an electric field with incident angle* θ, which we will denote by Γ_e. The term in parentheses in Eq. (57) can be thought of as the *transmission coefficient for an electric field with incident angle* θ, denoted τ_e. These expressions are valid only for the **E** field normal to the plane of incidence. Again, if $\theta = 0$, then $\theta'' = 0$, and Γ_e and τ_e reduce to the reflection and transmission coefficients, respectively, for an electric field normally incident on a dielectric boundary. Using Eqs. (52), (56), and (57), we can rewrite Eqs. (47) to (50) in terms of E_i:

$$\mathbf{E}_R = \Gamma_e E_i \exp\left[+j\beta_1(x \sin \theta + z \cos \theta)\right]\mathbf{a}_y \tag{58}$$

$$\mathbf{H}_R = \frac{\Gamma_e E_i}{\eta_1} \exp\left[+j\beta_1(x \sin \theta + z \cos \theta)\right](\cos \theta \, \mathbf{a}_x - \sin \theta \, \mathbf{a}_z) \tag{59}$$

$$\mathbf{E}_T = \tau_e E_i \exp\left[-j\beta_2(-x \sin \theta'' + z \cos \theta'')\right]\mathbf{a}_y \tag{60}$$

$$\mathbf{H}_T = \frac{\tau_e E_i}{\eta_2} \exp\left[-j\beta_2(-x \sin \theta'' + z \cos \theta'')\right]$$

$$(-\cos \theta'' \, \mathbf{a}_x - \sin \theta'' \, \mathbf{a}_z) \tag{61}$$

where

$$\sin \theta'' = \frac{v_2}{v_1} \sin \theta = \frac{\beta_1}{\beta_2} \sin \theta$$

As in the previous case, we have succeeded in describing the reflected and transmitted waves in terms of the incident wave. Again, the total fields in the region $z < 0$ have the characteristics of standing waves in the z direction and traveling waves in the x direction:

$$\mathbf{E} = E_i \exp(+j\beta_1 x \sin \theta)[\exp(-j\beta_1 z \cos \theta) + \Gamma_e \exp(+j\beta_1 z \cos \theta)]\mathbf{a}_y \quad (62)$$

$$\mathbf{H} = \frac{-E_i}{\eta_1} \exp(+j\beta_1 x \sin \theta)\{[\exp(-j\beta_1 z \cos \theta) - \Gamma_e \exp(+j\beta_1 z \cos \theta)] \cos \theta \, \mathbf{a}_x + [\exp(-j\beta_1 z \cos \theta) + \Gamma_e \exp(+j\beta_1 z \cos \theta)] \sin \theta \, \mathbf{a}_z\} \quad (63)$$

The total fields in the region $z > 0$ are the traveling waves described by Eqs. (60) and (61).

PROGRAMMED EXERCISE 3.8.1

UNDERSEA COMMUNICATION

An interesting practical problem is electromagnetic communication between submerged submarines, or between a submerged submarine and a receiving and transmitting station located above the surface of the sea.† We will consider a portion of the problem of transmitting a plane wave (this is an idealization) between a submerged submarine and a station above the surface. We will assume the media and coordinate system shown in Fig. 3.8.4.

We will assume that a plane wave at a frequency of 10 kHz, emanating from an antenna on the submarine, has field components $\mathbf{E}_i$ (in the plane of incidence) and $\mathbf{H}_i$. The assumed directions for the reflected and transmitted fields and the corresponding Poynting vectors are shown in the figure. The reader should verify each of the following true statements:

1. $\gamma_1 = 0.396 + j0.396$ (in the sea). $\lambda_1 = 15.8$ m (in the sea). $\lambda_2 = 3 \times 10^4$ m (in the air).
2. In seawater, skin depth is $\delta = 2.52$ m, and displacement current can be neglected in comparison with conduction current.
3. Attenuation of the field components in seawater is 8.68 dB/δ, or 86.8 dB for 25.2 m.
4. $\eta_1 = 0.099 + j0.099$ (in the sea). $\eta_2 = 377$ (in the air).

† This topic is discussed in more detail in Richard K. Moore, Communication in the Sea, *IEEE Spectrum*, November 1967.

PLANE-WAVE PROPAGATION

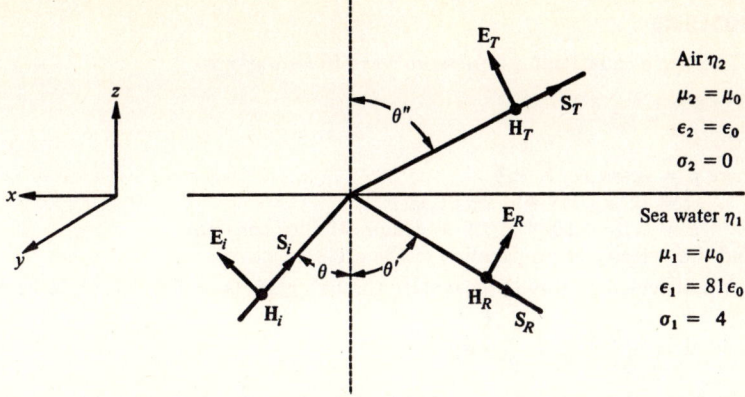

Fig. 3.8.4 A plane wave incident on the boundary between the sea and the air.

5. By applying the boundary conditions, we see that $\theta = \theta'$ and $\sin \theta'' = 1897 \sin \theta$. Therefore if the angle of incidence θ is greater than about 0.528 mrad, all energy is reflected back into the sea.
6. The fraction of power per unit area transmitted from the sea into the air

$$\frac{|\mathbf{S}_T|}{|\mathbf{S}_i|} = \frac{|\eta_1|}{\eta_2} = 3.71 \times 10^{-4}$$

does not include any attenuation due to propagation in the sea. Note that η_1 and $\mathbf{S}_i$ are complex because the sea is a conductor at 10 kHz.

From these calculations, we can conclude that seawater is a very poor medium for electromagnetic communication over distances normally considered short. Thus the path taken by the energy transmitted from one submarine and received by a distant submarine is:

1. Upward through the sea, within a cone having a total apex angle of 1.056 mrad.
2. Horizontally through the air, across the surface of the sea.
3. Downward through the sea, within a cone having a total apex angle of 1.056 mrad.

Although this up-over-and-down transmission path is very lossy, especially if the submarines are deeply submerged, losses are far less than for a direct path completely under water. The reader will probably find it interesting to investigate the advantages and disadvantages of using frequencies other than 10 kHz.

PROBLEMS

1. The wave equation for a pressure wave in seawater is

$$\frac{\partial^2 p}{\partial z^2} = \rho_v k \frac{\partial^2 p}{\partial t^2}$$

where p = pressure, N/m²
$\rho_v = 10^3$ kg/m³ = mass density
$k = 4.78 \times 10^{-10}$ m²/N = compressibility constant
Find the velocity of propagation of the pressure wave.

2. A positive-going wave propagating through glass ($\epsilon_r = 5.0$) has an **H** field

$$\mathbf{H} = \frac{2\sqrt{5}}{377}\mathbf{a}_x + \frac{\sqrt{5}}{377}\mathbf{a}_y$$

at a given point and instant. Find the corresponding **E** field at the same point and the same time.

3. Show that $U_E = U_H$ for a single negative-going wave.

4. For any given traveling wave, the energy density of the electric field U_E is the same as the energy density U_H of the magnetic field, so that $U_E - U_H = 0$. However, if we superimpose a negative-going wave and a positive-going wave and compare the energy densities of the total fields, we find that

$$U_E - U_H = \Delta U \neq 0$$

This shows that superposition is not valid for energy densities. Find ΔU for waves whose components lie in the xy plane.

5. From Maxwell's equations for a lossy medium, derive Eq. (1) of Sec. 3.4 for the curl of **H**.

6. The following constants were obtained for seawater at a frequency of $\omega = 4 \times 10^9$ rad/s:

$\mu_r \simeq 1.0 \qquad \epsilon_r \simeq 81 \qquad \sigma \simeq 4.36$

where $\mu_r = \mu/\mu_0$ and $\epsilon_r = \epsilon/\epsilon_0$. Find the attenuation constant.

7. Show that Eqs. (5) and (6) of Sec. 3.4 are valid if $E_x(z,t) = \text{Re}\,[E_x(z)e^{j\omega t}]$, even if $E_x(z)$ is complex.

8. Show that the expression for the intrinsic impedance of a lossy medium is given by Eq. (20) of Sec. 3.4.

9. We showed that a function of the form $E_x(z)e^{j\omega t}$ is a solution of the damped wave equation, provided $E_x(z)$ satisfies the ordinary differential equation $d^2E_x(z)/dz^2 = \gamma^2 E_x(z)$. This was a very special case because of the sinusoidal time function. Now assume a solution of the form $E_x(z)F(t)$, where $F(t)$ is an arbitrary time function. Find the differential equations which must be satisfied by $E_x(z)$ and $F(t)$.

10. Derive Eq. (22) in Sec. 3.4 for the curl of **H** with sinusoidal time variations.

11. With the following three assumptions, show that Eq. (23) of Sec. 3.4 for $E_x(z)$ in a conductor is obtained: (a) sinusoidal time variations; (b) $\sigma \gg \omega\epsilon$; (c) $\partial/\partial x = \partial/\partial y = 0$.

12. A source generates an **E** and an **H** field, and

$$\mathbf{E} = E_0 \cos\theta \, \mathbf{a}_x + E_0 \sin\theta \, \mathbf{a}_y$$

where $\theta = \omega(\sqrt{\mu\epsilon}\,z - t)$. Find (a) the **H** vector, (b) the direction and magnitude of power flow per unit area, (c) the type of polarization, and (d) the direction of rotation of the **E** and **H** vectors in a transverse plane.

PLANE-WAVE PROPAGATION

13. An elliptically polarized plane wave has the following E field components:

$$E_x = E_0 \cos(\omega t - \beta z) \qquad E_y = \sqrt{2}\, E_0 \cos(\omega t - \beta z - 45°) \qquad E_z = 0$$

Find the angle between the major axis of the ellipse and the x axis in the $z = 0$ plane.

14. Given $E_x = E_1 \cos \omega t$ and $E_y = 2E_1 \cos(\omega t + 90°)$ at the plane $z = 0$ and that the direction of propagation is a_z, derive (a) the H field vector; (b) the equation of the ellipse formed by the tip of the H field vector; (c) the direction of rotation of the H field vector.

15. Derive Eq. (10) in Sec. 3.6 for the group velocity from Eqs. (4) and (9).

16. For a particular medium, β is given approximately by $10^{-12}\omega^2 - 2 \times 10^{-8}\omega$ throughout the frequency range of 10^8 to 10^9 Hz. Find the phase velocity and the group velocity at each end of this frequency band.

17. Derive Eqs. (22) to (24) in Sec. 3.7 for the total fields obtained by reflection from a perfect conductor.

18. For a plane wave incident on a perfect dielectric, show that the transmitted and reflected E fields in Eq. (3) of Sec. 3.8 have vector components only in the x direction and that the transmitted and reflected H fields of Eq. (4) of Sec. 3.8 have vector components only in the y direction.

19. Derive (a) the reflection and (b) the transmission coefficients for a magnetic field at a dielectric boundary.

20. Show that the transmission coefficient τ for a plane wave transmitted normally from medium 1 to medium 2 is

$$\tau = \frac{2\sqrt{\epsilon_2}}{\sqrt{\epsilon_1} + \sqrt{\epsilon_2}}$$

where $\mu_1 = \mu_2$.

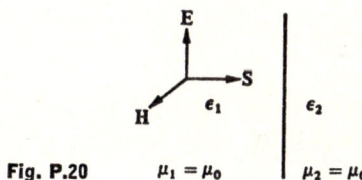

Fig. P.20

21. Suppose we are free to specify ϵ_2 of a medium to be sandwiched between fixed media. Find ϵ_2 to maximize the product $\tau_{12}\tau_{23}$ of the transmission coefficients. Is the product affected by the thickness of medium 2?

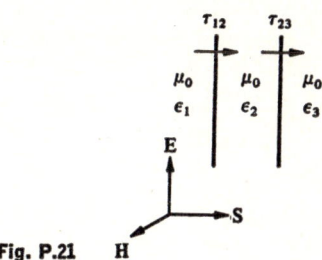

Fig. P.21

22. Let S_i be the Poynting vector of a sinusoidal plane wave normally incident on the surface of the sea from above the water. Let S_R and S_T be the Poynting vectors of the reflected and transmitted waves respectively. Find S_R and S_T in terms of S_i if $\sigma = 4$ and $\varepsilon_r = 81$.

23. A plane wave is obliquely incident on a dielectric boundary. (a) With **E** in the plane of incidence and $\mu_1 = \mu_2$, find the angle at which all energy is transmitted into the second medium even though $\eta_1 \neq \eta_2$. This angle is called the *Brewster angle*. Is it possible to find a similar angle if $\varepsilon_1 = \varepsilon_2$ but $\eta_1 \neq \eta_2$? (b) With **E** normal to the plane of incidence and $\varepsilon_1 = \varepsilon_2$, find the angle at which all energy is transmitted into the second medium. Can you find a similar angle if $\mu_1 = \mu_2$ but $\eta_1 \neq \eta_2$?

4
Transient Waves on Lossless Transmission Lines

4.1 COAXIAL TRANSMISSION LINE

Now we consider a case where an electromagnetic wave is constrained to travel down the inside of a conducting cylinder or tube in which a conducting rod concentric to the cylinder runs the full length of the cylinder, as shown in Fig. 4.1.1. It is assumed that the region between the rod and the cylinder is either free space or is filled with a dielectric medium. Conductors arranged with this geometry form the so-called *coaxial transmission line*. It will be assumed that at the far end of the transmission line the conducting rod and the cylinder are connected by a resistance.

Suppose that we close the switch of the coaxial transmission line of Fig. 4.1.1, thereby applying a voltage to the end of the coaxial line. Several questions immediately arise as a result of closing the switch. First, does current flow on the transmission line? And if current flows, how long will it be before an observer at the receiving end of the line observes the effect of the disturbance caused by closing the switch at the sending end? It would be reasonable to expect current to flow away from the battery on the conducting rod. Is this current flow continuous? For example, when the disturbance has traveled halfway down the line, how does the current get from the conducting rod to the conducting cylinder? If current flows, the battery supplies energy to the system, and some of this energy will be lost in the resistance at the receiving end. Can the transmission line store energy, and if so, how is the energy distributed along the transmission line? Rigorous answers to these questions will require the use of mathematical models; however, a brief qualitative description of the phenomena may be worthwhile as a guide through the mathematical maze.

When the switch is closed in Fig. 4.1.1, the conducting rod will be made positive with respect to the conducting cylinder because of the battery potential. Current will flow from the battery as a result of the charge redistribution, in much the same way that current flows from a battery when a capacitor is charged according to conventional circuit theory. An **H** field will encircle the conductor carrying the current, according to Ampere's law. Similarly, an **E** field will be established from the positive conducting rod to the negative conducting cylinder, as shown in Fig. 4.1.1b. Taking the vector product **E** × **H** according to Poynting's theorem, it is seen that the Poynting vector points down the transmission line, parallel to the center conducting rod. Thus we expect power (and consequently energy) to flow into the transmission line.

From Chap. 2 recall that the derivation of the wave equation does not depend on the particular geometry in which the fields exist. Thus the electromagnetic fields within a coaxial transmission line propagate down the line according to the wave equation, and closing the switch simply establishes an electromagnetic disturbance at the sending end of the line, in much the same way that the pebble dropped in water establishes a mechanical displacement initiating wave action. The disturbance appears as a transverse electromagnetic (TEM) wave, which propagates

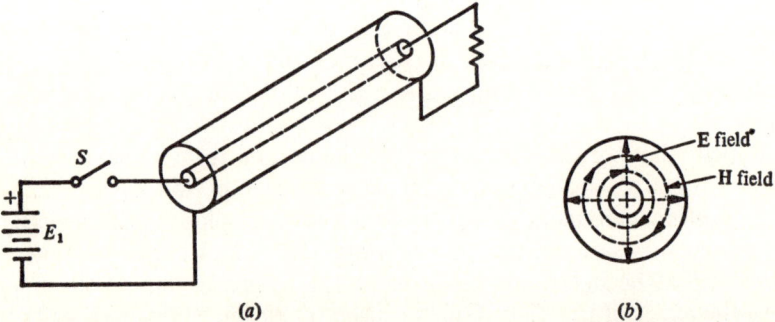

Fig. 4.1.1 The coaxial transmission line.

TRANSIENT WAVES ON LOSSLESS TRANSMISSION LINES

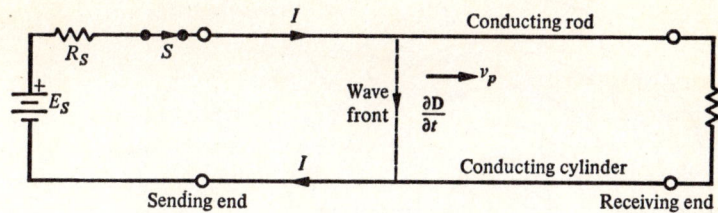

Fig. 4.1.2 Simplified drawing of the transmission line.

down the transmission line as a nonuniform plane wave. In this case the TEM wave is a step function, due to the manner in which energy was applied from the source, and an observer partway down the line would see this wavefront pass at the velocity of propagation v_p. As a result of this step function in the electromagnetic fields, an observer would detect a thin sheet of displacement-current density $\partial \mathbf{D}/\partial t$ flowing from the conducting rod to the conducting cylinder at the wavefront. One can trace a complete path for the current starting from the battery, then down the conducting rod in the center, through the displacement current, to the outer conducting cylinder, and back to the battery. This is shown in Fig. 4.1.2, which also illustrates a simplified way of drawing two-conductor transmission lines of different geometries. Because of the dc source, the electromagnetic fields in the region between the conductors will be constant after the wavefront has passed. In the following sections, we will proceed with a mathematical analysis of transient phenomena by first deriving the relationships between voltage, current, and the fields on a transmission line.

4.2 THE WAVE EQUATIONS FOR VOLTAGE AND CURRENT

The object of this section is to show that voltage and current on the transmission line are described by the wave equation. In elementary field studies it is shown that the static electromagnetic fields on a coaxial transmission line are distributed as sketched in Fig. 4.2.1. In this figure we assume a direct current into the paper in the center conductor and an equal current out of the paper in the outer conductor. Furthermore, we will assume that the conductors are perfect. The **E** field is radial from the center conductor and has a vector component only in the r direction. (A cylindrical coordinate system is assumed here, with the positive z direction into the paper.) It will be assumed that the time-varying fields have only the components E_r and H_ϕ, as did the static fields, and that they can vary with both time and distance along the transmission line.†

---

† For further discussion of this assumption, see R. B. Adler, L. J. Chu, and R. M. Fano, "Electromagnetic Energy Transmission and Radiation," sec. 2.2.2, John Wiley & Sons, Inc., New York, 1960.

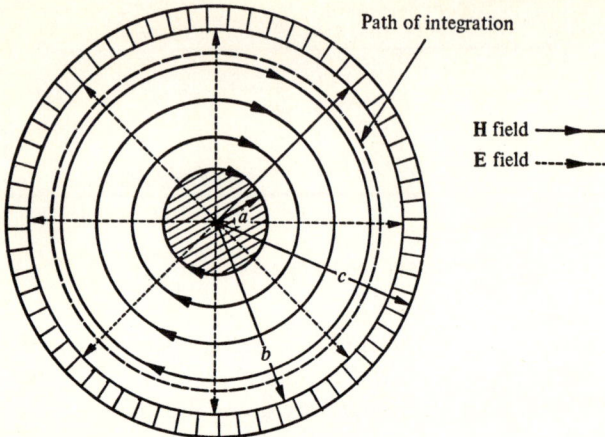

Fig. 4.2.1 Fields on the coaxial line.

We are thus assuming the functional forms

$$\mathbf{E} = E_r(r,z,t)\mathbf{a}_r \qquad \mathbf{H} = H_\phi(r,z,t)\mathbf{a}_\phi \tag{1}$$

From Fig. 4.2.1, the Poynting vector $\mathbf{S} = \mathbf{E} \times \mathbf{H}$ is in the positive z direction, so that power is flowing into the page. Thus the possibility exists of transmitting power down the transmission line with this electromagnetic field configuration. The derivation of the relationship between the **H** field and the current flowing in the center conductor begins with Maxwell's equation for the curl of **H** in the region between the conductors

$$\nabla \times \mathbf{H} = \varepsilon \frac{\partial E_r}{\partial t} \mathbf{a}_r \tag{2}$$

Conduction-current density **J** is zero, since $\sigma = 0$ in this region. The curl of **H** in cylindrical coordinates has the form

$$\nabla \times \mathbf{H} = \left(\frac{1}{r}\frac{\partial H_z}{\partial \phi} - \frac{\partial H_\phi}{\partial z}\right)\mathbf{a}_r + \left(\frac{\partial H_r}{\partial z} - \frac{\partial H_z}{\partial r}\right)\mathbf{a}_\phi$$
$$+ \frac{1}{r}\left(\frac{\partial r H_\phi}{\partial r} - \frac{\partial H_r}{\partial \phi}\right)\mathbf{a}_z \tag{3}$$

Since the right-hand side of Eq. (2) has a vector component only in the $\mathbf{a}_r$ direction, all components of the curl of **H** must be zero except the radial components. If we substitute Eqs. (1) into Eq. (3), we find that the following condition must be satisfied in order to have a circular **H** field as shown in Fig. 4.2.1:

$$\frac{\partial r H_\phi}{\partial r} = 0 \tag{4}$$

TRANSIENT WAVES ON LOSSLESS TRANSMISSION LINES

Using the product form of the derivative, an ordinary differential equation† for the **H** field is easily obtained and solved by separating variables

$$r\frac{dH_\phi}{dr} + H_\phi = 0 \tag{5}$$

$$H_\phi(r,z,t) = \frac{c_1(z,t)}{r} \tag{6}$$

Equation (6) shows that a circular **H** field can exist if the magnitude of **H** varies inversely with the radial distance. [Note that $c_1(z,t)$ must be a constant with respect to the variable r.] To relate c_1 to the current that flows in the center conductor, we integrate the **H** field about a circular path which is concentric to the center conductor and has a radius greater than a but less than b (see Fig. 4.2.1). According to Maxwell's equation, the line integral around the path equals the sum of the currents piercing any surface bounded by the path. For convenience in integration, we will use the surface defined by the plane of the paper, i.e.,

$$\oint_0^{2\pi} \mathbf{H} \cdot (r\,d\phi\,\mathbf{a}_\phi) = I(z,t) + \int \varepsilon \frac{\partial \mathbf{E}}{\partial t} \cdot (ds\,\mathbf{a}_z) \qquad a < r < b \tag{7}$$

The substitution of Eqs. (1) and (6) into Eq. (7) allows us to evaluate c_1. Since $\mathbf{a}_r \cdot \mathbf{a}_z = 0$, the displacement current is zero for the surface chosen

$$\oint_0^{2\pi} c_1\,d\phi = I(z,t) \tag{8}$$

$$c_1 = \frac{I(z,t)}{2\pi} \tag{9}$$

In these equations the current $I(z,t)$ is the conduction current which flows into the page in the center conductor, or in the positive z direction. In conclusion, the **H** field between conductors is related to the conduction current as follows:

$$H_\phi(r,z,t) = \frac{I(z,t)}{2\pi r} \tag{10}$$

By a similar development we can find the relationship between the **E** field and the applied voltage. Into Maxwell's equation

$$\nabla \times \mathbf{E} = -\frac{\partial \mathbf{B}}{\partial t} \tag{11}$$

† Given the total differential

$$df(r,z,t) = \frac{\partial f}{\partial r}dr + \frac{\partial f}{\partial z}dz + \frac{\partial f}{\partial t}dt$$

we can evaluate the partial derivative $\partial f/\partial r$ by dividing by dr. Note that dz/dr and dt/dr are zero because the variables are independent.

we substitute Eqs. (1) and (10)

$$\frac{\partial E_r}{\partial z}\mathbf{a}_\phi - \frac{1}{r}\frac{\partial E_r}{\partial \phi}\mathbf{a}_z = -\frac{\mu}{2\pi r}\frac{\partial I(z,t)}{\partial t}\mathbf{a}_\phi \tag{12}$$

Since no $\mathbf{a}_z$ component exists on the right-hand side of Eq. (12), the $\mathbf{a}_z$ component on the left must be zero. However, since we have assumed that E is not a function of ϕ, this condition gives us no new information. Next we integrate Eq. (12) with respect to z to obtain the functional form for E_r

$$E_r = -\frac{\mu}{2\pi r}\int \frac{\partial I(z,t)}{\partial t}dz = -\frac{\mu c_2(z,t)}{2\pi r} \tag{13}$$

By integrating E_r along a radial line from the outer conductor to the inner one, we can find the voltage $V(z,t) = V_{ab}$ between the two points, in terms of $c_2(z,t)$, and substitute it into Eq. (13). The result is

$$E_r(r,z,t) = \frac{V(z,t)}{r\ln(b/a)} \tag{14}$$

Thus the **E** and **H** fields [of Eqs. (10) and (14)] have the same variation in the r direction as the static fields.

The next objective is to relate the conduction current to the voltage that exists between the center conductor and the outer conductor. Substituting Eqs. (1) and (4) into the equation for the curl of **H** [Eq. (2)], we obtain a partial differential equation relating the **H** field and the **E** field

$$\boxed{\frac{\partial H_\phi}{\partial z} = -\varepsilon\frac{\partial E_r}{\partial t}} \tag{15}$$

Substituting Eqs. (10) and (14) into Eq. (15), we obtain a differential relationship between *voltage* and *current*

$$\frac{\partial I(z,t)}{\partial z} = -\left[\frac{2\pi\varepsilon}{\ln(b/a)}\right]\frac{\partial V(z,t)}{\partial t} \tag{16}$$

The constant in the brackets in Eq. (16) is the capacitance of a coaxial capacitor of unit length. To prove this, one calculates the capacitance as the ratio of charge to voltage ($C = Q/V$), and the result is the capacitance per meter of line length, as given by

$$C = \frac{2\pi\varepsilon}{\ln(b/a)} \quad \text{F/m} \tag{17}$$

On a transmission line this capacitance is uniformly distributed along the entire length of the line between the inner and outer conductors. Using

TRANSIENT WAVES ON LOSSLESS TRANSMISSION LINES

this capacitance, the differential equation relating the voltage and current can be rewritten

$$\boxed{\frac{\partial I(z,t)}{\partial z} = -C\frac{\partial V(z,t)}{\partial t}} \qquad (18)$$

Equation (18) is one of a pair of equations called the _telegrapher's equations_, a name which came about historically from studies of transmission lines used in telegraphy systems.

The second of the telegrapher's equations can be derived from Maxwell's equation for the curl of **E**

$$\nabla \times \mathbf{E} = -\mu\frac{\partial \mathbf{H}}{\partial t} \qquad (19)$$

The details of this derivation will be left as a problem, although the procedure is the same as followed above. Simplifying Eq. (19) by using the assumptions made earlier, we obtain a partial differential equation analogous to Eq. (15)

$$\boxed{\frac{\partial E_r}{\partial z} = -\mu\frac{\partial H_\phi}{\partial t}} \qquad (20)$$

Substituting Eqs. (10) and (14) into (20), we obtain

$$\frac{\partial V(z,t)}{\partial z} = -\left[\frac{\mu}{2\pi}\ln\frac{b}{a}\right]\frac{\partial I(z,t)}{\partial t} \qquad (21)$$

In this equation the constant in brackets is the _external inductance per unit length_, given by

$$L = \frac{\mu}{2\pi}\ln\frac{b}{a} \quad \text{H/m} \qquad (22)$$

Effects of the _internal_ inductance are illustrated in Prob. 8 and are negligible at the high frequencies used in most communications systems. Thus the second telegrapher's equation is

$$\boxed{\frac{\partial V(z,t)}{\partial z} = -L\frac{\partial I(z,t)}{\partial t}} \qquad (23)$$

From the telegrapher's equations it is relatively easy to show that the voltages and currents on the transmission line satisfy the wave equation. To show this for the voltage, we differentiate Eq. (23) with respect to z and Eq. (18) with respect to t

$$\frac{\partial^2 V(z,t)}{\partial z^2} = -L\frac{\partial^2 I(z,t)}{\partial z\, \partial t} \tag{24}$$

$$\frac{\partial^2 I(z,t)}{\partial t\, \partial z} = -C\frac{\partial^2 V(z,t)}{\partial t^2} \tag{25}$$

After interchanging the order of the derivatives on the left side of Eq. (25), the substitution of Eq. (25) into Eq. (24) yields the wave equation for voltage

$$\boxed{\frac{\partial^2 V(z,t)}{\partial z^2} = LC\frac{\partial^2 V(z,t)}{\partial t^2}} \tag{26}$$

A similar derivation using the telegrapher's equations yields the wave equation for the current

$$\boxed{\frac{\partial^2 I(z,t)}{\partial z^2} = LC\frac{\partial^2 I(z,t)}{\partial t^2}} \tag{27}$$

These wave equations for voltage and current are the same as those obtained in Sec. 3.1 for the transverse components of a plane wave (in rectangular coordinates) except that the constant LC replaces $\mu\varepsilon$. We derived functional forms for the solutions of these wave equations in Chap. 3 and need not repeat that work here to find the functional forms of the voltage and current waves. We can write the solutions of Eqs. (26) and (27) as before, with LC replacing $\mu\varepsilon$:

$$\boxed{V(z,t) = F_1(t - \sqrt{LC}\, z) + G_1(t + \sqrt{LC}\, z) + V_0} \tag{28}$$

$$\boxed{I(z,t) = F_2(t - \sqrt{LC}\, z) + G_2(t + \sqrt{LC}\, z) + I_0} \tag{29}$$

The constants V_0 and I_0 represent dc components. The functions F_1 and F_2 are waves traveling in the positive z direction with a velocity

$$\frac{dz}{dt} = v_p = \frac{1}{\sqrt{LC}} \tag{30}$$

and the functions G_1 and G_2 are negative-going waves with a velocity

$$\frac{dz}{dt} = -v_p = -\frac{1}{\sqrt{LC}} \tag{31}$$

TRANSIENT WAVES ON LOSSLESS TRANSMISSION LINES

The wave equations for the transverse field components E_r and H_ϕ can be derived from Eqs. (15) and (20), which are the fields forms of the telegrapher's equations

$$\frac{\partial^2 E_r}{\partial z^2} = \mu\varepsilon \frac{\partial^2 E_r}{\partial t^2} \tag{32}$$

$$\frac{\partial^2 H_\phi}{\partial z^2} = \mu\varepsilon \frac{\partial^2 H_\phi}{\partial t^2} \tag{33}$$

Therefore E_r and H_ϕ have the same functional forms as the transverse components of the plane wave described in Sec. 3.1. We can show that V, I, E_r, and H_ϕ all travel with the same velocity by substituting Eqs. (17) and (22) into Eq. (30)

$$\boxed{v_p = \frac{1}{\sqrt{LC}} = \frac{1}{\sqrt{\mu\varepsilon}}} \tag{34}$$

Thus the velocity of propagation depends only on the medium between the conductors.

The propagation of energy down the transmission line can be considered from two viewpoints. One viewpoint is in terms of voltages and currents and the wave equations (26) and (27), while the other viewpoint considers only electromagnetic fields and the wave equations (32) and (33).

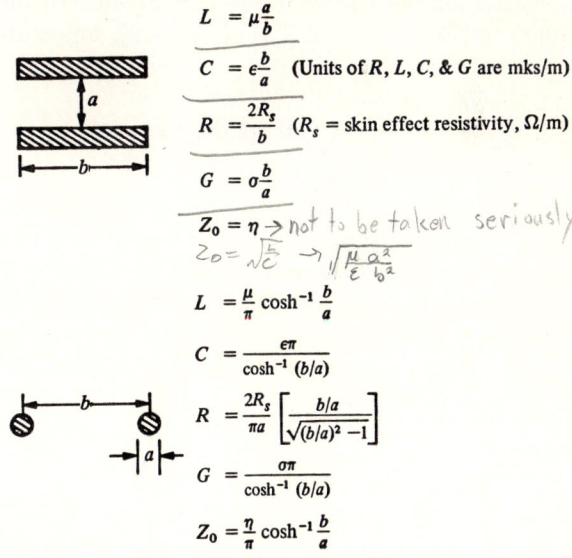

Fig. 4.2.2 Other transmission-line geometries.

The more fundamental viewpoint is in terms of the electromagnetic fields, since cases do exist† where there is no voltage and current analogy to the **E** and **H** fields. However, for the transmission lines to be considered, the voltage and current analogies exist, and working with voltage and current is usually more convenient than working with the electromagnetic fields. The telegrapher's equations and the wave equations were derived for the particular geometry of the lossless coaxial transmission line. However, the same equations can be obtained for a wide variety of transmission lines. Some of these other geometries‡ and their appropriate constants are shown in Fig. 4.2.2. The student will find it instructive to derive the parameters L and C, the telegrapher's equations, and the wave equations for lossless lines, starting from Maxwell's equations, for some of these other geometries. In most cases the same procedure can be followed as for the coaxial transmission line. Difficulties associated with finding the series resistance R and the parallel conductance G are discussed in the next section.

4.3 VOLTAGE, CURRENT, AND EQUIVALENT CIRCUITS

The telegrapher's equations relating voltage and current in the previous section have a simple interpretation in terms of an equivalent circuit. Consider an incremental section of transmission line z m long, as shown in Fig. 4.3.1. At a given instant of time there is a voltage drop ΔV and a current loss of ΔI in the incremental section of the transmission line. The equations for the voltage drop and the current loss are the telegrapher's equations [Eqs. (18) and (23) of Sec. 4.2], which are repeated below for an incremental section of line:

$$\frac{\Delta V}{\Delta z} = -L \frac{\Delta I}{\Delta t} = \text{voltage drop across section } \Delta z \tag{1}$$

$$\frac{\Delta I}{\Delta z} = -C \frac{\Delta V}{\Delta t} = \text{current loss in section } \Delta z \tag{2}$$

† The voltage and current analogy, as considered here, exists only for TEM fields.
‡ Transmission Lines, in "Reference Data for Radio Engineers," 4th ed., International Telephone and Telegraph Corporation, New York, 1956.

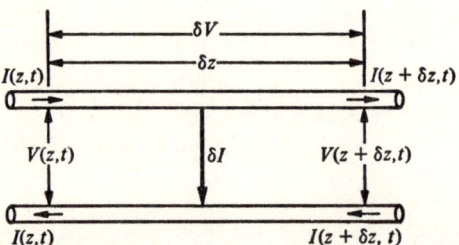

Fig. 4.3.1 An incremental section of a transmission line.

TRANSIENT WAVES ON LOSSLESS TRANSMISSION LINES

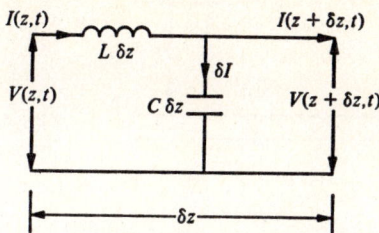

Fig. 4.3.2 The equivalent circuit of a lossless incremental section.

Equation (1) indicates that the incremental voltage drop $L\,\Delta I/\Delta t$ is due to a series inductance. Similarly, Eq. (2) shows that there is a shunt current $C\,\Delta V/\Delta t$ due to a shunt capacitance. Thus a simplified equivalent circuit can be drawn directly from the telegrapher's equations (Fig. 4.3.2). This equivalent circuit and the corresponding transmission line are said to be lossless because no energy-dissipative elements (resistances or conductances) appear. We have shown in Sec. 3.2 that plane waves propagate without distortion in a lossless medium. Since the fields in a lossless two-wire transmission line form a plane wave, such a line is also distortionless. The lossless transmission line is an ideal case which can never be achieved in actual practice, although many well-designed transmission lines today are so nearly lossless that the above equivalent circuit is valid for most analysis.

A more general equivalent circuit can be derived for a lossy line. For example, suppose that a coaxial transmission line is filled with salt water, so that some conduction current can flow from the inner conductor to the outer conductor. If the salt water has a conductivity of σ, Maxwell's equation for the curl of the **H** field is

$$\nabla \times \mathbf{H} = +\sigma \mathbf{E} + \varepsilon \frac{\partial \mathbf{E}}{\partial t} \tag{3}$$

Equation (3) can be reduced to an equation relating the voltage and current, as in Sec. 4.2,

$$\frac{\partial I}{\partial z} = -GV - C\frac{\partial V}{\partial t} \tag{4}$$

Equation (4) is one of the telegrapher's equations for a lossy line and should be compared with Eq. (2). Note that in Eq. (4) there is a conductance G which represents the conductance per unit meter of the salt water or the lossy medium inside the coaxial line. Because of the conductance term in Eq. (4), the equivalent circuit must be modified for the lossy line by placing a conductance G in parallel with the capacitance C, as shown in Fig. 4.3.3.

One additional modification is necessary to complete the lossy equivalent circuit. The conductors of the transmission line (the center con-

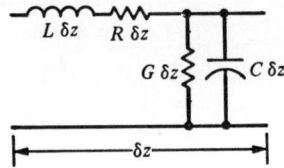

Fig. 4.3.3 The complete equivalent circuit of an incremental section.

ductor and the outer shield) are very good conductors, often being made of copper or aluminum, but they are not perfect conductors, as assumed in the lossless case. They have some finite resistivity ρ per unit length, and consequently the tangential **E** field is not zero. This causes a small voltage drop per unit length, due to the finite resistance of the conductors. These ideas are expressed by the addition of the term RI, where R has units of ohms per meter,

$$\frac{\partial V}{\partial z} = -RI - L\frac{\partial I}{\partial t} \tag{5}$$

One observes from Eq. (5) that it is necessary to add a resistance in series with the inductance in the equivalent circuit to account for the voltage drop due to the resistance of the conductors.† The complete equivalent circuit for the lossy line is shown in Fig. 4.3.3. It will be left as an exercise for the student to show that the damped-wave equations for the lossy line

$$\frac{\partial^2 V}{\partial z^2} = (RG)V + (RC + LG)\frac{\partial V}{\partial t} + LC\frac{\partial^2 V}{\partial t^2} \tag{6}$$

$$\frac{\partial^2 I}{\partial z^2} = (RG)I + (RC + LG)\frac{\partial I}{\partial t} + LC\frac{\partial^2 I}{\partial t^2} \tag{7}$$

can be derived from Eqs. (4) and (5). The lossy transmission line will be considered in more detail in Chap. 6, where it is shown that a properly designed lossy line can be nearly distortionless for a limited frequency range.

The immediate objective is to develop the relationship between the voltage and the current in terms of a parameter of the transmission line called the *characteristic impedance*. This parameter is analogous to the

---

† The existence of an E_z component changes the field distribution from that assumed in Fig. 4.2.1; consequently, Eq. (5) is an approximate equation, valid for lines with low-loss conductors. Carson and Schelkunoff have studied the validity of this equation and show that the difference between the exact analysis and the approximate one is extremely small (J. R. Carson, The Guided and Radiated Energy in Wire Transmission, *J. AIEE*, vol. 43, pp. 906–913, October 1924; S. A. Schelkunoff, The Electromagnetic Theory of Coaxial Transmission Lines and Cylindrical Shields, *Bell System Tech. J.*, vol. 13, pp. 532–579, October 1934).

intrinsic impedance of plane waves. Only the lossless line will be considered for the present. Consider a positive-going wave given by

$$V(z,t) = V^+\left(t - \frac{z}{v}\right) = V^+(y) \tag{8}$$

in which the argument has been written in terms of a new variable y, where

$$y = t - \frac{z}{v} \tag{9}$$

We now differentiate Eq. (8) with respect to z, using the chain rule, to obtain a derivative in terms of the new variable y

$$\frac{\partial V}{\partial z} = \frac{\partial V}{\partial y}\frac{\partial y}{\partial z} = -\frac{1}{v}\frac{\partial V}{\partial y} \tag{10}$$

As a result of the positive-going voltage wave there will be a positive-going current wave

$$I(z,t) = I^+\left(t - \frac{z}{v}\right) = I(y) \tag{11}$$

The differentiation of $I(z,t)$ in Eq. (11) with respect to time yields

$$\frac{\partial I}{\partial t} = \frac{\partial I}{\partial y}\frac{\partial y}{\partial t} = \frac{\partial I}{\partial y} \tag{12}$$

Equations (10) and (12) can be substituted into

$$\frac{\partial V}{\partial z} = -L\frac{\partial I}{\partial t} \quad \text{telegrapher's equation} \tag{13}$$

the telegrapher's equation for the voltage change on an incremental section, giving a differential equation in terms of the new variable y

$$-\frac{1}{v}\frac{\partial V}{\partial y} = -L\frac{\partial I}{\partial y} \tag{14}$$

Since Eq. (14) is a partial differential equation with only one independent variable, it can be reduced to an ordinary differential equation, which can be solved by direct integration. Here it is assumed that the initial conditions are zero, and consequently the constant of integration is zero (recall that the velocity of propagation is $v = 1/\sqrt{LC}$).

$$dV = vL\, dI = \sqrt{\frac{L}{C}}\, dI \tag{15}$$

Upon integrating Eq. (15) we find that the relationship between the voltage and the current on the lossless transmission line is

$$\boxed{\frac{V^+(t - z/v)}{I^+(t - z/v)} = \sqrt{\frac{L}{C}} = Z_0} \quad \text{lossless case} \tag{16}$$

The constant on the right-hand side of Eq. (16) has the units of ohms, is called the *characteristic impedance*, and is denoted by Z_0. The characteristic impedance, which is one of the most important parameters of the transmission line, is a constant determined by the physical characteristics of the transmission line and is independent of the functional form of the applied voltage and current. Because of this simple relationship between the voltage and the current, it is not necessary to solve the wave equation for both these variables. Once the positive-going voltage is known, the corresponding current can be found from Eq. (16), and conversely one can find the voltage if the current is known.

The characteristic impedance is simply related to the permeability and permittivity of the medium in which the wave propagates. This relationship is easily developed by the substitution of the equations for the inductance and capacitance for a *coaxial* transmission line as developed in Sec. 4.2

$$Z_0 = \sqrt{\frac{L}{C}} = \sqrt{\frac{\mu}{\varepsilon}} \frac{\ln(b/a)}{2\pi} = \eta \frac{\ln(b/a)}{2\pi} \tag{17}$$

We can always express the characteristic impedance of a lossless transmission line as the product of the intrinsic impedance and a constant which is a function of the geometry of the particular line

$$Z_0 = \eta(\text{const}) \tag{18}$$

where the constant is a function of the geometry of the line. This relationship is obvious in Eq. (17) for the geometry of the coaxial transmission line.†

The relationship between a negative-going voltage wave and the corresponding current wave can be developed by the same procedure as above. The result is

$$\frac{V^-(t + z/v)}{I^-(t + z/v)} = -Z_0 \tag{19}$$

---

† A variety of other geometries can be found in the section on characteristic impedance of transmission lines in "Reference Data for Radio Engineers," 4th ed., International Telephone and Telegraph Corporation, New York, 1956. In most of these tables the permeability does not appear directly because it has been replaced by the permeability of free space ($\mu_0 = 4 \times 10^{-7}$ H/m).

TRANSIENT WAVES ON LOSSLESS TRANSMISSION LINES 221

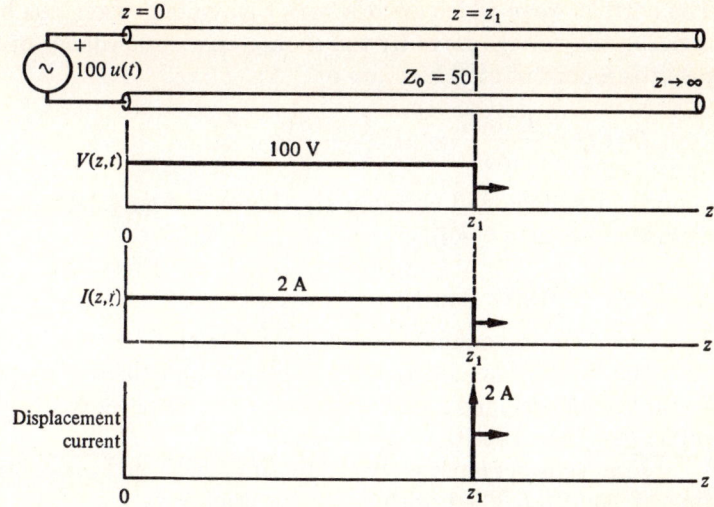

Fig. 4.3.4 Distribution of voltage and currents as a function of distance.

The negative sign in Eq. (19) should be interpreted as meaning that the current travels in the negative z direction. Otherwise, the result is the same for either negative-going or positive-going waveforms. The reciprocal of the characteristic impedance Y_0 is called the *characteristic admittance*

$$Y_0 = \frac{1}{Z_0} \tag{20}$$

The voltage and current relationships can now be summarized. It is known from the solution of the wave equation that the total voltage on the transmission line will be the sum of the positive-going waves and the negative-going waves. The same superposition relationship holds for the currents, but the currents can now be written in terms of the characteristic admittance and the voltage waves

$$\boxed{V(z,t) = V^+ + V^- \qquad I(z,t) = Y_0(V^+ - V^-)} \tag{21}$$

As a simple example of how one might use the characteristic impedance, suppose that 100 V from a battery is suddenly applied to a semi-infinite coaxial transmission line with a characteristic impedance of 50 Ω. The sudden application of the 100 V to the transmission line appears as a step function $U(t - z/v)$ which will travel down the transmission line as a function of the time and distance:

$$V^+(z,t) = 100 U\left(t - \frac{z}{v}\right) \quad \text{V} \tag{22}$$

The current wave which travels with this voltage wave can be found by dividing the voltage wave by the characteristic impedance of 50 Ω. The result is a current step function of 2 A:

$$I^+(z,t) = \frac{V^+(z,t)}{Z_0} = 2U\left(t - \frac{z}{v}\right) \quad \text{A} \tag{23}$$

Graphs of voltage and currents are shown in Fig. 4.3.4 as a function of distance at a time $t = t_1$.

4.4 LINES OF FINITE LENGTH

In this section two simple examples of a line of finite length are discussed, with the line terminated in a short circuit and then in an open circuit. In these examples, the boundary conditions on perfect conductors and the principle of superposition will be used.

First consider a perfectly conducting uncharged coaxial transmission line of length L filled with a lossless dielectric material. The line is shorted at $z = L$ by means of a perfectly conducting annulus, as shown in Fig. 4.4.1. At $t = 0$, a battery is connected at the sending end with the positive terminal on the center conductor and the negative terminal on the outer conductor. Then the fields between conductors are given by

$$\mathbf{E}_1^+(z,t) = \frac{V_0 U(t - z/v)}{r \ln (b/a)} \mathbf{a}_r \quad 0 \leq z \leq L \tag{1}$$

$$\mathbf{H}_1^+(z,t) = \frac{I_0}{2\pi r} U\left(t - \frac{z}{v}\right) \mathbf{a}_\phi \quad 0 \leq z \leq L \tag{2}$$

until the wave reaches the short circuit at $t = L/v$ [see Sec. 4.2, Eqs. (10) and (14)]. The superscript + indicates propagation in the positive z direction, the subscript 1 denotes the first positive-going wave, V_0 is the voltage of the battery, and $U(t - z/v)$ is the unit step function. By equating the argument of the unit step function to zero we can locate the position of the wavefront at any given time, until it impinges on the short circuit (at $t = L/v$). For all time greater than L/v, the fields which were established by the initial traveling wave remain constant as long as there is no change in the source voltage. Thus the unit step functions in these

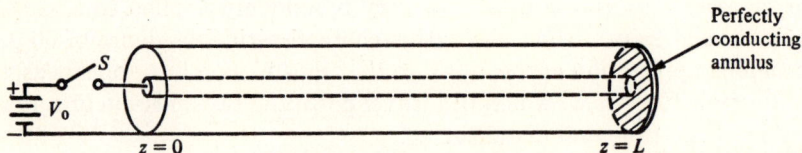

Fig. 4.4.1 The shorted transmission line.

equations are unity for all time greater than L/v, since z is not greater than L. We will show that the *total* fields on the line will change at $t = L/v$ because of the reflected wave.

When $t = L/v$, the wavefront has traveled the entire length of the line and impinges on the perfectly conducting annulus. Boundary conditions at the annulus require the electric field intensity to be zero. As discussed in Sec. 3.7, there must be a reflected electric-field-intensity vector of the same magnitude as the incident electric field but opposite direction. The magnetic field vector is likewise reflected at the conducting annulus but without reversal of direction. Thus the Poynting vector for the reflected plane wave indicates propagation in the $-z$ direction. The total fields on the line for $L/v < t < 2L/v$ are the sums of the corresponding fields in the initial wave and in the reflected wave. The expressions for these reflected fields† are given by

$$\mathbf{E}_1^-(z,t) = \frac{-V_0 U(t - 2L/v + z/v)}{r \ln(b/a)} \mathbf{a}_r \qquad 0 \le z \le L \qquad (3)$$

$$\mathbf{H}_1^-(z,t) = \frac{I_0}{2\pi r} U\left(t - \frac{2L}{v} + \frac{z}{v}\right) \mathbf{a}_\phi \qquad 0 \le z \le L \qquad (4)$$

To see that a wave has actually been reflected from the shorted termination, compare the direction of the Poynting vector $\mathbf{E} \times \mathbf{H}$ (the direction of power flow) for Eqs. (1) and (2) with that for Eqs. (3) and (4). Equations (1) and (2) describe the only positive-going waves for $0 < t < 2L/v$. At $t = 2L/v$, there are reflections from the sending-end termination which cause new positive-going waves $\mathbf{E}_2^+$ and $\mathbf{H}_2^+$ to propagate. This problem will be pursued in more detail in Sec. 4.5. Figure 4.4.2 illustrates the voltage and current waves on the line at $t = L/2v$ and $t = 3L/2v$. Voltage and current are ordinarily plotted, since $\mathbf{E}$ and $\mathbf{H}$ are functions of r.

As a second example, let us consider an uncharged lossless coaxial transmission line of length L with an open circuit at $z = L$. At $t = 0$, an arbitrary voltage $V(t)$ is connected at $z = 0$ with the positive terminal on

† Although the step-function notation is convenient for describing wave phenomena, we must exercise caution. For example, at $t = 3L/2v$, $U(t - z/v)$ has a value of unity for all $z < 3L/2$. Since z must be limited to the range $0 \le z \le L$, we interpret $U(3L/2v - z/v)$ as having a value of unity for all allowable values of z. As long as the battery remains connected to the line (Fig. 4.4.1), the initial positive-going fields remain constant on the line for $t > L/v$. Similarly, the first negative-going wave, launched from the receiving end as a reflection of the incident wave at $t = L/v$, remains constant on the line for $t > 2L/v$. Thus our interpretation of the notation for a unit step function is consistent with the behavior of the waves. The *total* electric and magnetic fields, however, are the sum of all waves on the line, and hence are modified with every reflection occurring at either end of the line.

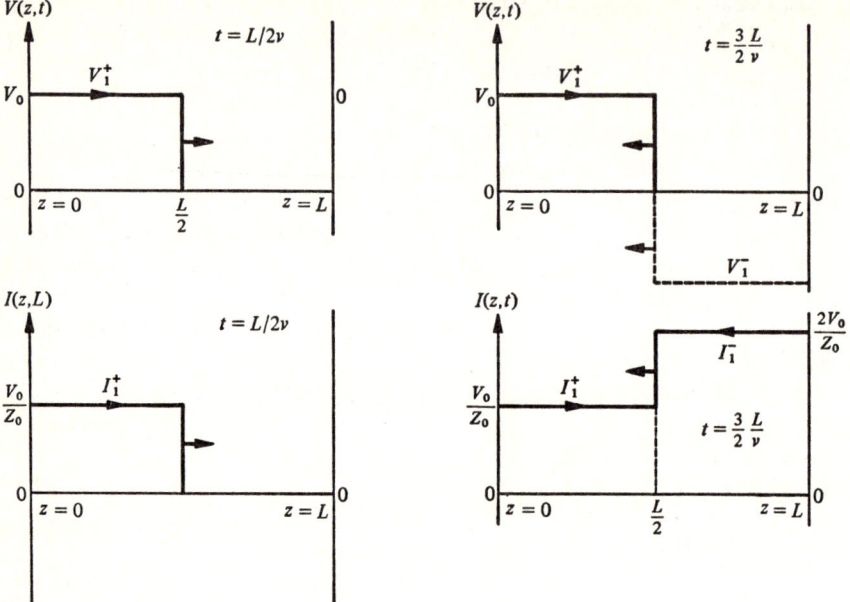

Fig. 4.4.2 Voltage and current on a short-circuited line.

the center conductor and the negative terminal on the outer conductor. For $t < L/v$, the total fields between conductors are given by

$$\mathbf{E}_1^+(z,t) = \frac{V(t - z/v)}{r \ln (b/a)} U\left(t - \frac{z}{v}\right) \mathbf{a}_r \tag{5}$$

$$\mathbf{H}_1^+(z,t) = \frac{I(t - z/v)}{2\pi r} U\left(t - \frac{z}{v}\right) \mathbf{a}_\phi \tag{6}$$

At $t = L/v$, however, the fields are incident at the open-circuit termination, and current here must be zero. Therefore the reflected magnetic field intensity† is of the same magnitude as the incident magnetic field intensity but of opposite sign and is given by

$$\mathbf{H}_1^-(z,t) = \frac{-I(t - 2L/v + z/v)}{2\pi r} U\left(t - \frac{2L}{v} + \frac{z}{v}\right) \mathbf{a}_\phi \tag{7}$$

† Consider a line integral $\oint \mathbf{H} \cdot d\mathbf{l}$ around the circumference of the open annulus at an incremental distance beyond the end of the line. No conductors pierce the plane defined by the path of the line integral, and so no conduction current flows through the plane. Since the displacement current in this region is also zero (assuming no fringing), an $\mathbf{H}^-$ field must exist such that $\oint \mathbf{H} \cdot d\mathbf{l} = 0$, and $\mathbf{H} = \mathbf{H}^+ + \mathbf{H}^- = 0$ at any point just beyond the end of the open-circuited line.

TRANSIENT WAVES ON LOSSLESS TRANSMISSION LINES

and the reflected electric field intensity is given by

$$\mathbf{E}_1^-(z,t) = \frac{V(t - 2L/v + z/v)}{r \ln(b/a)} U\left(t - \frac{2L}{v} + \frac{z}{v}\right) \mathbf{a}_r \qquad (8)$$

Again one should compare the Poynting vector for Eqs. (5) and (6) with that for Eqs. (7) and (8) to see that there actually has been a reversal of the direction of power flow from the open end of the line.

Figure 4.4.3 illustrates the voltage and current of a lossless open-circuited transmission line for $t = L/2v$ and $t = 3L/2v$, where the total voltage (or current) on the line is the sum of all voltage waves (or current waves) on the line. For this illustration the source voltage varies with time as shown in the figure.

As shown in Sec. 3.2, the electric and magnetic fields and the corresponding voltages and currents discussed in this section are traveling waves, because these waves propagate at the velocity

$$v = \frac{1}{\sqrt{LC}} = \frac{1}{\sqrt{\mu\varepsilon}}$$

in either the $+z$ or $-z$ direction.

In Sec. 4.3 it was shown that the ratio of voltage to current in a positive-going traveling wave is Z_0 and the ratio of voltage to current in

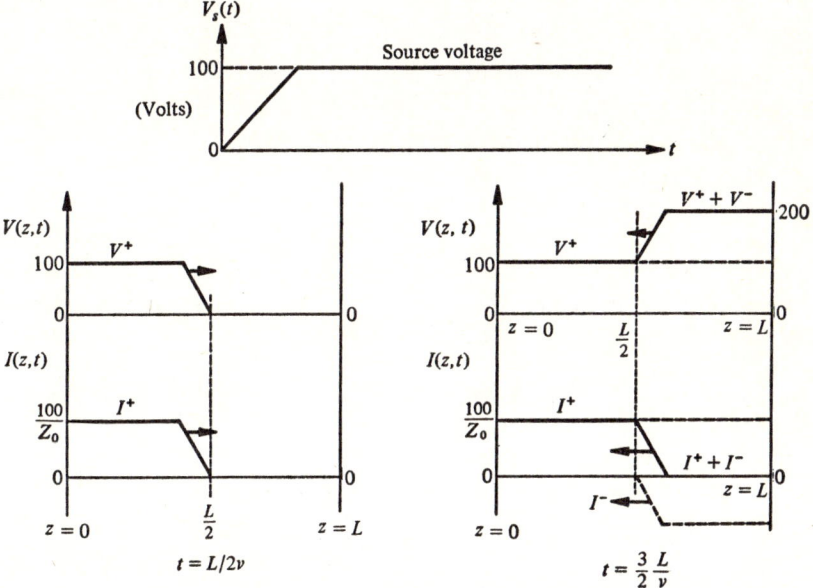

Fig. 4.4.3 Voltage and current on an open-circuited line.

a negative-going traveling wave is $-Z_0$. In the preceding two examples, the transmission line was lossless, and $Z_0 = \sqrt{L/C}$. Therefore the positive-going voltage and current waves have similar shapes and the same sign. However, the negative-going voltage and current waves have similar shapes but opposite sign, so that one appears as a reflection of the other. These relationships between voltage and current are illustrated in Figs. 4.4.2 and 4.4.3. Terminations other than an open or short circuit will be considered in the next section.

4.5 RESISTIVE TERMINATIONS

Between the extreme terminations of zero resistance or infinite resistance is the more general case of the termination of finite resistance. Suppose a lossless transmission line of length L is terminated with a resistance of value R, and suppose a voltage wave $V(z,t) = V^+(t - z/v)$ starts down the line at $z = 0$ and $t = 0$ with velocity v. For $t < L/v$, no reflection occurs, and so $V^-(t + z/v) = 0$. This can be shown by the following derivation. At $t = 0$, the voltage and current on the line are zero, except possibly at the ends of the line

$$V(z,0) = V^+\left(-\frac{z}{v}\right) + V^-\left(\frac{z}{v}\right) = 0 \qquad 0 < z < L \tag{1}$$

$$I(z,0) = Y_0\left[V^+\left(-\frac{z}{v}\right) - V^-\left(\frac{z}{v}\right)\right] = 0 \qquad 0 < z < L \tag{2}$$

Solving simultaneously, one finds that $V^+(-z/v) = 0$ for $0 < z < L$. As z varies over the range $0 < z < L$, the argument $-z/v$ varies over the range $-L/v < -z/v < 0$. For any argument y within the bounds $-L/v < y < 0$, the function $V(y)$ is zero. Therefore, $V^+(t - z/v) = 0$ for $-L/v < t - z/v < 0$. Adding z/v to the inequality, it can be seen that

$$V^+\left(t - \frac{z}{v}\right) = 0 \qquad \text{for } \frac{z - L}{v} < t < \frac{z}{v}$$

The first part of the inequality is true because the wave is not launched until $t = 0$, and the second part implies that the positive-going portion of the voltage wave is zero if the wave has had insufficient time to go a distance z down the line. Similarly, $V^-(+z/v) = 0$ for $0 < z < L$ implies that $V^-(t + z/v) = 0$ for $0 < t + z/v < L/v$, which is the same as $V^-(t + z/v) = 0$ for the range $-z/v < t < (L - z)/v$. The first part of this inequality is always true for $t < 0$, and the second part means that the negative-going voltage wave $V^-(t + z/v)$ originates at $z = L$. This can occur with the application of an external source or by the reflection of a positive-going wave. Since we assumed a voltage applied at the left end of the line in this example, there can be no negative-going wave until a reflection occurs at $t = L/v$. As an analogy, consider a pebble dropped into water. The resulting wave is launched from the point where the disturbance occurs, and no reflection occurs until the wave strikes a boundary.

We now derive some interesting relationships between the incident voltage wave, reflected voltage wave, and receiving-end voltage. At $z = L$, the boundary condition (Ohm's law),

$$V_r(t) = RI_r(t) \tag{3}$$

is true at all times at the receiving end. However, since the voltage and the current at the receiving end must also satisfy the wave equations, $V_r(t)$ and $I_r(t)$ can be written

$$V_r(t) = V(L,t) = V^+\left(t - \frac{L}{v}\right) + V^-\left(t + \frac{L}{v}\right) \tag{4}$$

$$I_r(t) = I(L,t) = Y_0\left[V^+\left(t - \frac{L}{v}\right) - V^-\left(t + \frac{L}{v}\right)\right] \tag{5}$$

From Eq. (5) it is seen that $V^-(t + L/v) = V^+(t - L/v) - Z_0 I_r(t)$. By substituting this negative-going voltage into Eq. (4) a relationship between receiving-end voltage, current, and incident voltage wave is obtained

$$V_r(t) = 2V^+\left(t - \frac{L}{v}\right) - Z_0 I_r(t) \tag{6}$$

By substituting the current from Eq. (3) into Eq. (6), we find the voltage across the resistor in terms of the positive-going voltage wave

$$V_r(t) = \frac{2RV^+(t - L/v)}{R + Z_0} \tag{7}$$

The reflected voltage wave, $V^-(t + L/v) = V_r(t) - V^+(t - L/v)$, is

$$V^-\left(t + \frac{L}{v}\right) = \frac{R - Z_0}{R + Z_0}\left[V^+\left(t - \frac{L}{v}\right)\right] \tag{8}$$

The coefficient of the positive-going wave in Eq. (8), called the *reflection coefficient* Γ_r, can be used to find V^- if V^+ is known

$$\boxed{\Gamma_r = \frac{V^-}{V^+} = \frac{R - Z_0}{R + Z_0}} \tag{9}$$

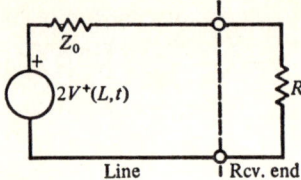
Fig. 4.5.1 Thévenin circuit at the receiving end.

A similar derivation can be given for a reflection coefficient for current, and it will be found that

$$\frac{I^-}{I^+} = -\Gamma_r = -\frac{R - Z_0}{R + Z_0} \tag{10}$$

Note that the same basic definition of the reflection coefficient used in Sec. 3.8 for plane waves

$$\Gamma = \frac{\text{reflected wave}}{\text{incident wave}} \tag{11}$$

has been applied for both voltage and current. Hereafter the terms Γ and "reflection coefficient" will refer to the *voltage* reflection coefficient unless otherwise designated. The reflection coefficient varies in range from -1 for a short circuit, to 0 if the termination resistance is equal to the characteristic impedance, to $+1$ if the termination is an open circuit. As an exercise, the reader can plot Γ_r as a function of R.

To find voltage and current at the receiving end, the transmission line can be replaced with a Thévenin equivalent circuit. Equation (7) is the equation of a simple voltage-dividing network with impedances Z_0 and R. The open-circuit voltage is $2V^+(t - L/v)$, and the short-circuit current is $2V^+(t - L/v)/Z_0$, which yields the circuit shown in Fig. 4.5.1.

Thévenin's circuit for the sending end can be derived in a similar manner. The reflected voltage wave $V^-(t - 2L/v + z/v)$ is incident at the sending end at $t = 2L/v$, and a portion of that negative-going wave is reflected back to form a second positive-going wave. Assuming a source voltage $V_s(t)$ and a source resistance R_s and using Thévenin's theorem, the circuit of Fig. 4.5.2 can be obtained. From the Thévenin circuit the sending-end current is

$$I_s(t) = \frac{V_s(t)}{R_s + Z_0} - \frac{2V^-(t)}{R_s + Z_0} \tag{12}$$

This current is also a solution of the wave equation

$$I_s(t) = I(0,t) = Y_0[V^+(0,t) - V^-(0,t)] \tag{13}$$

TRANSIENT WAVES ON LOSSLESS TRANSMISSION LINES

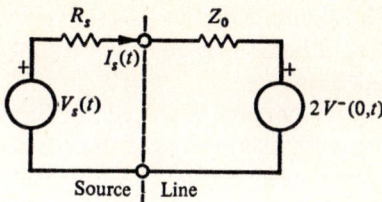

Fig. 4.5.2 Thévenin circuit at the sending end.

Therefore $V^+(0,t) = Z_0 I_s(t) + V^-(0,t)$, and substituting Eq. (12) for $I_s(t)$ gives

$$V^+(0,t) = \frac{V_s(t)}{1 + Y_0 R_s} + \frac{R_s - Z_0}{R_s + Z_0} V^-(0,t) = V_1^+(0,t) + V_2^+(0,t) \quad (14)$$

Equation (14)† represents the sum of all positive-going waves on the line for $t < 4L/v$. The first term of the right-hand side of Eq. (14) is the initial positive-going voltage wave on the line, is always a result of the source voltage, and is independent of the load resistance. This wave is denoted by V_1^+. The second term is a positive-going wave reflected at the sending end, due to the first negative-going wave's impinging on the source impedance, and is denoted by V_2^+. The sum of the two terms V_1^+ and V_2^+ is the total positive-going wave for $t < 4L/v$.

The coefficient of $V^-(0,t)$ in Eq. (14) is the reflection coefficient at the sending end, as defined by Eq. (11), and will be denoted Γ_s.

$$\Gamma_s = \frac{V_2^+}{V_1^-} = \frac{R_s - Z_0}{R_s + Z_0} \quad (15)$$

The power per unit of cross-sectional area traveling down the transmission line, in either the positive or negative direction, can be found from Poynting's vector. This vector relationship, $\mathbf{S} = \mathbf{E} \times \mathbf{H}$, where $\mathbf{E}$ and $\mathbf{H}$ are given by Eqs. (10) and (14) of Sec. 4.2, yields

$$\mathbf{S} = \frac{Y_0 [V^+(t - z/v)]^2}{2\pi r^2 \ln(b/a)} \mathbf{a}_z \quad (16)$$

for the power density traveling in the positive z direction. Similarly the power density in the negative-going wave is given by

$$\mathbf{S} = \frac{-Y_0 [V^-(t + z/v)]^2}{2\pi r^2 \ln(b/a)} \mathbf{a}_z \quad (17)$$

† The notation $V^+(z,t)$, with no subscript, denotes the sum of all the positive-going voltage waves on the line from $t = 0$ until the specified time. Similarly, $V^-(z,t)$ denotes the sum of all negative-going waves, and at any given time and position the total voltage is $V(z,t) = V^+(z,t) + V^-(z,t)$.

The sign is negative because either the magnetic field intensity is in the $-a_\phi$ direction or the electric field intensity is in the $-a_r$ direction, but not both.

A portion of the power flowing from the generator is reflected successively from the load impedance and the source impedance, depending upon the respective reflection coefficients, Γ_r and Γ_s. This reflected power, at either the load or the generator, represents power not absorbed by the load resistance or the generator resistance. Often the load resistance is selected to equal the characteristic impedance in order to eliminate power loss due to reflection. Then $\Gamma_r = 0$, and all power going down the line is dissipated in the load. In this case the load is said to be *matched* to the transmission line.

4.6 BOUNCE DIAGRAMS

For transient problems, a technique very useful for problem solving is the *bounce diagram*. Voltage or current can be plotted versus time and distance on the coordinate system indicated in Fig. 4.6.1. Ordinarily, a two-dimensional bird's-eye view (top view) of time versus distance is plotted, with numerical values of voltage written where applicable.

A simple example to illustrate the bounce-diagram technique will be given. Suppose at $t = 0$ a battery of V_0 V with internal resistance of $R_s = 30\ \Omega$ is connected to the sending end of a lossless transmission line, which has a characteristic impedance of $Z_0 = 50\ \Omega$. The transmission line shown in Fig. 4.6.2 has a length L and a phase velocity v, and the load has a resistance of $R_r = 100\ \Omega$. We see that $\Gamma_s = -\frac{1}{4}$ and $\Gamma_r = \frac{1}{3}$, and these values are written by the sending end and receiving end, respectively, in the bounce diagram for voltage shown in Fig. 4.6.2.

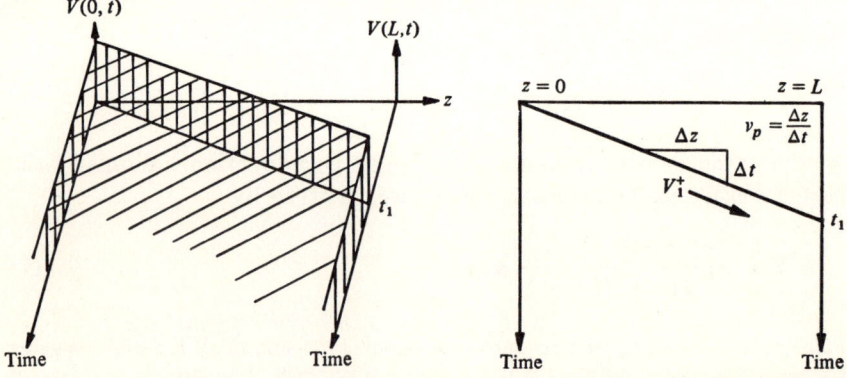

Fig. 4.6.1 The bounce diagram is a plot of voltage as a function of time and distance.

TRANSIENT WAVES ON LOSSLESS TRANSMISSION LINES

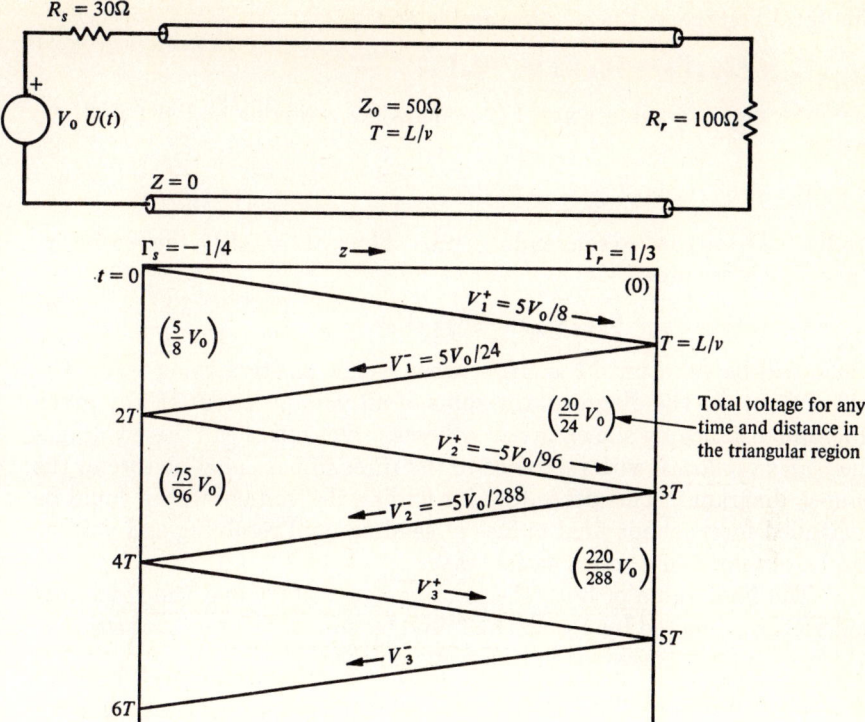

Fig. 4.6.2 The transmission line and its voltage bounce diagram.

From Thévenin's theorem, the initial positive-going voltage wave V_1^+ is found by voltage division between the characteristic impedance and the internal impedance of the generator. This voltage is

$$V_1^+ = \frac{50}{30 + 50} V_0 = \tfrac{5}{8} V_0$$

The first reflected voltage wave V_1^- from the receiving end is

$$V_1^- = \Gamma_r(\tfrac{5}{8} V_0) = \tfrac{5}{24} V_0$$

Therefore the total voltage at the receiving end just after $t = L/v$ is the sum of the incident and reflected voltage waves, which is

$$V_r = \tfrac{5}{8} V_0 + \tfrac{5}{24} V_0 = \tfrac{20}{24} V_0$$

The negative-going wave V_1^- is incident at the sending end at $t = 2L/v$, and the positive-going wave V_2^+ reflected from the sending-end resistance is

$$V_2^+ = \Gamma_s(\tfrac{5}{24} V_0) = -\tfrac{5}{96} V_0$$

The total voltage at the sending end is now

$$V_s = {}^{20}\!/_{24}V_0 - {}^{5}\!/_{96}V_0 = {}^{75}\!/_{96}V_0$$

This new positive-going wave V_2^+ reaches the receiving end at $t = 3L/v$, and

$$V_2^- = \tfrac{1}{3}(-{}^{5}\!/_{96}V_0) = -{}^{5}\!/_{288}V_0$$

is reflected back toward the sending end. Thus, at $t = 3L/v$, the receiving-end voltage becomes

$$V_r = {}^{25}\!/_{32}V_0 - {}^{5}\!/_{288}V_0 = {}^{220}\!/_{288}V_0$$

which will be valid until $t = 5L/v$, when a new positive-going wave V_3^+ will arrive. In the diagram, the sums of all voltage waves at any given time and distance are shown in parentheses. One might call these voltages the *plateau voltages*, with reference to the three-dimensional picture of the bounce diagram. This procedure for finding the voltage waves could be continued forever, but final values of sending- and receiving-end voltage can be obtained in a much easier way.

The final value of both the sending-end voltage and the receiving-end voltage, denoted by V_f, can be found by summing an infinite series in which each term corresponds to a wave on the line

$$V_f = \lim_{t \to \infty} V_r = \lim_{t \to \infty} V_s$$

$$= \frac{Z_0 V_0}{R_s + Z_0}(1 + \Gamma_r + \Gamma_r\Gamma_s + \Gamma_r^2\Gamma_s + \Gamma_r^2\Gamma_s^2 + \cdots) \quad (1)$$

Equation (1) can be written in closed form,† as in Eq. (3), by using the identity

$$\frac{1}{1-x} = 1 + x + x^2 + x^3 + \cdots \quad (2)$$

We apply the identity first to all positive-going waves (terms in which Γ_r and Γ_s appear raised to the same power), and then to all negative-going waves (terms in which Γ_r is raised to a higher power, by 1, than Γ_s);

$$V_f = \frac{Z_0 V_0}{R_s + Z_0}\underbrace{\left(\frac{1}{1 - \Gamma_r\Gamma_s}}_{\text{positive-going waves}} + \underbrace{\frac{\Gamma_r}{1 - \Gamma_r\Gamma_s}\right)}_{\text{negative-going waves}} = V^+(z, \infty) + V^-(z, \infty) \quad (3)$$

is the final value of voltage on the line. If $(R_s - Z_0)/(R_s + Z_0)$ is inserted for Γ_s and $(R_r - Z_0)/(R_r + Z_0)$ is inserted for Γ_r, a final value of voltage V_f for the sending end, the receiving end, and all points on the

† Any function written as a rational function is said to be written in closed form.

transmission line is seen to be independent of the characteristic impedance of the transmission line and is given by

$$V_f = V_s = V_r = \frac{R_r V_0}{R_s + R_r} \qquad (4)$$

Therefore, as t approaches infinity, the supply voltage is divided between the generator resistance and the load resistance according to Ohm's law, as if the transmission line were a small circuit.

In the present numerical example, the final value of line voltage is $100/130 \, V_0 = V_s = V_r$ from Eq. (4). From Eq. (3), as t approaches infinity, the total positive-going voltage is

$$V^+(z, \infty) = 50/80 \, V_0 (12/13) = 15/26 \, V_0 = 75/130 \, V_0$$

and the total negative-going voltage wave is

$$V^-(z, \infty) = \tfrac{1}{3} V^+(z, \infty) = 5/26 \, V_0 = 25/130 \, V_0$$

Current on the line is easily found, since the total positive-going current wave is

$$I^+(z, \infty) = \frac{V^+(z, \infty)}{Z_0}$$

and the total negative-going current wave is

$$I^-(z, \infty) = \frac{-V^-(z, \infty)}{Z_0}$$

Figure 4.6.3 shows the total positive-going and negative-going voltage waves as t approaches infinity.

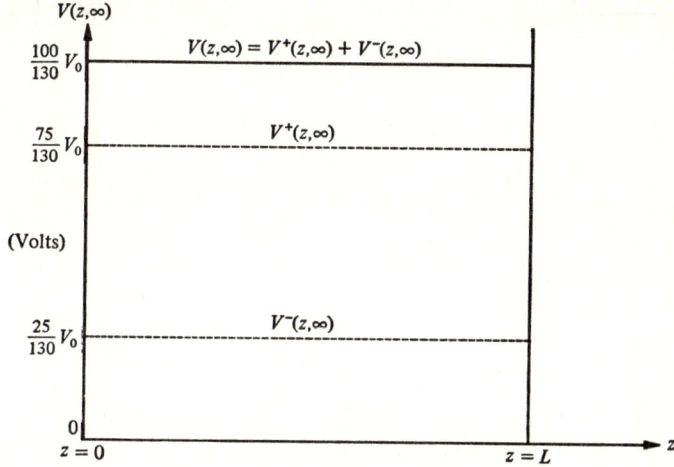

Fig. 4.6.3 Final values of the voltage waves.

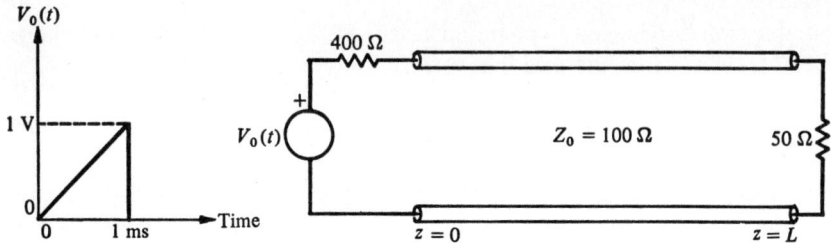

Fig. 4.7.1 An arbitrary waveform applied to a transmission line.

4.7 ARBITRARY WAVEFORMS

As an example of arbitrary waveforms, we now consider a pulse traveling down a lossless transmission line. A voltage pulse generator with a triangular waveform, as shown in Fig. 4.7.1, is applied to a transmission line of length L, which has a characteristic impedance of 100 Ω. The internal resistance of the pulse generator is 400 Ω, the load resistance is 50 Ω, and the transit time $T = L/v \gg 1$ ms, where v is the velocity of propagation.

At $t = 1.5$ ms the voltage on the transmission line appears as shown in Fig. 4.7.2. Note that the pulse appears to have been reversed when plotted as a function of distance rather than time.

To find the voltage or current at any point on the line at any time, it is convenient to use a bounce diagram, but we must remember that the peak voltage of the pulse occurs 1 ms after the beginning of the pulse. Since the pulse $V_0(t)$ can be synthesized as shown in Fig. 4.7.3, the ramp function $tU(t)$ starts propagating down the line at $t = 0$ and the ramp function $-(t - 1 \text{ ms})U(t - 1 \text{ ms})$ and the unit step function $-U(t - 1 \text{ ms})$ start propagating at $t = 1$ ms. The values of voltage given in the bounce diagram in Fig. 4.7.4 are the peak values and hence are valid only at the instant the peak of the pulse passes a certain point. The solid line in the

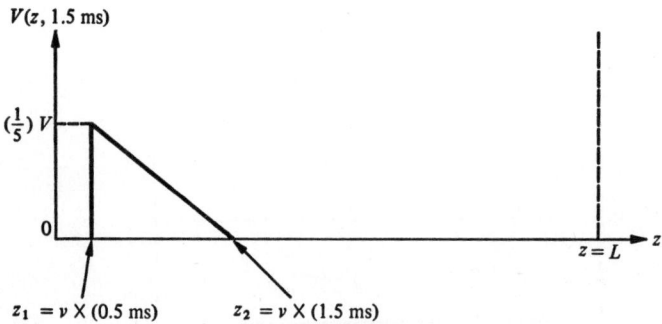

Fig. 4.7.2 Voltage as a function of distance at $t = 1.5$ ms.

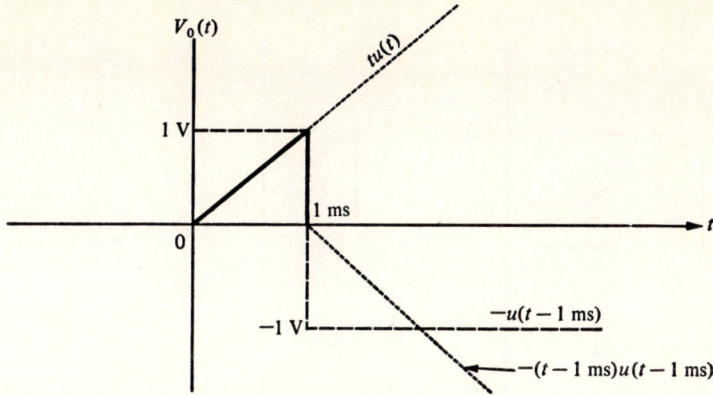

Fig. 4.7.3 Synthesis of the triangular pulse.

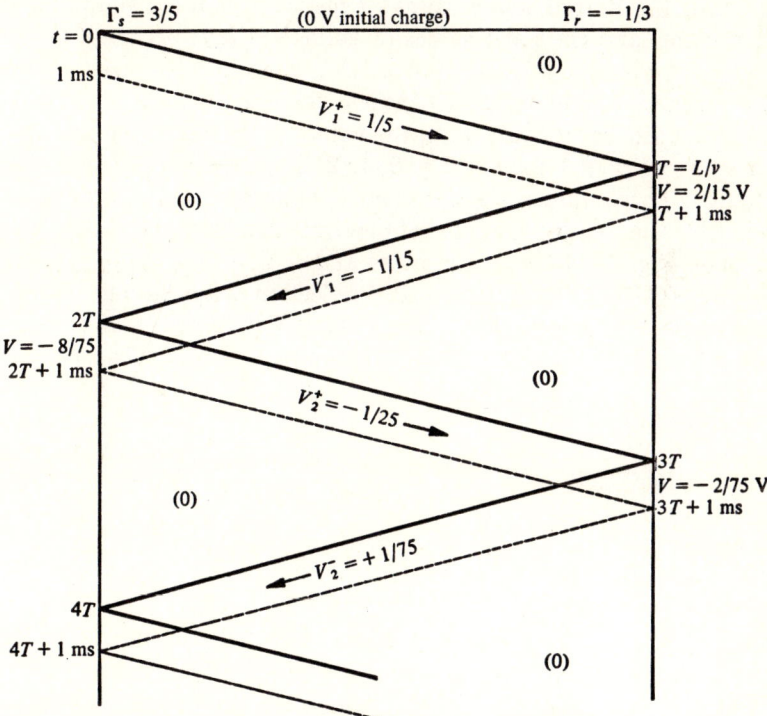

Fig. 4.7.4 Bounce diagram for a triangular voltage pulse.

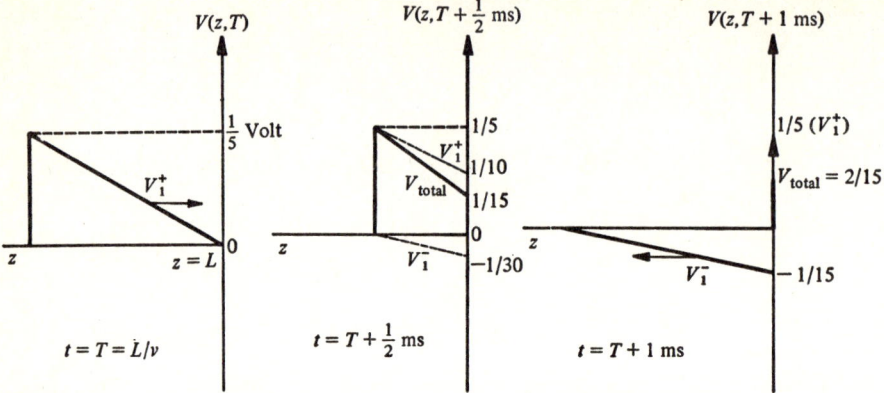

Fig. 4.7.5 Reflection from the termination at three different times.

bounce diagram represents the position at a given time of the wedge end of the first ramp function, and the dotted line represents the vertical end (the second ramp function and the step function) of the pulse. As expected, the dotted line is always 1 ms later than the solid line at any given point, and the peak values of voltage always occur on the dotted line.

Since the pulse length of 1 ms is much less than the line length, and since the pulse occurs only once, the values of voltage given at any of the times $t = nL/v + 1$ ms for $n = 0, 1, 2, 3, \ldots$ are valid only for that instant. These values are obtained by adding the peak value of one incident pulse to the peak value of the corresponding reflected pulse. For example, Fig. 4.7.5 shows the pulse at $t = L/v$, $t = L/v + 0.5$ ms, and $t = L/v + 1$ ms in various stages of reflection, after having traveled the length of the transmission line once. Thus $V(L,t)$ for $t < 4T$, as shown in Fig. 4.7.6, is the voltage that would be seen on an oscilloscope.

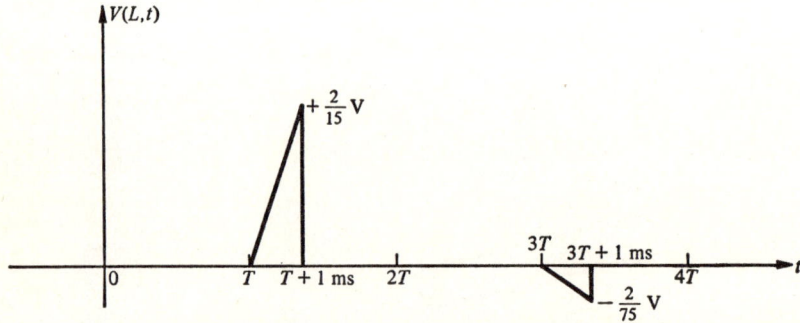

Fig. 4.7.6 The terminal voltage $V(L,t)$ as a function of time.

PROGRAMMED EXERCISE 4.7.1

AN IDEAL DIODE ON A LINE

A short-circuited transmission line of length L and characteristic impedance Z_0 is connected to a pulse generator, and an ideal diode is shunted across the line at $z = L/2$ as shown in Fig. 4.7.7. The reader should verify each of the following true statements:

1. At time $T/2$, the leading edge of the pulse is incident on the diode. The diode is back-biased by the pulse, and hence the pulse is transmitted entirely to the right, with no energy reflected back toward the generator.
2. At time T, the pulse is incident on the short circuit at $z = L$ and is reflected back toward the generator. At time $3T/2$, the voltage on the line is zero on the left half of the line and $-V_0$ on the right half, and the pulse, now negative-going, is incident on the diode.
3. Just after time $3T/2$, the diode is forward-biased by the pulse, and hence the (ideal) diode appears as a short circuit. Therefore the pulse reflected from the diode is positive-going and $+V_0$ V. During this reflection, the voltage across the diode is maintained at zero.
4. The diode thus effectively isolates the generator from incoming energy due to reflections, as long as the generator sends out only positive pulses.

The reader should attempt to answer the following questions pertinent to this problem.

1. Is the solution different if the diode is reversed?
2. Is the solution different if the diode is in series rather than in parallel?
3. What hazards are involved if the pulse is replaced by the sum of two step functions so that a bounce diagram can be used?

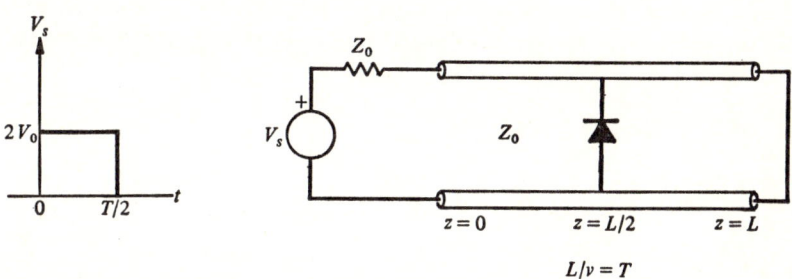

Fig. 4.7.7 A pulse applied to a shorted line, with an ideal diode in parallel with the line.

4. How is the solution different if the ideal diode is replaced by an actual diode with finite forward and back resistances?

4.8 NONZERO INITIAL CONDITIONS

To introduce the topic of nonzero initial conditions, we consider the charging and discharging of a transmission line. A line of length L can be charged to a voltage V_0 by means of the circuit shown in Fig. 4.8.1. Then, at $t = 2L/v$ after the battery is connected, the reflection from the open circuit arrives at the sending end and is not reflected there because the reflection coefficient Γ_s is zero. The battery can now be disconnected, leaving the line open-circuited at both ends and charged to V_0 V. Then $V^+ + V^- = V_0$ and $Y_0(V^+ - V^-) = 0$, so that $V^+ = V^- = V_0/2$ V.

If a resistor R_r is now connected at time $t = 0$ to the receiving end, current will flow through R_r and new current and voltage waves will be initiated due to the change in Γ_r. How the transmission line discharges can be determined by considering the reflection of the voltage wave on the line at various times, as shown in Fig. 4.8.2 for $\Gamma_r = \frac{1}{3}$ and $\Gamma_s = 1$. The voltage at the receiving end, just after the switch is closed, is

$$V_r = \frac{V_0}{2}(1 + \Gamma_r) = \frac{2V_0}{3} \tag{1}$$

which is valid until $t = 2T$. If we carefully continued to draw pictures like these, keeping track of the total positive- and negative-going waves, we could find the voltage or current at any point on the line and at any time.

A simpler approach, using superposition, permits use of the bounce diagram. At $t = 0^-$, just before R_r is connected, the voltage on the line is the sum of all the positive- and negative-going waves

$$V(z,0^-) = V^+(z,0^-) + V^-(z,0^-) = \frac{V_0}{2} + \frac{V_0}{2} \tag{2}$$

Since we are going to use superposition, we will consider the voltage waves $V^+(z,0^-)$ and $V^-(z,0^-)$ to remain constant at $V_0/2$ V. Thus at $t = 0^+$, a

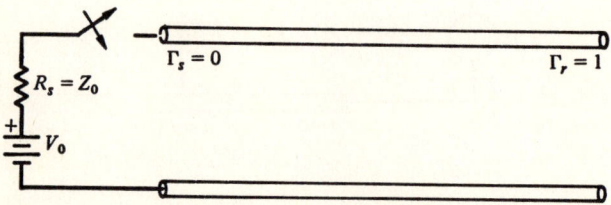

Fig. 4.8.1   Charging a transmission line.

TRANSIENT WAVES ON LOSSLESS TRANSMISSION LINES

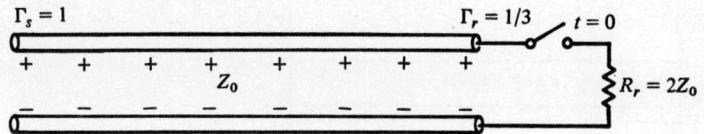

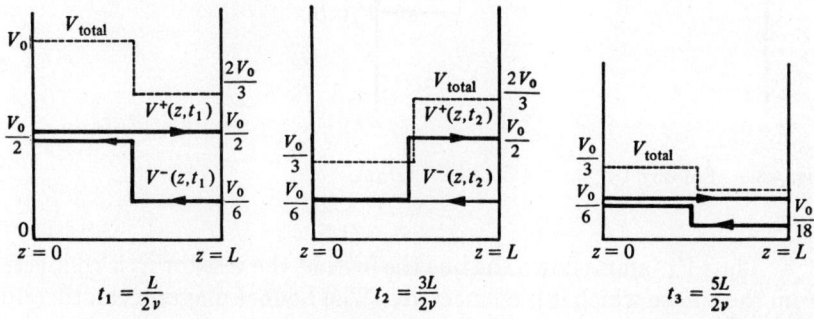

Fig. 4.8.2 Discharging a transmission line.

new negative-going wave $V_1^-(z,t)$ must be launched when the switch is closed, and so the total negative-going wave $V^-(z,t)$ (as shown in Fig. 4.8.2, at time t_1) is

$$V^-(z,t) = V^-(z,0^-) + V_1^-(z,t) = \frac{V_0}{2} + V_1^-(z,t) \quad (3)$$

The terminal voltage V_r, which is known from the previous example, is the sum of $V^+(z,0^-)$ and $V^-(z,t)$. Using the above equation for $V^-(z,t)$, we can write

$$V_r = V^+(z,0^-) + V^-(z,t) = \frac{V_0}{2} + \frac{V_0}{2} + V_1^-(z,t) \quad (4)$$

Therefore, solving for $V_1^-(z,t)$, we obtain

$$V_1^-(z,t) = V_r - V_0 = \frac{V_0}{2}(1 + \Gamma_r) - V_0 = \frac{2V_0}{3} - V_0 = -\frac{V_0}{3} \quad (5)$$

We can obtain a general expression for $V_1^-(z,t)$ by substituting $(R_r - Z_0)/(R_r + Z_0)$ for Γ_r in Eq. (5):

$$\boxed{V_1^-(z,t) = \frac{-Z_0 V_0}{R_r + Z_0}} \quad (6)$$

The wave $V_1^-(z,t)$ is illustrated in Fig. 4.8.3.

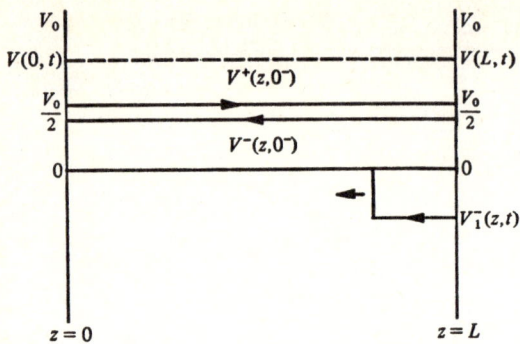

Fig. 4.8.3 Finding V_1^- for the bounce diagram.

Thus V_1^- starts down the line the instant the resistor R_r is connected from the end to which it is connected. The bounce diagram can therefore be used if it is understood that V_0 V is initially on the transmission line and if V_1^- is considered the initial disturbance on the line. Figure 4.8.4 shows the bounce diagram applicable for $R_r = 200\ \Omega$ and $Z_0 = 100\ \Omega$.

Another way of finding the wave V_1^- will now be considered. Suppose the line has been initially charged and has a static voltage of V_0 V

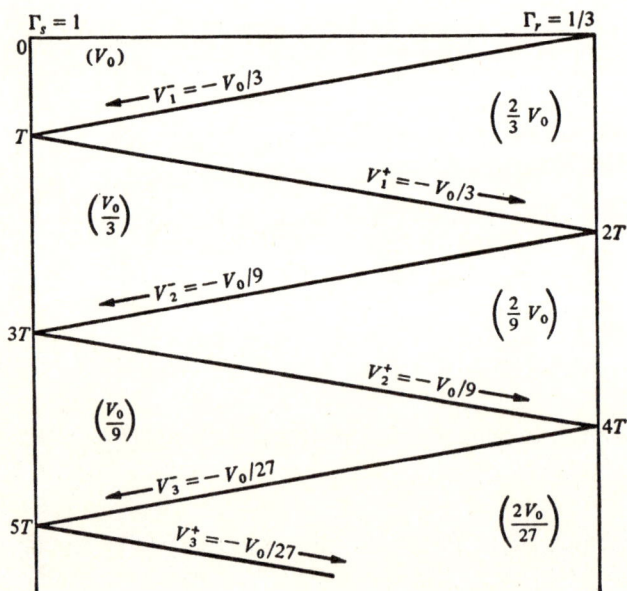

Fig. 4.8.4 Voltage bounce diagram for the discharging line.

and no net current along the entire length of the line. (Recall from Sec. 4.2 that a constant voltage is a solution of the wave equation.) An instant after R_r is connected, the total voltage on the line must be the superposition, or sum, of the initial voltage and the new wave V_1^-. Furthermore the voltage on the line must satisfy the boundary condition at the receiving end

$$V_0 + V_1^- = V_r = I_r R_r \tag{7}$$

The load current I_r must have the same value as the current I_1^-, for continuity of current, because there is no net current on the line initially. The negative-going voltage and current and I_r are related as follows:

$$I_r = I_1^- = -V_1^- Y_0 \tag{8}$$

By substituting I_1^- from Eq. (8) into (7) we can solve for V_1^-

$$V_1^- = \frac{-Z_0 V_0}{R_r + Z_0} \tag{9}$$

Once V_1^- is obtained, by either method, we can draw the bounce diagram for voltage to complete the solution. Similarly we can draw a bounce diagram for current, starting with $I_1^- = -V_1^-/Z_0$.

Using the bounce diagrams, the reader will find it instructive to plot voltage and current as a function of time at both the center of the line and the ends of the line.

For this example, we will show that the sum of the initial voltage and all the traveling waves is zero.

$$V_0 - V_0(\tfrac{1}{3} + \tfrac{1}{3} + \tfrac{1}{9} + \tfrac{1}{9} + \tfrac{1}{27} + \tfrac{1}{27} + \cdots) = 0$$

To show this, we sum the terms within the parentheses by writing the closed form for the series.

$$\frac{2}{3} + \frac{2}{9} + \frac{2}{27} + \cdots = 2\left[\frac{1}{3} + \left(\frac{1}{3}\right)^2 + \left(\frac{1}{3}\right)^3 + \cdots\right] = 2\frac{\tfrac{1}{3}}{1 - \tfrac{1}{3}} = 1$$

Thus the final value of voltage on the transmission line is zero, as expected.

As a second example, consider the transmission line shown in Fig. 4.8.5. We will assume that the switch has been closed for a long time prior to $t = 0$, so that a direct current $I_0 = V_0/2Z_0$ flows along the entire

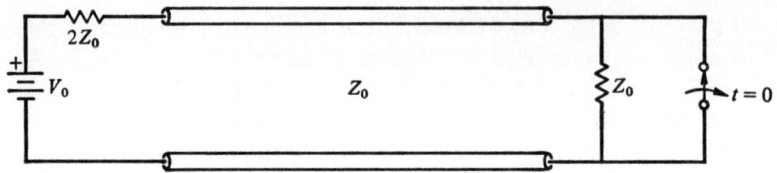

Fig. 4.8.5 A transmission line with a nonzero initial current.

length of the line and the line voltage is zero everywhere. The switch is opened at $t = 0$, and this disturbance (or sudden change in reflection coefficient) causes a new wave V_1^- to propagate toward the source. The immediate objective is to find V_1^- and to draw the bounce diagram for the voltage. To satisfy the voltage boundary conditions at the receiving end, we equate the line voltage V_1^- at the receiving end to V_r

$$V_1^- = V_r = I_r Z_0 \tag{10}$$

However, I_r can be considered to be the superposition of the initial current I_0 and the new current wave I_1^-. Hence

$$V_1^- = (I_1^- + I_0)Z_0 = (-Y_0 V_1^- + I_0)Z_0 \tag{11}$$

Solving Eq. (11) for V_1^- gives

$$V_1^- = \frac{I_0 Z_0}{2} \tag{12}$$

Since the initial current I_0 is $V_0/2Z_0$, we find that

$$V_1^- = \frac{V_0}{4} \tag{13}$$

Thus the bounce diagram can be drawn as shown in Fig. 4.8.6. Graphs of voltage as a function of distance at two different times, t_1 and t_2, are shown in Fig. 4.8.7. Voltage as a function of time, as seen at a point z_1 on the line, is sketched in Fig. 4.8.8. These graphs can be obtained directly from the bounce diagram as indicated by the dotted lines.

For a third example, the voltage on the lossless transmission line of Fig. 4.8.9 will be computed. The line is initially uncharged, and the initial current is zero. At $t = 0$ the switch is thrown to position 2, and at $t = 3T/2$ it is switched back to position 1. When the switch is changed to position 2, a voltage wave V_1^+ is launched from the sending end. A short time later, at $t = 3T/2$, the switch is returned to position 1 and a second wave V_2^+ is launched. Just before $t = 3T/2$, the voltage and current are not zero, and consequently we must account for nonzero initial conditions when V_2^+ is launched.

TRANSIENT WAVES ON LOSSLESS TRANSMISSION LINES

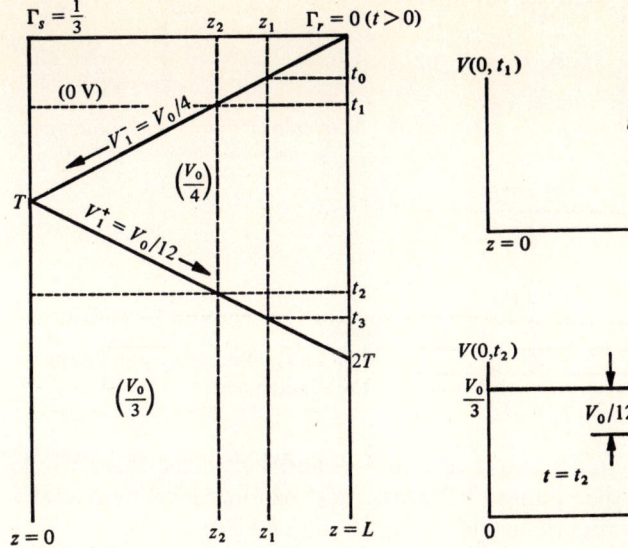

Fig. 4.8.6 The bounce diagram for the example in Fig. 4.8.5.

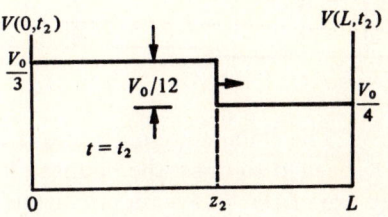

Fig. 4.8.7 Voltage distribution at times t_1 and t_2.

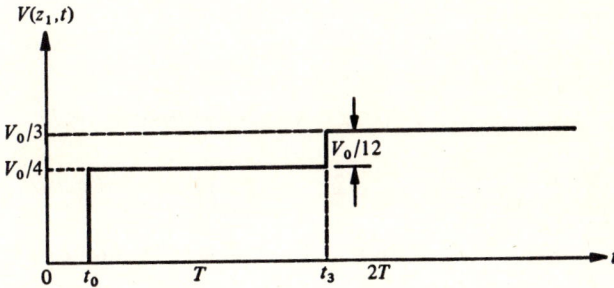

Fig. 4.8.8 The line voltage at point z_1.

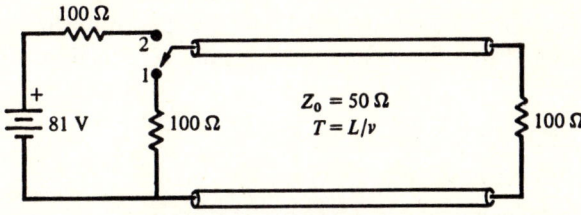

Fig. 4.8.9 A line with a switching transient.

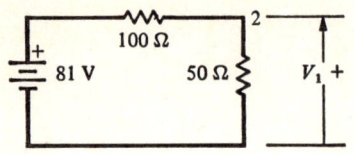

Fig. 4.8.10 Thévenin circuit at the sending end.

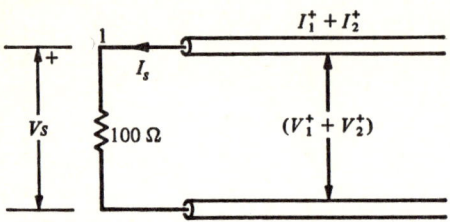

Fig. 4.8.11 Boundary conditions at the sending end.

The immediate objective is to find V_1^+ and V_2^+, since these waves are needed to draw the bounce diagrams. V_1^+ can be found most easily by use of the Thévenin circuit at the input (Fig. 4.8.10)

$$V_1^+ = {}^{50}\!/_{150}(81) = 27 \text{ V} \tag{14}$$

$$I_1^+ = \frac{V_1^+}{Z_0} = \frac{27}{50} \text{ A} \tag{15}$$

To find V_2^+, we can equate the line voltage and the terminal voltage at the sending end (the voltages and currents at the sending end are shown in Fig. 4.8.11)

$$V_1^+ + V_2^+ = V_s = I_s Z_s = -(I_1^+ + I_2^+)Z_s \tag{16}$$

$$V_1^+ + V_2^+ = -\left(I_1^+ + \frac{V_2^+}{Z_0}\right)Z_s \tag{17}$$

Since we know V_1^+ and I_1^+ from Eqs. (14) and (15), we can solve Eq. (12) for V_2^+

$$V_2^+ = -V_1^+ = -27 \text{ V}$$

We can now draw the bounce diagram for either voltage or current, and the diagram for voltage is shown in Fig. 4.8.12. A sketch of the voltage distribution on the line at $t = t_1$, taken from the voltages on the horizontal dotted line in the bounce diagram, is shown in Fig. 4.8.12.

PROGRAMMED EXERCISE 4.8.1

A PULSE GENERATOR

A transmission line 24 cm long has a characteristic impedance of 200 Ω. The velocity of propagation on the line is 3×10^8 m/s, and it is initially

TRANSIENT WAVES ON LOSSLESS TRANSMISSION LINES

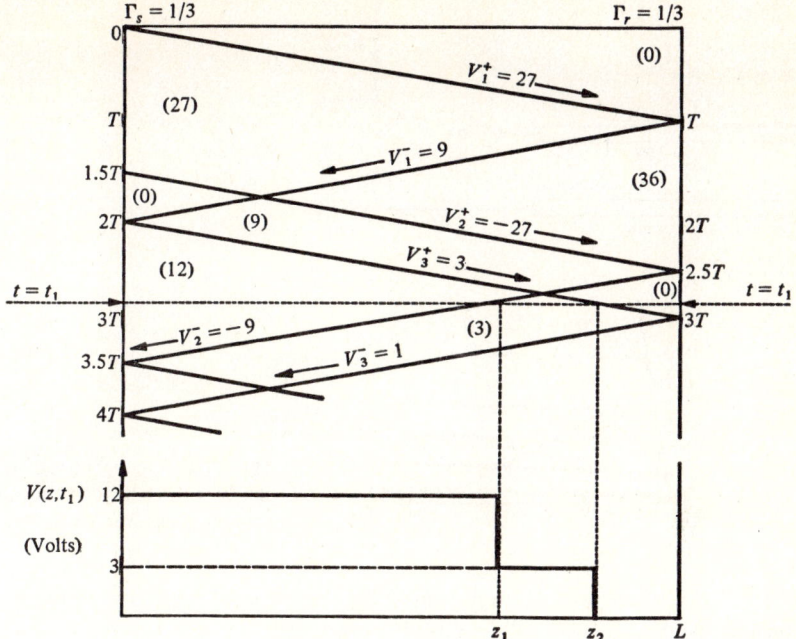

Fig. 4.8.12 Bounce diagram for a line with a switching transient.

charged to 0.4 V. At $t = 0$, the line is connected to the base of the NPN transistor as shown in Fig. 4.8.13. After the switch is closed, the dc input resistance of the transistor is 200 Ω (we will neglect the input capacitance). The reader should verify the following statements:

1. For $t < 0$, $V_0 = 9$ V, since the base current is zero during this time.
2. For $0 < t < 1.6$ ns, the base current is 1.0 mA and V_0 is 5 V.
3. For $t > 1.6$ ns, the transmission line is discharged, the base current is zero, and the collector voltage is $+9$ V. Thus a 4-V pulse of 1.6 ns duration has been generated. Note that the pulse width can be accurately controlled by the length of the transmission line and the pulse amplitude controlled by the load resistor and bias voltage.

4.9 DISCONTINUITIES ON A TRANSMISSION LINE

A discontinuity is said to exist at any point on the line from which waves are reflected. The unmatched terminations we considered previously are examples of discontinuities, but we now consider some other types. A discontinuity results at the junction of a transmission line composed of

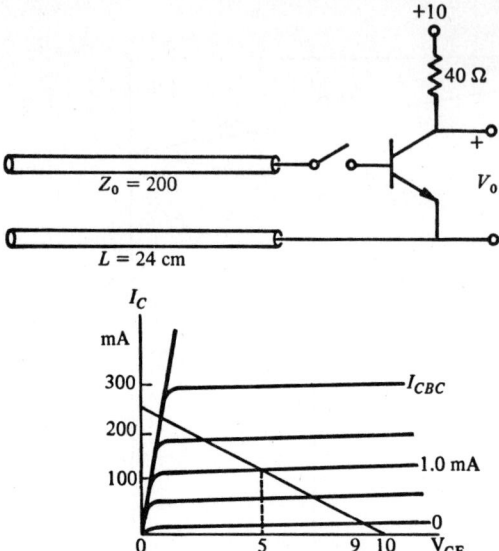

Fig. 4.8.13 A transmission line used to generate a pulse of known duration.

two (or more) sections each having different characteristic impedance. A discontinuity also results if a circuit composed of lumped elements is connected at some point on a uniform transmission line. Discontinuities also occur where the geometry of the line is changed, perhaps due to damage, or where the dielectric material changes, perhaps due to the intrusion of moisture. At discontinuities there will be a reflected wave and generally a transmitted wave due to every incident wave which impinges on the discontinuity. Bounce diagrams can be used, and total voltage or current on the transmission line at any point and at any time can be found by superposition.

A transmission-line network which can be used to illustrate three types of discontinuity is shown in Fig. 4.9.1. Reflected and transmitted waves will be calculated for this rather complicated discontinuity, and the results obtained include the simpler cases in which $R_1 = 0$, or $R_2 = \infty$, or both.

For a positive-going wave starting from the left end of the line, the Thévenin equivalent circuit is shown in Fig. 4.9.2. The reflection coeffi-

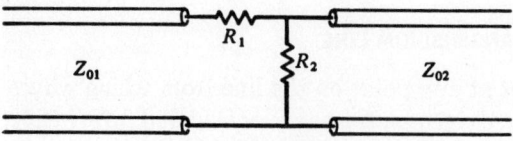

Fig. 4.9.1 Discontinuity on a line.

Fig. 4.9.2 From the equivalent circuit for a positive-going wave, we find that the apparent termination impedance is

$$Z_d = R_1 + \frac{R_2 Z_{02}}{R_2 + Z_{02}}$$

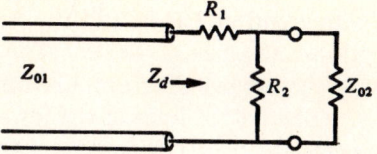

cient Γ_d at the discontinuity for a positive-going voltage wave is the same as if the network of resistors R_1, R_2, and Z_{02} terminated the line with an impedance Z_d, and is

$$\Gamma_d = \frac{Z_d - Z_{01}}{Z_d + Z_{01}} = \frac{R_1 + R_2 Z_{02}/(R_2 + Z_{02}) - Z_{01}}{R_1 + R_2 Z_{02}/(R_2 + Z_{02}) + Z_{01}} \quad (1)$$

The reflection coefficient at simpler discontinuities can be found from Eq. (1) by setting $R_1 = 0$, or letting R_2 approach infinity, or letting $Z_{02} = Z_{01}$, as appropriate.

The transmission coefficient τ_{12} is defined as the ratio of the wave transmitted into the second medium (due to a wave in the first medium incident at a discontinuity) to the incident wave in the first medium. For the discontinuity shown in Fig. 4.9.1, the voltage wave transmitted onto the second section of line is the voltage which will appear across the parallel combination of R_2 and Z_{02} in Fig. 4.9.2 due to a voltage wave incident from the first section. It should be noted that the total voltage at the left side of the discontinuity is $1 + \Gamma_d$ times the voltage wave incident at the discontinuity, i.e., the sum of the incident and reflected waves. Therefore the transmission coefficient can be computed

$$\tau_{12} = \frac{\text{transmitted voltage on 2d section}}{\text{incident voltage on 1st section}}$$

$$= (1 + \Gamma_d) \frac{R_2 Z_{02}/(R_2 + Z_{02})}{R_1 + R_2 Z_{02}/(R_2 + Z_{02})} \quad (2)$$

The transmission coefficient for simpler discontinuities can be found from Eq. (2) by appropriate substitutions, as discussed above for the reflection coefficient.

Reflection and transmission coefficients for a negative-going wave incident on the right side of the discontinuity (Fig. 4.9.1) can also be calculated. Because of the lack of symmetry, the results will be different from those given in Eqs. (1) and (2). These derivations will be left for exercises, however.

As an example of a discontinuity on a transmission line, a line of length L has a characteristic impedance of 100 Ω, and a resistance of 150 Ω is connected between conductors of the transmission line at $Z = L/3$. A battery of 3 V with an internal resistance of 200 Ω is connected to the sending end at $t = 0$, and the transmission line is terminated at $Z = L$ with a resistance of 400 Ω. The transmission line is shown in Fig. 4.9.3 with the bounce diagram for voltage below it.

The dotted lines in the lower part of the bounce diagram indicate a superposition of transmitted and reflected waves along the same path. The reader will find it instructive to continue the bounce diagram.

After a long period of time $(t \to \infty)$ the transients die, and the voltage approaches steady state, as shown in Sec. 4.6. The final values of voltage

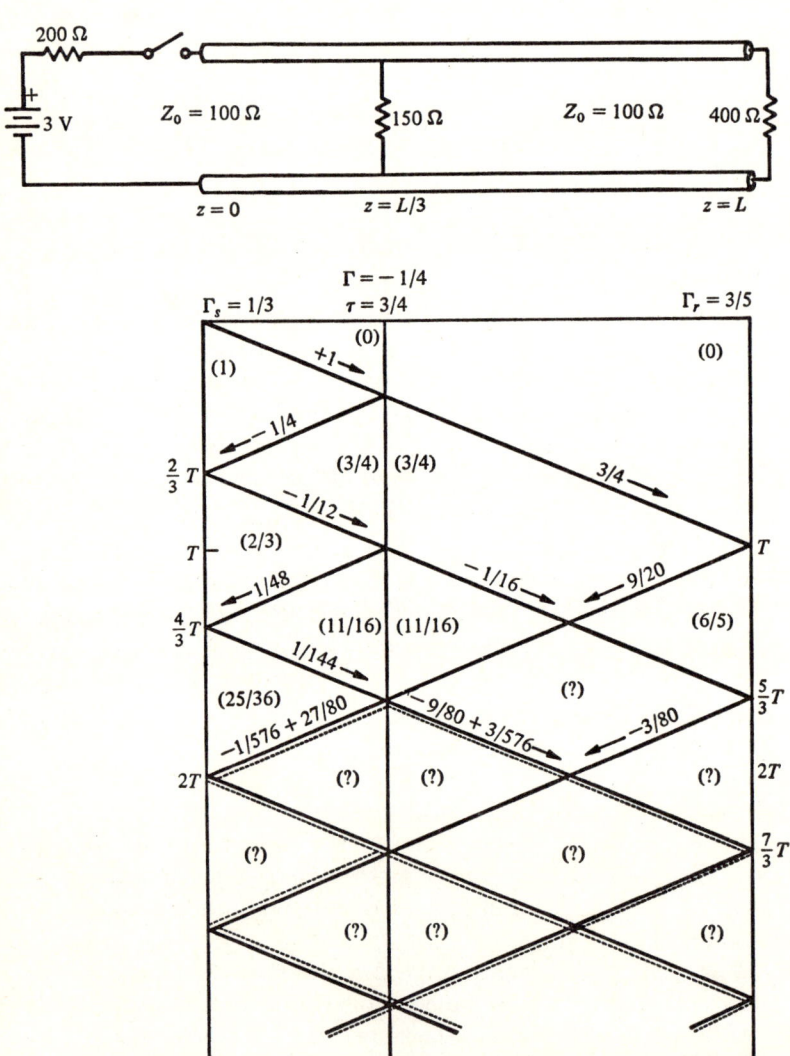

Fig. 4.9.3 Line with a discontinuity and the voltage bounce diagram.

TRANSIENT WAVES ON LOSSLESS TRANSMISSION LINES

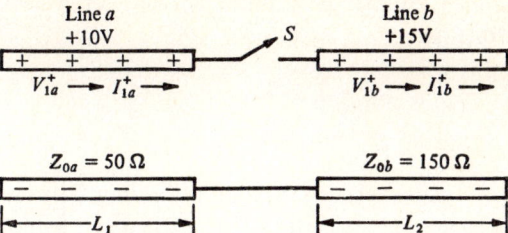

Fig. 4.9.4 Charged transmission lines.

and current can be obtained by treating the system as a dc circuit. Thus the final value of voltage V_f on the transmission line is the same as the final voltage across the 150-Ω resistance or the 400-Ω resistance and is

$$V_f = \frac{3 \dfrac{150(400)}{150 + 400}}{200 + \dfrac{150(400)}{150 + 400}} = \frac{18}{17} \text{ V}$$

Care must be taken to include the effect of all voltage waves. For example, between $z = 0$ and $z = L/3$, from $t = 5T/3$ to $t = 2T$, there are two negative-going waves. One is a reflection of the positive-going wave incident on the discontinuity, and one is transmitted through the discontinuity, as indicated by the dotted line.

The next example illustrates a case where a discontinuity and non-zero initial conditions must both be taken into account. Two ideal lossless transmission lines are connected together by the switch S (Fig. 4.9.4). Before the switch is closed, each line has reached a steady-state condition with 10 V on line a and 15 V on line b. The characteristic impedances are 50 and 150 Ω for lines a and b respectively. The line lengths are such that

$$\frac{L_1}{v_1} = \frac{L_2}{v_2}$$

so that the transit times are the same for the lines.

The switch is closed at $t = 0$, and the problem is to describe the voltage distribution on the lines after closing the switch. Voltage waves V_{1a}^- and V_{1b}^+ will be launched on lines a and b respectively, as a result of closing the switch. After closing the switch, the boundary conditions are as follows: (1) the voltage at the right end of line a must equal the voltage at the left end of line b, and (2) the currents at these ends must be equal:

$$10 + V_{1a}^- = 15 + V_{1b}^+$$
$$Y_{0a}(-V_{1a}^-) = Y_{0b}(V_{1b}^+) \tag{3}$$

These equations can be solved simultaneously for the two unknown voltages

$$V_{1a}^- = +1.25 \text{ V}$$
$$V_{1b}^+ = -3.75 \text{ V} \tag{4}$$

An alternate method of finding these waves originating at the switch is to use superposition. For example, from Eq. (6) of Sec. 4.8, a negative-going wave of -2.5 V will start at the switch and travel to the left on line a. The wave is due to the change in reflection coefficient when the switch is closed and the two 5-V waves on line a before the switch was closed. Since $\tau_{ba} = 0.5$, a voltage wave of 3.75 V is transmitted from line b to line a, and the total negative-going wave on line a is then

$$V_{1a}^- = -2.5 + 3.75 = 1.25$$

as in Eq. (4). Similarly,

$$V_{1b}^+ = -11.25 + 7.5 = -3.75$$

Drawing the bounce diagrams can now begin, as in Fig. 4.9.5. Note that the reflection coefficients at both the open-circuited ends are $+1$, so that $V_{1a}^+ = +1.25$ V and $V_{1b}^- = -3.75$ V. These waves arrive at the switch, where the reflection coefficient for line a is

$$\Gamma_a = \frac{Z_{0b} - Z_{0a}}{Z_{0b} + Z_{0a}} = +\frac{1}{2}$$

and the transmission coefficient (for a wave propagating from line a to line b) is

$$\tau_{ab} = 1 + \Gamma_a = +\tfrac{3}{2}$$

Similarly for line b the reflection coefficient is

$$\Gamma_b = -\Gamma_a = -\tfrac{1}{2}$$

and the transmission coefficient is

$$\tau_{ba} = 1 + \Gamma_b = +\tfrac{1}{2}$$

Now one finds V_{2a}^- as the sum, or superposition, of the reflected portion of the incident wave V_{1a}^+ and the transmitted portion of the incident wave V_{1b}^-

$$V_{2a}^- = \Gamma_a V_{1a}^+ + \tau_b V_{1b}^- = \tfrac{1}{2}(1.25) + \tfrac{1}{2}(-3.75) = -1.25 \text{ V}$$

TRANSIENT WAVES ON LOSSLESS TRANSMISSION LINES

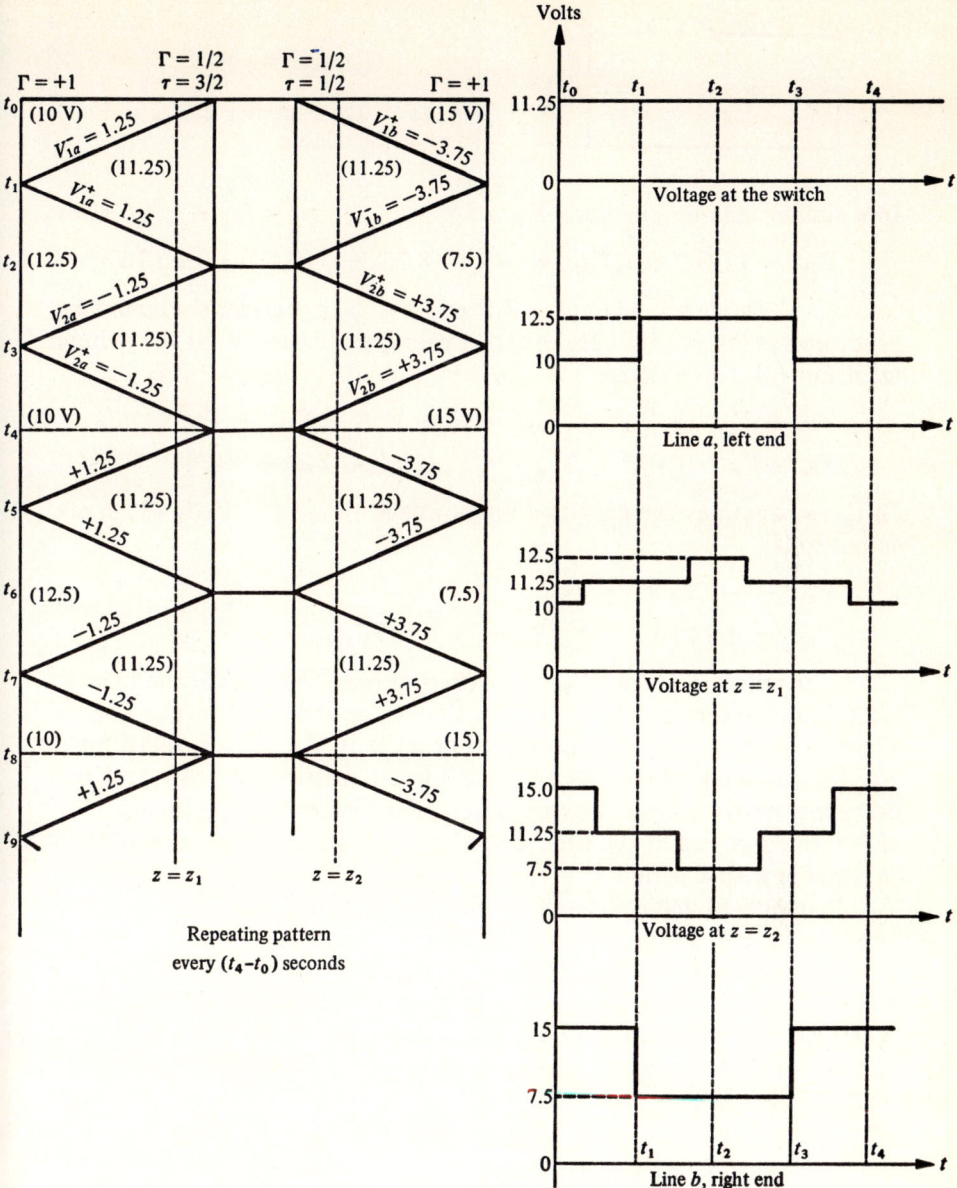

Fig. 4.9.5 Bounce-diagram solution for the charged-lines problem.

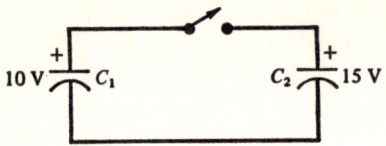

Fig. 4.9.6 The capacitor analogy of the steady-state solution.

In a similar manner one finds $V_{2b}{}^+$:

$$V_{2b}{}^+ = \Gamma_b V_{1b}{}^- + \tau_a V_{1a}{}^+ = -\tfrac{1}{2}(-3.75) + \tfrac{3}{2}(1.25) = +3.75 \text{ V}$$

An alternate way of finding $V_{2a}{}^-$ and $V_{2b}{}^+$ is to consider the boundary conditions at the switch. By equating voltages and currents at the switch, as in Eq. (3), one obtains

$$10 + V_{1a}{}^- + V_{1a}{}^+ + V_{2a}{}^- = 15 + V_{1b}{}^+ + V_{1b}{}^- + V_{2b}{}^+$$
$$Y_{0a}(-V_{1a}{}^- + V_{1a}{}^+ - V_{2a}{}^-) = Y_{0b}(V_{1b}{}^+ - V_{1b}{}^- + V_{2b}{}^+)$$

In these equations the only two unknowns are $V_{2a}{}^-$ and $V_{2b}{}^+$, which are found to be

$$V_{2a}{}^- = -1.25 \text{ V}$$
$$V_{2b}{}^+ = +3.75 \text{ V}$$

The waves $V_{2a}{}^-$ and $V_{2b}{}^+$ are reflected from the ends of the lines at time t_3 and return to the switch at time t_4, as indicated in the bounce diagram in Fig. 4.9.5. At this time, line a has 10 V uniformly distributed on the line, and line b has 15 V uniformly distributed. All previous traveling waves cancel, leaving a set of initial conditions identical to those which existed at the time the switch was closed. Thus the solution is periodic with a period $T = t_4$.

Sketches of voltage are shown for the open-circuited ends, for points near the switch, and at the switch in Fig. 4.9.5. Note that the voltage at the switch is constant at 11.25 V for all time after $t = 0$. This is the same voltage that one would find for the steady-state solution of the capacitor problem as indicated in Fig. 4.9.6. As an interesting exercise, the reader should consider the total energy before and after closing the switch.

4.10 REACTIVE TERMINATIONS

Transients on a transmission line which is terminated in a reactance provide a situation in which the reflection coefficient is useful as a concept but is of little help in solving problems. Problems of this nature are best approached by means of Thévenin's theorem. For example, let a transmission line of length L and characteristic impedance Z_0 be terminated with an inductor of value L_0 H having zero initial current. In the circuit

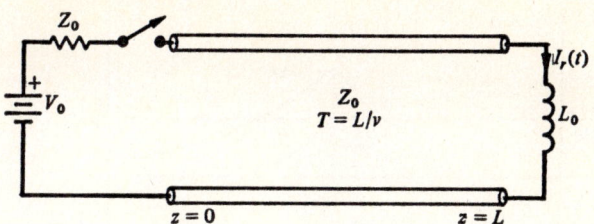

Fig. 4.10.1 Transmission line with inductive termination.

shown in Fig. 4.10.1, the switch is closed at $t = 0$, connecting the voltage source to the line through the source impedance. The voltage wave given by

$$V_1^+\left(t - \frac{z}{v}\right) = \frac{V_0}{2} U\left(t - \frac{z}{v}\right)$$

starts down the line at $t = 0$. From the equivalent circuit at the receiving end, as shown in Fig. 4.10.2, the differential equation of the current $I_r(t)$ is obtained by writing the loop equation

$$L_0 \frac{dI_r(t)}{dt} + Z_0 I_r(t) = V_0 U(t - T) \tag{1}$$

This well-known differential equation can be solved using Laplace transform techniques or by other standard means

$$I_r(t) = Y_0 V_0 U(t - T)\left(1 - \exp\frac{t - T}{Y_0 L_0}\right) \tag{2}$$

Now that $I_r(t)$ is known, the receiving-end voltage can be found from Fig. 4.10.2

$$V_r(t) = V_0 U(t - T) - Z_0 I_r(t) \tag{3}$$

Since $V_1^-(L,t) = V_r(t) - V_1^+(L,t)$, the negative-going voltage wave reflected from the inductor is

$$V_1^-\left(t - \frac{z}{v}\right) = V_0 U\left(t - 2T + \frac{z}{v}\right)\left[-\frac{1}{2} + \exp\left(-\frac{t - 2T + z/v}{Y_0 L_0}\right)\right] \tag{4}$$

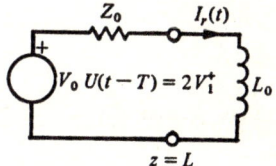

Fig. 4.10.2 Thévenin equivalent circuit at the receiving end.

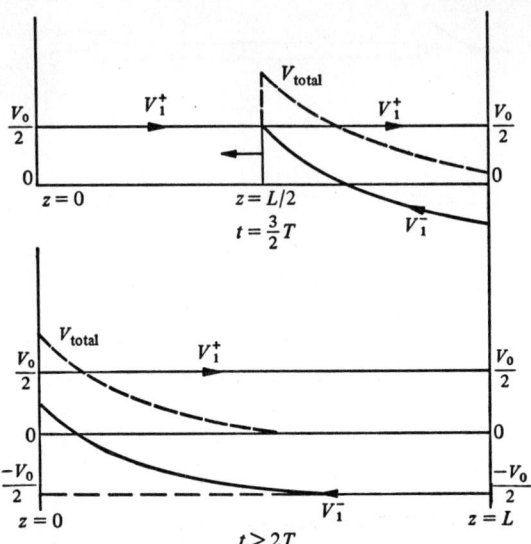

Fig. 4.10.3 Voltage on the line at two different times.

The positive-going and negative-going voltage waves are shown in Fig. 4.10.3 at $t = 3T/2$ and at time just after $t = 2T$. At the instant the positive-going voltage wave first impinges on the inductor, which has no initial current, the voltage wave is reflected as if the line were terminated in an open circuit (because current in an inductor cannot change instantaneously). When the current in the inductor has reached its final value, the voltage across the inductor is zero and the wave is reflected as if the inductor were a short circuit. Current waves on the line are found by dividing the positive-going voltage wave by Z_0 and negative-going voltage wave by $-Z_0$.

As a second example, let us consider an uncharged capacitor C at the receiving end of a transmission line of length L and characteristic impedance Z_0. At $t = 0$ a switch is closed, connecting the battery of voltage V_0 to the transmission line through a resistance of Z_0. This circuit is illustrated in Fig. 4.10.4. Qualitatively, it can be seen that at the instant

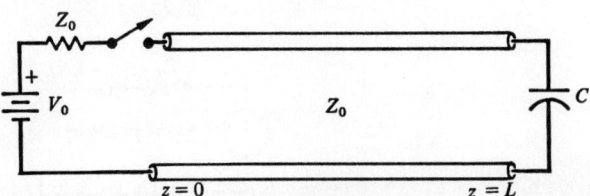

Fig. 4.10.4 Transmission line with capacitive termination.

TRANSIENT WAVES ON LOSSLESS TRANSMISSION LINES

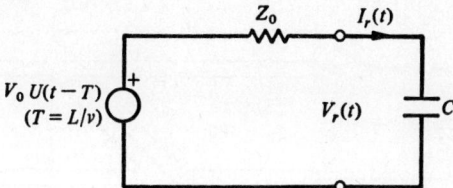

Fig. 4.10.5 Thévenin equivalent circuit at the receiving end.

the wavefront reaches the capacitor the uncharged capacitor will appear as a short circuit, so that the wave will be reflected with a reflection coefficient of -1 *at that instant*. As the capacitor charges, part of the incident wave is reflected, and finally, when the capacitor is totally charged, the entire incident wave is reflected with a reflection coefficient of $+1$, as if the capacitor were an open circuit.

The Thévenin equivalent circuit is shown in Fig. 4.10.5, and at $t = T$, the positive-going voltage wave, with magnitude of $V_0/2$, arrives at the receiving end.

From the equivalent circuit it is easily found that the receiving-end voltage is

$$V_r(t) = V_0 U(t - T)\left[1 - \exp\left(-\frac{t-T}{Z_0 C}\right)\right] \tag{5}$$

The negative-going voltage wave, reflected from the capacitor, is

$$V_1^-(L,t) = V_0 U(t - T)\left[\tfrac{1}{2} - \exp\left(-\frac{t-T}{Z_0 C}\right)\right] \tag{6}$$

at the receiving end, and an expression for the reflected wave at other points on the line is

$$V_1^-(z,t) = V_0 U\left(t - 2T + \frac{z}{v}\right)\left[\frac{1}{2} - \exp\left(-\frac{t - 2T + z/v}{Z_0 C}\right)\right] \tag{7}$$

The positive- and negative-going voltage waves on the line are shown at $t = 3T/2$ and at a time just after $t = 2T$ in Fig. 4.10.6.

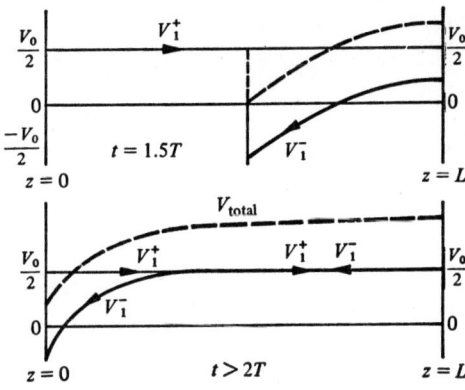

Fig. 4.10.6 Voltage on the line at two different times.

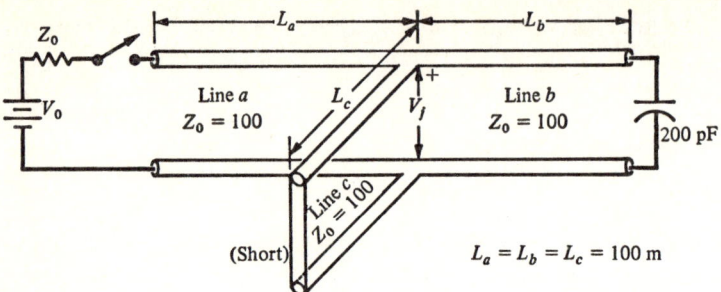

Fig. 4.10.7 An example of discontinuities and reactive terminations.

As another example, consider the three transmission lines shown in Fig. 4.10.7. All the lines are identical, with $Z_0 = 100$ Ω and a length of 100 m. For simplicity, suppose that the velocity of propagation on each of the lines is 3×10^8 m/s. When the switch is closed at $t = 0$, a wave V_{1a}^+ is launched, and it takes $T = \frac{1}{3}$ μs to reach the junction. As V_{1a}^+ impinges on the junction, a portion of the wave will be reflected back into line a and a portion will be transmitted into lines b and c. The objective of this example is to find the voltages as a result of these waves on all three lines.

As the switch closes, the initial wave on line a is

$$V_{1a}^+(z,t) = \frac{Z_0 V_0}{Z_0 + Z_0} U\left(t - \frac{z}{v}\right) = \frac{V_0}{2} U\left(t - \frac{z}{v}\right) \quad \text{V} \tag{8}$$

At $t = \frac{1}{3}$ μs, V_{1a}^+ arrives at the junction, where the effective load on line a is the parallel combination of the characteristic impedances of lines b and c. The Thévenin equivalent circuit for line a is shown in Fig. 4.10.8. With respect to line a, the reflection and transmission coefficient are

$$\Gamma = \tfrac{1}{3} \quad \text{and} \quad \tau = \tfrac{2}{3} \tag{9}$$

Consequently the junction voltage and the voltage reflected back on line a can be computed

$$V_j(t) = \tau V_{1a}^+ = \frac{V_0}{3} \text{V} \quad \text{for } \tfrac{1}{3} \leq t < 1.0 \text{ μs} \tag{10}$$

$$V_{1a}^- = \Gamma V_{1a}^+ = \frac{V_0}{6} U\left(t - \frac{2}{3}\text{μs} + \frac{z}{v}\right) \text{V} \tag{11}$$

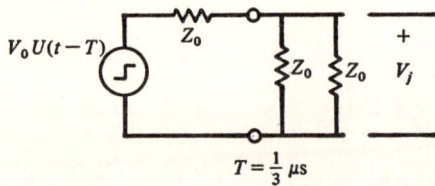

Fig. 4.10.8 The Thévenin equivalent circuit at the junction.

TRANSIENT WAVES ON LOSSLESS TRANSMISSION LINES

Waves will be launched on lines b and c as a result of the existence of a voltage at the junction

$$V_j = V_{1b}^+ = V_{1c}^+ = \frac{V_0}{3} \quad \text{for } \tfrac{1}{3} \le t < 1.0 \; \mu\text{s} \tag{12}$$

As V_{1c}^+ impinges on the short circuit at the end of line c, a reflected wave V_{1c}^- will be launched. Similarly a reflected wave V_{1b}^- will exist as a result of wave V_{1b}^+ impinging on the capacitor at the end of line b. The negative-going wave reflected from the capacitor can be calculated by the method given in the second example of this section in Eq. (7)

$$V_{1b}^-(z,t) = \tfrac{2}{3} V_0 U\left(t - 1\;\mu\text{s} + \frac{z}{v}\right)\left[\frac{1}{2} - \exp\left(-\frac{t - 1\;\mu\text{s} + z/v}{Z_0 C}\right)\right] \tag{13}$$

The two negative-going waves V_{1b}^- and V_{1c}^- reach the junction at the same time ($t = 1.0 \; \mu\text{s}$), where they are reflected and transmitted to form new waves on lines a, b, and c

$$V_{2a}^- = \tau V_{1b}^- + \tau V_{1c}^- \qquad t \ge 1.0 \; \mu\text{s} \tag{14}$$
$$V_{2b}^+ = \Gamma V_{1b}^- + \tau V_{1c}^- \qquad t \ge 1.0 \; \mu\text{s} \tag{15}$$
$$V_{2c}^+ = \tau V_{1b}^- + \Gamma V_{1c}^- \qquad t \ge 1.0 \; \mu\text{s} \tag{16}$$

Bounce diagrams can be used advantageously as a bookkeeping device to keep track of the various waves if one is careful to distinguish those waves which change exponentially from those which change as step functions. Bounce diagrams are shown for lines b and c in Fig. 4.10.9, where the hatched lines indicate those waves which change exponentially. Line a has been replaced by its Thévenin equivalent circuit and appears at the junction at the center of the diagram. Since there are no reflections from the source impedance, V_{1a}^+ is the only positive-going wave on line a.

The junction voltage and voltage distributions on lines b and c at times $t = \tfrac{1}{2}, \tfrac{3}{4}$, and $\tfrac{5}{4} \; \mu\text{s}$ are shown in Fig. 4.10.9. These voltages were computed using Eqs. (8) to (16) and the bounce diagrams. The time constant T_c for the exponential is critical for sketching the voltages:

$$T_c = Z_0 C = 100(200 \text{ pF}) = 0.02 \; \mu\text{s} \tag{17}$$

The exponential change will be essentially complete in five time constants, which is $0.10 \; \mu\text{s}$. Since the wave is traveling with a velocity of 3×10^8 m/s, the voltage changes exponentially for approximately 30 m of line.

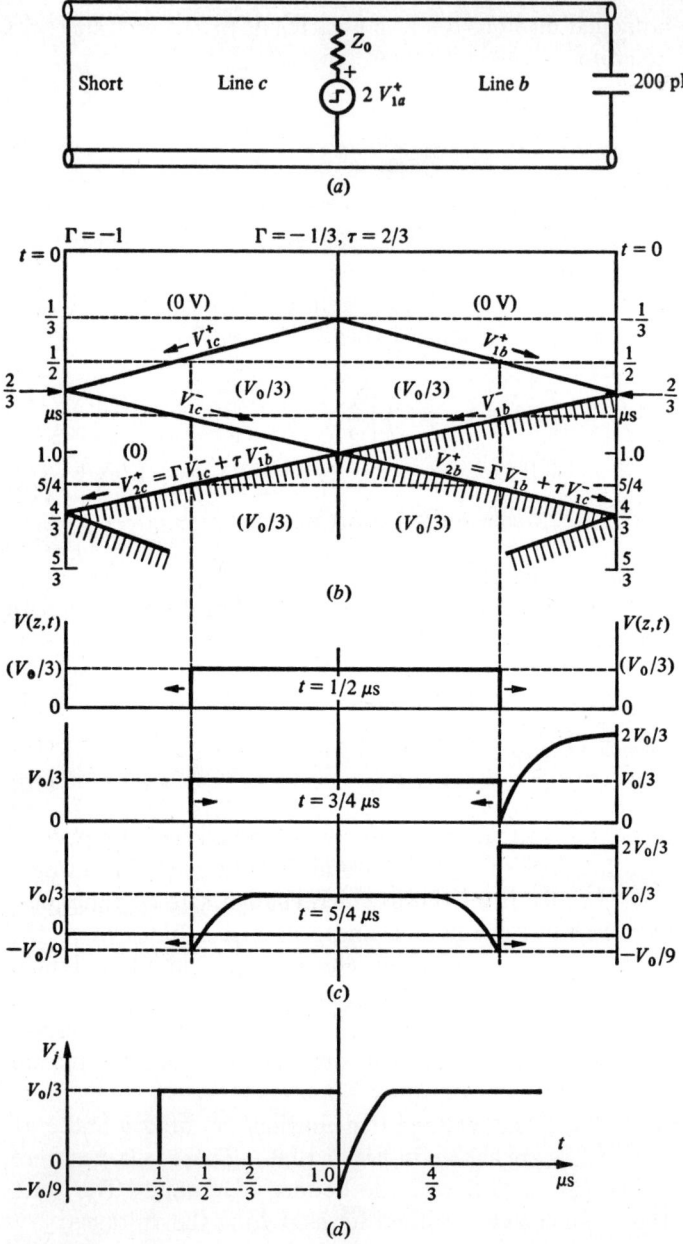

Fig. 4.10.9 Voltages for lines b and c. (*a*) Geometry of the lines. (*b*) Bounce diagrams. (*c*) Voltages versus distance. (*d*) Junction voltage.

Each line is 100 m long, and so one can scale the voltage sketches to show the approximate voltage distributions, as shown in Fig. 4.10.9c.

PROBLEMS

1. Find capacitance per unit length for a coaxial cable. The inner conductor has radius a, and the outer conductor has radius b.
2. Show that $E_r/H_\phi = \pm \eta = \pm \sqrt{\mu/\varepsilon}$ for the fields on coaxial cable.
3. Find $\mathbf{E} = E_r \mathbf{a}_r$ in terms of $V(z,t)$ from Maxwell's equations in differential form for coaxial line. Assume $\mathbf{H} = H_\phi \mathbf{a}_\phi$.
4. Show that the following functions are solutions of the wave equation:
   (a) $V(r,t) = (t - kr) \sin[\omega(t - kr)]$ for $k =$ constant.
   (b) $I(z,t) = U(t - kz) - e^{j\omega(t-kz)}$.
   (c) $P(a,b) = (a + kb)^2$ for $a = z$, $b = t$. (Then try $b = 1/t$.)
5. From Maxwell's equations derive an equivalent circuit for a differential length of transmission line assuming σ is not zero. Assume $\mathbf{H} = H_\phi \mathbf{a}_\phi$ and $\mathbf{E} = E_r \mathbf{a}_r$.
6. Derive Eq. (20) of Sec. 4.2, which is analogous to the telegrapher's equation given in Eq. (23) of Sec. 4.2.
7. The magnetic energy stored between conductors in a coaxial cable is given by $LI^2/2$ J/m, where L is the external inductance per meter. Another expression for this energy is the integral $\frac{\mu}{2}\int_v H^2\, dv$, where the integral is taken throughout the volume *between* conductors. Derive Eq. (22) of Sec. 4.2 for the external inductance per unit length from these expressions.
8. (a) Find $\mathbf{H}$ inside the conductors of coaxial line due to a low-frequency current, using the geometry shown in Fig. 4.2.1.
   (b) Using the volume integral of Prob. 7, find the energy per unit length of line stored in the conductors due to the magnetic field. This energy can be equated to $L_i I^2/2$, where L_i is the internal inductance per meter. Thus the total inductance per meter is the sum of the external inductance, found in the previous problem, and the internal inductance. Find the internal inductance L_i.
   (c) If the frequency is increased until the skin depth is negligible, what becomes of L_i?
9. Show that $1/\sqrt{LC} = 1/\sqrt{\mu\varepsilon}$ for coaxial cable, as claimed in Eq. (34) of Sec. 4.2.
10. Derive Eq. (4) of Sec. 4.3, one of the telegrapher's equations for a lossy line, from Eq. (3) for coaxial cable using the assumptions of Sec. 4.2.
11. The following voltage is applied at the left end ($z = 0$) of a lossless transmission line having characteristic impedance $Z_0 = 50\ \Omega$ and length $L = 300$ m. At $z = L$ is a short circuit, and the source acts as a short for reflected waves. Assume $v_p = 3 \times 10^8$ m/s on the line. Plot (a) $V(t)$ at both $z = 150$ m and $z = 270$ m for $0 \leq t \leq 5L/v_p$ and (b) $I(z)$ at both $t = 1.3\ \mu$s and $t = 2.6\ \mu$s.

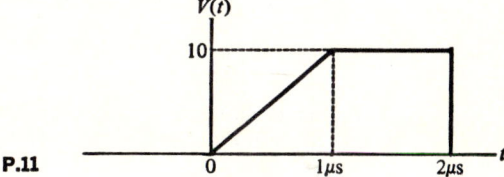

Fig. P.11

12. (a) Plot current versus time for $0 \leq t \leq 4L/v$ at $z = L/5$ by means of a bounce diagram.
(b) What are the final positive-going and negative-going voltage waves?

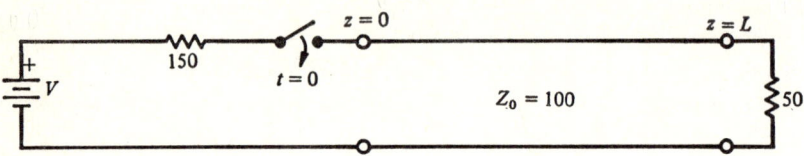

Fig. P.12

13. A battery having open-circuit voltage V_0 and internal resistance R is connected to an open-circuited transmission line with characteristic impedance Z_0. As t approaches infinity, find the total positive-going and negative-going voltage waves while the battery is still connected.

14. Draw the voltage bounce diagrams for $0 \leq t \leq 8$ μs and find the final values of voltages from a circuit-theory standpoint.

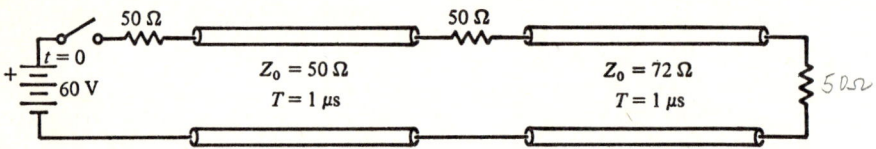

Fig. P.14

15. The voltage generator V_s produces a triangular pulse as shown. Draw the waveform for V_{in} and determine whether the circuit acts as an integrator or a differentiator. How can the differentiation or integration be improved, and at what expense?

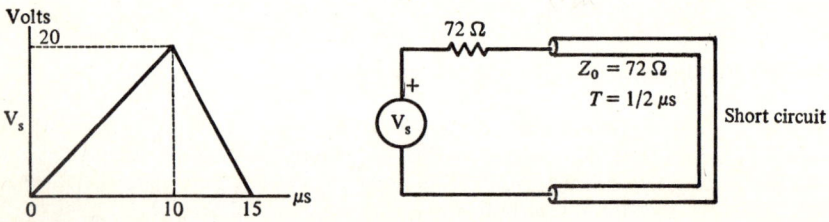

Fig. P.15

16. When the transistor in the circuit diagramed is held in saturation by the base current i_s, it appears as a short circuit. When the base current is zero, the transistor is cut off and appears as an open circuit. The base current is zero from $t_0 = 0$ to $t_1 = 1$ μs, as shown. (a) Find the current through R_L during the time interval $t_0 < t < t_1$. (b) Find the voltage at $z = L$ during the time interval $1.0 < t < 3.0$ μs.

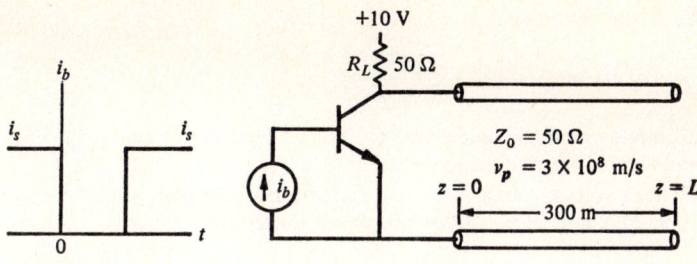

Fig. P.16

17. (a) Plot voltage at the receiving end for $0 \leq t \leq 4L/v_p$.
(b) Plot voltage versus z at $t = 3L/4v_p$ and at $t = 7L/4v_p$.

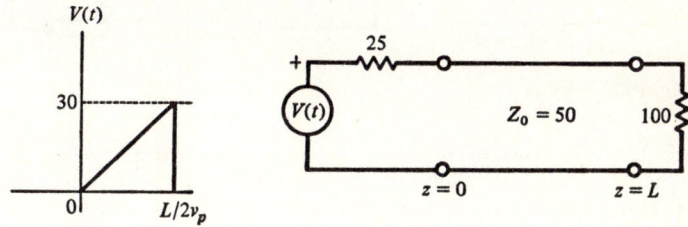

Fig. P.17

18. A transmission line with characteristic impedance Z_0 is shorted at both ends and has an initial current I_0 flowing down it. At $t = 0$, the receiving end is changed from a short circuit to a resistance $R_r = 3Z_0$. Plot V_r for $0 \leq t \leq 8T = 8L/v$.

19. If a battery of V_0^- V is suddenly connected to a lossless coaxial transmission line, show that current is continuous at the wavefront, and of proper magnitude, by finding displacement current.

20. For $t < 0$ and $t > 1$ ms, the switch is in position A, and for $0 < t < 1$ ms, the switch is in position B. The transmission line is lossless, and $l/v = 10$ ms. Plot voltage and current at the receiving end of the line for 15 ms, and plot voltage and current versus distance at $t = 10$, 10.5, and 11 ms.

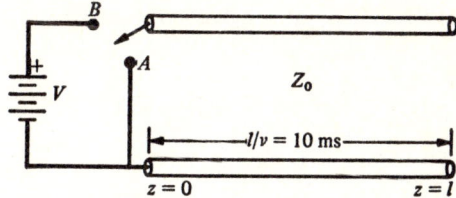

Fig. P.20

21. Repeat Prob. 14 if the transmission line is shorted at the right end.

22. A 1-V pulse of 2 ms duration starts down a line at $t = 0$. If $Z_0 = 100$, plot voltage and current waves and total voltage and current as a function of distance on the line at $t = 11.5$ ms if $l/v = 10$ ms and (a) $R_r = 50$; (b) $R_r = 150$.

23. If the sending-end current is given by

$$I_s(t) = \frac{V_s(t)}{R_s + Z_0} - \frac{2V^-(t)}{R_s + Z_0}$$

after one round trip, show that the second term is the sum of the first negative-going and second positive-going waves.

24. Plot voltage versus t for two round trips, and plot voltage versus z for $t = 1, 5, 10.5, 20.5,$ and 25 ms, given the circuit illustrated.

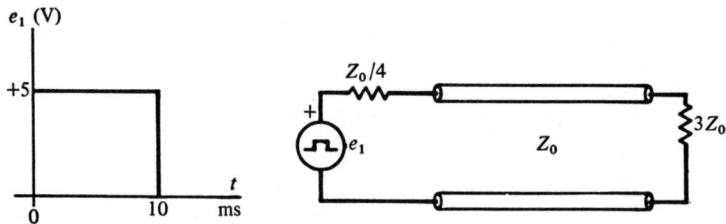

Fig. P.24

25. At $t = 0$, the switch is closed. (a) Draw bounce diagrams for voltage and current, until time $6l/v$. (b) Plot current versus time at $z = l/3$, and plot voltage versus distance at $t = 13L/v$. (c) Find the final values of V^+ and I^-.

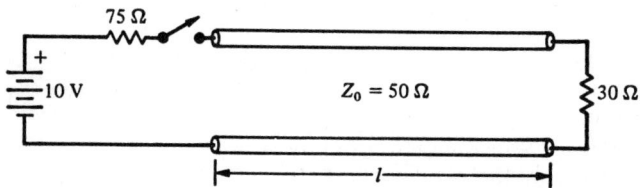

Fig. P.25

26. For the circuit shown, the switch is in position 2 for a long time, and then is in position 1 for 1 ms, and is then in position 2 for all time. (a) Draw bounce diagrams for voltage and current. (b) Plot sending-end and receiving-end voltage versus time. (c) Plot voltage at a distance of 1 km from the sending end versus time.

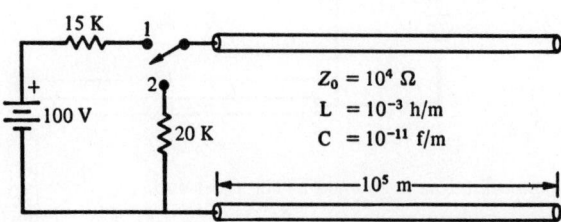

Fig. P.26

TRANSIENT WAVES ON LOSSLESS TRANSMISSION LINES

27. In the circuit shown, the switch is closed at $t = 0$. (a) Plot voltage and current at the sending end versus time. (b) Plot voltage and current at the center of the line versus time.

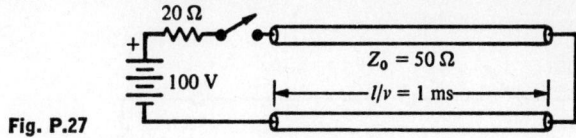

Fig. P.27

28. The voltage e_1 is applied to the circuit shown. Plot both positive- and negative-going voltages on the same graph at $t = L/v + 4$ ms.

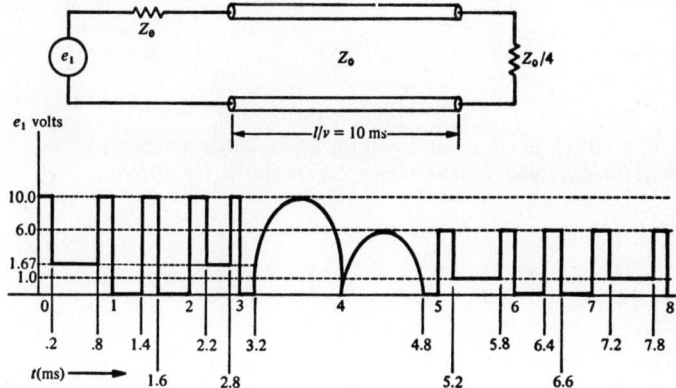

Fig. P.28

29. The line is initially uncharged, and the switch is closed at $t = 0$. (a) Plot sending- and receiving-end voltage for a time of $2L/v$. (b) Plot voltage across the 100 Ω resistor for a time of $2L/v$.

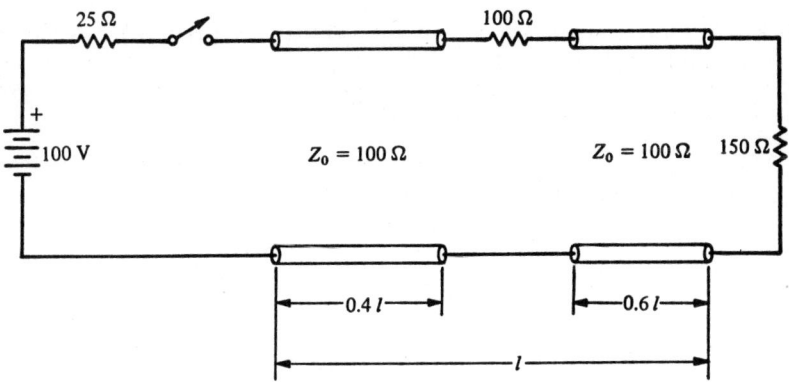

Fig. P.29

30. The line shown was charged to V V. At $t = 0$, both switches are closed. (a) Plot a voltage bounce diagram from $t = 0$ to $t = 3l/v$. (b) Plot voltage at $z = l/2$ versus time for $3l/v$.

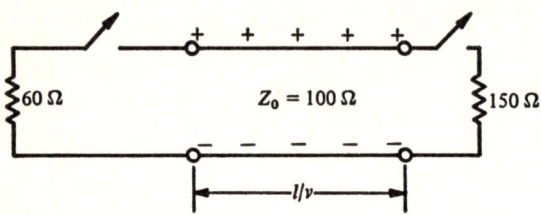

Fig. P.30

31. The switch is closed at $t = 0$ on a previously uncharged line. Plot voltage at $z = L$ for $4L/v$ and voltage across the resistor R for $4L/v$.

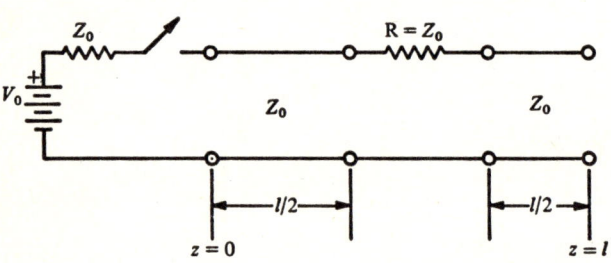

Fig. P.31

32. The line was charged to E V, and at $t = 0$ the switch is closed. Plot voltage across the ends of the line for $4L/v$.

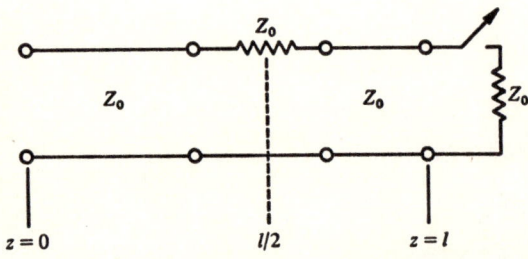

Fig. P.32

33. The switch is closed at $t = 0$ for the circuit shown. Plot receiving-end voltage for $3L/v$ and sending-end voltage for $4L/v$.

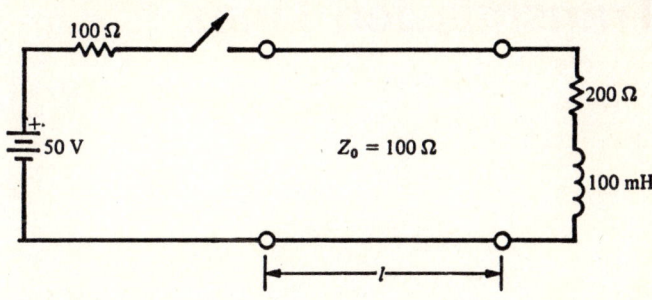

Fig. P.33

34. At $t = 0$, both switches are closed. For $t < 0$, $V_c = 100$ V. Plot sending- and receiving-end voltages.

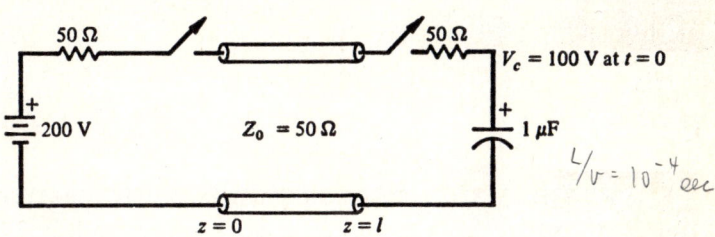

Fig. P.34

$L/v = 10^{-4}$ sec

5
Sinusoidal Waves on Lossless Transmission Lines

5.1 THE STEADY-STATE SOLUTION FROM TRANSIENT ANALYSIS

This chapter is a study of the transmission line under steady-state sinusoidal excitation. One branch of circuit theory, commonly called steady-state theory, has developed around linear circuits excited by sinusoidal generators. The sinusoid has a similarly important role in transmission-line theory.

The techniques developed in Chap. 4 for transient analysis on a lossless line can be applied to obtain steady-state solutions by allowing time to approach infinity. Inevitably, sums of infinite series will be needed to express steady-state solutions if one takes this approach. The problem is complicated somewhat by the fact that the source impedance and the load impedance (and consequently the reflection coefficients) can be complex. However, we will consider a special case where an uncharged transmission line with a resistive load R_r is excited by a sinusoidal generator, as shown in Fig. 5.1.1. We wish to find the expression for the voltage† on the line at any point for very large values of time.

† The measurable voltage is the real part of the product of the complex voltage phasor $V(z)$ and the rotational phasor $e^{j\omega t}$, as discussed in Sec. 3.4.

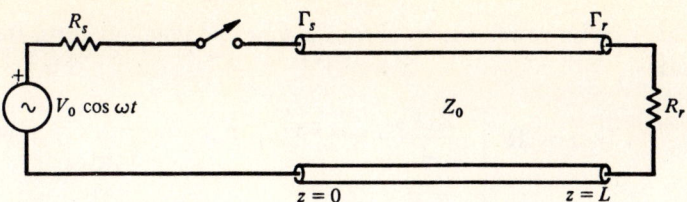

Fig. 5.1.1 The sinusoidally excited lossless transmission line.

Suppose that, at $t = 0$, the switch is closed in Fig. 5.1.1. The sinusoidal wave starts down the line and will be reflected every L/v s from R_r and R_s, alternately.

The final voltage distribution on the line (as $t \to \infty$) can be found by a transient analysis, as in Chap. 4. The first positive-going wave† down the line is

$$V_1{}^+ = \text{Re}\left(\frac{Z_0 V_0}{R_s + Z_0} e^{j(\omega t - \beta z)}\right) U\left(t - \frac{z}{v}\right) \tag{1}$$

where $\beta = 2\pi/\lambda = \omega/v$, as discussed in Sec. 3.4. The first negative-going wave on the line is

$$V_1{}^- = \text{Re}\left(\frac{Z_0 \Gamma_r}{R_s + Z_0} V_0 e^{j(\omega t + \beta z - 2\beta L)}\right) U\left(t - \frac{2L}{v} + \frac{z}{v}\right) \tag{2}$$

and, similarly, the second positive-going wave is

$$V_2{}^+ = \text{Re}\left(\frac{Z_0 \Gamma_s}{R_s + Z_0} \Gamma_r V_0 e^{j(\omega t - \beta z - 2\beta L)}\right) U\left(t - \frac{2L}{v} - \frac{z}{v}\right) \tag{3}$$

As t becomes very large, all voltage waves on the line can be written as an infinite sum of the positive- and negative-going waves

$$V(z,t) = \text{Re}\left[\frac{Z_0 V_0}{R_s + Z_0} \sum_{n=0}^{\infty} (\Gamma_s \Gamma_r)^n e^{j(\omega t - \beta z)} e^{-j2n\beta L}\right]$$
$$+ \text{Re}\left[\frac{Z_0 \Gamma_r V_0}{R_s + Z_0} \sum_{n=0}^{\infty} (\Gamma_s \Gamma_r)^n e^{j(\omega t + \beta z)} e^{-j2(n+1)\beta L}\right] \tag{4}$$

† It can be seen that Eq. (1) satisfies the boundary condition at the source if we set $z = 0$. Consequently, Eq. (1) is a unique solution of the wave equation for the first positive-going wave.

This can be rewritten to obtain a more convenient form for the infinite series.

$$V(z,t) = \text{Re}\left[\frac{Z_0 V_0}{R_s + Z_0}\left(e^{j(\omega t - \beta z)} + \Gamma_r e^{j(\omega t + \beta z)}\right)\sum_{n=0}^{\infty}(\Gamma_s \Gamma_r)^n e^{-j2n\beta L}\right] \quad (5)$$

$$V(z,t) = \text{Re}\left[\frac{Z_0 V_0}{R_s + Z_0}\left(e^{j(\omega t - \beta z)} + \Gamma_r e^{j(\omega t + \beta z)}\right)\frac{1}{1 - \Gamma_s \Gamma_r e^{-j2\beta L}}\right] \quad (6)$$

$$V(z,t) = \text{Re}\left(\frac{Z_0 V_0}{R_s + Z_0}\frac{e^{j(\omega t - \beta z)} + \Gamma_r e^{j(\omega t + \beta z)}}{1 - \Gamma_s \Gamma_r e^{-j2\beta L}}\right) \quad (7)$$

In Eq. (7) the sum of all positive-going waves is identified with the exponent $j(\omega t - \beta z)$, and the sum of all negative-going waves is identified with the exponent $j(\omega t + \beta z)$. This is more apparent if we write Eq. (7) in the form

$$V(z,t) = \text{Re}\,(\underbrace{C_1 e^{j(\omega t - \beta z)}}_{\text{positive-going wave}} + \underbrace{C_2 e^{j(\omega t + \beta z)}}_{\text{negative-going wave}}) \quad (8)$$

The coefficients C_1 and C_2 are in general complex numbers which can be determined from a knowledge of V_0, Z_0, R_s, Γ_r, and Γ_s in Eq. (7). The time functions and distance functions can be separated in Eq. (8) to yield a voltage which is more convenient for phasor notation

$$V(z,t) = \text{Re}\left[(C_1 e^{-j\beta z} + C_2 e^{+j\beta z})e^{j\omega t}\right] \quad (9)$$

In conclusion, the solution of the voltage wave equation in the sinusoidal steady-state condition is given by Eq. (9) for lossless lines. Bounce diagrams are not very useful for steady-state analysis, since an infinite number of reflections are generally required to establish a steady-state condition.

A more useful approach for steady-state analysis avoids the problems associated with the infinite-series solutions and applies for both lossless and lossy lines. The sinusoidal steady-state solutions of the wave equations for voltage and current are obtained in the same manner as those for the sinusoidal **E** and **H** fields in a plane wave, as discussed in Sec. 3.4.

5.2 STEADY-STATE SOLUTION OF THE WAVE EQUATION

The equivalent circuit of the transmission line was developed in Sec. 4.3, and the damped-wave equations for voltage and current on a lossy line were given by Eqs. (6) and (7) of that section. Since these damped-wave

SINUSOIDAL WAVES ON LOSSLESS TRANSMISSION LINES

equations are identical, it is convenient to assume a function $W(z,t)$ which can be used to represent either voltage or current

$$\frac{\partial^2 W}{\partial z^2} = (RG)W + (RC + GL)\frac{\partial W}{\partial t} + LC\frac{\partial^2 W}{\partial t^2} \tag{1}$$

It will be assumed that the function $W(z,t)$ has a sinusoidal time variation and can be separated into the product of a distance function $W(z)$ and a single-frequency time function $e^{j\omega t}$,

$$W(z,t) = W(z)e^{j\omega t} = \text{Re}\,[W(z)e^{j\omega t}] + j\,\text{Im}\,[W(z)e^{j\omega t}] \tag{2}$$

In general the function $W(z,t)$ is a complex variable and can be separated into real and imaginary parts. Any one of the functions—$W(z,t)$, its real part, or its imaginary part—can be substituted into Eq. (1). As in Sec. 3.4, any of these substitutions reduces the damped-wave equation to the following ordinary differential equation, which is independent of time:

$$\frac{d^2 W(z)}{dz^2} = [(R + j\omega L)(G + j\omega C)]W(z) = \gamma^2 W(z) \tag{3}$$

The propagation constant γ has the same significance in transmission-line theory as in the propagation of plane waves

$$\gamma = \sqrt{(R + j\omega L)(G + j\omega C)} = \text{propagation constant} \tag{4}$$

As in Sec. 3.4, the propagation constant can be factored into real and imaginary parts, called the *attenuation constant* and *phase constant* respectively:

$$\gamma = \alpha + j\beta \tag{5}$$

The solution of Eq. (3) is of the form

$$W(z) = C_1 e^{-\gamma z} + C_2 e^{+\gamma z} \tag{6}$$

The complex function $W(z)$ is either a voltage phasor $V(z)$ or a current phasor $I(z)$, and C_1 and C_2 are arbitrary complex constants determined by the boundary conditions. Letting $W(z) = V(z)$, we obtain a solution of the voltage wave equation

$$V(z,t) = \text{Re}\,[V(z)e^{j\omega t}] = \text{Re}\,[(V^+ e^{-\gamma z} + V^- e^{+\gamma z})e^{j\omega t}] \tag{7}$$

Since we are interested in a solution of the voltage wave equation, C_1 and C_2 have been replaced by V^+ and V^- respectively.† These constants have

---

† Note that this use of V^+ and V^- is more restrictive than in the previous chapter, where they were used to represent voltage waves as a function of time and distance. Throughout this chapter they represent constant-voltage phasors at $z = 0$, as determined by the boundary conditions.

been given these particular designations because of their association with traveling waves. Equation (7) is of the same form as equations of Sec. 3.4 for the transverse field components of a plane wave [e.g., see Eq. (13) of Sec. 3.4]. An equation of this form was also obtained in Sec. 5.1, Eq. (9), for the lossless transmission line (so that $\alpha = 0$, and $\gamma = j\beta$), using transient-analysis techniques.

Equation (7) can be rewritten to show the total positive- and negative-going waves:

$$V(z,t) = \text{Re} \, (\underbrace{V^+ e^{-\alpha z} e^{j(\omega t - \beta z)}}_{\text{positive-going wave}} + \underbrace{V^- e^{\alpha z} e^{j(\omega t + \beta z)}}_{\text{negative-going wave}}) \tag{8}$$

If we hold the phase angle constant ($\omega t - \beta z = \text{const}$) for the positive-going wave, we can find the velocity of propagation of the sinusoid by differentiation with respect to z, as was done in Sec. 3.2 for plane waves:

$$\frac{dz}{dt} = v = \frac{\omega}{\beta} \tag{9}$$

A similar analysis on the last term of Eq. (8) shows that the negative-going sinusoidal wave has a velocity of $-v$. Finally then, the steady-state solution for sinusoidal voltage on the transmission line may be written

$$V(z,t) = \text{Re} \, (V^+ e^{-\alpha z} e^{j\omega(t - z/v)} + V^- e^{\alpha z} e^{j\omega(t + z/v)}) \tag{10}$$

The current $I(z,t)$ must have the same form as Eq. (10) and can be obtained by replacing V^+ with I^+ and V^- with I^-. The voltage $V(z,t)$ and current $I(z,t)$ are traveling waves, since they have the same form as the transverse components of the plane wave discussed in Sec. 3.2. They will be discussed in more detail in the next section.

It will now be shown that the phasor voltage and phasor current are related by the characteristic impedance for either a positive- or negative-going wave, as was the case for the transient analysis. The steady-state solutions for voltage and current are

$$V(z,t) = \text{Re} \, [(V^+ e^{-\gamma z} + V^- e^{+\gamma z}) e^{j\omega t}] \tag{11}$$

$$I(z,t) = \text{Re} \, [(I^+ e^{-\gamma z} + I^- e^{+\gamma z}) e^{j\omega t}] \tag{12}$$

When these are substituted into the telegrapher's equation

$$\frac{\partial I(z,t)}{\partial z} = -GV(z,t) - C \frac{\partial V(z,t)}{\partial t} \tag{13}$$

one obtains an ordinary differential equation relating the phasor voltage and phasor current

$$\frac{dI(z)}{dz} = -(G + j\omega C) V(z) \tag{14}$$

SINUSOIDAL WAVES ON LOSSLESS TRANSMISSION LINES

Integration of Eq. (14) with respect to z yields the desired relationship between the voltage and current phasors

$$I(z) = I^+e^{-\gamma z} + I^-e^{+\gamma z} = \left(\frac{G + j\omega C}{\gamma}\right)(V^+e^{-\gamma z} - V^-e^{+\gamma z}) \qquad (15)$$

The constant in parentheses in Eq. (15) is the characteristic admittance Y_0

$$Y_0 = \frac{G + j\omega C}{\gamma} = \sqrt{\frac{G + j\omega C}{R + j\omega L}} \quad \text{mhos/m} \qquad (16)$$

The reciprocal of the characteristic admittance is the characteristic impedance Z_0. Note that the characteristic admittance is a function of frequency and is generally a complex number. From Eq. (15) one concludes that the voltage and current are related by the characteristic admittance, i.e.,

$$I(z,t) = \text{Re}\,[Y_0(V^+e^{-\gamma z} - V^-e^{+\gamma z})e^{j\omega t}] \qquad (17)$$

Throughout the remainder of this chapter lossless lines will be considered. In lossless lines, no energy can be dissipated, and consequently

$$R = 0 \qquad G = 0 \qquad (18)$$

If one substitutes Eqs. (18) into Eq. (4), it is found that the propagation constant γ has no real part, so that the attenuation constant α is zero for lossless lines:

$$\gamma = \alpha + j\beta = 0 + j\beta = j\omega\sqrt{LC} \qquad (19)$$

Since the attenuation constant α is zero, a wave does not diminish in amplitude as it travels down the lossless line.

If we assume that the line is lossless and that L and C are independent of frequency, the velocity of propagation is independent of frequency

$$v = \frac{\omega}{\beta} = \frac{1}{\sqrt{LC}} = \frac{1}{\sqrt{\mu\varepsilon}} \qquad (20)$$

Note that for the lossless line, the velocity of propagation is the same whether we are considering a transient wave, as in Chap. 4, or a steady-state sinusoidal wave. Substituting Eqs. (18) into Eq. (16), we find that the characteristic impedance is a constant for the lossless line, where L and C are determined at the frequency ω of the source

$$Z_0 = \sqrt{\frac{L}{C}} = \frac{1}{Y_0} \qquad (21)$$

This is the same characteristic impedance as obtained in the study of transients in Chap. 4.

If it is assumed that the line is lossless, voltage and current are given by

$$V(z,t) = \text{Re}\,[(V^+e^{-j\beta z} + V^-e^{j\beta z})e^{j\omega t}] \tag{22}$$

$$I(z,t) = \text{Re}\,[Y_0(V^+e^{-j\beta z} - V^-e^{j\beta z})e^{j\omega t}] \tag{23}$$

These equations were obtained from Eqs. (11) and (12). For future reference the principal equations of this section are summarized in Table 5.2.1 for both the lossy case and the lossless case.

Table 5.2.1 Summary of the steady-state equations

Lossy case

$\gamma = \sqrt{(R + j\omega L)(G + j\omega C)} = \alpha + j\beta$	propagation constant	(4)
$v = \dfrac{\omega}{\beta}$	velocity of propagation	(9)
$Y_0 = \dfrac{1}{Z_0} = \sqrt{\dfrac{G + j\omega C}{R + j\omega L}}$	characteristic admittance	(16)
$V(z,t) = \text{Re}\,[(V^+e^{-\gamma z} + V^-e^{\gamma z})e^{j\omega t}]$		(11)
$I(z,t) = \text{Re}\,[Y_0(V^+e^{-\gamma z} - V^-e^{\gamma z})e^{j\omega t}]$		(12)

Lossless case $(R = 0, G = 0)$

$\gamma = j\omega\sqrt{LC}$	propagation constant	(19)
$\alpha = 0$	attenuation constant	(19)
$\beta = \omega\sqrt{LC}$	phase constant	(19)
$v = \dfrac{1}{\sqrt{LC}}$	velocity of propagation	(20)
$Z_0 = \dfrac{1}{Y_0} = \sqrt{\dfrac{L}{C}}$	characteristic impedance	(21)
$V(z,t) = \text{Re}\,[(V^+e^{-j\beta z} + V^-e^{j\beta z})e^{j\omega t}]$		(22)
$I(z,t) = \text{Re}\,[Y_0(V^+e^{-j\beta z} - V^-e^{j\beta z})e^{j\omega t}]$		(23)

5.3 SINUSOIDAL TRAVELING WAVES AND STANDING WAVES

SINUSOIDAL TRAVELING WAVES

In this section, we first discuss traveling waves, phenomena discussed in Sec. 3.2 for a plane wave. To isolate a single traveling wave, a line terminated in the characteristic impedance Z_0 will be considered, as shown in Fig. 5.3.1. There are no reflections on this line.

The bounce-diagram technique of the previous chapter is not especially useful for lines under continuous sinusoidal excitation. However, in the simple case illustrated in Fig. 5.3.1, some information can be gained from a three-dimensional sketch of the bounce diagram, shown in Fig. 5.3.2. On the left-hand side of the bounce diagram the sinusoidal waveform travels down the line with a velocity of v and reaches the load end at time L/v. The sinusoidal load voltage can be seen on the right-hand side of the bounce diagram. There is no reflection at the load, and the load voltage has the same waveform as the source voltage. However, the voltage is delayed by the time that it takes the wave to travel down the line.

Now consider what an observer sees standing partway down the line, say at point A, as shown in Figs. 5.3.1 and 5.3.2. Since his position on the line is fixed, he observes a voltage as a function of time, as traced out on a cross section AA'. An observer at point A sees the same voltage waveform that an observer at the load sees, except with less time delay. In conclusion, an observer at any point on the transmission line observes a *sinusoidal function of time*, because the source voltage is sinusoidal.

Suppose that the transmission line has been under sinusoidal excitation for some time. At a given instant of time, say t_1, instantaneous measurements of voltage at every point on the line theoretically could be made. The voltage, as a function of position on the line, can be traced on a cross section BB' of the bounce diagram. The voltage is a sinusoidal function of *distance* because the source voltage is a sinusoidal function of *time*. During one period T of the source voltage, the waveform travels a distance of 1 wavelength, or λ m, as shown in Fig. 5.3.2. The sinusoidal wave in Fig. 5.3.2 is a traveling wave because it has the functional form $F(t \pm z/v)$, as shown in Eq. (10) of Sec. 5.2.

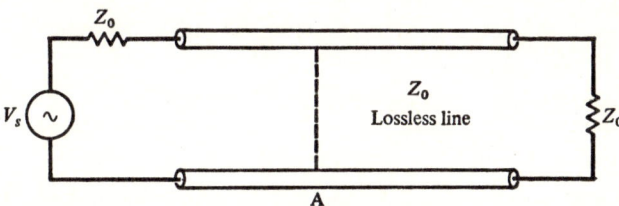

Fig. 5.3.1 The lossless matched line under sinusoidal excitation.

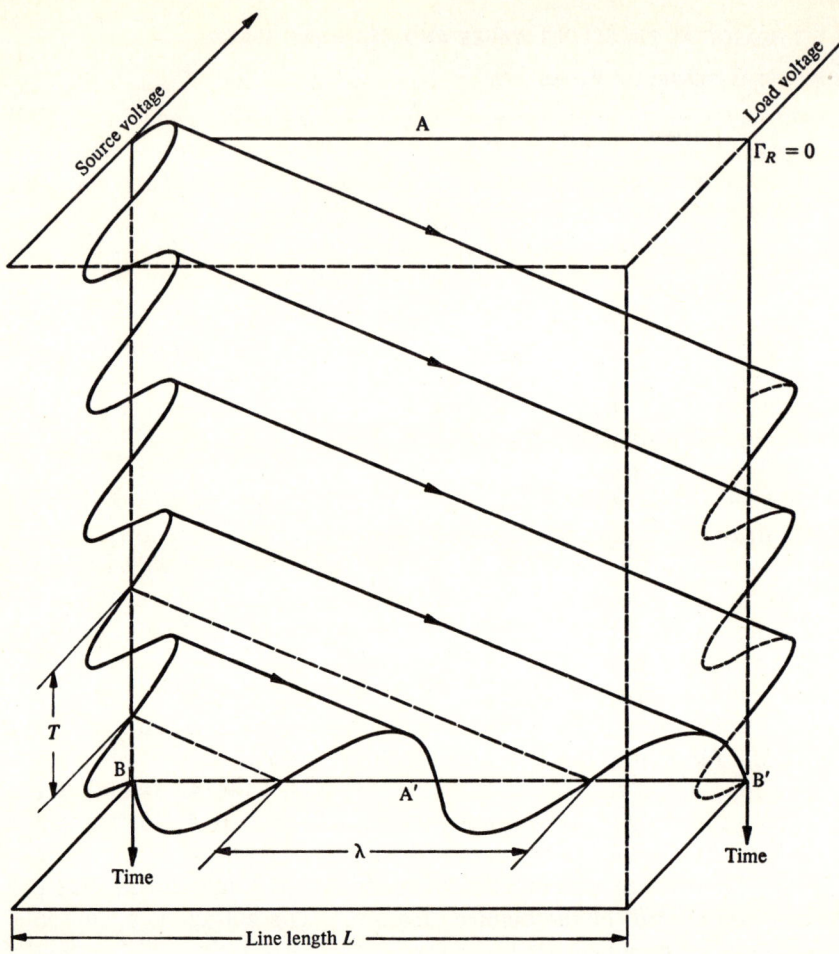

Fig. 5.3.2 A bounce diagram of traveling waves on a matched line.

We now review the relationships between the velocity of propagation v, the wavelength λ, and the phase constant β. When the velocity of propagation is constant (when μ and ε are frequency-independent on a lossless line), the distance traveled on the line in time t is

$$z = vt \tag{1}$$

However, the wave travels through 1 wavelength λ in time T_s, as shown in Fig. 5.3.2, so that one can obtain

$$\lambda = vT_s = \frac{v}{f} \qquad T_s = \frac{1}{f} \tag{2}$$

$$v = f\lambda \tag{3}$$

SINUSOIDAL WAVES ON LOSSLESS TRANSMISSION LINES

This relationship can be seen on the bounce diagram if one imagines increasing the frequency of the source and observing the effect on the wavelength λ. One further relationship will be noted in passing. If the equation for the velocity of propagation $v = \omega/\beta$ [Eq. (9) of Sec. 5.2] is substituted into Eq. (3), we obtain

$$\beta\lambda = 2\pi \tag{4}$$

The phase constant β expresses the phase shift in radians per meter. As shown by Eq. (4), a change of position of 1 wavelength on the line corresponds to a phase shift of 2π rad.

The equation of a positive-going sinusoidal traveling wave of voltage can be obtained from Eq. (22) of the previous section by letting $V^- = 0$

$$V(z,t) = \operatorname{Re}(V^+ e^{j(\omega t - \beta z)}) \tag{5}$$

Since V^+ is the positive-going voltage phasor at $z = 0$, it can be written in polar form,

$$V^+ = |V^+| e^{j\phi} \tag{6}$$

and therefore

$$V(z,t) = \operatorname{Re}(|V^+| e^{j\phi} e^{j(\omega t - \beta z)}) = |V^+| \cos(\omega t - \beta z + \phi) \tag{7}$$

The equation for the traveling wave of current $I(z,t)$ can be obtained from the voltage by multiplying by the characteristic admittance Y_0

$$I(z,t) = Y_0 |V^+| \cos(\omega t - \beta z + \phi) \tag{8}$$

These equations have been derived assuming a fixed reference point in time. If one wishes to make an arbitrary time shift t_1 to some new reference system, the variable t in Eqs. (7) and (8) can be replaced by the variable $t - t_1$. This corresponds to a simple change of phase of $-\omega t_1$ rad. Similarly, if one wishes to shift the zero reference point from the source to some other point on the line, this corresponds to replacing the variable z by the variable $z - z_1$. Thus a change in the origin with respect to distance corresponds to a change in phase by $-\beta z_1$ rad in Eqs. (7) and (8). In this manner Eqs. (7) and (8) can be generalized to account for any desired shift of the origin, either in time or distance, by the addition of an appropriate phase angle.

STANDING WAVES

It was found in the previous section that traveling waves exist on a sinusoidally excited transmission line when the line is terminated in its characteristic impedance. In this section we consider a lossless transmission line terminated in an open circuit. There will be reflections from

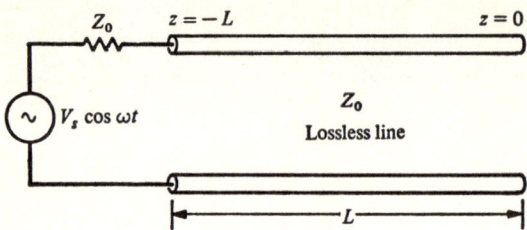

Fig. 5.3.3 The open-circuited line.

the open circuit, and therefore standing waves will develop. Similarly, in Sec. 3.7, standing waves which resulted from a plane wave incident on a perfect conductor were discussed. For convenience in applying the boundary conditions at the open circuit on the transmission line, *the origin will be shifted to the receiving end of the line* (Fig. 5.3.3). Under these conditions, the total positive-going voltage wave will be reflected from the open-circuited end, and consequently

$$V^+ = |V^+|e^{j\phi} = V^- \tag{9}$$

where ϕ is the angle of V^+ at the open circuit ($z = 0$). When Eq. (9) is substituted into the voltage equation

$$V(z,t) = \text{Re}\left[(V^+e^{-j\beta z} + V^-e^{+j\beta z})e^{j\omega t}\right]$$
$$= \text{Re}\left[|V^+|e^{j\phi}(e^{-j\beta z} + e^{+j\beta z})e^{j\omega t}\right] \tag{10}$$

we obtain for the voltage of an open-circuited line

$$V(z,t) = \text{Re}\,(2|V^+|e^{j\phi}\cos\beta z\, e^{j\omega t}) = \text{Re}\,[V(z)e^{j\omega t}] \tag{11}$$

The total phasor voltage

$$V(z) = 2|V^+|e^{j\phi}\cos\beta z \tag{12}$$

has an angle ϕ which is independent of position on the line.

At any point $z = z_1$, the voltage $V(z,t)$ can be represented as the projection of a rotating phasor $V(z)e^{j\omega t}$ on the real axis, as shown in Fig. 5.3.4a. The phasor $V(z)e^{j\omega t}$ rotates with an angular velocity of ω, and its projection on the real axis traces out a sinusoidal voltage $V(z,t)$ as a function of time. The phasor voltage $V(z)$ at each point on the line can be plotted from Eq. (12), as shown in Fig. 5.3.4b.

A *standing wave* is said to exist on the transmission line if two (or more) sinusoidal traveling waves of the same frequency are traveling in opposite directions. The phasor of a standing wave is the sum of the phasors of the traveling waves. The standing-wave phasor is a function of distance, as in Eq. (12). In the cases where the transmission line is terminated in an open circuit or a short circuit ($\Gamma_r = \pm 1$), the standing

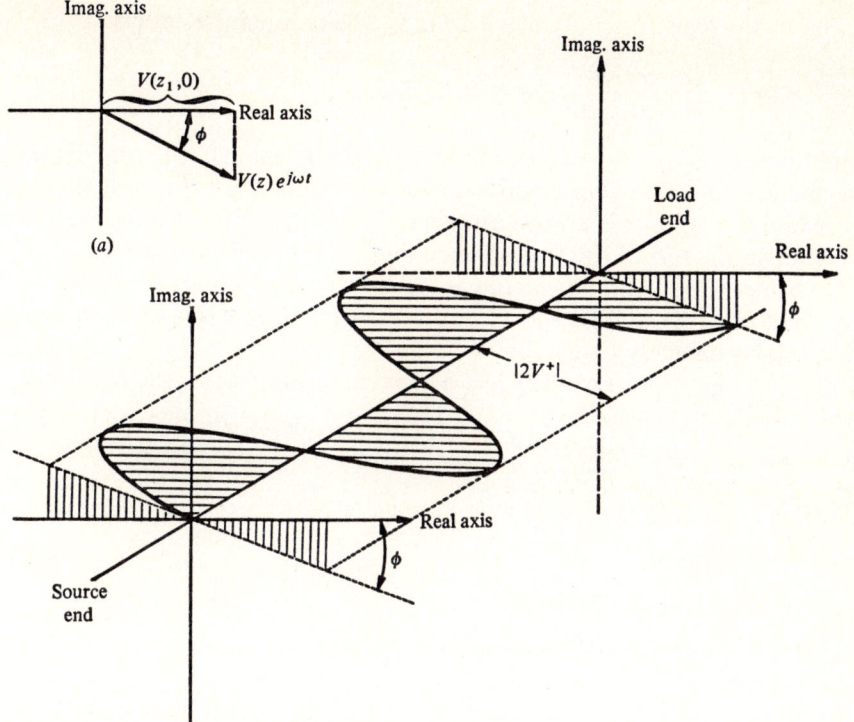

Fig. 5.3.4 Phasor diagrams of a complete standing wave. (a) The phasor $V(z)e^{j\omega t}$ at $z = z_1$ and $t = t_1$. (b) The voltage phasor versus distance at $t = t_1$.

waves that exist are examples of complete standing waves. A precise definition of complete standing waves will be given in the next section.

If one considers the measurable voltage as indicated in Eq. (11), taking the real part gives the voltage equation

$$V(z,t) = 2|V^+| \cos \beta z \cos (\omega t + \phi) \tag{13}$$

This voltage is periodic both in time and distance, as one would expect from the above diagrams. Because of the sinusoidal distance function $\cos \beta z$, there are points on the transmission line where the voltage is always zero, even though the source voltage is not zero. These points are called *null points* and occur in this example where

$$\beta z_n = -\frac{\pi}{2}, -\frac{3\pi}{2}, -\frac{5\pi}{2}, \ldots,$$

$$-\frac{(2n + 1)\pi}{2} \quad n = 0, 1, 2, 3, \ldots$$

The values of βz are negative because we have chosen the reference $z = 0$ to be at the load end. By solving for z_n, the null point z_n occurs at

$$z_n = -\frac{(2n+1)\pi}{2\beta} \qquad n = 0, 1, 2, 3, \ldots \tag{14}$$

and they are spaced ½ wavelength apart. Oddly enough, the conductors of the transmission line could be shorted together at the null points (because the voltage is always zero at these points) without affecting the voltage on the rest of the transmission line. This phenomenon can occur only because the transmission line is lossless and no power is consumed at the load end. It will be shown in a later section that this transmission line acts as an energy reservoir.

The phasor of Eq. (12) has a maximum amplitude whenever the cosine is unity, or when

$$\beta z_m = 0, -\pi, -2\pi, \ldots, -n\pi, \ldots$$

Therefore the points of maximum voltage occur at z_m:

$$z_m = \frac{-n\pi}{\beta} = \frac{-n}{2f\sqrt{LC}} \qquad n = 0, 1, 2, \ldots \tag{15}$$

The voltage at these points varies from the positive maximum of $2|V^+|$ to a minimum of $-2|V^+|$ as βz varies. However, at any point on the line, the phasor $V(z)e^{j\omega t}$ rotates with time, so that the instantaneous voltage $V(z,t)$ is a sinusoidal function of time.

If one could cut a narrow slot along the length of the transmission line and measure the voltage between the center conductor and the outer conductor with an rms-measuring voltmeter, one would obtain a plot as shown in Fig. 5.3.5. This same plot would be obtained for the magnitude

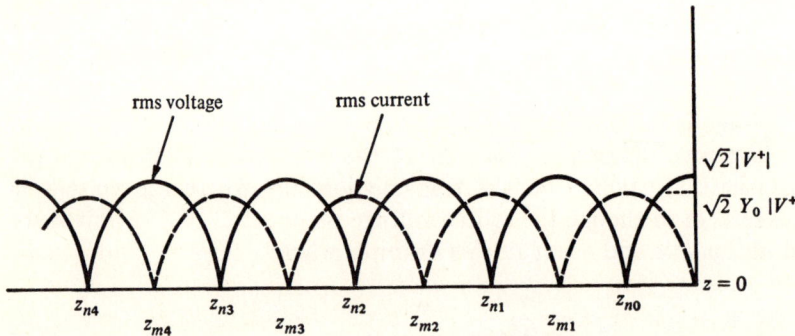

Fig. 5.3.5 A plot of rms voltage and current versus distance for an open-circuit termination, showing distance quadrature.

SINUSOIDAL WAVES ON LOSSLESS TRANSMISSION LINES

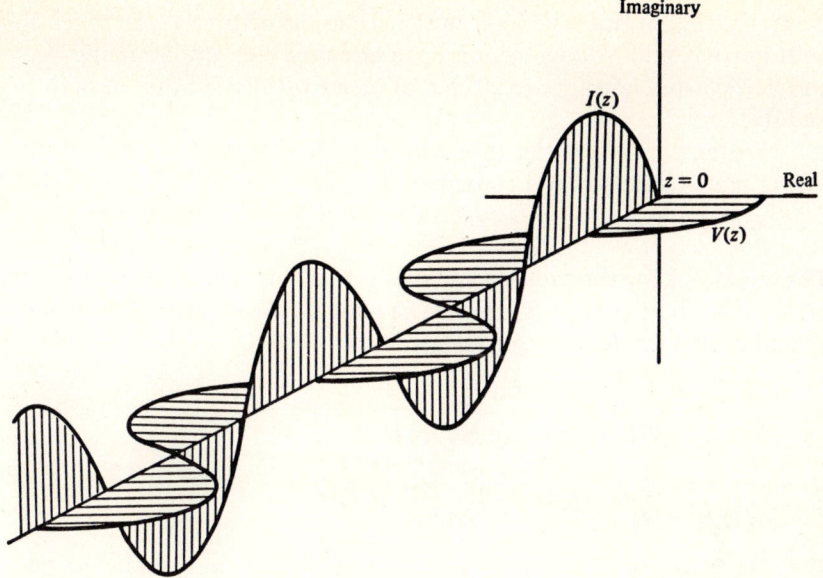

Fig. 5.3.6 Voltage and current phasors on an open-circuited line are 90° out of phase in both time and space.

of the total phasor voltage given in Eq. (12) except that the voltage scales would differ by $\sqrt{2}$.

It was shown in Sec. 5.2 that the current under sinusoidal excitation is given by

$$I(z,t) = \text{Re}\left[Y_0(V^+e^{-j\beta z} - V^-e^{+j\beta z})e^{j\omega t}\right] \tag{16}$$

In the case under consideration, where the line is open-circuited at the load end, the magnitude of the reflected wave equals the magnitude of the incident wave, as shown in Eq. (9). If we substitute this condition into Eq. (16), we obtain the equation for the current phasor, which is a function of distance

$$I(z) = -2jY_0|V^+|e^{j\phi}\sin\beta z \tag{17}$$

The imaginary number $j = \sqrt{-1}$ in Eq. (17) indicates that there is a 90° phase shift between the current phasor of Eq. (17) and the voltage phasor of Eq. (12) at every point on the transmission line. This relationship is shown in the phasor diagram of Fig. 5.3.6, where V^+ has been chosen so that its angle ϕ is zero. Taking the real part of $I(z)e^{j\omega t}$, as indicated in Eq. (16), gives the current as a function of time and distance

$$I(z,t) = 2Y_0|V^+|\sin\beta z \sin(\omega t + \phi) \tag{18}$$

Note that the current is periodic both in time and distance. In conclusion, both current and voltage on an open-circuited line are standing waves, and voltage and current are 90° out of phase with each other in both time and distance.

We now consider the case where the load end of the transmission line is short-circuited, and consequently

$$V^+ = -V^- = |V^+|e^{j\phi} \tag{19}$$

The equations for the voltage and current on this short-circuited line can be obtained by the same procedure as above. As an exercise, the reader should derive the following equations:

$$V(z) = -2j|V^+|e^{j\phi}\sin\beta z \tag{20}$$

$$V(z,t) = 2|V^+|\sin\beta z \sin(\omega t + \phi) \tag{21}$$

$$I(z) = 2Y_0|V^+|e^{j\phi}\cos\beta z \tag{22}$$

$$I(z,t) = 2Y_0|V^+|\cos\beta z \cos(\omega t + \phi) \tag{23}$$

Equations (20) and (22) are phasors of standing waves. As in the case of the open-circuited termination, there is a 90° phase shift between the voltage and the current in both time and distance. If one compares Eqs. (18) and (21), it is found that the voltage and current have interchanged roles as a result of switching the load from an open circuit to a short circuit. A plot of the rms voltage and current for the short-circuited line would appear as in Fig. 5.3.5 if one interchanged the role of voltage and current in that diagram.

The time-average power $\langle P \rangle$ at any point z on the transmission line over one period T of the source frequency is

$$\langle P \rangle = \frac{1}{T}\int_0^T I(z,t)V(z,t)\,dt \tag{24}$$

If one substitutes Eqs. (21) and (23) for the short-circuited line into the time-average power equation and performs the integration, one finds that the power $\langle P \rangle$ is zero at any point on the line. (The reader should verify this statement.) This means that there is no net power flow from the source. This is exactly what one would expect, because the load is a short circuit and consumes no power. Similarly, for the open-circuit case, one finds no net power flow. Under either of these conditions, the transmission line is an energy reservoir, and it dissipates no energy. This topic will be considered in more detail in a later section.

At any point on the transmission line, the ratio of the phasor voltage $V(z)$ to the phasor current $I(z)$ is the *impedance* $Z(z)$ at that point. This is the *driving-point impedance*, or two-terminal impedance, that one would measure if the line were cut at that point and the source applied there. This impedance is not a constant but varies from point to point, because the voltage and current phasors are functions of distance. Since the driving-point impedance $Z(z)$ is the ratio of the voltage and current phasors, it is generally different from the characteristic impedance Z_0. (The characteristic impedance is the ratio of a traveling voltage wave to the corresponding traveling current wave.)

Consider the specific case of an open-circuited lossless transmission line. The impedance $Z(z)$ can be found from the ratio of Eq. (12) to Eq. (17):

$$Z(z) = \frac{V(z)}{I(z)} = \frac{2|V^+|e^{j\phi} \cos \beta z}{-j2Y_0|V^+|e^{j\phi} \sin \beta z} \tag{25}$$

Simplification of this equation yields

$$Z(z) = jZ_0 \cot \beta z \tag{26}$$

and we can see that the impedance is imaginary at every point on the line. Thus the transmission line is inductive in some regions (where the impedance has a positive sign) and capacitive in other regions (where the sign of the impedance is negative). This imaginary impedance is consistent with the fact that the net power flow on the transmission line is zero. An impedance with a nonzero real part would mean a dissipation of energy at the load. A sketch of the impedance as a function of distance for the open-circuited line is shown in Fig. 5.3.7. The reader will find it instructive to sketch the impedance of the short-circuited line [the ratio of Eq. (20) to Eq. (22)] and to compare his sketch with Fig. 5.3.7.

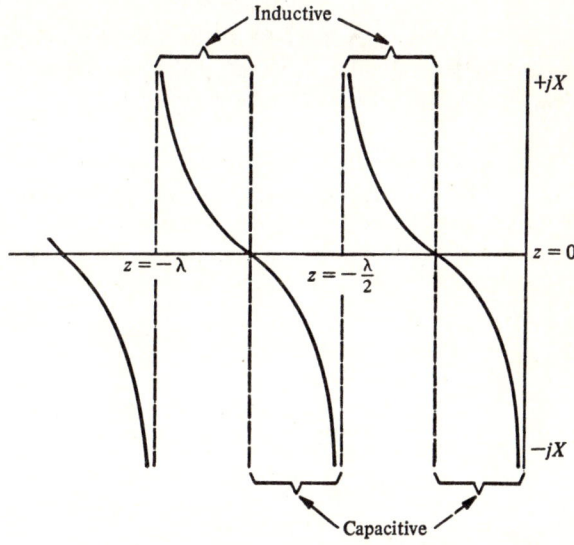

Fig. 5.3.7 Impedance versus distance for the open-circuited line.

We can imagine an interesting experiment with this lossless line. The voltage is always zero at the null points, for either the open-circuited line or the shorted line. Since the voltage is zero at these points, we can imagine chopping the line into half-wavelength sections, with cuts being made at each null point. Each section of line must be short-circuited at each end to maintain a zero voltage at that point and to allow current reflections. We could carry these short-circuited sections around in our coat pocket (assuming the burden was not too great), and reassemble them in any order at any time. Furthermore, an observer at either the source or the load ends of the line would be unable to detect that the intermediate portions of the transmission line had been removed, so long as no changes were made in either the load or the source.

PROGRAMMED EXERCISE 5.3.1

THE FERRANTI EFFECT

An interesting phenomenon, called the *Ferranti effect*, is illustrated in the following problem. A transmission line is terminated in an open circuit as shown in Fig. 5.3.8. Other data are

Characteristic impedance $= Z_0 = 50 \; \Omega$

Source voltage $= V_s(t) = 100 \cos (\omega t + \xi)$

Source impedance $= Z_s = 20 \; \Omega$

Line length $= L = \frac{5}{16}\lambda \quad$ m

The series of true statements now to be made about the transmission line of Fig. 5.3.8 is organized to guide the reader through a certain line of reasoning, as well as to illustrate the Ferranti effect and other phenomena. The reader should verify each statement.

1. The input impedance of the transmission line is an inductive reactance of 20.71 Ω, and the generator load is the sum of the source impedance and the input impedance, or $20.00 + j20.71 \; \Omega$.

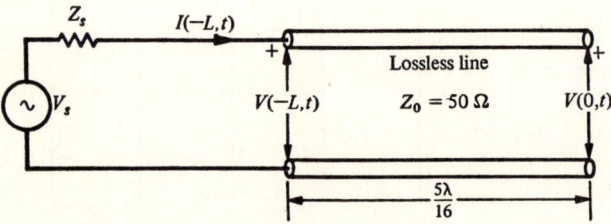

Fig. 5.3.8 A comparison of the voltages $V(0,t)$ and $V(-L,t)$ illustrates the Ferranti effect.

2. The voltage phasor $V(-L)$ and current phasor $I(-L)$ at the input to the line are found using the impedance above:

$$V(-L) = 73.7 \underline{/\xi + 44°} \quad \text{V} \tag{27}$$

$$I(-L) = 3.56 \underline{/\xi - 46°} \quad \text{A} \tag{28}$$

The voltage phasor leads the current phasor by 90° at the source end.

3. The sum of all the transient positive-going waves is the total positive-going wave $V^+ e^{-j\beta z} e^{j\omega t}$, whose coefficient V^+ is a complex constant. The term V^+, which appears in every voltage and current equation, must be of a proper magnitude and phase angle to satisfy the boundary condition at the sending end of the transmission line (and V^- must satisfy the boundary conditions at the load end). At the sending end the phasor voltage is

$$V(-L) = 73.7 \underline{/\xi + 44°} \quad \text{V}$$

where ξ is the phase angle of the generator, and so

$$V^+ = |V^+| e^{j\phi} = -96.3 \underline{/\xi + 44°} = 96.3 \underline{/\xi + 224°} \quad \text{V} \tag{29}$$

Consequently the phase angle ϕ of V^+ and the phase angle ξ of the generator are always related, in a manner which depends upon the nature of the load termination Z_r, the source impedance Z_s, and the length of the line. Similarly the magnitudes of V^+ and V_s are not independent.

4. The voltage $V(0,t)$ and current $I(0,t)$ at the load (open-circuit) end of the line are

$$V(0,t) = 192.6 \cos(\omega t + \xi + 224°) \quad \text{V} \tag{30}$$

$$I(0,t) = 0 \text{ A} \tag{31}$$

A comparison of $V(0,t)$ and $V(-L,t)$ shows that the voltage at the load end is 2.61 times greater than the voltage at the source end. Thus this transmission line acts like a voltage amplifier or transformer. This is called the *Ferranti effect*. The amount by which the voltage is increased depends upon the effective length† of the transmission line. If the line is $\frac{1}{4}$ wavelength long, the voltage ratio of $V(0,t)$ to $V(-L,t)$ is infinite because $V(-L,t) = 0$. By taking advantage of the Ferranti effect, the transmission line can be used as a voltage transformer whose transformation ratio varies with frequency. However, this effect can cause the voltage to exceed the maximum voltage rating of the insulation near the open circuit.

† The effective length of a line is measured in wavelengths, and to obtain the actual line length (in meters) we multiply the effective length by the number of meters per wavelength.

5. The voltage and current at $z = -\lambda/4$ are

$$V\left(\frac{-\lambda}{4}\right) = 0 \text{ V} \tag{32}$$

$$I\left(\frac{-\lambda}{4}\right) = 3.85\underline{/\xi + 314°} \quad \text{A} \tag{33}$$

The point $z = -\lambda/4$ is the first null point for the voltage. If the line were exactly $\frac{1}{4}$ wavelength long instead of $\frac{5}{16}$ wavelength, the generator would see a short circuit even though the load end is an open circuit. In this case the generator could be damaged by excessive current.

6. The open-circuited transmission line can be used for a current transformer as well as a voltage transformer. However, the average power transmitted at any point on the line is zero because of the open circuit at the load end. The voltage and current are orthogonal functions in both time and distance.

7. A duality relationship exists between the quarter-wavelength open-circuited line and the quarter-wavelength short-circuited line. The reader should explore this relationship.

5.4 IMPEDANCE TERMINATIONS AND THE GAMMA PLANE

VOLTAGE, CURRENT, IMPEDANCE, AND THE REFLECTION COEFFICIENT

In this section we study a lossless transmission line terminated in an arbitrary passive impedance Z_r, as shown in Fig. 5.4.1. A common example of this type is a transmission line leading from an antenna to a receiver. The antenna, if replaced by its Thévenin equivalent circuit, appears as the source in Fig. 5.4.1, and Z_r represents the input impedance to the receiver. The receiver impedance can be measured, but it can be rather difficult to determine the Thévenin equivalent of the antenna. Consequently we usually prefer to evaluate the boundary conditions at the load end, and so it is more convenient to choose the origin ($z = 0$) at the load end, as in Sec. 5.3.

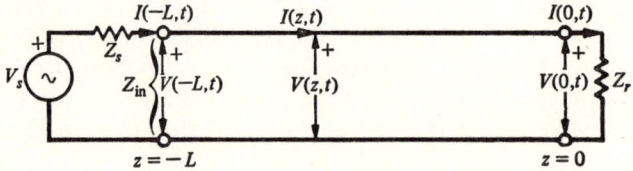

Fig. 5.4.1 A sinusoidally excited transmission line terminated in an arbitrary impedance.

In general, the impedance at any point on a lossless transmission line can be written as the ratio of the voltage phasor to the current phasor [see Eqs. (22) and (23) of Table 5.2.1]. This driving-point impedance is given by

$$Z(z) = \frac{V(z)}{I(z)} = \frac{V^+ e^{-j\beta z} + V^- e^{j\beta z}}{V^+ e^{-j\beta z} - V^- e^{j\beta z}} Z_0 \qquad (1)$$

To satisfy the boundary conditions at the load end, the line impedance $Z(0)$ must equal the load impedance Z_r

$$Z_r = Z(0) = \frac{V^+ + V^-}{V^+ - V^-} Z_0 \qquad (2)$$

One of the obvious advantages of shifting the origin to the load end is the simplicity of Eq. (2) with respect to Eq. (1). Furthermore, because of this shift of the origin, the reflection coefficient at the load (defined as the ratio of the reflected voltage wave to the incident voltage wave) has a particularly simple form

$$\Gamma_r = \left.\frac{V^- e^{+j\beta z}}{V^+ e^{-j\beta z}}\right|_{z=0} = \frac{V^-}{V^+} = |\Gamma_r| e^{j\rho} \qquad (3)$$

Since V^+ and V^- are phasors, Γ_r is usually complex, and the angle ρ is the angle between V^+ and V^- at the load. Normally, Eq. (3) is not convenient for the evaluation of the reflection coefficient because neither V^- nor V^+ is known. A much more convenient form can be obtained by substituting Eq. (3) into Eq. (2) and solving for Γ_r

$$\boxed{\Gamma_r = \frac{Z_r/Z_0 - 1}{Z_r/Z_0 + 1} = \frac{Z_r - Z_0}{Z_r + Z_0} = |\Gamma_r| e^{j\rho}} \qquad (4)$$

As a consequence of Eq. (4), one can easily evaluate a reflection coefficient if the characteristic impedance of the transmission line and the load impedance are known.

The impedance has been defined as a function of distance along the transmission line, as shown in Eq. (1). Similarly, the reflection coefficient also can be defined as a function of distance on the line

$$\boxed{\Gamma(z) = \frac{V^- e^{+j\beta z}}{V^+ e^{-j\beta z}} = \Gamma_r e^{j 2\beta z} = |\Gamma_r| e^{j(2\beta z + \rho)}} \qquad (5)$$

The generalized reflection coefficient $\Gamma(z)$ is a phasor whose magnitude is $|\Gamma_r|$, and $\Gamma(0)$ equals Γ_r.

We can now define precisely the term "complete standing wave." A standing wave is a complete standing wave when the magnitude of $\Gamma(z)$ is unity at every point on the line. For a lossless line, the reader can show that this occurs when the load is an open circuit or a short circuit or when Z_r has no real part.

By using the generalized reflection coefficient, we can express the impedance of Eq. (1) in the following convenient form:

$$Z(z) = \frac{1 + \Gamma(z)}{1 - \Gamma(z)} Z_0 \qquad (6)$$

The following example will illustrate how this equation can be used to find the input impedance of the transmission line, as seen at the source $(Z = -L)$.

Example 1 Find the input impedance Z_{in} for a line 7.5 cm long if $Z_r = 150 + j0\ \Omega$, $Z_0 = 50 + j0\ \Omega$, and $\lambda = 20$ cm.

Solution

$$\Gamma_r = \frac{Z_r - Z_0}{Z_r + Z_0} = \frac{3 - 1}{3 + 1} = \frac{1}{2}$$

$$\beta = \frac{2\pi}{\lambda} = 31.416 \text{ rad/m}$$

$$\Gamma(-L) = \Gamma_r e^{-j2\beta L} = \tfrac{1}{2} e^{-j2(31.416)(0.075)} = \tfrac{1}{2} e^{-j3\pi/2} = j0.5$$

$$Z_{in} = Z(-L) = \frac{1 + \Gamma(-L)}{1 - \Gamma(-L)} Z_0 = \frac{1 + j0.5}{1 - j0.5} 50 = 30 + j40\ \Omega$$

Thus the transmission line is an impedance transformer, transforming the load impedance of 150 Ω to an input impedance of $30 + j40\ \Omega$.

Voltages and currents can be expressed conveniently in terms of the $\Gamma(z)$ phasor also:

$$V(z) = V^+ e^{-j\beta z} + V^- e^{j\beta z} = [1 + \Gamma(z)] V^+ e^{-j\beta z} \qquad (7)$$

$$I(z) = Y_0(V^+ e^{-j\beta z} - V^- e^{j\beta z}) = [1 - \Gamma(z)] Y_0 V^+ e^{-j\beta z} \qquad (8)$$

However, before Eqs. (7) and (8) can be used for computations, V^+ must be known. The relationship between the sending-end voltage $V(-L)$ and V^+ can be obtained from Eq. (7):

$$V^+ = \frac{V(-L)}{[1 + \Gamma(-L)] e^{-j\beta(-L)}} \qquad (9)$$

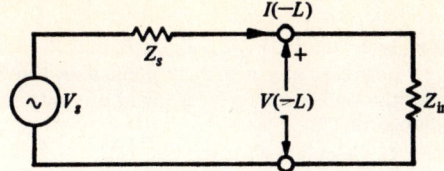

Fig. 5.4.2 Equivalent circuit at the source end.

Therefore the phasors V^+ and $V(-L)$ are interdependent. At the source, if the transmission line is replaced by its input impedance Z_{in}, one can solve for the voltage $V(-L)$ and current $I(-L)$ at the source end of the transmission line (Fig. 5.4.2):

$$V(-L) = \frac{Z_{in} V_s}{Z_{in} + Z_s} \tag{10}$$

$$I(-L) = \frac{V_s}{Z_s + Z_{in}} \tag{11}$$

Equations (9) and (10) can be substituted into Eqs. (7) and (8) to obtain the following equations for the voltage and current phasors, where V^+ has been eliminated:

$$V(z) = \frac{[1 + \Gamma(z)] Z_{in} V_s e^{-j\beta(z+L)}}{[1 + \Gamma(-L)](Z_s + Z_{in})} = |V(z)| e^{j\theta_1} \tag{12}$$

$$I(z) = \frac{[1 - \Gamma(z)] Y_0 Z_{in} V_s e^{-j\beta(z+L)}}{[1 + \Gamma(-L)](Z_s + Z_{in})} = |I(z)| e^{j\theta_2} \tag{13}$$

Although these computations can be carried out by hand, it may be more convenient to use a digital computer, especially if the voltage and current phasors are desired at several points. For the origin at the load end, the distance variable z in these equations is bounded by the interval $-L \leq z \leq 0$.

Because the transmission line is assumed to be lossless, the power received at the load end must be the same as the power sent into the transmission line at the source end. Consequently, we can show from Eq. (24) of Sec. 5.3 that the time-average power at any point on a lossless transmission line is a constant

$$\langle P \rangle = |V(z)| |I(z)| \cos(\theta_1 - \theta_2) \tag{14}$$

Usually the power can be evaluated most easily either at the source end or at the receiving end.

A typical problem and a procedure for solving it follows.

Example 2 Given V_s, Z_s, Z_0, λ, L, and Z_r, find the impedance, voltage, current, and power at any arbitrary point z on the line. (The arbitrary point could be at the source, or receiver, or at any intermediate point.)

Solution
1. Calculate Γ_r using Eq. (4).
2. Calculate $\Gamma(-L)$ and $\Gamma(z)$ using Eq. (5).
3. Find $Z(-L)$ and $Z(z)$ using Eq. (6).
4. Find $V(-L)$ using Eq. (10).
5. Calculate V^+ using Eq. (9).
6. Calculate $V(z)$ using Eq. (7).
7. Obtain $I(z)$ from the ratio $V(z)/Z(z)$ or from Eq. (8).
8. Obtain $\langle P \rangle$ from Eq. (14).

An alternate procedure, which avoids step 5, is to use Eqs. (12) and (13) to find $V(z)$ and $I(z)$. However, in those cases where computations at more than one point on the line are being considered, the first procedure is more efficient.

An example of the solution of the type of problem given above will now be discussed.

Example 3 Given

$Z_r = 100 + j57.74 \; \Omega$ $L = 43$ m
$Z_s = 10 + j120 \; \Omega$ $\lambda = 6$ m
$Z_0 = 100 \; \Omega$ $V_s = 261 \underline{/0°}$ V rms

Find the voltage, current, and power at the source end, at the load end, and 17.2 m from the load end. The problem will be solved first by using the above step-by-step procedure and then by a method using phasor diagrams.

Solution
1. Calculate Γ_r using Eq. (4):

$$\Gamma_r = \frac{Z_r - Z_0}{Z_r + Z_0} = \frac{100 + j57.74 - 100}{100 + j57.74 + 100} = 0.2774\underline{/73.9°} = 0.0769 + j0.2661$$

2. Calculate $\Gamma(-L)$ and $\Gamma(-17.2 \text{ m})$ using Eq. (5). The value for β can be obtained from the equation $\beta\lambda = 2\pi$:

$$\beta = \frac{2\pi}{\lambda} = \frac{\pi}{3} \quad \text{rad/m}$$

$$\Gamma(-L) = \Gamma_r e^{j2\beta(-L)} = \Gamma_r e^{-j(2\pi/3)(43)} = \Gamma_r e^{-j(2\pi/3 + 28\pi)}$$
$$= 0.2774\underline{/73.9°} - 120.0° = 0.2774\underline{/-46.1°} = 0.1922 - j0.2000$$

Similarly for $z_1 = -17.2$ m, one obtains

$$\Gamma(z_1) = \Gamma_r e^{-j2\beta(17.2)} = 0.2774\underline{/73.9°} - 264.0°$$
$$= 0.2774\underline{/-190.1°} = -0.2731 + j0.0504$$

SINUSOIDAL WAVES ON LOSSLESS TRANSMISSION LINES

3. Find $Z(-L)$ and $Z(z_1)$ using Eq. (6):

$$Z(-L) = Z_0 \frac{1 + \Gamma(-L)}{1 - \Gamma(-L)} = \frac{119.22 - j20}{0.8078 + j0.200}$$

$$= 134 - j58 = 145.5\underline{/-23.45°} \; \Omega$$

$$Z(z_1) = Z_0 \frac{1 + \Gamma(z_1)}{1 - \Gamma(z_1)} = \frac{72.7 + j5.04}{1.273 - j0.0504}$$

$$= 56.9 + j5.98 = 57.25\underline{/6.0°} \; \Omega$$

4. Find $V(-L)$ using Eq. (10):

$$V(-L) = \frac{Z_{in}V_s}{Z_{in} + Z_s} = \frac{145.5\underline{/-23.45°}}{144 + j62} \, 261\underline{/0°}$$

$$V(-L) = 242.0\underline{/-46.75°} \text{ V rms}$$

5. Calculate V^+ using Eq. (9):

$$V^+ = \frac{V(-L)}{[1 + \Gamma(-L)]e^{-j\beta(-L)}}$$

$$= \frac{242\underline{/-46.75°}}{(1.192 - j0.200)e^{-j(\pi/3)(-43)}} = 200\underline{/-97.15°} \text{ V}$$

6. Calculate the load voltage and $V(z_1)$ using Eq. (7):

$$V(0) = (1 + \Gamma_r)V^+ = (1.0769 + j0.2661)(200\underline{/-97.15°})$$

$$= 221.6\underline{/-83.24°} \text{ V rms}$$

This voltage appears across the load impedance Z_r. Similarly the voltage at point z_1 is

$$V(z_1) = [1 + \Gamma(z_1)]V^+e^{-j\beta z_1} = 145.3\underline{/218.7°} \text{ V rms}$$

7. The current at each point can be found from the ratio of voltage to impedance

$$I(-L) = \frac{V(-L)}{Z(-L)} = \frac{242\underline{/-46.75°}}{145.5\underline{/-23.45°}} = 1.663\underline{/-23.30°} \text{ A rms}$$

$$I(z_1) = \frac{V(z_1)}{Z(z_1)} = \frac{145.3\underline{/218.7°}}{57.25\underline{/6.0°}} = 2.540\underline{/212.7°} \text{ A rms}$$

8. The time average power $\langle P \rangle$ can be obtained from Eq. (14). At the source,

$$\langle P(-L) \rangle = |V(-L)| \, |I(-L)| \cos(\theta_1 - \theta_2)$$

$$= (242)(1.663) \cos(-46.75° + 23.30°) = 368 \text{ W}$$

At the point $z = z_1 = -17.2$ m,

$$\langle P(z_1) \rangle = 145.3(2.540) \cos(218.7° - 212.7°) = 368 \text{ W}$$

At the load,

$$\langle P(0) \rangle = 221.6(1.918) \cos(-83.24° + 113.24°) = 368 \text{ W}$$

As expected, the time-average power $\langle P \rangle$ is the same at any point on the line because the line is lossless.

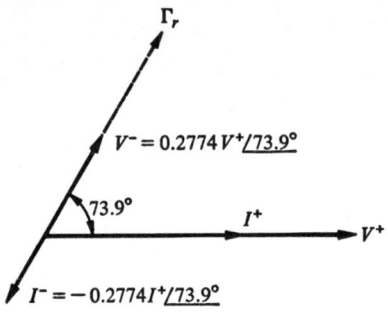

Fig. 5.4.3 Phasor diagram at $z = 0$, with V^+ as the reference.

The transmission line of the preceding example will now be used to illustrate the phasor-diagram method. When using phasor diagrams to solve problems like Example 3, it is convenient to define a positive-going voltage phasor $V^+ e^{-j\beta z}$ and a negative-going voltage phasor $V^- e^{+j\beta z}$. The sum of these two phasors at any point z_1 on the line is the total phasor voltage $V(z_1)$. Similarly, positive-going and negative-going current phasors can be defined as $I^+ e^{-j\beta z}$ and $I^- e^{+j\beta z}$ respectively, so that the total phasor current $I(z)$ is the sum of these two phasors at any point.

For Example 3 we will find $V(z_1)$, $I(z_1)$, $Z(z_1)$, $\Gamma(z_1)$, and $\langle P(z_1) \rangle$ for $z_1 = -17.2$ m. As before, $\Gamma_r = 0.2774/73.9°$, and so the phasor diagram at $z = 0$ appears as shown in Fig. 5.4.3. Since the angle of V^+ with respect to the source voltage V_s (our assumed reference) is unknown, V^+ is drawn as the reference for the present, so that the angle of V^- is the angle of Γ_r. It will be possible to determine the angle of V^+ with respect to V_s later, and then the diagram in Fig. 5.4.3 can be rotated by this angle to reestablish V_s as the reference phasor.

To find V^+, rotate the positive-going phasor $V^+ e^{-j\beta z}$ back to the origin through $-\beta L = (2\pi/6)(43)$ rad, or $60°$, counterclockwise and then rotate the negative-going phasor $(V^- e^{+j\beta z})$ $60°$ clockwise. Since $I^+ e^{-j\beta z} = Y_0 V^+ e^{-j\beta z}$, the positive-going voltage and current phasors are always in phase. Similarly, the negative-going voltage and current phasors are $180°$ out of phase. These phasors are shown at $z = -L$ in Fig. 5.4.4. Then by summing the voltage phasors,

$$V(-L) = V^+ e^{+j\beta L} + V^- e^{-j\beta L}$$
$$= V^+[0.5 + j0.866 + 0.2774(0.97 + j0.2404)]$$
$$= V^+(1.21/50.4°)$$

Similarly, by summing the current phasors,

$$I(-L) = I^+[0.5 + j0.866 + 0.2774(-0.97 - j0.2404)]$$
$$= I^+(0.832/73.85°)$$

SINUSOIDAL WAVES ON LOSSLESS TRANSMISSION LINES

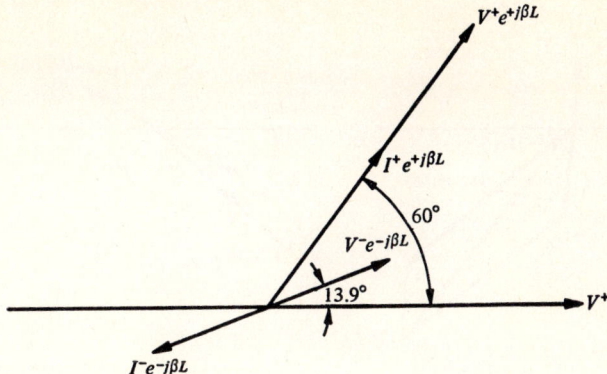

Fig. 5.4.4 Phasor diagram at $z = -L$, with V^+ as the reference.

Therefore

$$Z(-L) = \frac{V(-43 \text{ m})}{I(-43 \text{ m})} = \frac{V^+(1.21\underline{/50.4°})}{I^+(0.832\underline{/73.85°})} = 145.5\underline{/-23.45°}$$

where $V^+/I^+ = Z_0$. Since the quantities $V_s = 261\underline{/0°}$ and $Z_s = 10 + j120$ are given, another expression for $V(-L)$ can be obtained (as in step 4 of Example 3) by means of the circuit shown in Fig. 5.4.5. Therefore

$$V(-L) = 261\underline{/0°} \, \frac{145.5\underline{/-23.45°}}{10 + j20 + 134 - j58}$$

where we use $V_s = 261\underline{/0°}$ to reestablish V_s as the reference phasor. Equating the two expressions for $V(-L)$, it is found that

$$V^+ = 200\underline{/-97.15°}$$

as in step 5 of Example 3. Since V_s was the reference phasor, the V^+ phasor is seen to be complex. In order to establish V_s as the reference phasor in Figs. 5.4.3 and 5.4.4, we rotate both diagrams 97.15° clockwise.

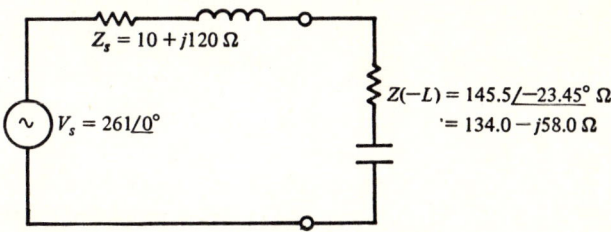

Fig. 5.4.5 Equivalent circuit at $z = -L$.

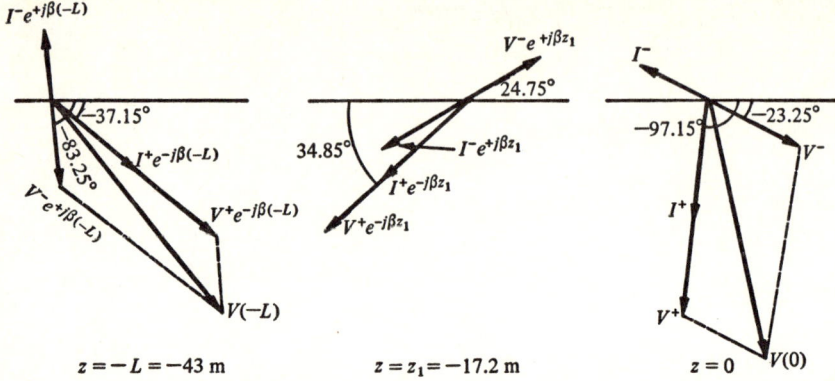

Fig. 5.4.6 Phasor diagrams with V_s as the reference.

These diagrams and a phasor diagram at $z_1 = -17.2$ m are shown in Fig. 5.4.6.

From the diagram at $z = z_1$, $V(z_1)$, $I(z_1)$, $Z(z_1)$, $\Gamma(z_1)$, and $\langle P(z_1)\rangle$ can be found:

$$V(z_1) = V^+ e^{-j\beta z_1} + V^- e^{+j\beta z_1}$$
$$= 200[-0.82 - j0.57 + 0.2774(0.908 + j0.4185)]$$
$$= -113.6 - j90.8 = 145.3\underline{/218.7°}$$
$$I(z_1) = I^+ e^{-j\beta z_1} + I^- e^{+j\beta z_1}$$
$$= 2[-0.82 - j0.57 + 0.2774(-0.908 - j0.4185)]$$
$$= -2.144 - j1.372 = 2.54\underline{/212.7°}$$

Therefore

$$Z(-17.2 \text{ m}) = \frac{V(z_1)}{I(z_1)} = \frac{145.3\underline{/218.7°}}{2.54\underline{/212.7°}} = 57.25\underline{/6°} = 56.9 + j5.98$$

$$\Gamma(-17.2 \text{ m}) = \frac{V^- e^{+j\beta z_1}}{V^+ e^{-j\beta z_1}} = \frac{0.2774\underline{/24.75°}}{1\underline{/-145.15°}} = 0.2774\underline{/169.9°}$$

$$\langle P(z_1)\rangle = |V(z_1)|\,|I(z_1)|\cos(\theta_1 - \theta_2)$$
$$= 145.3(2.54)\cos(218.7° - 212.7°) = 368 \text{ W}$$

The procedures used in these examples are not difficult in theory, but the actual computations become laborious if one is concerned with many points on the transmission line. Two techniques can be used to avoid some of the laborious calculations. The first is the use of a computer

SINUSOIDAL WAVES ON LOSSLESS TRANSMISSION LINES

Table 5.4.1 A lossless line with a real load $Z_r = 3Z_0$

LOSSLESS TRANSMISSION LINE NO. 2

GIVEN DATA

CHARACTERISTIC IMPEDANCE (OHMS)	100.0000
LINE LENGTH (METERS)	7.0000
CENTER FREQUENCY (MHZ)	50.0000
VELOCITY (M/S)	3.00000E 08
SOURCE VOLTAGE (VOLTS)	120.0000
LOAD IMPEDANCE (REAL+IMAG,OHMS)	300.0000 0
SOURCE IMPEDANCE (RE+IM,OHMS)	100.0000 0
NP, NF, FD (MHZ) ARE 22 0	0

COMPUTED DATA

FREQUENCY OF OPN. (MHZ)	5.00000E 01	
STANDING WAVE RATIO	3.00000E 00	
WAVE LENGTH (METERS)	6.00000E 00	
REFLECTION COEFF. (MAG+ANGLE)	5.00000E-01	0
INPUT ADMITTANCE (RE+IM, MHOS)	1.00000E-02	1.15470E-02
V-PLUS (MAG+ANGLE, VOLTS)	6.00000E 01	-6.00000E 01
LOAD POWER (WATTS+VARS)	2.70000E 01	4.65661E-10

LENGTH METERS	IMPEDANCE MAG.	ANGLE	LINE VOLTS MAG.	ANGLE	LINE AMPS MAG.	ANGLE	POWER VARS	ANGLE GAMMA
0	300.00	0	90.00	-60.00	.30	-60.00	0	0
-0.33	204.10	-40.60	85.19	-53.08	.42	-12.48	-23.14	-40.00
-0.67	115.01	-52.71	71.59	-44.37	.62	8.33	-35.45	-80.00
-1.00	65.47	-49.11	51.96	-30.00	.79	19.11	-31.18	-120.00
-1.33	37.64	-24.51	33.42	2.12	.89	26.64	-12.31	-160.00
-1.67	37.64	24.51	33.42	57.88	.89	33.36	12.31	-200.00
-2.00	65.47	49.11	51.96	90.00	.79	40.89	31.18	-240.00
-2.33	115.01	52.71	71.59	104.37	.62	51.67	35.45	-280.00
-2.67	204.10	40.60	85.19	113.08	.42	72.48	23.14	-320.00
-3.00	300.00	-0.00	90.00	120.00	.30	120.00	-0.00	-360.00
-3.33	204.10	-40.60	85.19	126.92	.42	167.52	-23.14	-400.00
-3.67	115.01	-52.71	71.59	135.63	.62	-171.67	-35.45	-440.00
-4.00	65.47	-49.11	51.96	150.00	.79	-160.89	-31.18	-480.00
-4.33	37.64	-24.51	33.42	-177.88	.89	-153.36	-12.31	-520.00
-4.67	37.64	24.51	33.42	-122.12	.89	-146.64	12.31	-560.00
-5.00	65.47	49.11	51.96	-90.00	.79	-139.11	31.18	-600.00
-5.33	115.01	52.71	71.59	-75.63	.62	-128.33	35.45	-640.00
-5.67	204.10	40.60	85.19	-66.92	.42	-107.52	23.14	-680.00
-6.00	300.00	-0.00	90.00	-60.00	.30	-60.00	-0.00	-720.00
-6.33	204.10	-40.60	85.19	-53.08	.42	-12.48	-23.14	-760.00
-6.67	115.01	-52.71	71.59	-44.37	.62	8.33	-35.45	-800.00
-7.00	65.47	-49.11	51.96	-30.00	.79	19.11	-31.18	-840.00

program[†] to perform all the calculations indicated above. Second, many graphical aids for transmission-line computations have been developed, one of the most ingenious and useful of which is that developed by P. H. Smith.[‡] The Smith chart is valuable not only for computational purposes but also as an aid to understanding transmission-line phenomena. In the

[†] Fortran package SANTLINE, in Appendix B, has been included for the convenience of those who have access to a computer.
[‡] P. H. Smith, Transmission Line Calculator, *Electronics*, vol. 12, pp. 29-31, January 1939; see also *Electronics*, vol. 17, p. 130, January 1944.

Table 5.4.2 A lossless line with a complex load $\quad Z_r = Z_0 + j0.5774Z_0$

LOSSLESS TRANSMISSION LINE NO. 1

GIVEN DATA

CHARACTERISTIC IMPEDANCE (OHMS)	100.0000	
LINE LENGTH (METERS)	7.0000	
CENTER FREQUENCY (MHZ)	50.0000	
VELOCITY (M/S)	3.00000E 08	
SOURCE VOLTAGE (VOLTS)	120.0000	
LOAD IMPEDANCE (REAL+IMAG,OHMS)	100.0000	57.7400
SOURCE IMPEDANCE (RE+IM,OHMS)	100.0000	0
NP, NF, FD (MHZ) ARE 22 4	.5000	

COMPUTED DATA

FREQUENCY OF OPN. (MHZ)	5.00000E 01	
STANDING WAVE RATIO	1.76768E 00	
WAVE LENGTH (METERS)	6.00000E 00	
REFLECTION COEFF. (MAG+ANGLE)	2.77372E-01	7.38966E 01
INPUT ADMITTANCE (RE+IM,MHOS)	6.31556E-03	2.73503E-03
V-PLUS (MAG+ANGLE, VOLTS)	6.00000E 01	-6.00000E 01
LOAD POWER (WATTS+VARS)	3.32303E 01	1.91872E 01

LENGTH	IMPEDANCE		LINE VOLTS		LINE AMPS		POWER	ANGLE
METERS	MAG.	ANGLE	MAG.	ANGLE	MAG.	ANGLE	VARS	GAMMA
0	115.47	30.00	66.57	-46.10	.58	-76.10	19.19	73.90
-0.33	157.92	18.53	74.40	-32.83	.47	-51.36	11.14	33.90
-0.67	176.07	-3.66	76.57	-21.32	.43	-17.67	-2.12	-6.10
-1.00	145.30	-23.42	72.54	-9.52	.50	13.90	-14.39	-46.10
-1.33	103.56	-30.95	63.35	4.80	.61	35.75	-19.92	-86.10
-1.67	73.10	-25.90	51.96	25.00	.71	50.90	-16.14	-126.10
-2.00	57.73	-8.21	44.03	54.79	.76	63.00	-4.80	-166.10
-2.33	60.62	14.81	45.65	89.23	.75	74.42	8.79	-206.10
-2.67	80.91	28.79	55.39	115.94	.68	87.16	18.26	-246.10
-3.00	115.47	30.00	66.57	133.90	.58	103.90	19.19	-286.10
-3.33	157.92	18.53	74.40	147.17	.47	128.64	11.14	-326.10
-3.67	176.07	-3.66	76.57	158.68	.43	162.33	-2.12	-366.10
-4.00	145.30	-23.42	72.54	170.48	.50	-166.10	-14.39	-406.10
-4.33	103.56	-30.95	63.35	-175.20	.61	-144.25	-19.92	-446.10
-4.67	73.10	-25.90	51.96	-155.00	.71	-129.10	-16.14	-486.10
-5.00	57.73	-8.21	44.03	-125.21	.76	-117.00	-4.80	-526.10
-5.33	60.62	14.81	45.65	-90.77	.75	-105.58	8.79	-566.10
-5.67	80.91	28.79	55.39	-64.06	.68	-92.84	18.26	-606.10
-6.00	115.47	30.00	66.57	-46.10	.58	-76.10	19.19	-646.10
-6.33	157.92	18.53	74.40	-32.83	.47	-51.36	11.14	-686.10
-6.67	176.07	-3.66	76.57	-21.32	.43	-17.67	-2.12	-726.10
-7.00	145.30	-23.42	72.54	-9.52	.50	13.90	-14.39	-766.10

next section we develop and use the Smith chart and a variation of it, the Z-θ chart.

COMPUTER SOLUTIONS OF SOME TYPICAL PROBLEMS

Computer package SANTLINE (Appendix B) has been used to solve typical problems following the procedure outlined in Example 2. Four examples are included in Tables 5.4.1 to 5.4.4. The same transmission line is used in each case, but the load impedance is varied as shown.

The voltage and current from Table 5.4.1 are plotted in Fig. 5.4.7. Although the standing waves of voltage and current are not sinusoidal, the

SINUSOIDAL WAVES ON LOSSLESS TRANSMISSION LINES

Table 5.4.3 A lossless line with an imaginary load $Z_r = jZ_0$

LOSSLESS TRANSMISSION LINE NO. 3

GIVEN DATA

CHARACTERISTIC IMPEDANCE (OHMS)	100.0000	
LINE LENGTH (METERS)	7.0000	
CENTER FREQUENCY (MHZ)	50.0000	
VELOCITY (M/S)	3.00000E 08	
SOURCE VOLTAGE (VOLTS)	120.0000	
LOAD IMPEDANCE (REAL+IMAG,OHMS)	0	100.0000
SOURCE IMPEDANCE (RE+IM,OHMS)	100.0000	0
NP, NF, FD (MHZ) ARE 22	0	0

COMPUTED DATA

FREQUENCY OF OPN. (MHZ)	5.00000E 01	
STANDING WAVE RATIO	2.00004E 06	
WAVE LENGTH (METERS)	6.00000E 00	
REFLECTION COEFF. (MAG+ANGLE)	9.99999E-01	9.00000E 01
INPUT ADMITTANCE (RE+IM,MHOS)	5.35891E-09	2.67949E-03
V-PLUS (MAG+ANGLE, VOLTS)	6.00000E 01	-6.00000E 01
LOAD POWER (WATTS+VARS)	0	7.19999E 01

LENGTH	IMPEDANCE		LINE VOLTS		LINE AMPS		POWER	ANGLE
METERS	MAG.	ANGLE	MAG.	ANGLE	MAG.	ANGLE	VARS	GAMMA
0	100.00	90.00	84.85	-15.00	.85	-105.00	72.00	90.00
-0.33	214.45	90.00	108.76	-15.00	.51	-105.00	55.16	50.00
-0.67	1143.01	90.00	119.54	-15.00	.10	-105.00	12.50	10.00
-1.00	373.21	-90.00	115.91	-15.00	.31	75.00	-36.00	-30.00
-1.33	142.81	-90.00	98.30	-15.00	.69	75.00	-67.66	-70.00
-1.67	70.02	-90.00	68.83	-15.00	.98	75.00	-67.66	-110.00
-2.00	26.79	-90.00	31.06	-15.00	1.16	75.00	-36.00	-150.00
-2.33	8.75	90.00	10.46	165.00	1.20	75.00	12.50	-190.00
-2.67	46.63	90.00	50.71	165.00	1.09	75.00	55.16	-230.00
-3.00	100.00	90.00	84.85	165.00	.85	75.00	72.00	-270.00
-3.33	214.45	90.00	108.76	165.00	.51	75.00	55.16	-310.00
-3.67	1143.01	90.00	119.54	165.00	.10	75.00	12.50	-350.00
-4.00	373.20	-90.00	115.91	165.00	.31	-105.00	-36.00	-390.00
-4.33	142.81	-90.00	98.30	165.00	.69	-105.00	-67.66	-430.00
-4.67	70.02	-90.00	68.83	165.00	.98	-105.00	-67.66	-470.00
-5.00	26.79	-90.00	31.06	165.00	1.16	-105.00	-36.00	-510.00
-5.33	8.75	90.00	10.46	-15.00	1.20	-105.00	12.50	-550.00
-5.67	46.63	90.00	50.71	-15.00	1.09	-105.00	55.16	-590.00
-6.00	100.00	90.00	84.85	-15.00	.85	-105.00	72.00	-630.00
-6.33	214.45	90.00	108.76	-15.00	.51	-105.00	55.16	-670.00
-6.67	1143.01	90.00	119.54	-15.00	.10	-105.00	12.50	-710.00
-7.00	373.20	-90.00	115.91	-15.00	.31	75.00	-36.00	-750.00

slope is continuous everywhere, in contrast to the slope of a complete standing wave. Furthermore, the voltage is at a maximum when the current is at a minimum, and vice versa. The impedance and the magnitudes of the voltage and the current phasors are all periodic with a period of $\lambda/2$. As an exercise, the reader should sketch the tabulated data from Tables 5.4.2 to 5.4.4 and compare the results with the graphs of Fig. 5.4.7. It is interesting to compare where the minima and maxima of the impedance, voltage, and current for all these cases occur. The matched case, where the load impedance equals the characteristic

Table 5.4.4 A lossless line with no load $Z_r = 0$

LOSSLESS TRANSMISSION LINE NO. 4

GIVEN DATA

CHARACTERISTIC IMPEDANCE (OHMS)	100.0000	
LINE LENGTH (METERS)	7.0000	
CENTER FREQUENCY (MHZ)	50.0000	
VELOCITY (M/S)	3.00000E 08	
SOURCE VOLTAGE (VOLTS)	120.0000	
LOAD IMPEDANCE (REAL+IMAG,OHMS)	0	0
SOURCE IMPEDANCE (RE+IM,OHMS)	100.0000	0
NP, NF, FD (MHZ) ARE 22	0	0

COMPUTED DATA

FREQUENCY OF OPN. (MHZ)	5.00000E 01	
STANDING WAVE RATIO	2.00004E 06	
WAVE LENGTH (METERS)	6.00000E 00	
REFLECTION COEFF. (MAG+ANGLE)	9.99999E-01	-1.80000E 02
INPUT ADMITTANCE (RE+IM,MHOS)	6.66629E-09	-5.77350E-03
V-PLUS (MAG+ANGLE, VOLTS)	6.00000E 01	-6.00000E 01
LOAD POWER (WATTS+VARS)	3.59985E-09	0

LENGTH	IMPEDANCE		LINE VOLTS		LINE AMPS		POWER	ANGLE
METERS	MAG.	ANGLE	MAG.	ANGLE	MAG.	ANGLE	VARS	GAMMA
0	.00	.04	.00	-59.96	1.20	-60.00	.00	-180.00
-0.33	36.40	90.00	41.04	30.00	1.13	-60.00	46.28	-220.00
-0.67	83.91	90.00	77.13	30.00	.92	-60.00	70.91	-260.00
-1.00	173.21	90.00	103.92	30.00	.60	-60.00	62.35	-300.00
-1.33	567.13	90.00	118.18	30.00	.21	-60.00	24.63	-340.00
-1.67	567.13	-90.00	118.18	30.00	.21	120.00	-24.63	-380.00
-2.00	173.21	-90.00	103.92	30.00	.60	120.00	-62.35	-420.00
-2.33	83.91	-90.00	77.13	30.00	.92	120.00	-70.91	-460.00
-2.67	36.40	-90.00	41.04	30.00	1.13	120.00	-46.28	-500.00
-3.00	.00	5.34	.00	125.34	1.20	120.00	.00	-540.00
-3.33	36.40	90.00	41.04	-150.00	1.13	120.00	46.28	-580.00
-3.67	83.91	90.00	77.13	-150.00	.92	120.00	70.91	-620.00
-4.00	173.21	90.00	103.92	-150.00	.60	120.00	62.35	-660.00
-4.33	567.13	90.00	118.18	-150.00	.21	120.00	24.63	-700.00
-4.67	567.13	-90.00	118.18	-150.00	.21	-60.00	-24.63	-740.00
-5.00	173.21	-90.00	103.92	-150.00	.60	-60.00	-62.35	-780.00
-5.33	83.91	-90.00	77.13	-150.00	.92	-60.00	-70.91	-820.00
-5.67	36.40	-90.00	41.04	-150.00	1.13	-60.00	-46.28	-860.00
-6.00	.00	10.56	.00	-49.44	1.20	-60.00	.00	-900.00
-6.33	36.40	90.00	41.04	30.00	1.13	-60.00	46.28	-940.00
-6.67	83.91	90.00	77.13	30.00	.92	-60.00	70.91	-980.00
-7.00	173.21	90.00	103.92	30.00	.60	-60.00	62.35	-1020.00

impedance, has not been considered. The reader will find it instructive to solve this problem and compare the results with the data of the above tables.

THE GAMMA PLANE

The variables $\Gamma(z)$, $V(z)$, and $I(z)$ are all complex numbers and can be plotted on a complex plane using the real and imaginary axes. We now consider the plot of $\Gamma(z)$ in the complex plane. From the equation for $\Gamma(z)$,

$$\Gamma(z) = \Gamma_r e^{+j2\beta z} = |\Gamma_r| e^{j\rho} e^{+j2\beta z}$$

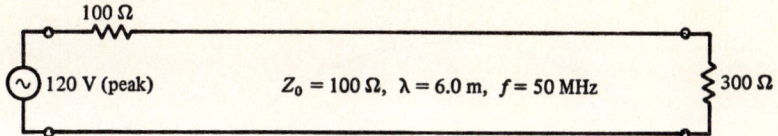

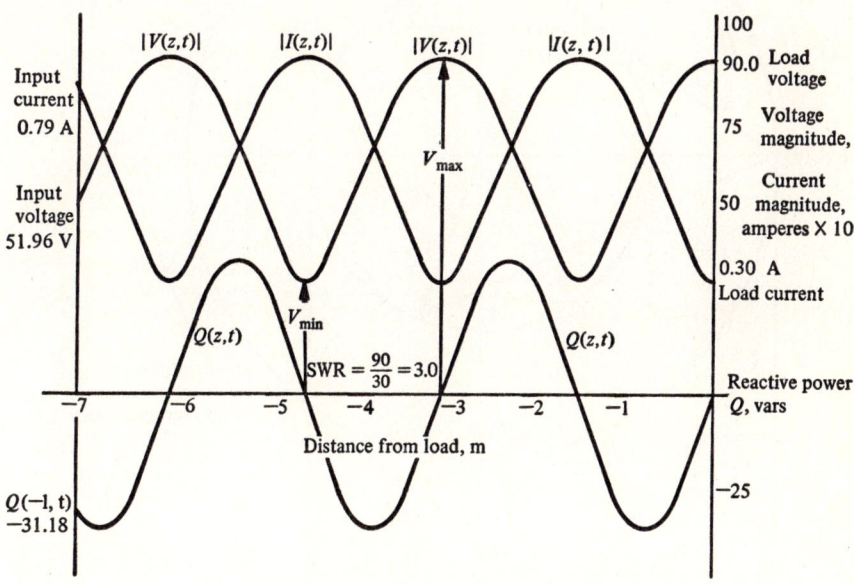

Fig. 5.4.7 Voltage, current, and power for line 2 (see Table 5.4.1).

we see that the magnitude of $\Gamma(z)$ is independent of position on the line, and at $z = 0$, $\Gamma(z)$ has a phase angle of ρ. As we move in the negative z direction on the transmission line, the gamma phasor rotates in the clockwise direction (because of the term $e^{-j2\beta z}$) and traces out a circle of radius $|\Gamma_r|$ centered at the origin. For convenience, we call this complex plane the *gamma plane*.

We can take advantage of the gamma-plane geometry for finding the magnitudes of voltage and current phasors. It is helpful to normalize the magnitude of the voltage and current phasors from Eqs. (7) and (8), as follows:

$$|V_n(z)| = \frac{|V(z)|}{|V^+|} = |1 + \Gamma(z)| \tag{15}$$

$$|I_n(z)| = \frac{|I(z)|}{|Y_0||V^+|} = |1 - \Gamma(z)| \tag{16}$$

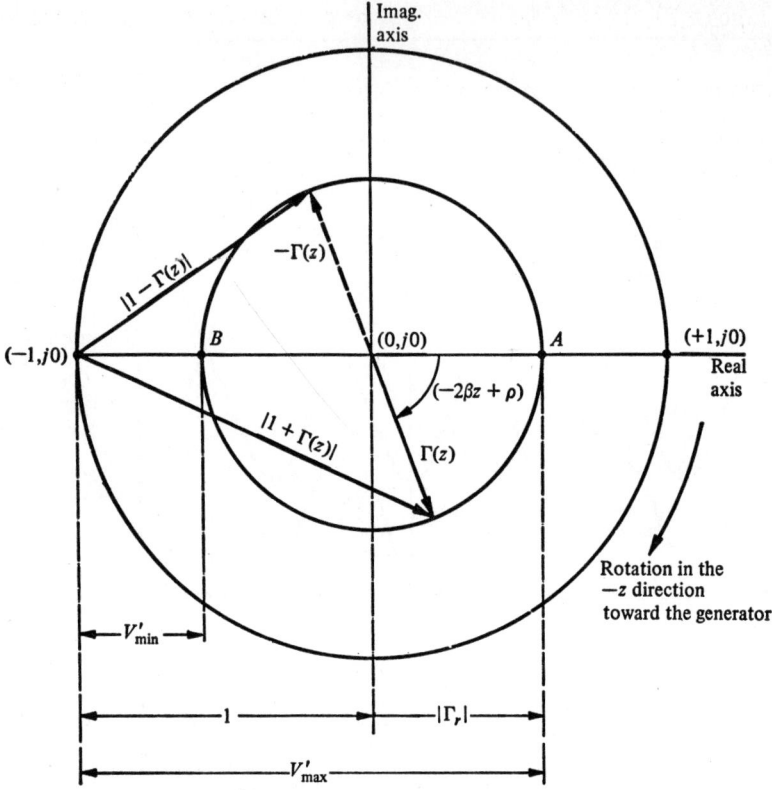

Fig. 5.4.8 The complex gamma plane.

In these equations, the subscript n denotes that the phasors have been normalized. A phasor from the point $(-1,j0)$ in the gamma plane to the tip of the gamma phasor has a magnitude of $|1 + \Gamma(z)|$ and hence represents the normalized voltage phasor of Eq. (15). Similarly, a current phasor can be drawn which corresponds to the normalized current phasor of Eq. (16). These phasors are shown in Fig. 5.4.8. Note that the tips of the voltage and current phasors are always diametrically opposite and lie on the circle of radius $|\Gamma_r|$. (Why?)

As the $\Gamma(z)$ phasor rotates, the voltage phasor rotates about the circle of radius $|\Gamma_r|$ as a function of the distance on the transmission line. We find that voltage maxima occur at point A and voltage minima occur at point B in Fig. 5.4.8. The length of these voltage phasors at the maxima and minima depends upon the length of the gamma phasor. However, the magnitude of the gamma phasor is never greater than unity on a lossless transmission line. For this reason, we are presently interested only in the small region of the entire complex plane bounded by the circle of unit radius.

Suppose we use a voltmeter to locate a point z_A on the transmission line where the voltage is maximum. Then, by moving the voltmeter $\frac{1}{4}$ wavelength in either direction from z_A, we find a point z_B where the voltage is a minimum. The voltages at these two points are

$$|V(z_A)| = |V_{max}| = (1 + |\Gamma_r|)|V^+| \qquad (17)$$
$$|V(z_B)| = |V_{min}| = (1 - |\Gamma_r|)|V^+| \qquad (18)$$

An important parameter, the *standing-wave ratio* (SWR), is defined as the ratio of the maximum voltage to the minimum voltage

$$\boxed{\text{SWR} = \frac{|V_{max}|}{|V_{min}|} = \frac{1 + |\Gamma_r|}{1 - |\Gamma_r|}} \qquad (19)$$

The SWR is important because it is easily measured and is useful in mathematical analysis. Some of its important properties are:

1. SWR ≥ 1.
2. If SWR $= 1$, then $\Gamma_r = 0$ and the line is matched.
3. If SWR $\to \infty$, then $|\Gamma_r| = 1$ and no power is consumed at the load.

By using Eq. (6) for impedance as a function of distance we can derive some simple relationships between the SWR and the minimum and maximum values of impedance

$$\frac{Z_{max}}{Z_0} = \frac{1 + \Gamma(z_A)}{1 - \Gamma(z_A)} = \frac{1 + |\Gamma_r|}{1 - |\Gamma_r|} = \text{SWR} \qquad (20)$$

$$\frac{Z_{min}}{Z_0} = \frac{1 + \Gamma(z_B)}{1 - \Gamma(z_B)} = \frac{1 - |\Gamma_r|}{1 + |\Gamma_r|} = \frac{1}{\text{SWR}} \qquad (21)$$

$$Z_{max} Z_{min} = Z_0^2 \qquad (22)$$

Naturally, the proof of these equations will be left to the reader. In the next section, the gamma plane and some of the equations of this section will be used to develop the Smith chart and the Z-θ chart.

5.5 THE SMITH CHART AND THE Z-θ CHART

In this section, graphical techniques for solving transmission-line problems are developed. Before presenting the Smith chart, it is convenient to discuss the generalized reflection coefficient and to define the normalized impedance. The generalized reflection coefficient, defined previously, is here repeated:

$$\Gamma(z) = \frac{V^-(z)}{V^+(z)} = \frac{V^- e^{j\beta z}}{V^+ e^{-j\beta z}} = \Gamma_r e^{j2\beta z} \tag{1}$$

Since this expression can also be written

$$\Gamma(z) = |\Gamma_r| e^{j(\rho + 2\beta z)} \tag{2}$$

it is evident that $\Gamma(z)$ and Γ_r have the same magnitude and that if $z = 0$, then $\Gamma(z) = \Gamma_r$.

We now discuss an interpretation of the generalized reflection coefficient. Suppose that a line is terminated in Z_r, thereby fixing Γ_r. If the load end of the line is chopped off and reterminated at a point $z = -h$, the standing-wave pattern on the remainder of the line is the same as before if the new terminal impedance is

$$Z(-h) = Z_0 \frac{1 + \Gamma(-h)}{1 - \Gamma(-h)} \tag{3}$$

The *normalized impedance* is defined by

$$Z_n(z) = \frac{Z(z)}{Z_0} = \frac{1 + \Gamma_r e^{j2\beta z}}{1 - \Gamma_r e^{j2\beta z}} \tag{4}$$

The $Z_n(z)$ plane is a complex plane in which the impedance $Z_n(z)$ is plotted in either rectangular or polar coordinates. Since realizable impedances have nonnegative real parts, we will restrict our interest to the right half of the $Z_n(z)$ plane. Written in terms of the generalized reflection coefficient, the normalized impedance can be expressed as

$$Z_n(z) = \frac{1 + \Gamma(z)}{1 - \Gamma(z)} \tag{5}$$

If Eq. (5) is solved for the generalized reflection coefficient in terms of the normalized impedance, we obtain

$$\Gamma(z) = \frac{Z_n(z) - 1}{Z_n(z) + 1} \tag{6}$$

Equations (5) and (6) define a bilinear transformation† between the $\Gamma(z)$ plane and the $Z_n(z)$ plane. We now consider how an impedance transforms from the right half of the $Z_n(z)$ plane to the interior of the unit circle on the $\Gamma(z)$ plane. This transformation leads to the Smith chart.

† R. V. Churchill, "Complex Variables and Applications," 2d ed., McGraw-Hill Book Company, New York, 1960; see chaps. 8 and 9 for a discussion of conformal mapping and fig. 12 of app. II.

THE SMITH CHART

To construct the Smith chart (see Fig. 5.5.1) it is necessary to plot $Z_n(z) = r + jx$, for loci of constant r and then of constant x, on the $\Gamma(z)$ plane. Two restrictions, of course, are that the real part of the normalized impedance is never negative and the magnitude of the generalized reflection coefficient never exceeds unity. To construct the Smith chart

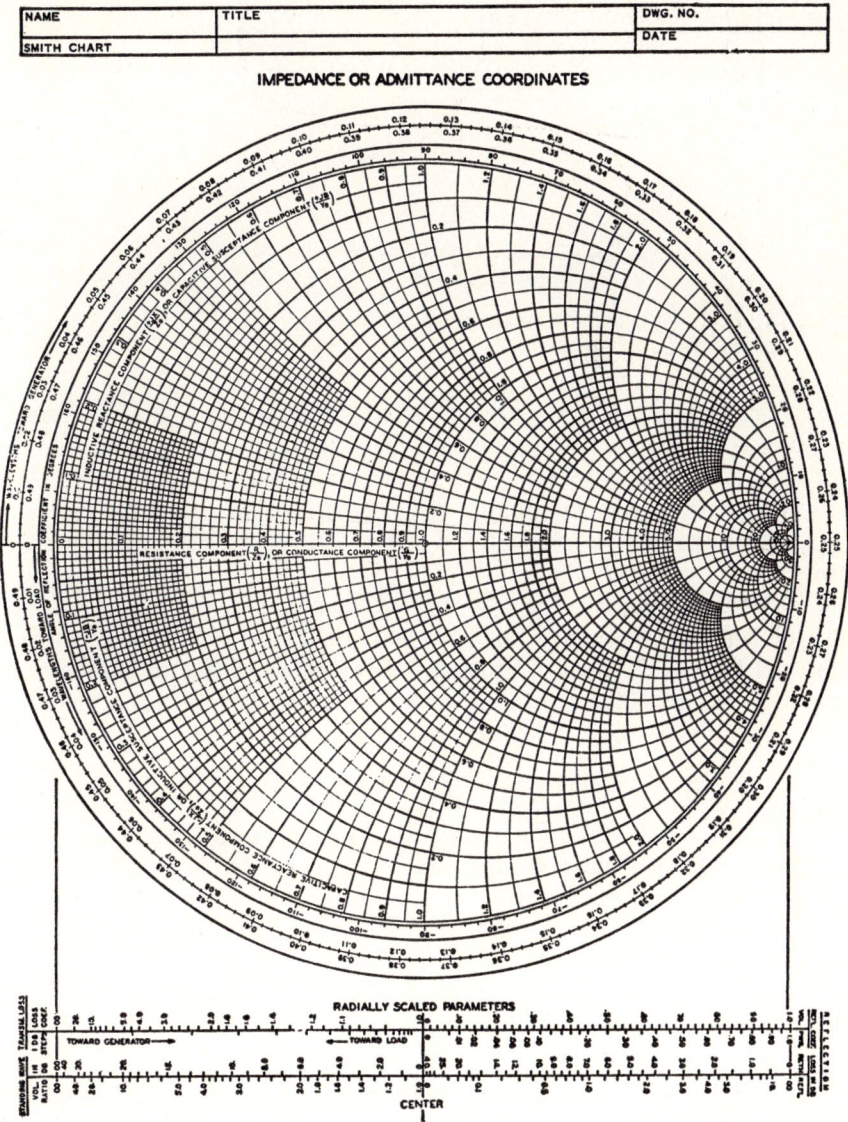

Fig. 5.5.1 The Smith chart.

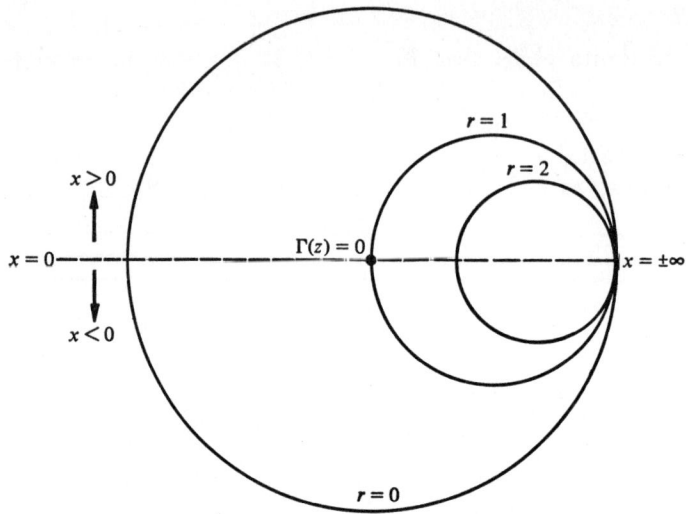

Fig. 5.5.2 Loci of constant resistance on the $\Gamma(z)$ plane.

we must (1) map $Z_n(z) = r_0 + jx$ onto the $\Gamma(z)$ plane and (2) map $Z_n(z) = r + jx_0$ onto the $\Gamma(z)$ plane, where r_0 and x_0 are constants. Loci for $r_0 = 0, 1, 2$ on the $\Gamma(z)$ plane are shown in Figure 5.5.2, where the reactance varies from minus infinity to plus infinity. Loci for $x_0 = 0$, ± 1, ± 2 on the $\Gamma(z)$ plane are shown in Fig. 5.5.3 as the resistance r varies from zero to infinity.

The Smith chart is the transformed impedance plane (on which the loci of constant resistance and constant reactance are plotted) superimposed on the $\Gamma(z)$ plane. Proof that these loci are either circles or circle segments will be left as an exercise. Any point on the Smith chart can be expressed in either of two types of coordinate system:

1. The impedance coordinates are read from the loci of the constant resistance and reactance curves. Any impedance is expressed in rectangular form $r + jx$, although the coordinate system of the transformed impedance is not a cartesian system.
2. The coordinate system of the reflection coefficient $\Gamma(z)$ is the ordinary polar coordinate system of the $\Gamma(z)$ plane. Grid lines of this system do not appear explicitly on the interior of the Smith chart, but the coordinates can be read from scales on the periphery of the chart.

Thus if a particular impedance is known, we can find the corresponding reflection coefficient, and vice versa.

SINUSOIDAL WAVES ON LOSSLESS TRANSMISSION LINES

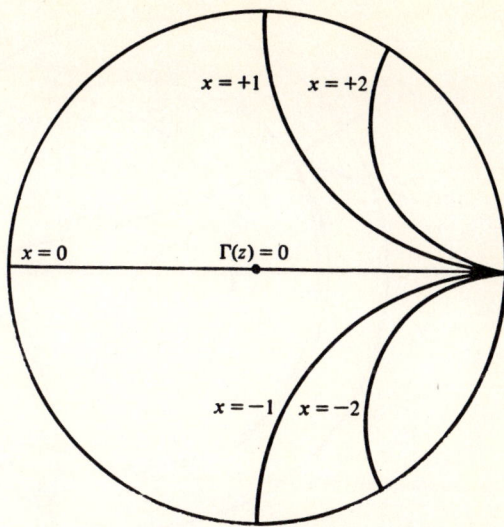

Fig. 5.5.3 Loci of constant reactance on the $\Gamma(z)$ plane.

Since the magnitude of $\Gamma(z)$ is $|\Gamma_r|$ on a lossless line, then as z varies, $\Gamma(z)$ traverses (on the Smith chart) a circle of radius $|\Gamma_r|$ centered about the origin of the $\Gamma(z)$ plane. Since $\Gamma(z) = \Gamma_r e^{j2\beta z}$, angular distance on the line is measured in radians by the factor $2\beta z$ (where $\beta = 2\pi/\lambda$), from which it can be seen that $\Gamma(z)$ and the normalized impedance repeat every half wavelength on the line. Moving from the load end toward the sending end, distance becomes increasingly negative and $2\beta z$ also becomes increasingly negative. Since $\rho + 2\beta z$ is the angle of $\Gamma(z)$, traveling down the line toward the sending end corresponds to a clockwise rotation on the $\Gamma(z)$ plane.

As an example of the use of the Smith chart consider a lossless transmission line of length 0.7λ with characteristic impedance 50 Ω and terminated with an impedance $15 - j70$ Ω. The impedance at the sending end (or at any point along the line) can be found on the Smith chart. The immediate objective is to find the impedance at the sending end. The first step is to find the normalized load impedance $Z_n(0)$, which is $0.3 - j1.4$, and to locate this point on the Smith chart as shown in Fig. 5.5.4. The line from the center of the Smith chart (the origin) to the point $Z_n(0)$ is the Γ_r phasor. Its magnitude and angle are measured in the polar coordinate system of the gamma plane as follows. From the *linear scale* on the lower right-hand corner of the Smith chart, $|\Gamma_r| = 0.82$, and from the *angular scale* around the periphery of the chart, $\rho = -69.8°$. To find the sending-end impedance, the next step is to rotate the gamma phasor

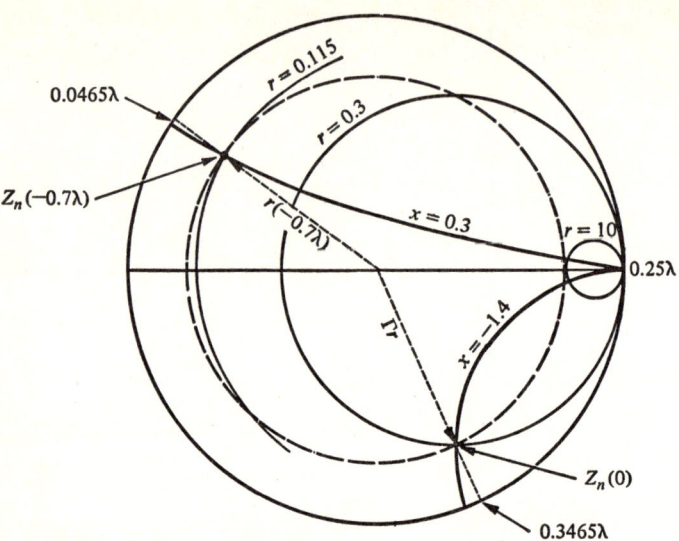

Fig. 5.5.4 Use of Smith chart for a complex termination.

0.7λ [which is $2(2\pi/\lambda)(0.7\lambda) = 2.8\pi$ rad] clockwise around the Smith chart. This rotation is facilitated by the wavelength scale around the periphery of the chart. The line from the origin through $Z_n(0)$ intercepts the outer wavelength scale at 0.3465λ. The sending-end normalized impedance is located at $0.7\lambda + 0.3465\lambda = 1.0465\lambda$ (which is not on the wavelength scale because one complete rotation is 0.5λ). However, by subtracting 1.0λ from 1.0465λ, we obtain 0.0465λ. This point is found on the periphery of the chart, and a line is drawn from it to the center of the chart. After rotating the gamma phasor through 0.7λ, the phasor $\Gamma(-0.7\lambda)$ lies along this line. The coordinates of the tip of the gamma phasor $\Gamma(-0.7\lambda)$ correspond either to the sending-end impedance $Z_n(-0.7\lambda)$ or to the reflection coefficient $\Gamma(-0.7\lambda)$, depending on which set of coordinates is desired. It is seen that normalized impedance is $Z_n(-0.7\lambda) = 0.115 + j0.3$, so that the sending-end impedance is $Z(-0.7\lambda) = 5.75 + j15 \ \Omega$.

The relationship between the SWR and $|\Gamma(z)|$ is given by

$$\text{SWR} = \frac{1 + |\Gamma(z)|}{1 - |\Gamma(z)|} = \frac{1 + |\Gamma_r|}{1 - |\Gamma_r|} \tag{7}$$

and the normalized impedance is given by

$$Z_n(z) = \frac{1 + |\Gamma_r|e^{j(\rho + 2\beta z)}}{1 - |\Gamma_r|e^{j(\rho + 2\beta z)}} \tag{8}$$

SINUSOIDAL WAVES ON LOSSLESS TRANSMISSION LINES

It can be seen that the SWR can be obtained by reading the impedance coordinates when $\rho + 2\beta z$, the angle of $\Gamma(z)$ (from the polar coordinate system), is equal to zero. Thus *the maximum resistive value of normalized impedance is also the standing-wave ratio,* which can be read directly from the Smith chart. The magnitude of the reflection coefficient can easily be obtained when the SWR is known by rearranging Eq. (7), to give

$$|\Gamma_r| = \frac{\text{SWR} - 1}{\text{SWR} + 1} \tag{9}$$

In the previous example, the maximum normalized impedance is $r = 10.0$, so that SWR $= 10.0$, and then $|\Gamma_r| = |\Gamma(z)| = 9.0/11.0 = 0.82$, as before. Using the scale marked in degrees around the periphery of the chart, it is seen that $\Gamma_r = 0.82/\!-69.8°$ and that $\Gamma(-0.7\lambda) = 0.82/146.5°$.

As another example, let a line of length $3\lambda/8$ be terminated in a short circuit. The impedance and reflection coefficient at the sending end are to be found. After locating $Z_n(0) = 0$ on the chart, we find that $\Gamma_r = \Gamma(0) = 1/180°$, which is on the outer periphery of the chart, as shown in Fig. 5.5.5. Rotation of the gamma phasor through a distance of $3\lambda/8$ toward the generator (clockwise through $3\pi/2$ rad) locates a point three-fourths of the way around the circle. The impedance coordinates

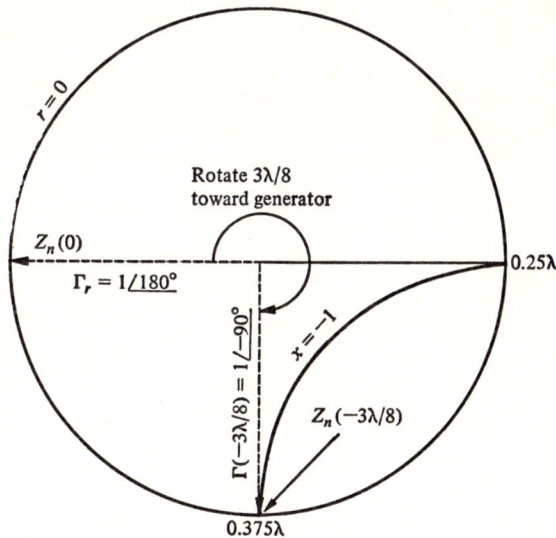

Fig. 5.5.5 Use of the Smith chart for a short-circuit termination.

of this point are $Z_n(-3\lambda/8) = -j1$, and the polar coordinates are $\Gamma(-3\lambda/8) = 1/\underline{-90°}$. Thus the input impedance of a short-circuited line of length $3\lambda/8$ is a capacitive reactance of the same magnitude as the characteristic impedance, $Z(-3\lambda/8) = -jZ_0$ Ω.

From Fig. 5.4.8 we found that voltage is maximum when the angle of the gamma phasor is equal to zero. This is the point where the normalized impedance $Z_n(z)$ is maximum and where the SWR is read

$$\text{SWR} = \frac{1+\Gamma(z)}{1-\Gamma(z)}\bigg|_{\xi=0} = Z_n(z)\bigg|_{\xi=0} = \text{maximum value of } Z_n(z)$$

where $\xi = \rho + 2\beta z$. Similarly the voltage and impedance are both at a minimum when the angle of the gamma phasor is π rad.

$$\frac{1}{\text{SWR}} = \frac{1+\Gamma(z)}{1-\Gamma(z)}\bigg|_{\xi=\pi} = Z_n(z)\bigg|_{\xi=\pi} = \text{minimum value of } Z_n(z)$$

where $\xi = \rho + 2\beta z$. Consequently, the voltage has a maximum where $\rho = -2\beta z$, or $\xi = 0$, and a minimum where $\rho + 2\beta z = \pi$. Thus if the SWR is known and the distance between the termination and a voltage minimum or maximum is also known, the normalized termination impedance itself can be read from the Smith chart. In fact, the normalized impedance at any point on the transmission line can be read from the corresponding point on the $\Gamma(z)$ circle.

As an example, let SWR $= 1.5$, and let the first voltage minimum be 0.37λ from the load. The SWR $= 1.5$ circle [the $\Gamma(z)$ circle] intersects the $\xi = 0$ axis at $Z_n(z) = 1.5$ and the $\xi = \pi$ axis at $Z_n(z) = 0.667$, as shown in Fig. 5.5.6. The normalized impedance at the point on the line corresponding to minimum voltage is therefore $Z_n(-0.37\lambda) = 0.667 + j0$. The normalized load impedance is found by rotating the gamma phasor from the point of minimum voltage 0.37λ toward the load (counterclockwise), and we find that $Z_n(0) = 0.95 + j0.4$.

The Smith chart can be used in terms of admittance rather than impedance, and also for finding admittance if impedance is known. Since

$$Z_n(z) = \frac{1+|\Gamma_r|e^{j\xi}}{1-|\Gamma_r|e^{j\xi}}$$

the normalized admittance at any point must be

$$Y_n(z) = \frac{1-|\Gamma_r|e^{j\xi}}{1+|\Gamma_r|e^{j\xi}} = \frac{1+|\Gamma_r|e^{j(\xi+\pi)}}{1-|\Gamma_r|e^{j(\xi+\pi)}}$$

Therefore if coordinates of impedance are plotted on the Smith chart, the corresponding admittance is found at the point 180° around the circle

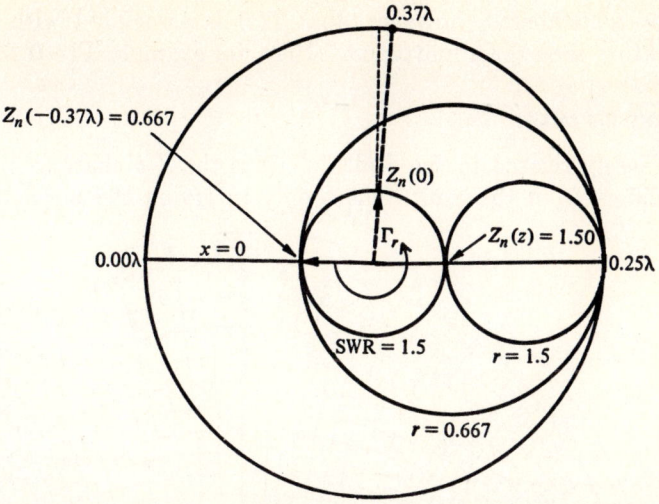

Fig. 5.5.6 Use of the Smith chart when the position of a voltage minimum is known.

of radius $|\Gamma_r|$. Note that this is a simple way of inverting any complex number.

As an example, let $Z_n(0) = 0.6 + j0.1$, and let the line length be 0.15λ. The problem is to find the sending-end admittance. From the Smith chart (see Fig. 5.5.7) it is seen that the receiving-end admittance is $Y_n(0) = 1.63 - j0.25$, located 180° from $Z_n(0)$. After rotating 0.15λ toward the generator, it is seen that $Y_n(-0.15\lambda) = 0.7 - j0.32$. It must

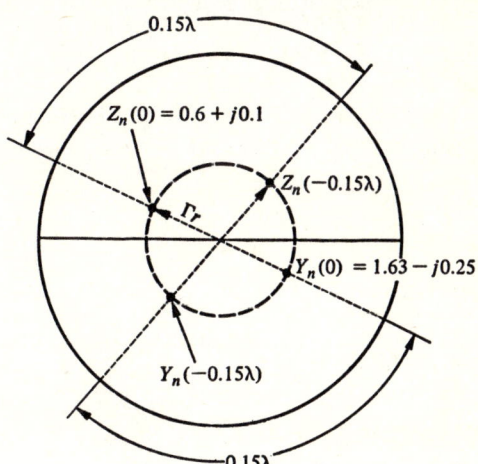

Fig. 5.5.7 Use of the Smith chart to find admittance.

be remembered, however, that $\Gamma(z)$ is associated with the impedance rather than the admittance. Here, for example, $\Gamma(-0.15\lambda) = 0.25/\underline{58°}$.

THE Z-θ CHART

Closely related to the Smith chart is the Z-θ chart (see Fig. 5.5.8). It differs from the Smith chart only in giving the normalized impedance

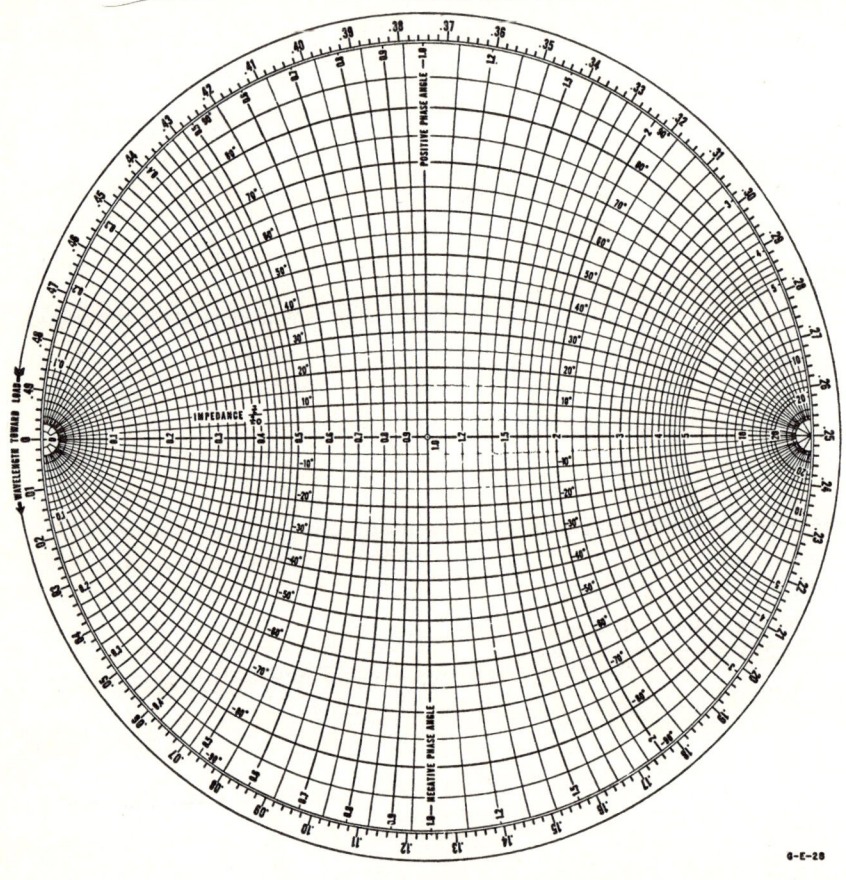

Fig. 5.5.8 The Z-θ chart.

SINUSOIDAL WAVES ON LOSSLESS TRANSMISSION LINES

coordinates in polar form, $Ae^{j\theta}$, rather than in rectangular form, $r + jx$. The transformation between the $\Gamma(z)$ plane and the $Z_n(z)$ plane is given by

$$\Gamma(z) = \frac{Z_n(z) - 1}{Z_n(z) + 1} = \frac{Ae^{j\theta} - 1}{Ae^{j\theta} + 1} \tag{10}$$

The Z-θ chart is composed of the loci of constant amplitude A and of constant angle θ superimposed on the gamma plane. Any point on the Z-θ chart can be expressed in either coordinate system—the normalized impedance coordinates in polar form or the reflection-coefficient coordinates in polar form. By substituting a few values for A and θ into Eq. (10), the reader can verify that the loci shown on the Z-θ chart are correct.

The Z-θ chart can be used just like the Smith chart, as the reader can show for all the preceding examples. Indeed, the same derivations would yield the same relationships between $\Gamma(z)$, SWR, $Y_n(z)$, and $Z_n(z)$ for the Z-θ chart as for the Smith chart. The only difference between the two charts is the coordinate system used to express $Z_n(z)$ and $Y_n(z)$.

The Z-θ chart is particularly convenient if for some reason the impedance or admittance at some point on the line is restricted to a particular angle or amplitude. As an example, suppose it is undesirable to draw more than 0.36 A from a 100-V rms source having negligible internal impedance. A line with $Z_0 = 100$ Ω is to be connected to a load of $Z_r = 450\underline{/30°}$ Ω. What line lengths are permissible? The minimum value of $|Z(-L)|$ allowable is

$$|Z(-L)| = \frac{100}{0.36} = 277.8 \text{ Ω}$$

The allowable line lengths must be such that $Z_n(-L)$ lies to the right of the $|Z_n| = 2.778$ locus on the Z-θ chart, as shown in Fig. 5.5.9. The normalized load impedance is

$$Z_{nr} = 4.5\underline{/30°}$$

and the shortest permissible range for line length is, from Fig. 5.5.9, $L = 0$ to $L = 0.0665\lambda$, the next longer permissible range is $L = 0.4715\lambda$ to $L = 0.5665\lambda$, and the next allowable range is $L = 0.9715\lambda$ to $L = 1.0665\lambda$. In conclusion, there are an infinite number of allowable ranges for line length, obtained by successively adding 0.5λ to the limits of the second permissible range of L. The reader will appreciate the value of the Z-θ chart if he attempts to solve this problem on the Smith chart.

5.6 VOLTAGE, CURRENT, AND POWER FROM THE SMITH CHART

The Smith chart can be used to find voltage or current at any point on the line if either voltage or current is known at another point. Since V^+ is

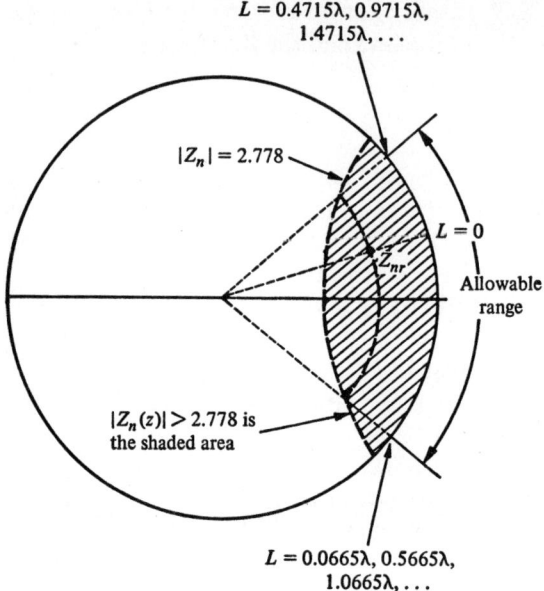

Fig. 5.5.9 The Z-θ chart for the example.

independent of distance along the line, and since $V(z) = V^+e^{-j\beta z}[1 + \Gamma(z)]$,

$$|V^+| = \frac{|V(z_1)|}{|1 + \Gamma(z_1)|} = \frac{|V(z_2)|}{|1 + \Gamma(z_2)|} \tag{1}$$

where z_1 and z_2 are any two points on the line. This can be rewritten

$$\frac{|V(z_1)|}{|V(z_2)|} = \frac{|1 + \Gamma(z_1)|}{|1 + \Gamma(z_2)|} \tag{2}$$

Similarly, the expression

$$\frac{|I(z_1)|}{|I(z_2)|} = \frac{|1 - \Gamma(z_1)|}{|1 - \Gamma(z_2)|} \tag{3}$$

can be obtained for the ratio of current magnitudes at any two points on the line. The length of $|1 + \Gamma(z)|$ and $|1 - \Gamma(z)|$ can be measured from the Smith chart for a particular $\Gamma(z)$, as shown in Fig. 5.6.1. It is often convenient to use the linear scale at the bottom of the chart to estimate line length. Equations (1) to (3) are valid for either rms or peak values. Peak values of voltage and current will be used unless otherwise specified.

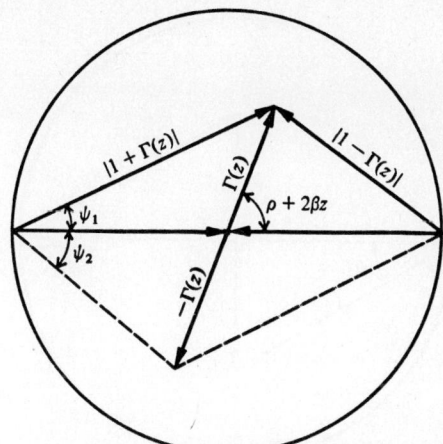

Fig. 5.6.1 Finding $|1 + \Gamma(z)|$ and $|1 - \Gamma(z)|$ from the Smith chart.

Example 1 Let $Z_0 = 50\ \Omega$ and $Z_r = 150\ \Omega$. If $V_r = 30$ V, the problem is to find the magnitude of voltage and current at $z = -3\lambda/8$. The normalized load impedance is $3 + j0$, which can be located on the Smith chart, as shown in Fig. 5.6.2. It can be seen that

$$|\Gamma(z)| = \tfrac{1}{2} \quad \text{and} \quad \Gamma\left(\frac{-3\lambda}{8}\right) = \tfrac{1}{2}\underline{/90°}$$

The ratio $|V(-3\lambda/8)|/|V(0)|$ is the same as the ratio of the magnitudes of the corresponding phasors shown in Fig. 5.6.2. The magnitudes of the phasors are given in parentheses in the figure. Thus

$$\frac{|V(-3\lambda/8)|}{|V(0)|} = \frac{\sqrt{1 + (\tfrac{1}{2})^2}}{1.5} = \frac{\sqrt{5}}{3}$$

$$\left|V\left(\frac{-3\lambda}{8}\right)\right| = 10\sqrt{5}\ \text{V}$$

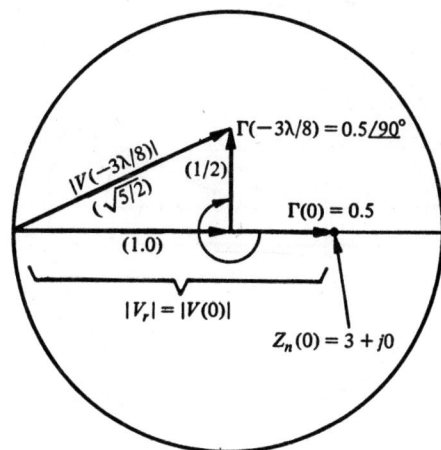

Fig. 5.6.2 Finding $|V(-3\lambda/8)|$ from $|V_r|$.

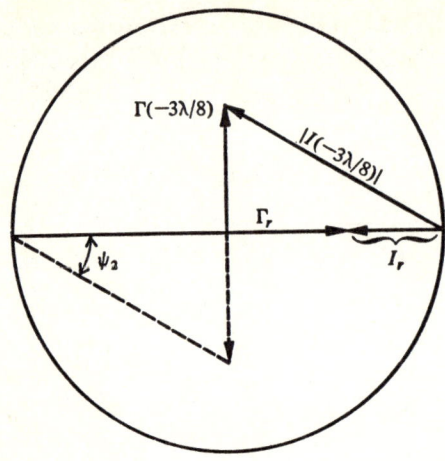

Fig. 5.6.3 Finding $|I(-3\lambda/8)|$ from $|I_r|$.

Similarly, since $|I_r| = \frac{30}{150} = \frac{1}{5}$ A, the current at $z = -3\lambda/8$ is found as illustrated in Fig. 5.6.3. Thus

$$\left|I\left(\frac{-3\lambda}{8}\right)\right| = \sqrt{5}\,(\tfrac{1}{5})\text{ A}$$

Example 2 We will find $|V_r|$, given that $\Gamma_r = 0.75\underline{/30°}$ and $|V(-\tfrac{7}{8}\lambda)| = 10$ V; the voltage ratio can be scaled from the Smith chart, as illustrated in Fig. 5.6.4:

$$\frac{|V_r|}{10} = \frac{1.7}{0.905}$$

so that

$$|V_r| = 18.8$$

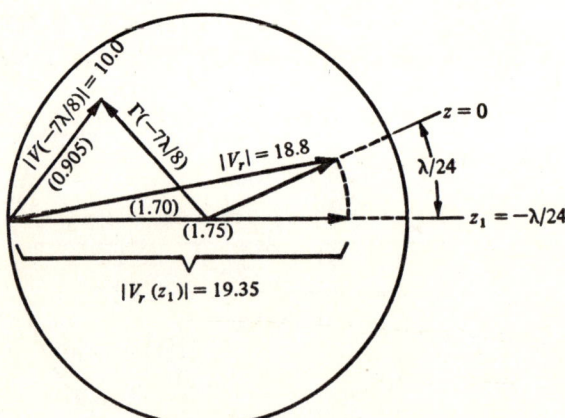

Fig. 5.6.4 Finding $|V_r|$ and $|V^+|$ from $|V(-7\lambda/8)|$.

In addition, the magnitude of V^+ can be found by choosing a point z_1 on the line where the angle of $\Gamma(z)$ is zero and finding the voltage there. For Example 2 such a point is located at $z_1 = -\lambda/24$, so that $\Gamma(-\lambda/24) = 0.75/\underline{0°}$ and

$$\left|V\left(\frac{-\lambda}{24}\right)\right| = \frac{10(1.75)}{0.905} = 19.35$$

Since the magnitude of V^+ is just

$$|V^+| = \frac{|V(z_1)|}{|1 + \Gamma(z_1)|}$$

here

$$|V^+| = \frac{19.35}{1.75} = 11.05 \text{ V}$$

The Z-θ chart and the Smith chart can be used to find the time-average complex power $\langle S \rangle$, where

$$\langle S \rangle = \langle P \rangle + jQ = \tfrac{1}{2} V(z) I^*(z) \tag{4}$$
$$= \tfrac{1}{2} |V(z)| |I(z)| (\cos \theta + j \sin \theta) \tag{5}$$

in which θ is the phase angle between $V(z)$ and $I(z)$ and the asterisk denotes the complex conjugate. Therefore, the time-average power is

$$\langle P \rangle = \tfrac{1}{2} |V(z)| |I(z)| (\cos \theta) \tag{6}$$

and the reactive power is

$$Q = \tfrac{1}{2} |V(z)| |I(z)| (\sin \theta) \tag{7}$$

In Eqs. (4) to (7) we have assumed that $|V(z)|$ and $|I(z)|$ are *peak* values of the phasors. If the phasors are given in *rms* values, the factor of $\tfrac{1}{2}$ can be removed. Since we can find the magnitudes of voltage and current by the methods given above, we can find $\langle S \rangle$ if we can find the power-factor angle θ. Let

$$1 + \Gamma(z) = |1 + \Gamma(z)| e^{j\psi_1} \tag{8}$$
$$1 - \Gamma(z) = |1 - \Gamma(z)| e^{j\psi_2} \tag{9}$$

The complex power is

$$\langle S \rangle = \tfrac{1}{2} V(z) I^*(z)$$
$$= \tfrac{1}{2} \{V^+ e^{-j\beta z}[1 + \Gamma(z)]\} \{Y_0 V^+ e^{-j\beta z}[1 - \Gamma(z)]\}^* \tag{10}$$
$$= \tfrac{1}{2} Y_0 |V^+|^2 |1 + \Gamma(z)| |1 - \Gamma(z)| e^{j(\psi_1 - \psi_2)} \tag{11}$$

Therefore the power-factor angle

$$\theta = \psi_1 - \psi_2 \tag{12}$$

where the angles ψ_1 and ψ_2 are shown in Fig. 5.6.1.

Example 3 Apply the above results to Example 1 to calculate complex power at the point $z = -3\lambda/8$. Data from that example are

$Z_r = 150\ \Omega$

$Z_0 = 50\ \Omega$

$|V_r| = 30\ \text{V}$

$\left|V\left(\dfrac{-3\lambda}{8}\right)\right| = 10\sqrt{5} = 22.4\ \text{V}$

$\left|I\left(\dfrac{-3\lambda}{8}\right)\right| = \dfrac{1}{\sqrt{5}} = 0.448\ \text{A}$

From Figs. 5.6.2 and 5.6.3, one finds that

$\psi_1 = 26.6°\qquad \psi_2 = -26.6°$

and so $\theta = 53.2°$. Consequently the complex power from Eq. (5) is

$\langle S \rangle = \tfrac{1}{2}(10\sqrt{5})\dfrac{1}{\sqrt{5}}(\cos 53.2° + j\sin 53.2°)$

$= \langle P \rangle + jQ = 3 + j4\ \text{VA}$

The time-average power is a constant on a lossless line. For example, at the load one finds

$\langle P \rangle = \dfrac{1}{2}\dfrac{(30^2)}{150} = 3\ \text{W}$

which agrees with the result at $z = -3\lambda/8$.

5.7 IMPEDANCE MATCHING WITH STUB TUNERS

It is often desirable to eliminate negative-going waves between the generator and the load. If there are no negative-going waves, all power sent down the line is dissipated in the load. Otherwise, standing waves occur, and energy is stored on the line. If the termination impedance is equal to the characteristic impedance, there are no reflections, and the load is said to be matched to the line. However, if the load is not equal to the characteristic impedance, it is necessary to go through a procedure known as *impedance matching* or *tuning*. In most practical cases, impedance matching is accomplished by attaching short sections of transmission line (called *tuning stubs*) to the main line near the load, thus causing the impedance of the main line to change. The objective is to select the stub lengths and locations so that the impedance $Z(z)$ on the main line equals Z_0 in the region between the generator and the stub. Of course, we obtain the same result if $Y(z) = Y_0$. It is easier to work with admittance than impedance for stub-tuning problems because the stub impedances appear in *parallel* with the main-line impedance. We will use admittances in solving the two most common types of impedance-matching problems, namely for single- and double-stub tuning.

SINGLE-STUB TUNING

The first step of the single-stub-tuning procedure is to find a point $z = -d$ on the line where the real part of $Y(z)$ equals the characteristic admittance, i.e., where

$$\operatorname{Re}[Y(-d)] = Y_0 = \frac{1}{Z_0}$$

At that point, we cancel $\operatorname{Im}[Y(-d)]$ (using a section of shorted, lossless transmission line) by adding a susceptance which is the conjugate of the susceptance on the transmission line. This situation is illustrated in Fig. 5.7.1, where the tuning stub (the shorted section of transmission line) is attached to the main transmission line at a point $z = -d$. Admittance on the main line to the right of $z = -d$ will be denoted $Y(-d^+)$ and admittance to the left of that point by $Y(-d^-)$. These admittances can be written in the rectangular forms

$$Y_1 = Y(-d^+) = Y_0 + jB_1$$
$$Y_t = jB_t \quad \text{input admittance of the stub}$$
$$Y_2 = Y(-d^-) = Y_1 + Y_t$$

At the point $z = -d$, the input susceptance B_t of the shorted tuning stub is chosen (by adjusting the length of the stub) to cancel the susceptance B_1 of the transmission line ($B_t = -B_1$), so that the admittance Y_2 is the same as the characteristic admittance. Therefore no energy is reflected back to the generator, and $\text{SWR} = 1.0$ for that portion of the line between the generator and the stub. Since no power can be dissipated in the shorted line, all power sent down the transmission line is dissipated in Z_r after a steady-state condition is attained. However, some energy is stored in the stub and on the main line, where $\text{SWR} > 1.0$.

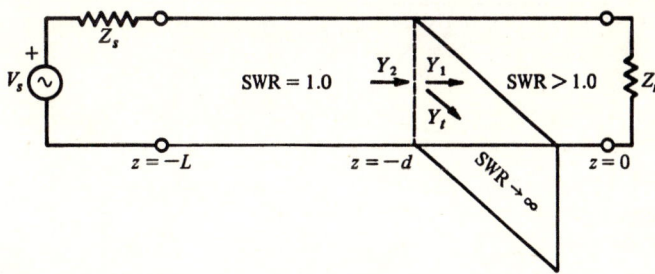

Fig. 5.7.1 A shorted tuning stub on a transmission line.

As an example of the single-stub-tuning technique, let the receiving-end admittance be $Y_r = 0.011 + j0.0054$ mho, and let the characteristic admittance be $Y_0 = 0.02$ mho. The normalized receiving end admittance is $Y_n(0) = 0.55 + j0.27$, and the SWR is 2.0 (from the Smith chart). Thus $\Gamma_r = 0.33\underline{/-41°}$, where the angle of the reflection coefficient must be found from the angle associated with the receiving-end impedance rather than admittance. The first step is to find a location (usually the one nearest the load) where $\text{Re}\,[Y(z)] = Y_0$, or $\text{Re}\,[Y_n(z)] = 1.0$. This is at the intersection of the $Y_n(z) = 1 + jb$ and SWR = 2.0 circles, as shown in Fig. 5.7.2. These circles intersect at the two points $z = -d_1$ and $z = -d_2$, within the first half wavelength from the load. At $z = -d_1$, the normalized admittance $Y_n(-d_1)$ has a positive imaginary part, which is a capacitive susceptance. For this case, the transmission line is matched by adding an equal inductive susceptance in parallel at this point. For the second case, at $z = -d_2$, the normalized admittance $Y_n(-d_2)$ has a negative imaginary part, which is an inductive susceptance. For this case, the transmission line is matched by adding an equal capacitive susceptance in parallel at $z = -d_2$.

The next steps are to find numerical values for d_1 and d_2 and to find the normalized admittances at these points from the Smith chart. For the first case, it is a distance of $d_1 = 0.096\lambda$ between the termination $Y_n(0)$ and the point where the real part of the admittance is unity. At this

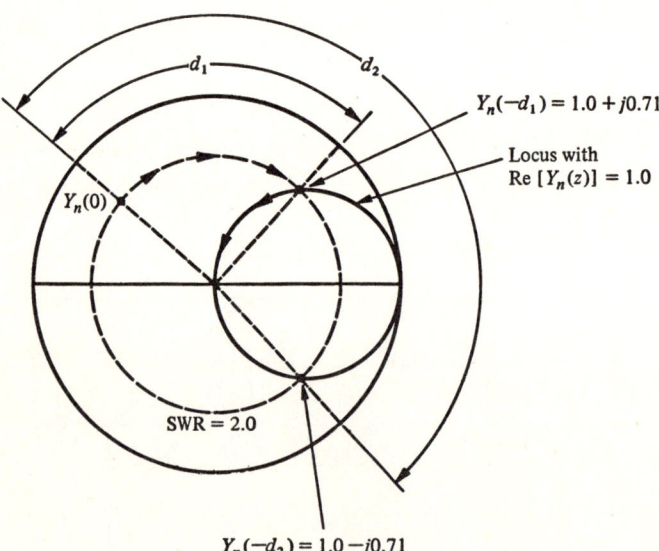

Fig. 5.7.2 Two possible solutions for single-stub tuning.

point, we see from the Smith chart that the normalized admittance is

$$Y_n(-d_1) = 1.0 + j0.71$$

Therefore the normalized susceptance to be added by the stub is

$$B_{tn} = -0.71$$

For the second case, the distance from the termination $Y_n(0)$ to the point $z = -d_2$, where the real part of the admittance is unity, is $d_2 = 0.292\lambda$. At that point, the admittance is

$$Y_n(-d_2{}^+) = 1.0 - j0.71$$

The susceptance to be added by the tuning stub to match the line is therefore

$$B_{tn} = +0.71$$

The remaining problem for either of these cases is to find the length of the tuning stub required to obtain the desired value of B_{tn}.

For a tuning stub, which is a section of short-circuited transmission line,† the receiving-end admittance is infinity. For the first case, where the added susceptance is -0.71, it can be seen from Fig. 5.7.3 that the distance from the short circuit to the sending end of the shorted line is

† Theoretically, an open-circuited section of line could also be used for a tuning stub. However, since energy will be lost at the open end due to fringing fields, it is easier to obtain a "good" short circuit than a "good" open circuit.

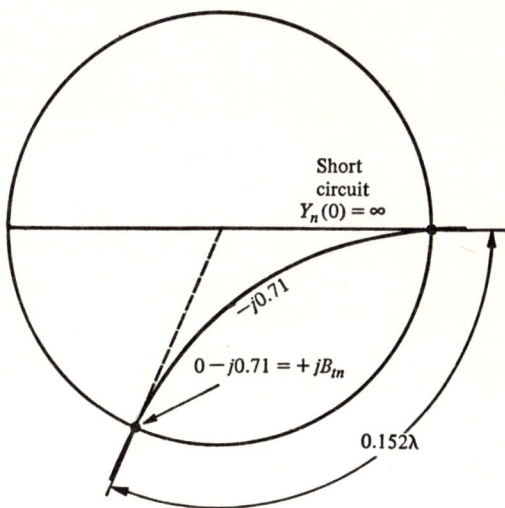

Fig. 5.7.3 Input admittance of the tuning stub for the first case.

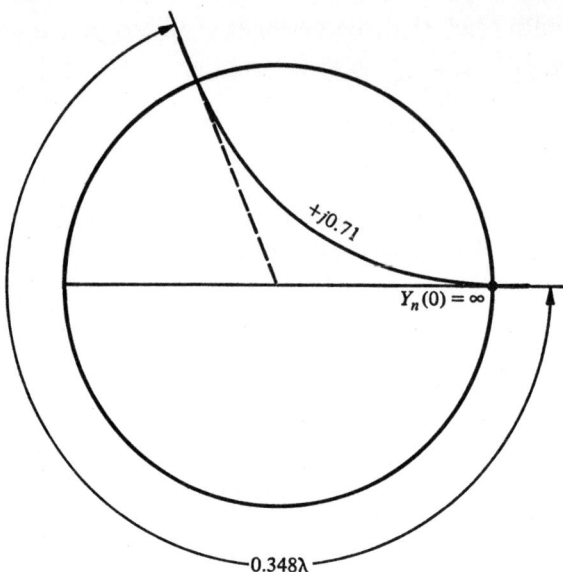

Fig. 5.7.4 Input admittance of the tuning stub for the second case.

0.152λ. This segment of shorted line, placed in parallel with the main transmission line at a distance $d_1 = 0.096\lambda$ from the receiving end, guarantees that the impedance at the point $z = -d_1$ is the characteristic impedance, and therefore the line is matched for all $z < -d_1$.

For the second case, the normalized susceptance to be added by the stub is $+0.71$. It can be seen from Fig. 5.7.4 that a length 0.348λ of shorted transmission line placed in parallel with the primary transmission line at the point d_2 guarantees that the line is matched for $z < -d_2$.

A disadvantage of matching impedances with a single shorted section of transmission line is that if the load is changed slightly, the length of transmission line between the load and the tuning stub will also have to be changed. This necessitates cutting the transmission line and replacing it with a section of line of different length, which is often inconvenient.

DOUBLE-STUB TUNING

An alternate solution of the impedance-matching problem is to use two tuning stubs, where the distance between the stubs is fixed and the distance between the load and the nearest tuning stub is also fixed. Matching the load is accomplished by changing the length of one or both of the shorted stubs rather than changing the distance between the shorted stubs and the load. Changing the length of the shorted sections of line is facilitated

SINUSOIDAL WAVES ON LOSSLESS TRANSMISSION LINES

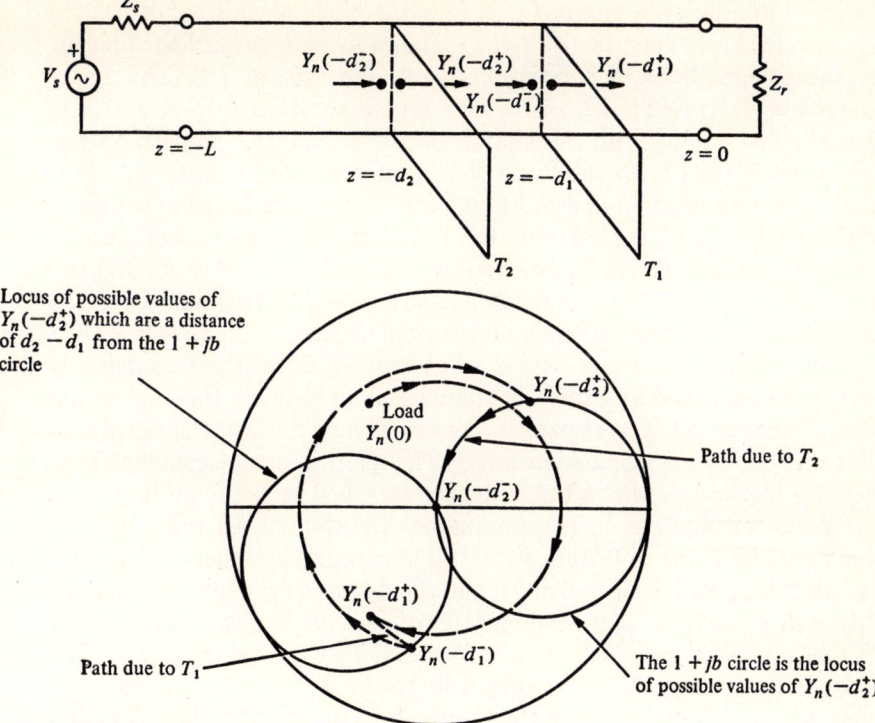

Fig. 5.7.5 Transmission line with two shorted tuners and the solution by Smith chart.

by using commercially available variable-length stub tuners. With this procedure large variations in load impedance can be matched without changing the distance between tuning stubs or between the tuning stubs and the load. This method of impedance matching is illustrated in Fig. 5.7.5.

We will assume that the load impedance and the distances d_1 and d_2 are known. The normalized load admittance $Y_n(0)$ is located on the chart, and then the admittance $Y_n(-d_1^+)$ is found by a rotation corresponding to a distance d_1 around a circle of constant SWR. This admittance, just to the right of the tuning stub T_1 in Fig. 5.7.5, depends only on Y_r and d_1. Since we do not know the length of the stub T_1, we cannot find $Y_n(-d_1^-)$. However, by starting at $z = -d_2$, we can find the allowable locus of $Y_n(-d_1^-)$. To obtain a match, it is required that

$$Y_n(-d_2^-) = 1.0 + j0$$

and consequently

$$Y_n(-d_2^+) = 1.0 + jb$$

The locus of allowable points $1.0 + jb$ is the circle of unit conductance. This circular locus must be rotated a distance $d_2 - d_1$ toward the load to find the circular locus of permissible points for $Y_n(-d_1^-)$.† This locus of points is illustrated in Fig. 5.7.5 for a stub separation $d_2 - d_1$ of 0.31λ. Since the first tuning stub can change only susceptance, the path on the chart from $Y_n(-d_1^+)$ to the locus of allowable points for $Y_n(-d_1^-)$ is a path of constant conductance. From T_1 to T_2, the path is one of constant SWR, and $Y_n(-d_2^+)$ is always on the $1 + jb$ circle. The tuning stub T_2 then adds just enough susceptance so that $Y_n(-d_2^-) = 1.0 + j0$, and the load is matched to the transmission line at $z = -d_2$. It should be noted that there are two types of path on the Smith chart. The first is a path of constant SWR, and this occurs when moving down the transmission line between the load and the first tuning stub or between the two tuning stubs. The second type of path is due to adding a finite susceptance at a point on the main transmission line. This path is one of constant conductance and occurs where the stubs are attached to the main line.

As an example of this technique, let the normalized receiving-end admittance be $Y_n(0) = 0.20 + j0.30$. The distance between the load and the first tuning stub is $d_1 = 0.150\lambda$, and the distance between tuning stubs is $d_2 - d_1 = 0.375\lambda$. This configuration is shown in Fig. 5.7.6. From the Smith chart it is seen that the SWR is 5.5 on the section nearest the load, as shown in Fig. 5.7.7. Going from the load $d_1 = 0.150\lambda$ toward the generator, we follow a path on the Smith chart in which SWR = 5.5, and we find that $Y_n(-d_1^+) = 1.40 + j2.20$. At this point, the tuning stub T_1 adds enough susceptance for $Y_n(-d_1^-)$ to be on the locus of points which is a distance of $d_2 - d_1 = 0.375\lambda$ from the $1 + jb$ circle. This is illustrated in Fig. 5.7.7, from which it can be seen that there are two points on the circle of conductance 1.40 which intersect the circular locus of points 0.375λ from the unit-conductance circle. Either of these two points could be chosen, but for this example we choose $Y_n(-d_1^-) = 1.40 - j0.08$. Thus the tuning stub T_1 must be chosen so that the change in normalized susceptance is $\Delta b = -0.08 - 2.20 = -2.28$; therefore T_1 must have an inductive susceptance of -2.28 at its input terminal.

Going from d_1^- to d_2^+ along the uniform transmission line between the tuning stubs, the path on the Smith chart is one of a constant SWR = 1.42, which is a different value from that existing between the termination and the first tuning stub. After going $\frac{3}{8}$ wavelength toward the generator from the first tuning stub, we see that the admittance is

$$Y_n(-d_2^+) = 1.0 + j0.35$$

---

† The locus of $Y_n(-d_1^-)$ can be found by rotating the center of the $1.0 + jb$ circle a distance $d_2 - d_1$ and drawing a circle of the same diameter about that point. This circle is the locus of $Y_n(-d_1^-)$.

SINUSOIDAL WAVES ON LOSSLESS TRANSMISSION LINES

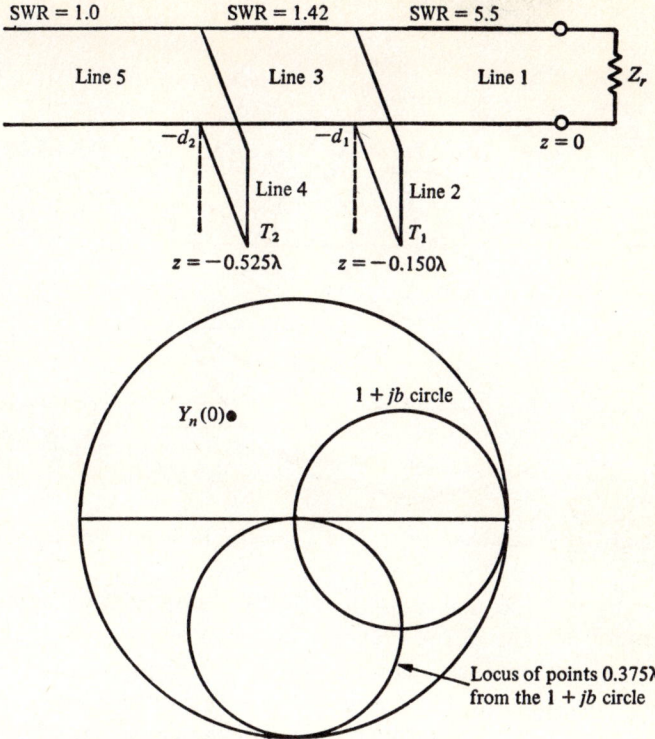

Fig. 5.7.6 An example of double-stub tuning.

Thus the susceptance added by the tuning stub T_2 must be $\Delta b = -0.35$. Therefore $Y_n(-d_2^-) = 1.0 + j0$, and there are no negative-going waves from $z = -d_2$ to $z = -L$. The remaining problem is to find the lengths of the two stubs so that the input susceptance to the first tuning stub is -2.28 and to the second tuning stub is -0.35. These lengths can be obtained from the Smith chart, as indicated in Fig. 5.7.8. It can be seen that the length of the first tuning stub is 0.0655λ, and the length of the second tuning stub is 0.1966λ.

If we had chosen the other point, $Y_n(-d_1^-) = 1.4 - j1.93$, we would have obtained

$Y_n(-d_2^+) = 1 + j1.67$

$\Delta b_{T_1} = -4.13 \qquad \Delta b_{T_2} = -1.67$

$L_{T_1} = 0.038\lambda \qquad L_{T_2} = 0.086\lambda$

One difficulty with double-stub tuning is that not all loads can be matched for a particular placement of the tuning stubs. For example,

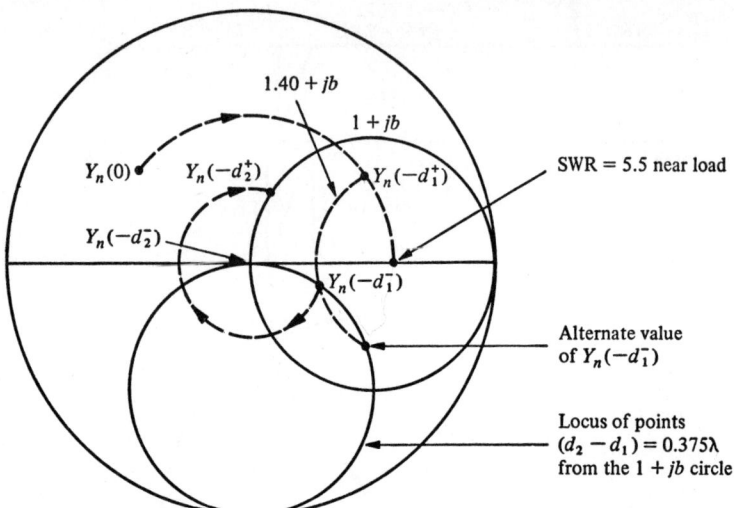

Fig. 5.7.7 Smith chart solution of a problem in double-stub tuning.

let $d_1 = 0.150$ and $d_2 - d_1 = 0.375\lambda$, with $Y_n(0) = 0.3 + j0.7$, as illustrated in Fig. 5.7.9. Then $Y_n(-d_1{}^+) = 4.8 + j0$, and no amount of added susceptance will allow $Y_n(-d_1{}^-)$ to be on the locus of points $\frac{3}{8}$ wavelength from the $1 + j0$ circle. It can be seen that no admittance $Y_n(-d_1{}^+)$ that lies within region 1, as shown in Fig. 5.7.9, can be matched

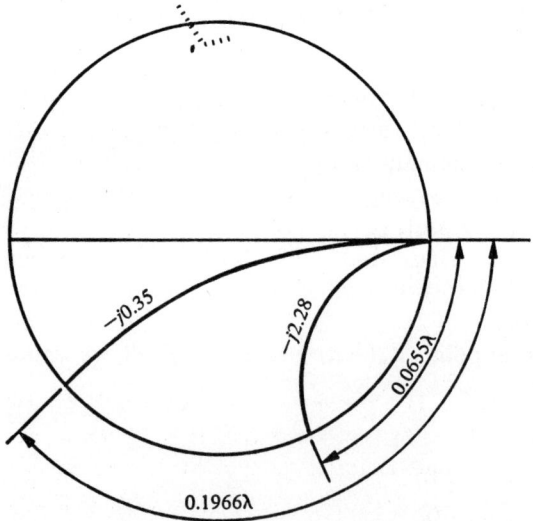

Fig. 5.7.8 Use of the Smith chart to find the lengths of the tuning stubs.

SINUSOIDAL WAVES ON LOSSLESS TRANSMISSION LINES

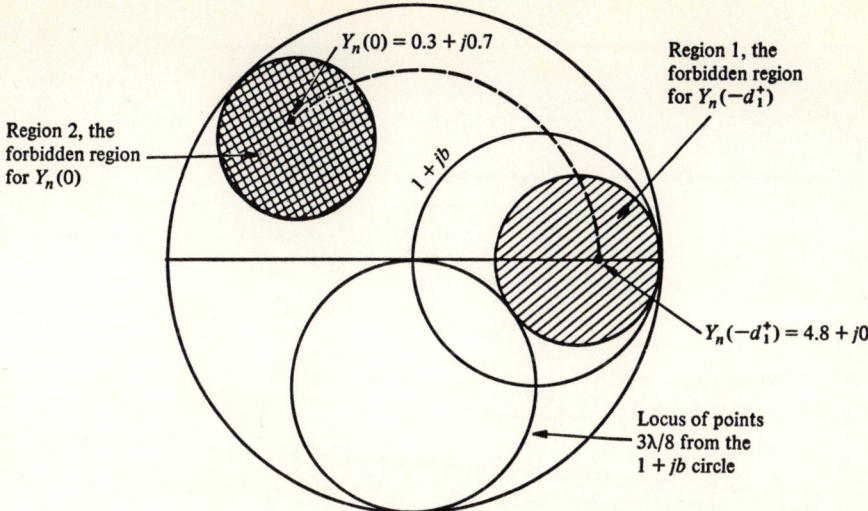

Fig. 5.7.9 An example of a load that cannot be matched with a given geometry.

with this configuration of tuning stubs. The forbidden region for $Y_n(-d_1^+)$, region 1, is determined by the circle of constant conductance which is tangent to the circle which is $3\lambda/8$ from the $1 + jb$ circle. If region 1 is rotated back to the load, we obtain a forbidden region for the load admittance, shown as region 2.

The load admittance can be moved out of the forbidden region by changing either the length d_1 or the length $d_2 - d_1$. The diameter of the forbidden region can be reduced by making $d_2 - d_1$ greater than $3\lambda/8$. However, it can be shown that if $d_2 - d_1$ is increased much beyond $3\lambda/8$, the match becomes increasingly frequency-dependent. For many applications, the resulting narrow bandwidth may be undesirable. Consequently, it is usually preferable to change the length d_1 to move the load admittance out of the forbidden region.

5.8 THE QUARTER-WAVELENGTH TRANSFORMER

In cases where the load impedance is nearly constant, it is sometimes most convenient to match the load to the line with a *quarter-wavelength transformer*. This consists of a quarter-wavelength section of line, of appropriate characteristic impedance Z_{02}, inserted between the load and the original transmission line of characteristic impedance Z_{01}, as shown in Fig. 5.8.1. We now derive a technique to eliminate reflections on the main line by finding the required characteristic impedance Z_{02} of the

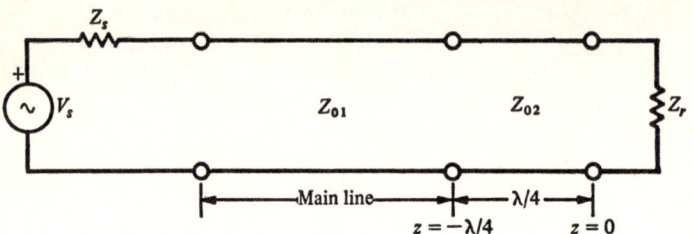

Fig. 5.8.1 The quarter-wavelength transformer.

quarter-wavelength section. If the load impedance is Z_r, the normalized impedance at $z = -\lambda/4$ is given by

$$Z_n\left(-\frac{\lambda}{4}\right) = \frac{Z(-\lambda/4)}{Z_{02}} = \frac{1 + \Gamma_r \exp\left(j\frac{4\pi}{\lambda}\frac{\lambda}{4}\right)}{1 - \Gamma_r \exp\left(j\frac{4\pi}{\lambda}\frac{\lambda}{4}\right)} = \frac{1 - \Gamma_r}{1 + \Gamma_r} \qquad (1)$$

The impedance at $z = -\lambda/4$ is

$$Z\left(-\frac{\lambda}{4}\right) = Z_{02}\frac{1 - \Gamma_r}{1 + \Gamma_r} \qquad (2)$$

If there are to be no reflections, $Z(-\lambda/4) = Z_{01}$. Therefore, the relationship between Z_{02}, Z_{01}, and Z_r is given by

$$\frac{Z_{01}}{Z_{02}} = \frac{1 - \Gamma_r}{1 + \Gamma_r} = \frac{2Z_{02}}{2Z_r} = \frac{Z_{02}}{Z_r} \qquad (3)$$

From Eq. (3) we see that Z_{02} is the geometric mean of Z_{01} and Z_r, which can also be written

$$Z_{02} = \sqrt{Z_{01}Z_r} \qquad (4)$$

It can be verified on the Smith chart that this value of characteristic impedance for the quarter-wavelength transformer does indeed match the load impedance Z_r to the impedance Z_{01}.

Theoretically, none of the impedances in Eq. (4) are required to be real. Practically, however, there are only a limited number of different values of characteristic impedance commercially available in transmission lines, and these values are generally real. Suppose a complex load impedance must be matched to a transmission line with a real characteristic impedance. Then a quarter-wavelength section having a real characteristic impedance can be used if the proper length of transmission line is

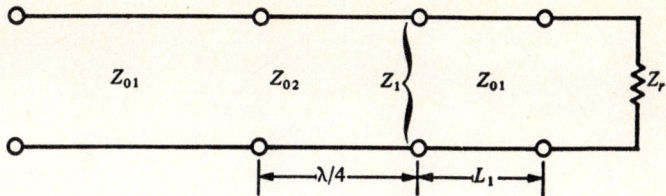

Fig. 5.8.2 The quarter-wavelength transformer used to match a complex load impedance to a real characteristic impedance.

placed between the quarter-wavelength transformer and the load. This situation is illustrated in Fig. 5.8.2, where the length L_1 of transmission line is chosen so that the impedance Z_1 is real. The characteristic impedance of the quarter-wavelength transformer is then given by $Z_{02} = \sqrt{Z_{01}Z_1}$.

A disadvantage of matching with one quarter-wavelength transformer is its frequency sensitivity. Often this can be improved by using two or more quarter-wavelength sections. As an example, the apparent termination impedance Z_1 will be compared for a 400-Ω load resistance matched (at one frequency) to a transmission line with a characteristic impedance of 50 Ω by means of a single quarter-wavelength transformer and then by means of two quarter-wavelength transformers. These two cases are illustrated in Fig. 5.8.3.

When two quarter-wavelength transformers are used, it will be assumed that the quarter-wavelength transformer nearest the load resis-

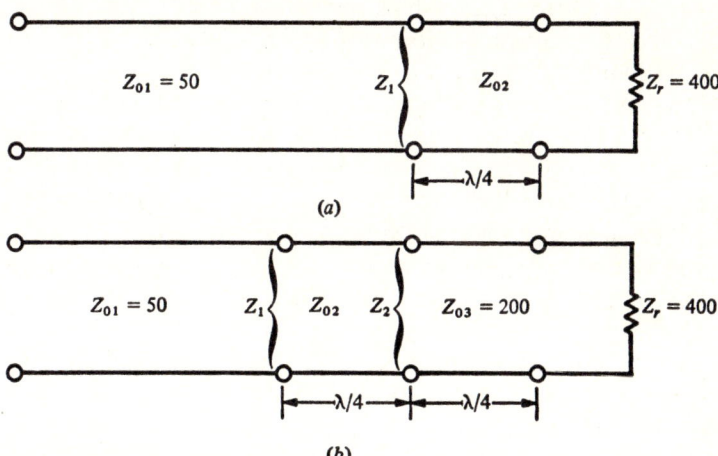

Fig. 5.8.3 Impedance matching with (a) one section and (b) two sections of quarter-wavelength transformer.

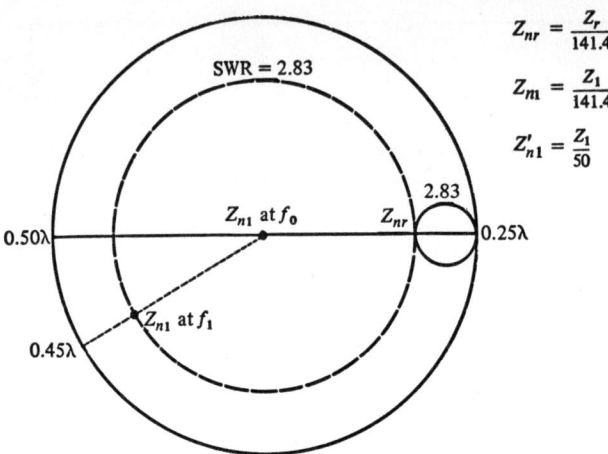

Fig. 5.8.4 Finding the apparent termination impedance Z_1 where one quarter-wavelength transformer is used.

tance has a characteristic impedance of $Z_{03} = 200\ \Omega$. Also, the frequency f_1 used to calculate the apparent termination impedance Z_1 will be four-fifths the frequency f_0 at which the lines are matched. Thus the actual length of the quarter-wavelength transformers is $\frac{4}{5}(\lambda/4) = \lambda/5$ instead of $\lambda/4$ at the frequency f_1.

If one section of quarter-wavelength transformer is used, it is seen that the characteristic impedance of this section must be $Z_{02} = 141.4\ \Omega$ at the frequency f_0 at which matching is to occur. At a frequency $f_1 = 4f_0/5$, the apparent termination impedance Z_1 of the line using the single quarter-wavelength transformer is $Z_1 = 53.8 - j39.6$. The SWR is 2.83 on the transformer and 2.1 on the line. This can be obtained from the Smith chart, as shown in Fig. 5.8.4. If two quarter-wavelength transformers are used, the characteristic impedance Z_{02}, from Fig. 5.8.3, of the first quarter-wavelength transformer is $Z_{02} = 70.7\ \Omega$. For this case, both transformers must be considered to be $\frac{1}{5}$ wavelength at the frequency f_1. The input impedance for the first transformer (the apparent termination impedance) is $Z_1 = 35.5 + j0.71$. This impedance must be found in two steps, shown in Fig. 5.8.5.

In conclusion, we can compare the normalized values Z'_{n1} obtained for the apparent termination impedance Z_1 for the two cases. These values of Z'_{n1} are normalized with respect to the 50-Ω characteristic impedance of the transmission lines. We see that when two quarter-wavelength transformers are used, the apparent termination impedance Z_1 lies closer to the center of the Smith chart than the apparent termination impedance

SINUSOIDAL WAVES ON LOSSLESS TRANSMISSION LINES

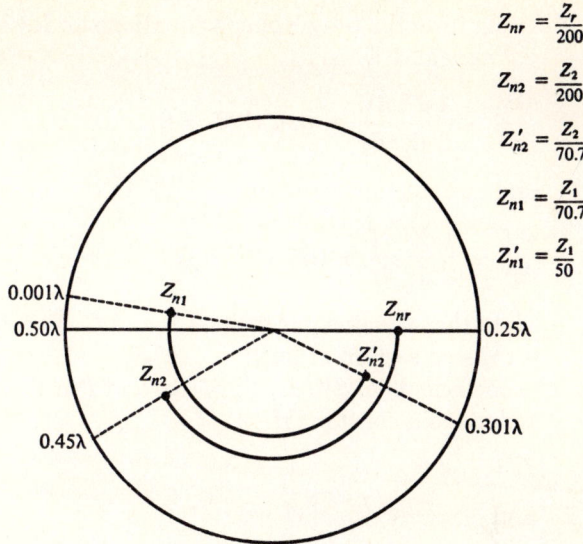

Fig. 5.8.5 Finding the apparent termination impedance Z_1 where two quarter-wavelength transformers are used.

found when only one quarter-wavelength transformer is used. Since the reflection coefficient and the SWR have smaller magnitudes on the main line (SWR = 1.3) when two quarter-wavelength transformers are used, we have been able to decrease the frequency sensitivity of the apparent termination by using two quarter-wavelength sections. Although some general statements regarding frequency sensitivity could be derived, this example illustrates that a problem exists and that the use of two (or more) quarter-wavelength sections of line can decrease frequency sensitivity. Bandwidth will be discussed in more detail in Sec. 5.10.

PROGRAMMED EXERCISE 5.8.1

THE EXPONENTIALLY TAPERED LINE

For matching fixed loads in which bandwidth must be maximized, it is often convenient to use a tapered line. If the diameter of the center (or outer) conductor of coaxial cable is tapered, the characteristic impedance varies continuously along the tapered section. Ideally, the taper is several wavelengths long at the lowest frequency of interest, and it then provides a transition from one impedance level to another with essentially no reflected energy due to the taper. The reader should verify each of the following true statements.

1. Starting from the telegrapher's equations for lossless lines, we find that voltage must satisfy

$$\frac{d^2V}{dz^2} - \frac{1}{L}\frac{dL}{dz}\frac{dV}{dz} + (\omega^2 LC)V = 0 \tag{5}$$

and current must satisfy

$$\frac{d^2I}{dz^2} - \frac{1}{C}\frac{dC}{dz}\frac{dI}{dz} + (\omega^2 LC)I = 0 \tag{6}$$

Note that the line parameters L and C are functions of distance on a tapered section of line.

2. The coefficients of dV/dz, V, dI/dz, and I in Eqs. (5) and (6) will each reduce to a constant if

$$L = L_0 e^{-\tau z} \tag{7}$$

and

$$C = C_0 e^{+\tau z} \tag{8}$$

(Derivational hint: set the coefficient of dV/dz equal to $-\tau$ and separate variables in the resulting differential equation.)

3. The sign of each of the exponents in Eqs. (7) and (8) could be reversed. However, the signs actually chosen imply that the center conductor (assuming coaxial cable is used) is larger at $z = 0$ than at any point $z < 0$.

4. After substituting Eqs. (7) and (8) into Eqs. (5) and (6), we obtain

$$\frac{d^2V}{dz^2} + \tau\frac{dV}{dz} + \omega^2 L_0 C_0 V = 0 \tag{9}$$

and

$$\frac{d^2I}{dz^2} - \tau\frac{dI}{dz} + \omega^2 L_0 C_0 I = 0 \tag{10}$$

5. The solutions of these differential equations with constant coefficients are, respectively,

$$V = V^+ e^{-\tau z/2} e^{-j\beta z} + V^- e^{-\tau z/2} e^{+j\beta z} \tag{11}$$

and

$$I = I^+ e^{+\tau z/2} e^{-j\beta z} + I^- e^{+\tau z/2} e^{+j\beta z} \tag{12}$$

where

$$\beta = \sqrt{\omega^2 L_0 C_0 - \frac{\tau^2}{4}} \tag{13}$$

Thus β is real for frequencies high enough to ensure that

$$\omega > \sqrt{\frac{\tau^2}{4L_0C_0}} = \omega_c \tag{14}$$

where ω_c is called the *cutoff frequency*. For $\omega < \omega_c$, β becomes imaginary (thus $j\beta$ is real), and both voltage and current are attenuated in the tapered section. Normal operation of the tapered line, however, is with $\omega > \omega_c$, so that normally there is no attenuation.

6. The characteristic impedance seen by the positive-going wave is

$$Z_0^+(z) = \frac{j\omega L}{\tau/2 + j\beta} = \frac{j\omega L_0 e^{-\tau z}}{\tau/2 + j\beta} \tag{15}$$

and for a negative-going wave is

$$Z_0^-(z) = \frac{j\omega L}{\tau/2 - j\beta} = \frac{j\omega L_0 e^{-\tau z}}{\tau/2 - j\beta} \tag{16}$$

The reader should answer the following questions:†

1. What is I^+ in terms of V^+ and $Z_0^+(z)$, and I^- in terms of V^- and $Z_0^-(z)$ in Eq. (12)?
2. If $\omega \gg \omega_c$, what are approximate expressions for β, $Z_0^+(z)$, and $Z_0^-(z)$?
3. If a tapered section of line is an integral number of half wavelengths long, how are impedance, voltage, and current transformed?

5.9 THE SYSTEMS APPROACH USING THE DIGITAL COMPUTER

THE COMPUTER VERSUS CONVENTIONAL TECHNIQUES

In our prior studies, we have developed analytical and graphical techniques for solving a wide variety of electromagnetic field and transmission-line problems, and one might question the need for computers. However, there are at least two important types of transmission-line problems in which the digital computer can be valuable to us: (1) problems concerned with bandwidth (or with problems covering a wide range of frequencies) and (2) almost any lossy-line problem. The decision whether or not to use the computer is normally made by comparing the effort (or time) one must expend using conventional techniques with that required using computer techniques.

The initial effort needed to write a computer program can be large, but once written, it can be used to solve many problems with little additional effort. That initial effort has been expended for the reader in the

† For further discussion of tapered lines see Walter C. Johnson, "Transmission Lines and Networks," pp. 201–208, McGraw-Hill Book Company, New York, 1950.

computer package SANTLINE (systems analysis of transmission lines) to be described.† It is assumed that the reader has a knowledge of elementary FØRTRAN IV and has access to a computer capable of using that language. The investment of a minimal amount of time using SANTLINE will be amply rewarded in additional knowledge of transmission lines and in the scope of practical problems one can solve. After successfully solving a few simple transmission-line problems on the computer, we will be able to decide whether to apply computer techniques or solve the problem conventionally. We should not blindly use the computer to the exclusion of graphical and analytical techniques, or vice versa, since each technique has its advantages.

The computer package SANTLINE is based upon the sinusoidal steady-state analysis developed in Chaps. 5 and 6. Transient problems, such as those considered in Chap. 4, have not been programmed, since the techniques previously presented are adequate for most cases on lossless lines. Transient problems on dissipative lines can be solved by Fourier analysis and will not be considered in this book.

THE SYSTEMS APPROACH

The purpose in writing SANTLINE was to free the user from the computational details (thus avoiding the problem of selecting the proper equations) so that he could concentrate on the transmission-line system as a whole. Of course, this same economy of effort is what makes the Smith chart so attractive compared with equation solving. We have previously derived and used all the equations we need for an efficient computer program. These equations, valid for either lossless or lossy lines, are summarized in Table 5.9.1. SANTLINE uses these equations in the subprograms (FUNCTIØNS or SUBRØUTINES) to be described. Consequently, SANTLINE can be applied to either lossless or lossy transmission lines. The FUNCTIØNS and SUBRØUTINES have been written and can be called into use in our main program as necessary. We must write a main program (which we will call TLINE) for each given problem, so that we are not released from the obligation of knowing how to solve the problem. We must also learn how and when to call the various subprograms.

To illustrate with a specific example, suppose we have three sections of line connected in cascade and loaded with an admittance $Y_r = G_r + j\omega C_r$, as shown in Fig. 5.9.1. We will assume that the three characteristic impedances Z_{01}, Z_{02}, and Z_{03} (in ohms), the lengths l_1, l_2, and l_3 (in meters), the propagation constants γ_1, γ_2, and γ_3, and the values of G_r and C_r are all known at the frequency FPZ0 (the frequency at which the propagation

† We use capital letters to denote words or symbols in the FØRTRAN language.

SINUSOIDAL WAVES ON LOSSLESS TRANSMISSION LINES

Table 5.9.1 Summary of the steady-state equations

Dissipative lines

$$\gamma = \sqrt{(R + j\omega L)(G + j\omega C)} = \alpha + j\beta \tag{1}$$

$$Y_0 = \frac{1}{Z_0} = \sqrt{\frac{G + j\omega C}{R + j\omega L}} \tag{2}$$

$$v = \frac{\omega}{\beta} \tag{3}$$

$$V(z,t) = \text{Re}\,[(V^+ e^{-\gamma z} + V^- e^{+\gamma z})e^{j\omega t}] \tag{4}$$

$$I(z,t) = \text{Re}\,[Y_0(V^+ e^{-\gamma z} - V^- e^{+\gamma z})e^{j\omega t}] \tag{5}$$

$$\langle S(z) \rangle = \tfrac{1}{2} V(z)\, I^*(z) = \langle P(z) \rangle + jQ(z) \tag{6}$$

$$V^+ = \frac{V(-L)}{[1 + \Gamma(-L)]e^{-\gamma(-L)}} \tag{7}$$

$$\Gamma(z) = \frac{V^- e^{+\gamma z}}{V^+ e^{-\gamma z}} = \Gamma_r e^{2\gamma z} = \frac{Z_r - Z_0}{Z_r + Z_0} e^{2\gamma z} \tag{8}$$

$$Y(z) = Y_0 \frac{1 - \Gamma(z)}{1 + \Gamma(z)} \tag{9}$$

$$\text{SWR}(z) = \frac{1 + |\Gamma(z)|}{1 - |\Gamma(z)|} = \frac{1 + |\Gamma_r|e^{2\alpha z}}{1 - |\Gamma_r|e^{2\alpha z}} \tag{10}$$

Lossless lines

Each of the above equations is valid for lossless lines, if we set

$$R = 0 \quad G = 0 \tag{11}$$

so that

$$\alpha = 0 \tag{12}$$

and

$$\gamma = j\beta \tag{13}$$

---

constant and Z_0 are known). When computations are made at frequencies F other than FPZ0, SANTLINE will recompute γ and Z_0 for the new frequency. The frequency at which calculations are to be made is the stored value of the common variable F, which must be specified by the programmer before the calculations are performed. Suppose we want to find the admittances Y_1, Y_2, and Y_3, shown in Fig. 5.9.1, at some given

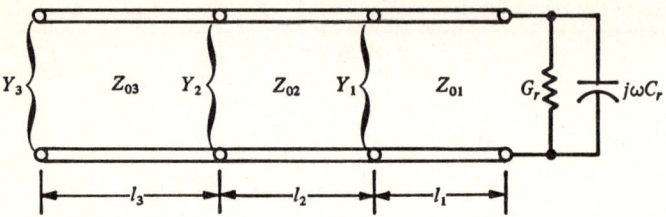

Fig. 5.9.1 A loaded three-line transmission system.

frequency F. A FUNCTIØN subprogram called CYIN has been written to find the complex admittance Y_{in} for any line given its terminating admittance $Y_r = $ YTERM. This subprogram, designated

FUNCTIØN CYIN(NLINE,YTERM)

uses Eq. (9) of Table 5.9.1 (repeated below)

$$Y_{in}(l,Y_0,F,Y_r) = Y_0 \frac{1 - \Gamma(l)}{1 + \Gamma(l)} \tag{1}$$

to compute the complex input admittance CYIN. The computer retrieves the length l and the characteristic admittance to form a common data block (the permanent store), where the programmer has previously stored the line parameters. The variable NLINE is a line identification number used for designation purposes, since there is normally more than one section of line involved in a given problem.

To find the admittances Y_1, Y_2, and Y_3 for lines 1, 2, and 3 in Fig. 5.9.1, we write the following computational part of the main program TLINE:

$$\begin{aligned}
&\text{YTERM} = \text{CMPLX(GR,BR)} \\
&\text{Y1} = \text{CYIN(1,YTERM)} \\
&\text{Y2} = \text{CYIN(2,Y1)} \\
&\text{Y3} = \text{CYIN(3,Y2)}
\end{aligned} \tag{2}$$

The complex admittance YTERM is determined by the numbers G_r and B_r ($B_r = 2\pi F C_r$) for the particular load.† The subprogram CYIN is used three times, in the same way and in the same order that we would use Eq. (1) to compute Y_1, Y_2, and Y_3 by hand. For comparison, consider how you would work the problem on the Smith chart. If these admittances

---

† The method of specifying complex numbers varies. Consequently the first line of (2) will not be the same on every machine. See "complex numbers" in the user's manual for your computer. SANTLINE was originally written for use on an IBM 360. However, the form Z = (X,Y), where Z has been declared complex, is commonly used with other computers.

are all that are required, we could add the appropriate input-output statements to the program to obtain the desired information.

The function CYIN also computes the reflection coefficient and the standing-wave ratio at both ends of the line, by using Eqs. (8) and (10) of Table 5.9.1. It assumes an input voltage of $1.0 + j0.0$ V and computes voltage, current, and power at each end of the line by using Eqs. (4) to (6) of Table 5.9.1. (Another function CVLØAD can be used to adjust the numbers for voltage, current, and power for any given source voltage.) All these variables computed by CYIN are returned to a common data block, TLDATA, for use by other subprograms or for other computations. The data can be extracted from the common data block and printed as necessary. Thus from FØRTRAN statements (2) we can obtain much more information than just the admittances Y_1, Y_2, and Y_3.

The function CYIN is one of 12 subprograms, many of which can be used for different purposes, at the programmer's option. These subprograms are given in detail in Appendix B. As the programmer becomes more familiar with these subprograms, he will find himself freed from most of the algebraic details of problem solving, so that he can concentrate on the system under investigation.

THE SANTLINE COMPUTER PACKAGE

The SANTLINE† computer package can be divided into four parts:

1. The main program TLINE
2. A group of subprograms used primarily to transfer data between storage areas
3. Four utility subprograms used for certain elementary computations
4. Five subprograms used for most of the computational work and problem solving

These four parts are shown in Fig. 5.9.2, along with two storage areas which are shared by the main program and most of the subprograms. All the subprograms have been written for the user and are listed in Appendix B. The four parts of SANTLINE and the use of the two storage areas will now be discussed. The main program TLINE consists of:

1. The control and accounting cards required by the programmer's computer center.
2. A declarations package, which has been written for the programmer and is listed in Appendix B.
3. The programmer's package, to be written by the programmer so as to solve the given problem. He must design the input and out-

---

† The SANTLINE package was developed by James C. Wells, whose work is gratefully acknowledged.

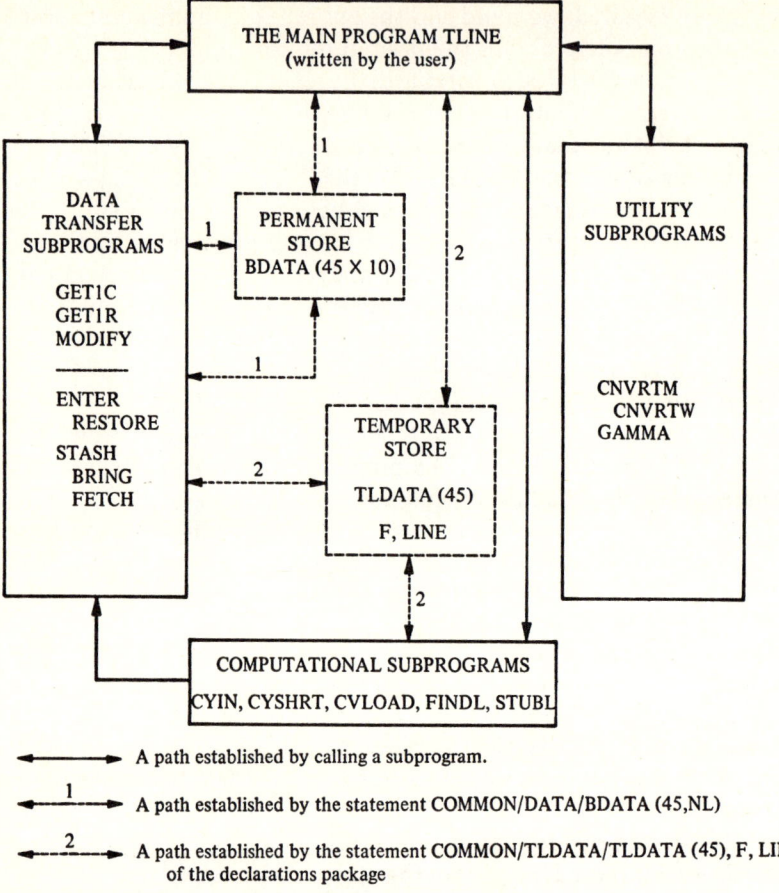

Fig. 5.9.2 The major parts of the SANTLINE package showing the calling and common connections.

put, supply the data cards, and use the subprograms, as for any FØRTRAN program. The ordering of the complete deck is shown in Fig. 5.9.3.

The declaration package consists of the following declarations statements:

1. TYPE CØMPLEX list
2. TYPE REAL list
3. DIMENSIØN list
4. EQUIVALENCE list
5. CØMMØN list

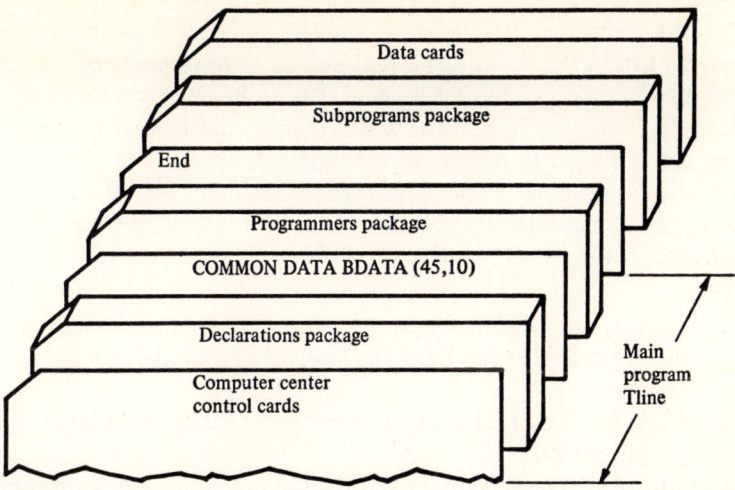

Fig. 5.9.3 The card deck for SANTLINE.

This package is necessary to make the main program compatible with the subprograms and to provide the declaration statements needed by the subprograms. It need not be of particular concern to the programmer as long as it is included in his TLINE deck. Additional declaration statements can be added by the programmer as necessary in his TLINE deck, but none should be removed from the declarations package. The first card following the declarations package is the CØMMØN statement for the permanent store

   CØMMØN/DATA/BDATA(45,NL)

The user replaces the symbol NL with an integer specifying the number of transmission-line sections involved in the problem, up to a maximum of NL = 10.

After the declarations package comes the programmer's package, mentioned previously; some examples are given in Sec. 5.10. Next comes the subprogram package with all the subprograms of Fig. 5.9.2 plus any others the programmer includes. The input data cards are last in the deck; they are written by the programmer to be compatible with his package. In addition, some control cards may be required at the end of the deck by the programmer's computer center.

Of the two storage areas shown in Fig. 5.9.2, the temporary store is simplest. It consists of a 45-word column labeled TLDATA(I), where I = 1, 2, 3, . . . , 45, plus the two words F and LINE. This 47-word column (vector) allows us to store 47 items of information about a given section of transmission line while we are working with that section. These

47 items, defined in Table 5.9.2, include items such as the line length in meters [TLDATA(7)] and the real and imaginary parts of the characteristic admittance [TLDATA(9) and TLDATA(10)]. The word F designates the frequency in hertz at which the calculations are made, and LINE is an integer used to designate the particular section of transmission line under investigation. The first eight words consists of initial data and are normally supplied by the programmer in his main program.† These eight words are:

TLDATA(1) and TLDATA(2) A two word A-format array with which the programmer can label the line, using a total of 8 letters, numbers, and symbols

TLDATA(3) The real part of the characteristic impedance at the frequency FPZ0

TLDATA(4) The imaginary part of the characteristic impedance at the frequency FPZ0

TLDATA(5) The attenuation constant in nepers per meter at the frequency FPZ0

TLDATA(6) The phase constant in radians per meter at the frequency FPZ0

TLDATA(7) The line length in meters

TLDATA(8) The frequency FPZ0 (not FPZØ) in hertz at which the above constants are known

The remaining words in the TLDATA vector are computed using the subprogram CYIN (and others, in some cases). Discussion of the computation of the remaining words will be delayed until CYIN has been described. The complete TLDATA array is defined in Table 5.9.2, however.

The information in the TLDATA store can be transferred to the permanent storage (BDATA array) until needed. This frees the temporary store for use on other sections of transmission line. The statement CØMMØN/TLDATA/TLDATA(45),F,LINE, which appears in the declarations package allows all the computational subprograms and some of the data-transfer subprograms to share the information in the tem-

---

† The real and imaginary parts of the propagation constant, TLDATA(5) and TLDATA(6) respectively, can be computed by the function CPRØP(A,VF,FPZ0). It returns the complex number $\gamma = \alpha + j\beta$ when the power-attenuation constant A (in decibels per meter), the velocity factor VF (the ratio of the velocity of propagation to the velocity of light in free space), and the frequency FPZ0 are known. Since manufacturers' catalogs normally give A at FPZ0 (but seldom α or β) and often VF = 1.0, the function CPRØP is quite useful. We obtain TLDATA(5) from the real part of CPRØP and TLDATA(6) from the imaginary part.

SINUSOIDAL WAVES ON LOSSLESS TRANSMISSION LINES 337

Table 5.9.2 Definitions of words and element assignments
for the common array TLDATA

ELEMENT(S) IN TLDATA	VARIABLE NAME(S)	EXPLANATION
TLDATA(1) TLDATA(2)	LINEID	THIS TWO-WORD ARRAY CONTAINS A LABEL OR DESCRIPTION OF THE LINE PRESENTLY IN THE ARRAY TLDATA. THE INFORMATION IS IN A-FORMAT (DECLARED REAL).
TLDATA(3)	Z0	THE KNOWN CHARACTERISTIC IMPEDANCE OF THE LINE (AT A FREQUENCY FPZ0) IN OHMS.
	Z0X	THE REAL PART OF Z0, STORED IN TLDATA(3).
TLDATA(4)	Z0Q	THE IMAGINARY PART OF Z0.
TLDATA(5)	P0	THE KNOWN PROPAGATION CONSTANT OF THE LINE (AT THE FREQUENCY FPZ0) IN NEPERS/METER.
	P0X	THE REAL PART OF P0, STORED IN TLDATA(5).
TLDATA(6)	P0Q	THE IMAGINARY PART OF P0.
TLDATA(7)	LENGTH	THE LENGTH OF THE LINE SEGMENT IN METERS (THIS VARIABLE IS DECLARED REAL).
TLDATA(8)	FPZ0	THE FREQUENCY, IN HERTZ, AT WHICH BOTH Z0 AND PLINE ARE KNOWN.
TLDATA(9)	YLINE	THE CHARACTERISTIC ADMITTANCE OF THE LINE AT THE FREQUENCY OF OPERATION, IN MHOS (COMPUTED BY FETCH).
	YLINEX	THE REAL PART OF YLINE.
TLDATA(10)	YLINEQ	THE IMAGINARY PART OF YLINE.
TLDATA(11)	VPLUS	CONSTANT POSITIVE-GOING VOLTAGE WAVE (V+) (COMPUTED BY CYIN OR CYSHRT, MODIFIED BY CVLOAD).
	VPLUSX	REAL PART OF VPLUS, STORED IN TLDATA(11).
TLDATA(12)	VPLUSQ	IMAGINARY PART OF VPLUS.
TLDATA(13)	IPLUS	CONSTANT POSITIVE-GOING CURRENT WAVE (I+) (COMPUTED BY CYIN OR CYSHRT, MODIFIED BY CVLOAD).
	IPLUSX	REAL PART OF IPLUS (DECLARED REAL).
TLDATA(14)	IPLUSQ	IMAGINARY PART OF IPLUS (DECLARED REAL).
TLDATA(15)	YIN	INPUT ADMITTANCE OF THIS LINE SEGMENT WHEN TERMINATED BY YLOAD AT FREQUENCY F (COMPUTED BY CYIN OR CYSHRT).
	YINX	REAL PART OF YIN, STORED IN TLDATA(15).
TLDATA(16)	YINQ	IMAGINARY PART OF YIN.
TLDATA(17)	RCFI	THE REFLECTION COEFFICIENT AT THE INPUT END OF THIS LINE SEGMENT (COMPUTED BY CYIN OR CYSHRT).
	RCFIX	REAL PART OF RCFI, STORED IN TLDATA(17).
TLDATA(18)	RCFIQ	IMAGINARY PART OF RCFI.
TLDATA(19)	VIN	THE VOLTAGE AT THE INPUT END OF THE LINE (COMPUTED BY CYIN OR CYSHRT, MODIFIED BY CVLOAD).
	VINX	REAL PART OF VIN, STORED IN TLDATA(19).
TLDATA(20)	VINQ	IMAGINARY PART OF VIN.

porary store. These connections are shown by the dotted paths in Fig. 5.9.2.

Three words in the temporary store are reserved for frequencies. The first, FPZ0 or TLDATA(8), is the frequency at which the characteristic impedance and propagation constant are known, as previously explained. Since we often want to perform calculations at several fre-

Table 5.9.2 Definitions of words and element assignments for the common array TLDATA   (Continued)

TLDATA(21)	IIN	THE CURRENT AT THE INPUT END OF THE LINE, ASSUMING VIN IS 1.0 + J(0.0), IN AMPS (SET TO 1.0 + J(0.0) BY CYIN OR CYSHRT, AND SET TO TRUE VALUE BY CVLOAD).
	IINX	REAL PART OF IIN (DECLARED REAL).
TLDATA(22)	IINQ	IMAGINARY PART OF IIN (DECLARED REAL).
TLDATA(23)	PIN	POWER AT THE INPUT END OF THE LINE IN VOLT-AMPERES, ASSUMING VIN IS 1.0 + J(0.0) (COMPUTED BY CYIN OR CYSHRT, MODIFIED BY CVLOAD).
	PINX	REAL PART OF PIN IN WATTS.
TLDATA(24)	PINQ	IMAGINARY PART OF PIN IN VARS.
TLDATA(25)	YLOAD	THE TERMINATING ADMITTANCE FOR THIS LINE SECTION IN MHOS (COPIED FROM YTERM BY CYIN, AND SET TO 1.0E300 BY CYSHRT).
	YLOADX	REAL PART OF YLOAD, STORED IN TLDATA(25).
TLDATA(26)	YLOADQ	IMAGINARY PART OF YLOAD.
TLDATA(27)	RCFL	THE REFLECTION COEFFICIENT AT THE LOAD END OF THIS LINE SEGMENT (COMPUTED BY CYIN OR CYSHRT).
	RCFLX	REAL PART OF RCFL, STORED IN TLDATA(27).
TLDATA(28)	RCFLQ	IMAGINARY PART OF RCFL.
TLDATA(29)	VLOAD	VOLTAGE SEEN AT LOAD END OF LINE ASSUMING VIN IS 1.0 + J(0.0) VOLTS (COMPUTED BY CYIN OR CYSHRT, AND MODIFIED BY CVLOAD).
	VLOADX	REAL PART OF VLOAD, STORED IN TLDATA(29).
TLDATA(30)	VLOADQ	IMAGINARY PART OF VLOAD.
TLDATA(31)	ILOAD	CURRENT AT LOAD END ASSUMING VIN IS 1.0 + J(0.0) AMPERES (COMPUTED BY CYIN OR CYSHRT, AND MODIFIED BY CVLOAD).
	ILOADX	REAL PART OF ILOAD (DECLARED REAL).
TLDATA(32)	ILOADQ	IMAGINARY PART OF ILOAD (DECLARED REAL).
TLDATA(33)	PLOAD	POWER AT THE LOAD END OF THE LINE, ASSUMING VIN IS 1.0 + J(0.0), IN VOLT-AMPERES (COMPUTED BY CYIN OR CYSHRT, AND MODIFIED BY CVLOAD).
	PLOADX	REAL PART OF PLOAD IN WATTS.
TLDATA(34)	PLOADQ	IMAGINARY PART OF PLOAD IN VARS.
TLDATA(35)	ZLINE	THE ACTUAL CHARACTERISTIC IMPEDANCE OF THE LINE AT THE FREQUENCY, F, AT WHICH THE CALCULATIONS WERE MADE (COMPUTED BY FETCH).
	ZLINEX	REAL PART OF ZLINE, STORED IN TLDATA(35).
TLDATA(36)	ZLINEQ	IMAGINARY PART OF ZLINE.
TLDATA(37)	PLINE	THE ACTUAL PROPAGATION CONSTANT OF THE LINE AT THE FREQUENCY, F, AT WHICH THE CALCULATIONS WERE MADE (COMPUTED BY FETCH).
	ALPHA	REAL PART OF PLINE, STORED IN TLDATA(37).
TLDATA(38)	BETA	IMAGINARY PART OF PLINE.

quencies, the variable F is provided as a working frequency. It is not stored in the BDATA array. Of course, F can be equal to FPZ0 if it happens that we know Z_0 and γ at that frequency, but generally they are not the same. For example, we can use the variable F in a DØ loop to investigate response over a range of frequencies. It is not necessary that FPZ0 be one of the frequencies in the DØ loop. The third word reserved for frequency is FCØMP, or TLDATA(45), which is copied into the

Table 5.9.2 Definitions of words and element assignments for the common array TLDATA (Continued)

TLDATA(39)	R	THE RESISTANCE, IN OHMS/METER, OF THE TRANSMISSION LINE AT FREQUENCY FPZ0 (COMPUTED BY ENTER).
TLDATA(40)	L	THE INDUCTANCE OF THE TRANSMISSION LINE IN HENRYS/METER AT FREQUENCY FPZ0 (COMPUTED BY ENTER).
TLDATA(41)	G	THE SHUNT CONDUCTANCE OF THE TRANSMISSION LINE IN MHOS/METER AT FREQUENCY FPZ0 (COMPUTED BY ENTER).
TLDATA(42)	C	THE SHUNT CAPACITANCE OF THE TRANSMISSION LINE IN FARADS/METER AT FREQUENCY FPZ0 (COMPUTED BY ENTER).
TLDATA(43)	SWRI	VOLTAGE STANDING WAVE RATIO AT THE INPUT (COMPUTED BY CYIN OR CYSHRT).
TLDATA(44)	SWRL	VOLTAGE STANDING WAVE RATIO AT THE LOAD (COMPUTED BY CYIN OR CYSHRT).
TLDATA(45)	FCOMP	FREQUENCY AT WHICH CALCULATIONS WERE MADE, IN HERTZ.

BDATA array to provide a permanent record of the frequency at which the elements of the TLDATA array were calculated. When the computations are being performed, F = FCØMP, but the frequency F can be changed without changing FCØMP.

The permanent store (BDATA array) is an array of 45 × NL words, with one 45-word column vector assigned to each section of transmission line. This store is specified by the statement CØMMØN/DATA/BDATA(45,NL). For the system given in Fig. 5.9.1, only three vectors for the permanent store would be used since there are three transmission lines; thus the symbol NL would be replaced by the integer 3. In this fashion the programmer can control the size of the BDATA array. The 45 words of each column of the BDATA array are identical to those of the TLDATA array and are identified in Table 5.9.2. As numerical values are computed for a given section of line, they are stored temporarily in the TLDATA array and later transferred to a column of the BDATA array reserved for the given section of line.

The primary functions of the permanent store are:

1. To receive information from TLDATA which was read into the computer initially
2. To receive (from TLDATA) and store the new information after it is computed
3. To store information on each section of line not pertaining to computations in progress
4. To transfer data to the temporary store, as required

The programmer can obtain a printed output from either the permanent or the temporary storage areas, according to his instructions in the main program.

The five data-transfer subprograms listed in Fig. 5.9.2 are used primarily to enter the data into the array BDATA, or to transfer data between the two storage areas. In addition, ENTER, MØDIFY, and FETCH do some computation. These five subprograms will be discussed in the following paragraphs.

SUBRØUTINE ENTER(NLINE) The values of R, L, G, and C of transmission lines normally are not known directly from manufacturer's data, but they can be found from the characteristic impedance Z_0 and the propagation constant γ. From the initial data in the temporary store, the subroutine ENTER(NLINE) computes R, L, G, and C from

$$R + j\omega L = Z_0 \gamma \qquad G + j\omega C = \frac{Z_0}{\gamma} \tag{3}$$

and then stores R, G, $2\pi L$, $2\pi C$, and the entire TLDATA array in the permanent store, in the column specified by the integer NLINE. The values of L and C are multiplied by 2π before being stored in the BDATA array, since they are more useful for computational purposes in this form.

An alternate entry to the subroutine ENTER is provided† in the form ENTRY RESTØR(NLINE). RESTØR copies a column from the BDATA array into the TLDATA array, and the BDATA array column is specified by the number NLINE. The numbers stored for L and C in the BDATA array are divided by 2π, so that the true values of L and C are stored in the TLDATA array. Thus RESTØR is used to transfer data from the BDATA array to the TLDATA array. ENTER transfers data from the TLDATA array to the BDATA array.

SUBRØUTINE STASH The subroutine STASH does no computation on the data, but it copies any data from the TLDATA array into the BDATA array in a column specified by the value of the parameter LINE (in the TLDATA array). Thus STASH is similar to ENTER, but it does no computation.

There is an entry to STASH called BRING. It copies the data from the BDATA array (in the column specified by the value of LINE) into the TLDATA array, with no computation. Hence STASH and BRING are a transform pair for data transformation. A second entry to STASH is called FETCH. It performs the same function as BRING, except that certain values are recomputed at the frequency F (of the TLDATA array). For more detail see the subroutine STASH in Appendix B.

† For CDC processors the argument (NLINE) is deleted, and the form is ENTRY RESTØR. The argument of the original subroutine applies to all entry names for these processors.

FUNCTIØN GET1C(NLINE,NUMBER) GET1C is used to obtain any single complex number (real and imaginary parts) stored in the BDATA array, as designated by the line identification number NLINE and by NUMBER, an integer between 1 and 45 which identifies the real part of a complex number in the TLDATA array. For example, if one wants the reflection coefficient at the load end of line 2 in Fig. 5.9.1, one writes

$\quad$ RCØEFL = GET1C(2,27)

which will be a complex number. The NUMBER 27 is for the real part of the reflection coefficient, as shown in Table 5.9.2. Word number 28, the imaginary part of the reflection coefficient, will also be returned. See Appendix B for further details.

FUNCTIØN GET1R(NLINE,NUMBER) This function is the same as GET1C except that it is used to obtain any single *real* number. For example, if we want the inductance of line 2 from the BDATA array, we write WL = GET1R(2,40) and obtain 2π times the inductance; that is WL = $2\pi L$. Recall that the values of L and C are multiplied by 2π before being stored in the BDATA array. Of course the inductance can be obtained from WL by division by 2π. The FUNCTIØN GET1R returns a number from the BDATA array without change.

SUBRØUTINE MØDIFY(NLINE,NUMBER,VALUE,MØDE) This subroutine allows the programmer to change the number stored in any individual word in the BDATA array (permanent store) without changing any other words, except as noted below. NUMBER is an integer (from 3 through 45) used to specify the word to be changed. VALUE is the new value to be inserted into the word. For example, if we want to change LENGTH of line 2 in the permanent store to 5.3 m, we write

$\quad$ CALL MØDIFY(2,7,5.3,1)

since LENGTH = TLDATA(7), as shown in Table 5.9.2. MØDE indicates how many words are to be changed, starting with the one specified by NUMBER. If R, L, G, or C is changed, then the values of Z_0 and γ are changed accordingly. Otherwise only the words specified by NUMBER are changed. If VALUE is an array, then MØDE can be greater than unity and more than one word will be changed. For example, suppose we want to change the value of the load admittance

$\quad$ YLØAD = YLØADX + jYLØADQ

in line 3. We could write

$\quad$ DIMENSIØN VALUE(2)
$\quad$ VALUE(1) = YLØADX
$\quad$ VALUE(2) = YLØADQ

where YLØADX and YLØADQ are real numbers to be inserted into line 3 of the BDATA array, starting with word 25 in Table 5.9.2. Next we write

   CALL MØDIFY(3,25,VALUE,2)

and the numerical values in words 25 and 26 will be changed to the new values as specified.

The computational subprograms are CYIN, CYSHRT, CVLØAD, FINDL, and STUBL. They will be described in the next few paragraphs.

FUNCTIØN CYIN(NLINE,YTERM) This is the major computational subprogram. First, it copies the data from a column (identified by NLINE) of the BDATA array into the TLDATA array by using FETCH. Then, using the input data (usually the first eight words) in the TLDATA array, CYIN computes values for most of the remaining words of the TLDATA vector. One can specifically identify the words from Table 5.9.2. Then it uses STASH to copy the TLDATA vector into the original column of the BDATA array.

CYIN assumes that a voltage of $1.0 + j0.0$ V is applied at the input terminals of a section of line, identified in the BDATA array by NLINE, and loaded at the far end by a complex admittance YTERM in mhos. It computes the following variables at each end of the line:

1. Reflection coefficient
2. Standing-wave ratio
3. Voltage
4. Current
5. Power

All the computations are done at the frequency F, which is stored as FCØMP via TLDATA(45) in the BDATA array.

In addition, CYIN computes the input admittance and returns this complex number to the main program. An example was given in Eq. (2). All these operations are done for the programmer when he calls the function, without additional work on his part.

FUNCTIØN CYSHRT(NLINE) This subprogram is identical with CYIN (it does the same data-transfer operations and computations) except that it assumes that the termination of the line is a short circuit. This FUNCTIØN was designed for use with shorted stubs. It is more economical with machine time than CYIN in such cases, and it avoids problems associated with an infinite admittance and standing-wave ratio. It should be used wherever shorted terminations occur.

FUNCTIØN CVLØAD(NLINE,CVIN) If a generator voltage and an internal impedance are given at the source end, the voltages, currents, and powers obtained from CYIN (and CYSHRT) must be corrected. It is necessary to use CYIN first, because we cannot find the true input voltage until we know the input impedance (which we obtain using CYIN). We encountered this same difficulty when using the Smith chart. The subprogram CVLØAD is used to correct the voltage, current, and power when the actual complex input voltage is CVIN.

CVLØAD uses FETCH to copy the data from the BDATA column, identified by NLINE, into the TLDATA array. Then it computes the correct values of V^+, I^+, voltage, current, and power at both ends of the line and uses STASH to copy the corrected values into the BDATA array.

The FUNCTIØN returns the value of the complex voltage at the load end to the main program. As a general procedure in solving a problem, we start from the load end and use CYIN and CYSHRT in progressing to the source end, thereby filling the BDATA array with most of the necessary information. Once the input impedance is known, we can use CVLØAD to work back to the load end, section by section.

For example, in Fig. 5.9.1, we found the input admittances for the three sections of transmission line by using CYIN in Eqs. (2). If we assume that the generator voltage is V1 and its impedance is Z1, then the complex input voltage is

$$\text{CVIN} = V1/(1.0+Z1*Y3) \tag{4}$$

where Y3 is the input admittance from Eq. (2) (see Fig. 5.9.4). We can use CVLØAD to compute the voltages at the load end of each of the three sections as follows:

$$\begin{align} V3 &= \text{CVLØAD}(3,\text{CVIN}) \\ V2 &= \text{CVLØAD}(2,V3) \\ V1 &= \text{CVLØAD}(1,V2) \end{align} \tag{5}$$

where V1, V2, and V3 have been declared complex in the main program. In conclusion, Eqs. (2), (4) and (5) are the essential elements of the computer program for the given three-section system used for illustration.

FUNCTIØN FINDL(NLINE,YTERM,ITER) Given the termination admittance YTERM of the line identified by the number NLINE, the function

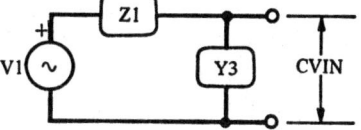

Fig. 5.9.4 The complex voltage at the input CVIN is determined by the generator parameters and the input admittance.

FINDL computes the shortest length (in meters) such that the input admittance YIN to that line has a real part equal to the characteristic admittance (or equal to the real part of the characteristic admittance for lossy lines). For example, on a lossless line (arbitrarily called line 1) at an unknown distance D (meters) from the load end, assume YIN $= Y_0 + jB$. Then D is given by

$$D = \text{FINDL}(1,\text{YTERM},3)$$

An iteration technique, involving 30 iterations at most, is used to find the length to an accuracy of $\frac{1}{2000}$ wavelength. A smaller number of iterations can be used if less accuracy is needed. The parameter ITER determines the accuracy and also the number of iterations. If ITER = 1, 2, or 3, the distance will be found to the nearest 0.05, 0.005, or 0.0005 wavelength respectively. If ITER is greater than 3, more than 30 iterations can occur, but the improved accuracy probably is not warranted for most problems.

FUNCTIØN STUBL(NLINE,YINST) Given an admittance

$$\text{YINST} = G + jB$$

the function STUBL computes the minimum length (in meters) of a lossless shorted stub (identified as NLINE in the BDATA array) so that the input admittance to the stub is YIN $= -jB$. The stub of length STUBL can be used to cancel the imaginary part of YINST when connected in parallel with YINST. The length is based upon the assumption that the stub is lossless (see Appendix B for further details).

Next, the utility subprograms will be described. They are simple FUNCTIØN subprograms that have proved useful often enough to be included in SANTLINE.

FUNCTIØN CNVRTM(WAVEL,VF,F) Given the length in wavelengths (WAVEL) of a section of line, CNVRTM converts the length to meters at the frequency F for the given velocity factor VF. (See FUNCTIØN CPRØP for a definition of VF.) For example, if a line is 0.8931 wavelength long at the known frequency F, the length of the line in meters (call it DIST) is

$$\text{DIST} = \text{CNVRTM}(0.8931,1.0,\text{F})$$

where we have assumed that VF = 1.0.

Another entry to this function can be used to convert a known length in meters to the number of wavelengths at the frequency F. For this entry,

$$\text{ENTRY CNVRTW(WAVEL,VF,F)}$$

SINUSOIDAL WAVES ON LOSSLESS TRANSMISSION LINES

the parameter WAVEL must be given in meters. Using the above example, if we write

$$WL = CNVRTW(DIST, 1.0, F)$$

we find that WL is 0.8931 wavelength.

FUNCTION CPROP(A,VF,F) This function computes the propagation constant $\gamma = \alpha + j\beta$, given A, VF, and F, where

A = power attenuation, db/m at frequency F

$$VF = \text{velocity factor} = \frac{\text{velocity of light on transmission line}}{\text{velocity of light in free space}}$$

F = frequency at which A and VF are known

Table 5.9.3   A description of the subprograms of SANTLINE and a glossary of terms

FUNCTION CYIN(NLINE,YTERM)

```
CYIN..........GIVEN THE TERMINATING ADMITTANCE, CYIN COMPUTES THE
              INPUT ADMITTANCE OF A SEGMENT OF TRANSMISSION LINE
              WHOSE PARAMETERS HAVE BEEN PREVIOUSLY STORED IN THE
              STORAGE ARRAY BDATA.  THE REFLECTION COEFFICIENTS
              AND STANDING WAVE RATIOS AT BOTH ENDS ARE COMPUTED.
              ASSUMING AN INPUT VOLTAGE OF 1.0+J(0.0), THE
              VOLTAGES, CURRENTS, AND POWERS AT EACH END ARE ALSO
              COMPUTED.  THESE ARE THEN PLACED IN THE STORAGE
              ARRAY, BDATA, AND THE FUNCTION EXITS.  THE FREQUENCY
              AT WHICH THE CALCULATIONS WERE MADE (CONTAINED IN
              THE COMMON VARIABLE F) IS STORED VIA TLDATA(45).
NLINE.........THE COLUMN NUMBER OF THE LINE IN THE ARRAY BDATA
              WHICH IS TO BE AFFECTED.
YTERM.........THE TERMINATING ADMITTANCE IN MHOS.
```

FUNCTION CYSHRT(NLINE)

```
CYSHRT........CYSHRT COMPUTES THE INPUT ADMITTANCE OF A LINE
              SEGMENT ASSUMING THAT IT IS SHORTED.  IN ADDITION,
              IT PERFORMS ALL OTHER FUNCTIONS OF CYIN.
NLINE.........THE COLUMN NUMBER OF THE SEGMENT TO USE.  IF IT IS
              LESS THAN 1, THE LINE SEGMENT PRESENTLY IN THE
              COMMON ARRAY TLDATA IS USED.
```

FUNCTION CVLOAD(NLINE,CVIN)

```
CVLOAD........ONCE CYIN HAS CALCULATED THE VOLTAGES, CURRENTS, AND
              POWERS AT THE LOAD AND INPUT ENDS AND THE VALUES
              OF VPLUS AND IPLUS ASSUMING THAT VIN IS 1.0+J(0.0),
              CVLOAD ADJUSTS THESE VALUES TO REFLECT THE ACTUAL
              VALUES FOR A GIVEN INPUT VOLTAGE CVIN.
CVIN..........THE ACTUAL VALUE OF THE INPUT VOLTAGE IN VOLTS.
NLINE.........THE COLUMN NUMBER OF THE LINE IN THE ARRAY BDATA
              WHICH IS TO BE AFFECTED.
```

Table 5.9.3 A description of the subprograms of SANTLINE and a glossary of terms (Continued)

```
FUNCTION FINDL(NLINE,YTERM,ITER)

FINDL.........FINDL DETERMINES THE DISTANCE FROM THE LOAD TO THE
              POINT ON A TRANSMISSION LINE AT WHICH THE REAL PART
              OF THE INPUT ADMITTANCE IS APPROXIMATELY EQUAL TO THE
              REAL PART OF THE CHARACTERISTIC ADMITTANCE.
NLINE.........THE NUMBER OF THE LINE TO BE USED IN THE SEARCH.
YTERM.........THE COMPLEX LOAD ADMITTANCE.
ITER..........ITER CONTROLS THE DEGREE OF APPROXIMATION.  IF ITER
              IS 1, THE DISTANCE WILL BE FOUND TO THE NEAREST 0.05
              WAVELENGTH; IF 2, TO THE NEAREST 0.005 WAVELENGTH;
              IF 3, TO THE NEAREST 0.0005 WAVELENGTH, ETC.


FUNCTION STUBL(NLINE,YINST)

STUBL.........THIS FUNCTION COMPUTES THE LENGTH OF A SECTION OF
              SHORTED, LOSSLESS TRANSMISSION LINE NEEDED TO GIVE
              AN INPUT SUSCEPTANCE EQUAL TO THE IMAGINARY PART OF
              YINST.  ALTHOUGH THE LINE TO BE USED MAY NOT
              ACTUALLY BE LOSSLESS IT IS TREATED AS SUCH.
NLINE.........THE LINE NUMBER OF THE LINE SEGMENT TO BE USED FOR
              THE STUB.  IF NLINE IS LESS THAN 1, THE LINE
              SEGMENT WHOSE DATA IS IN THE COMMON ARRAY TLDATA
              WILL BE USED FOR THE STUB.
YINST.........THE COMPLEX ADMITTANCE WHOSE IMAGINARY PART IS TO BE
              PRODUCED BY THE SHORTED STUB.


SUBROUTINE ENTER(NLINE)

ENTER.........ONCE THE PROGRAMMER HAS INITIALIZED THE REQUIRED
              ELEMENTS IN THE COMMON ARRAY TLDATA, ENTER COMPUTES
              THE VALUES OF R,L,G AND C (L AND C ARE MULTIPLIED
              BY 2*PI) AND THEN THE CONTENTS OF TLDATA ARE COPIED
              INTO THE ARRAY BDATA IN COMMON BLOCK DATA.  THE
              COLUMN NUMBER FOR THE BDATA ARRAY IS SPECIFIED BY THE
              PARAMETER NLINE.  IF NLINE IS LESS THAN 1, ONLY THE
              DATA IN THE TLDATA ARRAY IS AFFECTED.  NOTHING
              WILL BE COPIED FROM BDATA.  ENTER USES STASH.
```

This function is useful when A and VF can be obtained from manufacturers' data sheets and α and β cannot.

The various subprograms of SANTLINE have been described and the BDATA and TLDATA arrays discussed. For convenience, a summary of the subprograms appears in Table 5.9.3. They are also described with the computer listings of the subprograms in Appendix B.

In the next section we illustrate the use of SANTLINE with a few examples. Although the description of SANTLINE has been fairly long and involved, we will find it rather simple to use.

5.10 APPLICATIONS OF THE COMPUTER

In this section we illustrate some applications of the digital computer and SANTLINE with a few examples. In addition we consider some effects of a varying frequency on matching, SWR, and on other variables. It is

Table 5.9.3 A description of the subprograms of SANTLINE and a glossary of terms (Continued)

```
ENTRY RESTOR
RESTOR........RESTOR COPIES A COLUMN OF BDATA FOR THE LINE
              SPECIFIED BY THE PARAMETER NLINE INTO THE COMMON
              ARRAY TLDATA.  IN ADDITION, THE VALUES OF L AND
              C CONTAINED IN BDATA ARE DIVIDED BY 2*PI TO REFLECT
              THEIR TRUE VALUES.  RESTOR USES BRING.


SUBROUTINE STASH

STASH.........STASH COPIES THE CONTENTS OF THE COMMON ARRAY TLDATA
              INTO A COLUMN IN BDATA AS SPECIFIED BY THE COMMON
              VARIABLE LINE.  IF LINE IS LESS THAN 1 THE CALL
              IS IGNORED.
ENTRY BRING
BRING.........BRING COPIES A COLUMN FROM BDATA INTO THE COMMON
              ARRAY TLDATA.  THE COLUMN NUMBER IS SPECIFIED BY
              THE PARAMETER LINE.  IF THE VALUE OF LINE IS
              LESS THAN 1 THE CALL IS IGNORED.
ENTRY FETCH
FETCH.........FETCH PERFORMS THE SAME FUNCTION AS BRING EXCEPT
              THAT NEW VALUES FOR ZLINE,PLINE, AND YLINE ARE
              COMPUTED AT THE FREQUENCY OF OPERATION F.
              IF LINE IS LESS THAN 1, THE CALL IS IGNORED.
              FETCH USES THE FUNCTION CSQRT.


SUBROUTINE MODIFY(NLINE,NUMBER,VALUE,MODE)

MODIFY........MODIFY ALLOWS THE PROGRAMMER TO CHANGE THE VALUES OF
              SELECTED PARAMETERS OF ANY LINE SEGMENT WITHOUT
              CHANGING ANY OTHERS (EXCEPT FOR CHANGING R, L, G AND
              C TO REFLECT CHANGES IN Z0, P0 OR FPZ0).  NLINE IS
              THE COLUMN-NUMBER IN THE BDATA ARRAY.  NUMBER IS
              THE ELEMENT IN THE BDATA ARRAY WHICH IS TO BE
              MODIFIED, OR THE FIRST OF SEVERAL ELEMENTS TO BE
              MODIFIED.  IF NLINE IS LESS THAN 1, THE CALL IS
              IGNORED.
VALUE.........THE NEW VALUE TO BE INSERTED INTO THE
              BDATA ARRAY, OR IF MORE THAN ONE VALUE IS
              TO BE INSERTED, THE NAME OF THE ARRAY CONTAINING
              THE NEW VALUES.  (COMPLEX VARIABLES ARE
              CONSIDERED AS TWO-ELEMENT REAL ARRAYS BY MODIFY).
```

important to note that the example programs in this section can be used for lines with losses, a topic considered in Chap. 6. With the aid of the computer, we can solve lossless or lossy problems of this type with equal ease.

Example 1 TLINE1: A simple test program In this first example we discuss a simple main program, TLINE1, which computes voltages, currents, powers, etc., for the system shown in Fig. 5.10.1. (The computer program TLINE1 is listed in Table 5.10.1 with the declarations package omitted. It should be inserted as indicated before using TLINE1; see Sec. 5.9 for discussion.) This example serves a dual purpose: (1) it is an introductory example using SANTLINE, and (2) it can be used to test the user's SANTLINE deck against the known results.†

† TLINE1 does not test *all* features of SANTLINE, however.

Table 5.9.3 A description of the subprograms of SANTLINE and a glossary of terms (Continued)

MODE..........THE LENGTH OF THE REAL ARRAY VALUE (1 FOR A REAL
              VARIABLE, 2 FOR A COMPLEX VARIABLE).
NOTE   IF THE COMBINATION OF NUMBER AND MODE WOULD ALLOW
              CHANGING DATA NOT CONCERNED WITH THE DESIRED LINE
              SEGMENT, NO MODIFICATION IS ALLOWED.

FUNCTION GET1C(NLINE,NUMBER)

GET1C.........THIS COMPLEX FUNCTION RETURNS ANY COMPLEX NUMBER
              FROM THE BDATA ARRAY FOR THE COLUMN AND ROW
              SPECIFIED BY THE PARAMETERS NLINE AND NUMBER.
NLINE.........THE COLUMN NUMBER OF THE WORD IN THE BDATA ARRAY
              WHICH IS TO BE COPIED.
NUMBER........SPECIFIES THE ELEMENT IN THE GIVEN COLUMN IN THE
              BDATA ARRAY CORRESPONDING TO THE VARIABLE
              TO BE RETURNED.
NOTE   NO CHECK IS MADE TO ASSURE THAT NUMBER ADDRESSES AN
              ACTUAL COMPLEX DATA ITEM. THIS IS LEFT TO THE
              PROGRAMMER. IF NUMBER IS LESS THAN 1 OR EXCEEDS 44
              OR NLINE IS LESS THAN 1, A VALUE OF 0.0+J0.0
              IS RETURNED.

FUNCTION GET1R(NLINE,NUMBER)

GET1R.........THIS REAL FUNCTION RETURNS ANY REAL DATA ITEM FOR
              THE LINE SEGMENT SPECIFIED BY THE PARAMETERS
              NLINE AND NUMBER (THE COLUMN AND ROW NUMBERS
              OF THE BDATA ARRAY).
NLINE.........THE COLUMN NUMBER OF THE WORD IN THE BDATA ARRAY
              WHICH IS TO BE COPIED.
NUMBER........SPECIFIES THE ELEMENT IN THE COMMON ARRAY BDATA
              CORRESPONDING TO THE VARIABLE WHOSE VALUE IS TO BE
              RETURNED.
NOTE   IF NUMBER IS LESS THAN 1 OR EXCEEDS 45 OR NLINE IS
              LESS THAN 1, A VALUE OF 0.0 IS RETURNED.

Otherwise, one would not take the time to write a computer program for such a simple system.

The first READ statement obtains an integer KØDE which allows one to analyze any number of lines as long as KØDE is not zero; when KØDE is zero, the program halts. The variable LINE is an integer for line identification, and

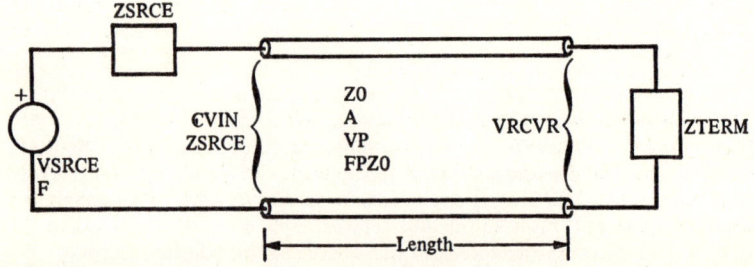

Fig. 5.10.1 The system for program TLINE1.

Table 5.9.3 A description of the subprograms of SANTLINE and a glossary of terms (Continued)

```
FUNCTION CPROP(A,VF,F)

CPROP.........THIS COMPLEX FUNCTION COMPUTES THE
              PROPAGATION CONSTANT P0 FROM THE POWER
              ATTENUATION IN DB/METER, THE VELOCITY FACTOR (RATIO
              OF THE VELOCITY OF PROPAGATION TO THE SPEED OF LIGHT
              IN A VACUUM) AND THE FREQUENCY AT WHICH THESE TWO
              ARE KNOWN (FPZ0).
A.............POWER ATTENUATION IN DB PER METER AT FREQUENCY F.
VF............THE VELOCITY FACTOR.
F.............THE FREQUENCY AT WHICH BOTH A AND VF ARE KNOWN.


FUNCTION CNVRTM(WAVEL,VF,F)

CNVRTM........THIS FUNCTION COMPUTES THE LENGTH OF A LINE SEGMENT
              IN METERS FROM THE LENGTH IN WAVELENGTHS, THE
              VELOCITY FACTOR (RATIO OF THE VELOCITY OF
              PROPAGATION TO THE SPEED OF LIGHT IN A VACUUM) AND
              THE FREQUENCY AT WHICH THESE ARE KNOWN.
WAVEL.........LENGTH OF THE LINE SEGMENT IN WAVELENGTHS AT
              FREQUENCY F.
VF............VELOCITY FACTOR AT FREQUENCY F.
F.............FREQUENCY AT WHICH WAVEL AND VF ARE KNOWN
ENTRY CNVRTW
CNVRTW........THIS FUNCTION ENTRY CONVERTS A LENGTH IN METERS TO
              A LENGTH IN WAVELENGTHS.
WAVEL.........FOR THIS ENTRY, THIS PARAMETER IS THE LENGTH IN
              METERS TO BE CONVERTED TO WAVELENGTHS.
              ALL OTHER PARAMETERS ARE AS FOR CNVRTM.
```

Z0, FPZ0, LENGTH, and LINEID supply initial data† for six of the first eight elements of TLDATA. The variables A, VF, and FPZ0 are used with the subprogram CPRØP to obtain the propagation constant and supply the initial data for TLDATA(5) and TLDATA(6). The first READ statement obtains initial values for the frequency at which computations are to occur and the source voltage, while the second READ statement initializes the load impedance. Next the subroutine ENTER is used to compute R, L, G, and C and to transfer the data to the BDATA array. At this point, all the necessary information about the transmission line is properly stored, and we are ready to begin the system analysis.

The analysis is done with two statements. The first uses the function CYIN to compute most of the unknown data of the TLDATA array. Then the function CVLØAD is used to compute the voltage, current, and power, as explained in Sec. 5.9. Since the voltage at the source end of the line is given by

CVIN = VSRCE/(1.0 + ZSRCE*YSEND)

this computation is included in the calling statement for CVLØAD; we have assumed ZSRCE = 100.0 + j0.0. (For some compilers, it may be necessary to compute CVIN before using it in the calling statement.) The final computation (ZSEND = 1.0/YSEND) is done for the convenience of having both the input admittance and impedance.

† The method of "initializing" complex variables is not standardized. Use the manufacturer's recommended method for your computer.

The program is completed with the appropriate WRITE statements and their FORMAT statements. The transmission line shown in Fig. 5.4.7 has been analyzed to provide a specific illustration, and the results of using TLINE1 are given in Table 5.10.2. The data cards for this specific line are listed in Table 5.10.1.

A DISCUSSION OF BANDWIDTH

In most applications of microwaves and transmission lines we are concerned with a band of frequencies, as in AM, FM, and TV broadcasts. The frequency content (spectrum) of the signals varies instant by instant but is confined to some frequency band by equipment design or limitations (often specified by federal regulations). Our work in Chap. 5 has been limited to single-frequency analyses (where the frequency is constant) except in one or two cases. Now we consider some of the problems that arise when frequency changes.

Assume that a microwave system, such as a quarter-wave transformer or a double-stub tuner, has been designed so that the system is matched at some frequency ω_0, which we call the *center frequency*. If the source changes from frequency ω_0 to some other frequency ω_1, the following parameters will change (assuming lossless lines):

Load impedance Unless the load Z_r is composed of a pure resistance (it seldom is), the load impedance will change with frequency. On the Smith chart, the point Z_{rn} will change correspondingly. In general, its trace will be some curve across the Smith chart, as determined by the frequency and the characteristics at the load.

Wavelength Since the wavelength is the ratio of the velocity of propagation to the frequency, we can write

$$\frac{\lambda_1}{\lambda_0} = \frac{\omega_0}{\omega_1} \frac{v_1}{v_0} \tag{1}$$

where v_0 and v_1 are the velocities of propagation at the frequencies ω_0 and ω_1 respectively. In most cases the velocity ratio is assumed to be unity, so that

$$\lambda_1 = \frac{\omega_0}{\omega_1} \lambda_0 = \frac{f_0}{f_1} \lambda_0 \quad \text{m/c} \tag{2}$$

In any case, the wavelength changes as the frequency changes.

Effective line lengths The actual length l, in meters, of a transmission line is independent of the frequency of the source, of course. But if we

Table 5.10.1 Program TLINE1

```
C       THIS PROGRAM COMPUTES AND WRITES THE VARIABLES OF THE TLDATA
C       ARRAY FOR THE SYSTEM SHOWN IN FIGURE 5.10.1.

C       ******** INSERT DECLARATIONS PACKAGE HERE ********
        COMMON/DATA/BDATA(45,10)
        COMPLEX CYIN,CVLOAD,CPROP,VSRCE,YSEND,ZTERM,VRCVR,ZSEND,YTERM
10      READ(5,20)KODE,LINE,F,VSRCE,Z0
20      FORMAT(I2,I5,3X,5E10.0)
        IF(KODE.EQ.0) GO TO 90
30      READ(5,40)A,VF,FPZ0,XTERMD,QTERMD,LENGTH,LINEID
40      FORMAT(6E10.0,4X,2A4)
        P0=CPROP(A,VF,FPZ0)
        CALL ENTER(LINE)
        ZTERM=CMPLX(XTERMD,QTERMD)
        YTERM=1.0/ZTERM
        YSEND=CYIN(LINE,YTERM)
        VRCVR=CVLOAD(LINE,VSRCE/(1.0+100.0*YSEND))
        ZSEND=1.0/YSEND
        CALL RESTOR(LINE)
        WRITE(6,60)LINE,LINEID,F,LENGTH,Z0,P0,FPZ0,R,L,G,C,ZLINE,PLINE,
       A YLOAD,RCFL,RCFI,YIN,VPLUS,IPLUS,VLOAD,ILOAD,VIN,IIN,PLOAD,PIN,
       B SWRL,SWRI,ZSEND,VSRCE,VRCVR,ZTERM,YSEND
60      FORMAT('1',9X,'DATA FOR LINE ',I4,5X,2A4//10X,'F=',6X,1PE9.2/10X,'
       ALENGTH= ',E9.2/10X,'Z0=',5X,E9.2,' +J',E9.2/10X,'P0=',5X,F9.2,' +J
       B',E9.2/10X,'FPZ0=',3X,E9.2/10X,'R=',6X,E9.2/10X,'L=',6X,E9.2/10X,'
       CG=',6X,E9.2/10X,'C=',6X,E9.2/10X,'ZLINE=',2X,E9.2,' +J',E9.2/10X,'
       DPLINE=',2X,E9.2,' +J',E9.2/10X,'YLOAD=',2X,E9.2,' +J',F9.2/10X,'RC
       EOEFL=',1X,E9.2,' +J',E9.2/10X,'RCOEFI=',1X,E9.2,' +J',F9.2/10X,'YI
       FN=',4X,E9.2,' +J',E9.2/10X,'VPLUS=',2X,F9.2,' +J',E9.2/10X,'IPLUS=
       G',2X,E9.2,' +J',E9.2/10X,'VLOAD=',2X,E9.2,' +J',E9.2/10X,'ILOAD=',
       H2X,E9.2,' +J',E9.2/10X,'VIN=',4X,E9.2,' +J',E9.2/10X,'IIN=',4X,E9.
       I2,' +J',E9.2/10X,'PLOAD=',2X,E9.2,' +J',E9.2/10X,'PIN=',4X,E9.2,'
       J+J',E9.2/10X,'SWRL=',3X,E9.2/10X,'SWRI=',3X,E9.2/10X,'ZSEND=',2X,E
       K9.2,' +J',E9.2/10X,'VSRCE=',2X,E9.2,' +J',E9.2/10X,'VRCVR=',2X,E9.
       L2,' +J',E9.2/10X,'ZTERM=',2X,E9.2,' +J',E9.2/10X,'YSEND=',2X,E9.2,
       M' +J',E9.2)
        GO TO 10
90      CALL EXIT
        END
```

THE DATA CARDS FOR PROGRAM TLINE1 ARE

```
1   1        5E7      120.0       0.0      100.0     0.0    LINE 1
    0.0      1.0      5E7         300.0    0.0       7.0    LINE 1
```

express its length as products

$$l = \kappa_0 \lambda_0 = \kappa_1 \lambda_1 \quad \text{m} \tag{3}$$

then the constants κ_0 and κ_1 have units of cycles and represent the length of the line in cycles or wavelengths (sometimes called the electrical lengths) at the frequencies ω_0 and ω_1 respectively. Thus

$$\kappa_1 = \frac{\lambda_0}{\lambda_1} \kappa_0 = \frac{f_1}{f_0} \kappa_0 \tag{4}$$

ELECTROMAGNETIC WAVE PROPAGATION

Table 5.10.2 Results for the line of Fig. 5.4.7

```
DATA FOR LINE    1      LINE 1

F=        5.00E 07
LENGTH=   7.00E 00
Z0=       1.00E 02 +J 0.0
P0=       0.0      +J 1.05E 00
FPZ0=     5.00E 07
R=        0.0
L=        3.33E-07
G=        0.0
C=        3.33E-11
ZLINE=    1.00E 02 +J 0.0
PLINE=    0.0      +J 1.05E 00
YLOAD=    3.33E-03 +J 0.0
RCOEFL=   5.00E-01 +J 0.0
RCOEFI=  -2.50E-01 +J-4.33E-01
YIN=      1.00E-02 +J 1.15E-02
VPLUS=    3.00E 01 +J-5.20E 01
IPLUS=    3.00E-01 +J-5.20E-01
VLOAD=    4.50E 01 +J-7.79E 01
ILOAD=    1.50E-01 +J-2.60E-01
VIN=      4.50E 01 +J-2.60E 01
IIN=      7.50E-01 +J 2.60E-01
PLOAD=    2.70E 01 +J 0.0
PIN=      2.70E 01 +J-3.12E 01
SWRL=     3.00E 00
SWRI=     3.00E 00
ZSEND=    4.29E 01 +J-4.95E 01
VSRCE=    1.20E 02 +J 0.0
VRCVR=    4.50E 01 +J-7.79E 01
ZTERM=    3.00E 02 +J 0.0
YSEND=    1.00E-02 +J 1.15E-02
```

and so the effective line length in wavelengths changes with frequency. On the Smith chart, the constants κ_1 and κ_0 are measured on the wavelength scale (outer periphery). To see the results of a change in effective line length, consider a line which is ¼ wavelength long ($\kappa_0 = 0.25$) at 300 MHz. Then it is ½ wavelength long ($\kappa_1 = 0.50$) at 600 MHz, and the change corresponds to an additional half revolution on the Smith chart.

Standing-wave ratio Suppose that a microwave system is matched at the center frequency ω_0 so that the SWR is unity on the main line. If the frequency of the source changes, the load impedance changes and the effective line lengths change. Stubs or transformers that matched the line at the frequency ω_0 are no longer the correct length for matching. As a consequence the SWR on the main line will increase.

Voltage, current, power As the above four parameters change, the rms values of voltage, current, and power at any given point on the line will change.

Other parameters can also change with frequency, such as the velocity of propagation (unless the line is air-filled or essentially free space). If the line has dissipation, the characteristic impedance, propagation constant, attenuation constant, and phase constant can also change with frequency. The generator impedance and its emf are other parameters which can change. Although graphical analysis of variable-frequency systems can be very laborious, SANTLINE can be used effectively, even with dissipative lines, as will be demonstrated in the examples which follow in this section and in Sec. 6.5.

It is a common practice to define the bandwidth of microwave systems in terms of the SWR (rather than using the half-power frequencies as in circuit theory) because of the ease with which SWR can be determined and measured. As frequency varies from the center frequency ω_0, where matching occurs, the SWR increases as shown in the sketch of Fig. 5.10.2. If one arbitrarily selects an SWR greater than unity, usually two frequencies ω_1 and ω_2 can be found such that $\omega_1 < \omega_0 < \omega_2$, where the SWR has the specified value. The bandwidth B is defined for that value of SWR as

$$B = \omega_2 - \omega_1 \tag{5}$$

The bandwidth is commonly defined in terms of a SWR of either 1.5 or 3.0. If SWR = 5.83, then $|\Gamma_r| = 0.707$ so that the frequencies ω_1 and ω_2 are the half-power frequencies, and then the bandwidth is defined as in circuit theory. However, it is more common to use SWR = 1.5 for bandwidth determination, so that a system is designed to confine the signal to this region where attenuation (due to mismatch) and distortion are small.

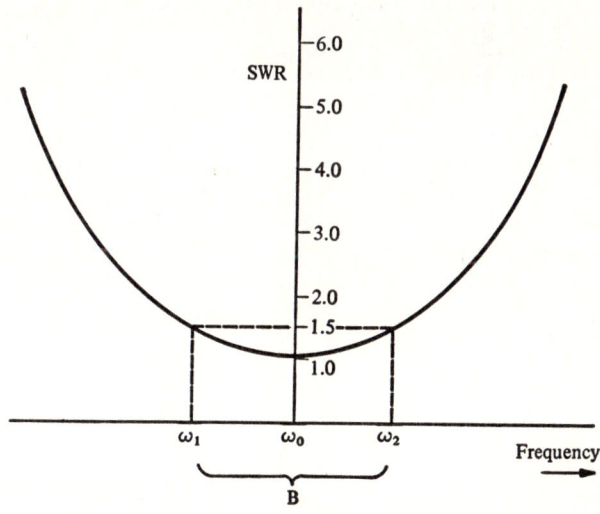

Fig. 5.10.2 The bandwidth of a microwave system is found from the SWR curve.

Example 2 The bandwidth of a double-stub tuner The double-stub tuner of Fig. 5.7.6 was analyzed at its center frequency (300 MHz), and the stub lengths were found, using the Smith chart and the techniques described in Sec. 5.7. The results are shown in the table. The load in this problem was the parallel

No.	Line	SWR at 300 MHz	Length, m
1	Main line	1.0	
2	Stub	∞	0.1966
3	Three-eighth section	1.42	0.3750
4	Stub	∞	0.0655
5	Stub to load	5.5	0.1500

connection of a 250-Ω resistor and a 3.18-pF (or $10/\pi$-pF) capacitor. The normalized load admittance was $0.2 + j0.3$ mho per mho, assuming 50-Ω lossless lines. This system has been analyzed using the computer program of Table 5.10.3 for a frequency range of 200 to 500 MHz. The results are shown in Fig. 5.10.3.

Table 5.10.3 Program double-stub bandwidth

```
C       THIS PROGRAM WAS USED TO FIND SWR AS A FUNCTION OF
C       FREQUENCY FOR THE DOUBLE STUB TUNER PROBLEM WHICH WAS
C       SOLVED IN SECTION 5.7 USING THE SMITH CHART.

C       ******** INSERT DECLARATIONS PACKAGE HERE ********
        COMPLEX CYIN,CYSHRT,CPROP,YSEND,YTERM,CMPLX
        REAL LINEL(5)
        COMMON/DATA/BDATA(45,5)
        READ(5,10)FPZ0,Z0,LINEL,YTERM,F,FMAX,FSTEP
10      FORMAT(8E10.0)
        P0=CPROP(0.0,1.0,FPZ0)
        CALL ENTER(0)
        DO 20 LINE=1,5
        LENGTH=CNVRTM(LINEL(LINE),1.0,FPZ0)
20      CALL STASH
        WRITE(6,30)
30      FORMAT('1',10X,'DOUBLE STUB TUNER BANDWIDTH PROBLEM...'///16X,'FRE
       AQ',5X,'YTERMX',4X,'YTERMQ',4X,'SWR(1)',4X,'SWR(3)',4X,'SWR(5)'//)
40      YSEND=CYIN(1,CYSHRT(2)+CYIN(3,CYSHRT(4)+CYIN(5,YTERM)))
        SWR1=GET1R(1,43)
        SWR3=GET1R(3,43)
        SWR5=GET1R(5,43)
        WRITE(6,60)F,YTERM,SWR1,SWR3,SWR5
60      FORMAT(10X,1P3E10.2,0P3F10.3)
        F=F+FSTEP
        YTERM=CMPLX(REAL(YTERM),AIMAG(YTERM)*(1.0+FSTEP/F))
        IF(F.LE.FMAX) GO TO 40
        CALL EXIT
        END
C       THE DATA CARDS FOR PROGRAM DSTUB2 ARE
           3E8       50.0       0.0       10.0      0.1966    0.3750    0.0655
           .004      .002       1E8       5E8       2E7
```

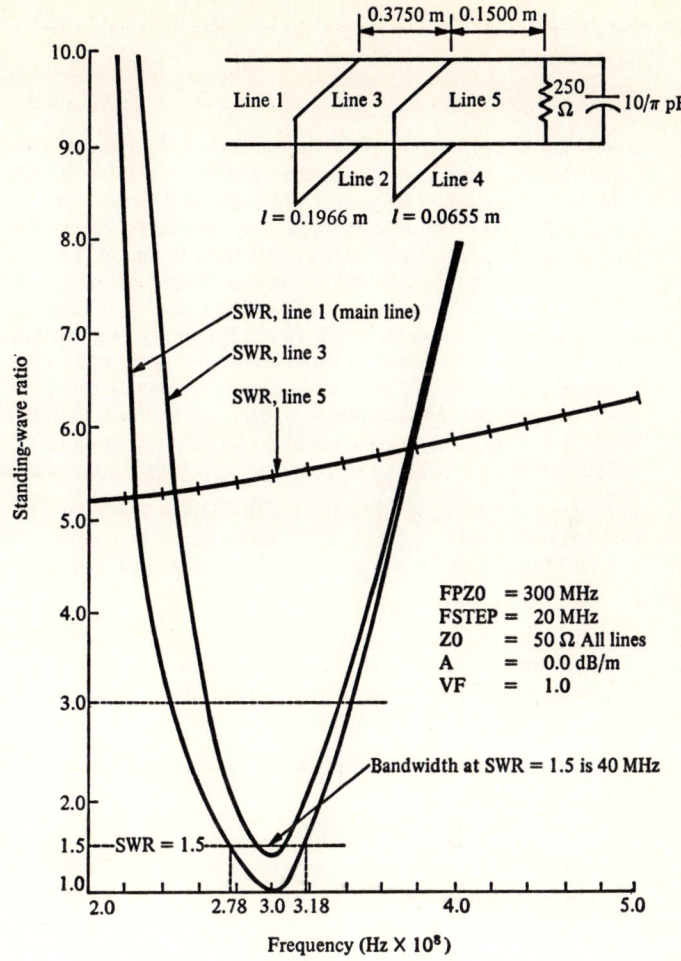

Fig. 5.10.3 SWR and bandwidth of a double-stub tuner.

Over the specified frequency range, the SWR of line 5 changes very little (from 5.2 to 6.3), and this change is due to the increasing susceptance of the capacitor. However, the SWR of lines 1 and 3 changes significantly. The minimum SWR occurs at 300 MHz, as expected, but on the main line, the SWR is 21.0 at 200 MHz and 11.6 at 500 MHz. The bandwidth (for SWR ≤ 1.5) is approximately 40 MHz. Note that tuning improves the SWR over the range of frequencies where the SWR of the main line (line 1) is less than the SWR of line 5, i.e., over a range from about 225 to 375 MHz. At frequencies above or below these two frequencies, the SWR is greater than it would be if the load were connected directly to line 1 with no tuning.

Example 3 The bandwidth of a quarter-wave transformer The use of a quarter-wavelength section of line as an impedance transformer was discussed in Sec. 5.8.

We showed that a load Z_r can be matched to a main line of characteristic impedance Z_{01} if

$$Z_{02} = \sqrt{Z_r Z_{01}} \tag{6}$$

where Z_{02} is the characteristic impedance of the quarter-wavelength section. Thus at the frequency at which the transformer section is truly ¼ wavelength long, i.e., at the center frequency, the SWR on the main line will be unity. At frequencies *above* the center frequency, the transformer section will appear longer than ¼ wavelength, and the SWR on the main line will be greater than unity. At frequencies *below* the center frequency, the transformer section appears shorter than at the center frequency, and again the SWR will be greater than unity. If we graph SWR versus frequency, we can measure the bandwidth from those frequencies where SWR = 1.5, using Eq. (5).

We could find the SWR as a function of frequency using the Smith chart or the Z-θ chart by the techniques of Sec. 5.8, but the process is tedious.† In Table 5.10.4 we have written a simple computer program using SANTLINE to obtain SWR as a function of frequency. This program was used to find the

† A more complete technique using the Smith chart is available, but the work is still somewhat tedious; see H. J. Reich, P. F. Ordung, H. L. Krauss, and J. G. Skalnik, "Microwave Theory and Techniques," D. Van Nostrand Company, Inc., New York, 1953.

Table 5.10.4   Program QWTX

```
C       THIS PROGRAM COMPUTES THE STANDING WAVE RATIO AS A FUNCTION OF
C       FREQUENCY FOR THE QUARTER WAVE TRANSFORMER OF FIGURE 5.8.3(A).

C       ******** INSERT DECLARATIONS PACKAGE HERE ********

        COMMON/DATA/BDATA(45,2)
        COMPLEX CYIN,CPROP,YSEND,YTERM
        WRITE(6,10)
10      FORMAT('1',9X,38HBANDWIDTH OF QUARTER-WAVE TRANSFORMER.///18X,4HFR
       AEQ,8X,4HSWR1,8X,4HSWR2//)
        FPZO=3.0E8
        P0=CPROP(0.0,1.0,FPZ0)
        DO 30 LL=1,2
        READ(5,20)Z0,LENGTH
20      FORMAT(3E10.0)
30      CALL ENTER(LL)
        F=1.0E8
        YTERM=CMPLX(1.0/400.0,0.0)
        DO 50 I=1,21
        YSEND=CYIN(1,CYIN(2,YTERM))
        SWR1=GET1R(1,43)
        SWR2=GET1R(2,43)
        WRITE(6,40)F,SWR1,SWR2
40      FORMAT(10X,1PE12.3,0P2F12.3)
50      F=F+2.0E7
        CALL EXIT
        END

C       THE DATA CARDS FOR PROGRAM QWTX ARE

            50.0         0.0          5.0
          1.4142E2       0.0          0.25
```

Table 5.10.5 Frequency response of the quarter-wave transformer of Fig. 5.8.3a

Frequency	SWR1	SWR2
1.000E 08	6.438	2.828
1.200E 08	5.838	2.828
1.400E 08	5.190	2.828
1.600E 08	4.521	2.828
1.800E 08	3.857	2.828
2.000E 08	3.221	2.828
2.200E 08	2.634	2.828
2.400E 08	2.111	2.828
2.600E 08	1.664	2.828
2.800E 08	1.294	2.828
3.000E 08	1.000	2.828
3.200E 08	1.294	2.828
3.400E 08	1.664	2.828
3.600E 08	2.111	2.828
3.800E 08	2.634	2.828
4.000E 08	3.221	2.828
4.200E 08	3.857	2.828
4.400E 08	4.521	2.828
4.600E 08	5.190	2.828
4.800E 08	5.838	2.828
5.000E 08	6.438	2.828

bandwidth of the system shown in Fig. 5.8.3a. Data cards for that system are listed in Table 5.10.4, and the computed results are given in Table 5.10.5. The computer could be programmed to search by iteration for the points where SWR = 1.5, but that has not been done here. Instead the computed data have been graphed in Fig. 5.10.4. We find that the bandwidth is 64 MHz and the SWR is unity at 300 MHz, the center frequency.

Example 4 Frequency response of two quarter-wave sections In Sec. 5.8 we discussed how the insertion of more than one quarter-wave section between the load and the main line can be used to improve the bandwidth of a matching system. One of the problems encountered was that the characteristic impedance of one of the sections had to be chosen somewhat arbitrarily. Also, the problem of finding bandwidths was even more tedious than for a single quarter-wave section. We are now in a position to use the computer for finding the bandwidth of a given set of quarter-wave sections and also to find the conditions for optimum bandwidth by varying the characteristic impedances.

Figure 5.8.3b shows a system where a load Z_r is to be matched to a characteristic impedance Z_{01} using two quarter-wave sections with characteristic impedances of Z_{02} and Z_{03}. From the figure, and by using Eq. (4) of Sec. 5.8, we can show that

$$Z_{02} = \sqrt{\frac{Z_1}{Z_r}} Z_{03} \tag{7}$$

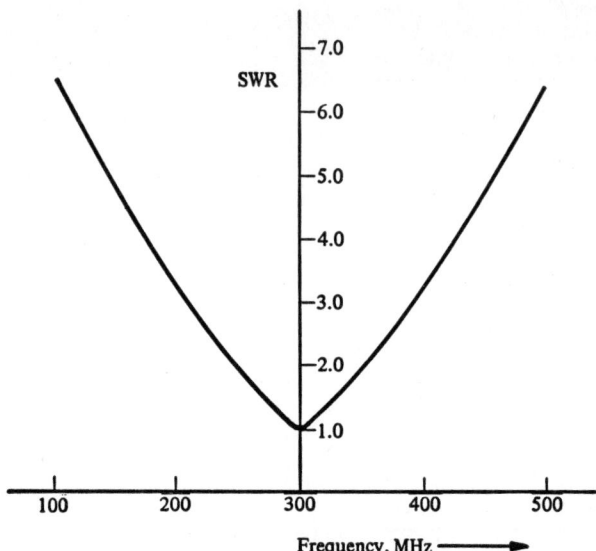

Fig. 5.10.4 SWR and bandwidth of a quarter-wave transformer.

It is necessary to satisfy this relationship between Z_{02} and Z_{03} if matching is to occur.

The computer program of Table 5.10.6 computes SWR as a function of frequency for each section of line of the double quarter-wave transformer. Given a value for Z_{03}, it computes Z_{02} from Eq. (7). For example, when $Z_{03} = 200\ \Omega$, as in Fig. 5.8.3b, the program computes $Z_{02}(70.7\ \Omega)$ and the SWR for the main line and each quarter-wave section over a specified frequency range. In order to compare the double quarter-wave transformer of Sec. 5.8 with the single quarter-wave transformer of that section, the computed results have been plotted in Fig. 5.10.5. Note that the bandwidth of the double transformer is 136 MHz, compared with 64 MHz for the single transformer. The SWR on the main line is unity at 300 MHz for either system.

The computer program of Table 5.10.6 computes SWR as a function of frequency, but also Z_{03} is allowed to vary in incremental steps of 50 Ω, and Z_{02} is computed using Eq. (7). For each pair of values of Z_{02} and Z_{03}, a table of the three SWRs versus frequency is obtained. The computer can be used to plot any of these curves if a digital plotter is available. (The program as listed does not include this feature, since software plotter techniques vary widely.) From these tables and curves, we can plot the bandwidth (for SWR ≤ 1.5) as a function of Z_{03}. The results are given in Fig. 5.10.6 for the system of Fig. 5.8.3b. We find that the optimum values for maximum bandwidth are $Z_{03} = 235\ \Omega$ and $Z_{02} = 83\ \Omega$. The optimum bandwidth is 170 MHz, compared with 64 MHz for the single-section transformer of the previous example. Also, by changing Z_{03} from 200 to 235 Ω (with the appropriate change in Z_{02}) the bandwidth was increased from 136 to 170 MHz. Note from Fig. 5.10.6 that it is also possible

Table 5.10.6 Program QWTX2

```
C     THIS PROGRAM COMPUTES THE STANDING WAVE RATIO AS A FUNCTION OF
C     FREQUENCY FOR THE DOUBLE QUARTER WAVE TRANSFORMER SHOWN
C     IN FIGURE 5.8.3(B).

C     ******** INSERT DECLARATIONS PACKAGE HERE ********

      COMMON/DATA/BDATA(45,3)
      COMPLEX CYIN,CPROP,YSEND,Z2,Z3,YTERM
      FPZ0=3.5E8
      Z0=50.0
      P0=CPROP(0.0,1.0,FPZ0)
      LENGTH=5.0
      CALL ENTER(1)
      LINE=2
      LENGTH=0.25
      CALL STASH
      LINE=3
      CALL STASH
      Z3=50.0
      YTERM=CMPLX(1.0/400.0,0.0)
      DO 40 NR=1,8
      CALL MODIFY(3,1,Z3,2)
      Z2=Z3/2.8284
      CALL MODIFY(2,1,Z2,2)
      WRITE(6,10)Z3,Z2
10    FORMAT(1H1,9X,42HDOUBLE QUARTER-WAVE TRANSFORMER BANDWIDTH.////10X
     A,5HZ03=  ,1PE9.2,3H +J,E9.2,5X,5HZ02=  ,E9.2,3H +J,E9.2///18X,4HFREQ
     B,8X,4HSWR1,8X,4HSWR2,8X,4HSWR3/)
      F=1.0E8
      DO 30 NF=1,21
      YSEND=CYIN(1,CYIN(2,CYIN(3,YTERM)))
      SWR1=GET1R(1,43)
      SWR2=GET1R(2,43)
      SWR3=GET1R(3,43)
      WRITE(6,20)F,SWR1,SWR2,SWR3
20    FORMAT(10X,1PE12.3,0P3F12.3)
30    F=F+2.0E7
40    Z3=Z3+50.0
      CALL EXIT
      END

C     THERE ARE NO DATA CARDS FOR PROGRAM QWTX2.
```

to decrease bandwidth to even less than could be obtained from a single transformer, depending upon the choice of Z_{02} and Z_{03}.†

One of the interesting characteristics of lossless quarter-wave transformers with resistive loads is that the SWR is a periodic function of frequency. For example, if a section is $\frac{1}{4}$ wavelength at 300 MHz in a matched system, then at 900 MHz it will be $\frac{3}{4}$ wavelength and the system will again be matched. Similarly, matching will occur at frequencies where the length is $5\lambda/4$, $7\lambda/4$, $9\lambda/4$, etc. At even multiples of $\lambda/4$, the impedance transformation due to one section corresponds to one revolu-

† See also Quarter-wave Matching Sections in "Reference Data for Radio Engineers," 4th ed., International Telephone and Telegraph Corporation, New York, 1956.

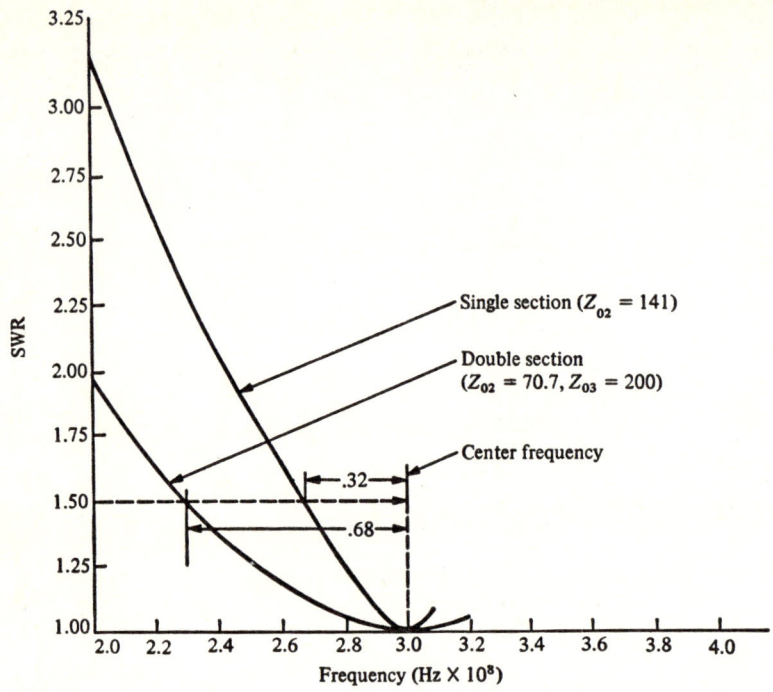

Fig. 5.10.5 SWR and bandwidth of quarter-wave transformers (see Fig. 5.8.3).

tion on the Smith chart. Consequently, it appears as though there were no transformer in the system.

In most practical cases, the load impedance is frequency-dependent and the lines are lossy, especially at high frequencies. Therefore the SWR is seldom a periodic function of frequency in practice.

In the above examples we have considered systems using one and two quarter-wave sections for matching, and we noted that bandwidth could be improved as we increased the number of sections. Can we obtain a similar increase if we add more sections? The answer is that we can obtain some improvement by adding more sections, although we exponentially approach a limiting bandwidth. In practice, little improvement is obtained if more than four sections are used. One could investigate these cases and find the optimum system using the computer. Another technique sometimes used for matching is to vary the characteristic impedance of the matching transformer along its length. This can be done by changing the medium inside the coaxial cable or waveguide or by changing its dimensions. The most common example of this type is the exponential

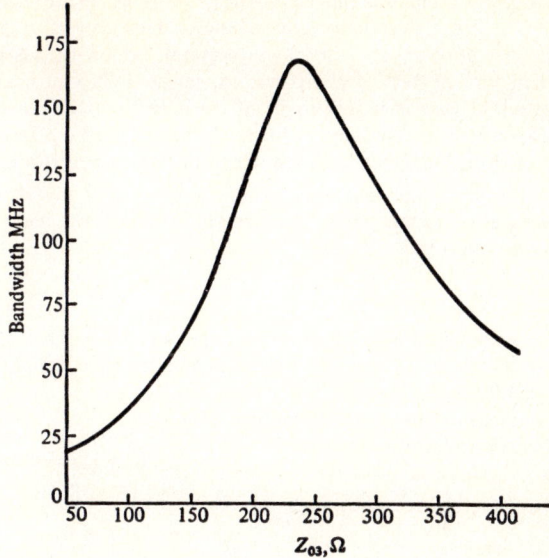

Fig. 5.10.6 Bandwidth of the double quarter-wave transformer.

section where the characteristic impedance is

$$Z_0(z) = Z_0 e^{-\Upsilon z}$$

Here Z_0 and Υ are constants selected to obtain a match between a given load and a main line (see Programmed Exercise 5.8.1). One can approximate the exponential line by using several quarter-wave sections where the change in characteristic impedance from one section to another follows the exponential equation approximately.

Many special cases of matching systems to bandwidth requirements have been investigated and are discussed in the various texts and in the literature of microwave systems.

PROBLEMS

1. Use the equivalent circuit to derive the telegrapher's equations for the leakage-free noninductive cable having the incremental equivalent circuit illustrated.

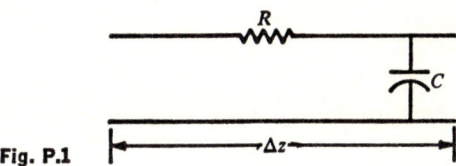

Fig. P.1

2. A secret telephone line between the boys' dorm and the girls' dorm normally terminates in its characteristic impedance. However, the girls often hang purses, coat hangers, clothes, etc., on their terminal, which causes great variation in the load. When you protest, a girl phys. ed. major tells you that (a) the voltage V^+ in the sinusoidal case is independent of load, (b) for this reason the transmitted power is independent of how they terminate their end, and (c) that engineers are intelligent and make good husbands. Prove which of her statements are true.

3. Consider a sonar system where a sinusoidal pressure wave propagates through seawater. Assume that on a straight-line path the wave acts as a wave on a lossless transmission line. Define

p = pressure, N/m² (analogous to voltage)
v = velocity of particle, m/s (analogous to current)
k = compressibility constant (analogous to capacitance)
 = 4.90×10^{-10} m²/N for seawater
ρ = mass density (analogous to inductance)
 = 10^3 kg/m³ for seawater

The telegrapher's equations are

$$\frac{\partial v}{\partial z} = -k \frac{\partial p}{\partial t} \qquad \frac{\partial p}{\partial z} = -\rho \frac{\partial v}{\partial t}$$

Find the velocity of propagation and the characteristic impedance for seawater.

4. The constants R, L, C, and G of a transmission line are

$R = 3.00$ Ω/km $\qquad L = 4.00$ mH/km
$C = 16.00 \times 10^{-3}$ F/km $\qquad G = 12.00$ mho/km

(a) Calculate α, β, γ, and Z_0. (b) Sketch the variation of α and β versus frequency. (c) Calculate the voltage 100 m from the source if the line is terminated in Z_0 and the source voltage is 10 V rms at $\omega = 10^3$ rad/s.

5. On a 60-c power line, a short circuit occurs suddenly at a distance of 750 km (about 466 mi) from the generating station. The circuit breaker at the station will open if the current exceeds 120 A. Prove that if the breaker does not open during the transient period, it will not open. How many microseconds does the breaker have to react to the transient?

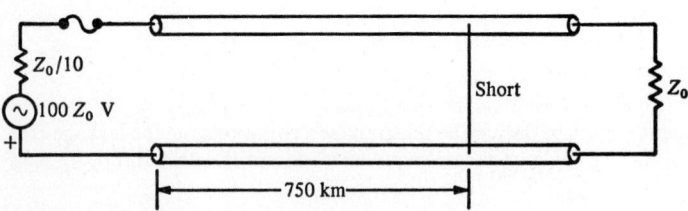

Fig. P.5

$C = 0.8 \text{ mfd}$

6. The generator in the system shown has a frequency of 1000 Hz. Find the rms current flowing through the capacitor.

SINUSOIDAL WAVES ON LOSSLESS TRANSMISSION LINES

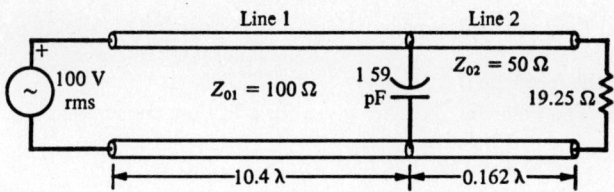

Fig. P.6

7. The Γ_r plane diagram for the load end of a lossless line is shown. (a) Find the distance to the first voltage maximum and minimum in wavelengths and sketch the voltage phasors. (b) Repeat for current.

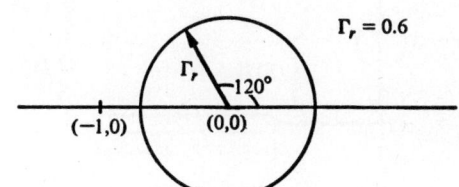

Fig. P.7

8. A transmission line is terminated in a capacitance C_0, and a sinusoidal generator of frequency f_0 is connected to the sending end. Measurements show that the maximum voltage on the line is V_m V (rms) and the capacitor voltage is $0.707 V_m$.
  (a) Where does the first voltage minimum (nearest C_0) occur?   (**Ans.** $z = -\lambda/8$.)
  (b) Where does the first current minimum occur?   (**Ans.** $z = -3\lambda/8$.)
  (c) Sketch the impedance and rms voltage and current for 2 wavelengths from the load.

9. The "lumped" capacitors and inductors of normal electronic circuits perform poorly at high frequencies. However, a transmission line terminated in a short circuit (called a shorted stub) can be used for an inductor or a capacitor at high frequencies. Suppose that 50-Ω coaxial cable ($Z_0 = 50\ \Omega$) is available for use at a frequency of 10^9 Hz. Assume $v_p = 3 \times 10^8$ m/s. How many centimeters long should a shorted stub be to have an input impedance of (a) $Z_{\text{in}} = j50\ \Omega$; (b) $Z_{\text{in}} = j500\ \Omega$; (c) $Z_{\text{in}} = -j50\ \Omega$? Calculate the equivalent value of capacitance or inductance in each case. Make a sketch of reactance similar to the one in Fig. 5.3.7.

10. Repeat Prob. 9 using an open-circuited stub.

11. Suppose the accuracy in cutting the length of the tuning stub of Prob. 9 were ± 1 cm. What range of impedances could one expect for parts (a) and (b) of Prob. 9?

12. Prove that (a) $Z_{\text{max}} = \text{SWR}(Z_0)$; (b) $Z_{\text{min}} = Z_0/\text{SWR}$; (c) $Z_0 = \sqrt{Z_{\text{max}} Z_{\text{min}}}$.

13. The known data on a lossless line are

  $\text{SWR} = 4.3$
  $Z_0 = 100\ \Omega$
  $v_p = 2 \times 10^8$ m/s

  First voltage null at 0.918 m from load
  Second voltage null at 2.218 m from load

(a) Find the load impedance. (b) By connecting a variable reactance (real part zero) in parallel with the load, what is the minimum SWR that can be obtained? (c) What is the frequency? d) VSWR if X series

14. The following data are given for a lossless transmission line:

$\omega = 10^9$ rad/s $Z_0 = 20\ \Omega$
$\beta = 4$ rad/m   Line length $= 1.961$ m

(a) How many wavelengths long is the line? (b) If $Z_r = 20 + j30\ \Omega$, what is Γ_r? (c) If $Z_r = 60 + j0\ \Omega$, what is Z_{in}? (d) If $\Gamma_r = -0.25/0°$, what are Z_r and Z_{in}? (e) How far from the load is the first voltage maximum in cases (b) to (d)?

15. A 50-Ω line $3\lambda/4$ long is connected between an antenna and a receiver as illustrated. (a) Find $\Gamma(-L)$ and Z_{in}. (b) Find V_{in} and I_{in}. (c) Find $\langle P \rangle$ dissipated in the load.

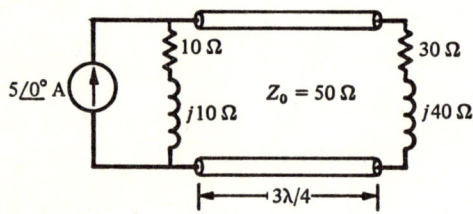

Fig. P.15

16. A 100-Ω transmission line is terminated in an unknown load. However, the measured SWR is 2.60, and the first voltage minimum is 0.140λ from the load. Find the load impedance.

17. A 50-Ω transmission line is terminated in an impedance of $25 - j25\ \Omega$ and is fed by a source of 100 V rms with an internal impedance of $25 + j25\ \Omega$. Find the rms current at the midpoint of the line if the line is $\frac{1}{2}$ wavelength long.

18. Prove that maximum power is obtained at the load if the load is adjusted so that the input impedance of a transmission line is the complex conjugate of the source impedance.

19. A 50-Ω line is 0.290λ long and is terminated in an impedance at $4.0 + jX\ \Omega$, where the reactance X is variable. The reactance is adjusted to obtain maximum time-average power at the load. The generator has an open-circuit voltage of 100 V rms and an internal impedance of $20 + j100\ \Omega$. Find the value of X for maximum power. Compute the time-average power for $X = 5, 10,$ and $15\ \Omega$ and compare your answers.

20. A 50-Ω line transmits a power $\langle P \rangle = 20$ W to a load. What maximum rms voltage must the cable withstand if the SWR ≤ 2.50?

21. A 50-Ω quarter-wavelength section of transmission line is shorted at the load. The voltage at the input end of the line is

$$V_{in} = 10 \cos \omega_0 t + 10 \cos 3\omega_0 t \quad \text{V}$$

Sketch the rms voltage at each frequency and find the total rms voltage at (a) one-third the distance from the load to the source, (b) two-thirds the distance, and (c) at the input to the line.

22. (a) Prove that, for complex power $\langle S \rangle = \langle P \rangle + jQ$,

$$\langle S \rangle = (V^+)^2 \frac{Y_0}{2}(1 - |\Gamma_r|^2) + jY_0(V^+)^2 b(z)$$

where $\Gamma(z) = a(z) + jb(z)$. (b) Show that $\langle P \rangle = Y_0|V_{max}||V_{min}|$.

SINUSOIDAL WAVES ON LOSSLESS TRANSMISSION LINES

23. The double-stub tuning method is used to match a load of $100 + j60\ \Omega$ to a $100\text{-}\Omega$ line. The distance between the stubs is $3\lambda/8$, and the first stub is connected across the load. Show that the minimum possible lengths for the shorted stubs are $L_{T_1} = 0.092\lambda$ and $L_{T_2} = 0.066\lambda$.

24. Two shorted stubs are to be used to match a load of $25 - j50\ \Omega$ to a $50\text{-}\Omega$ line. The stubs are separated by a distance of $3\lambda/8$, and the first stub is located $\lambda/8$ m from the load. (a) On a Smith chart locate and crosshatch the forbidden region for the load. (b) If the load lies in the forbidden region, add a third shorted stub (at the load) to move the load out of the forbidden region. Find the lengths of all the stubs.

25. A $50\text{-}\Omega$ line is terminated in an unknown impedance. Measurements show that SWR $= 5.0$ and the first voltage minimum is 0.1λ from the load. (a) Find the load impedance. (b) Reduce the SWR to unity by the connection of a lumped susceptance across the line as near as possible to the load. Find the value of the susceptance and its location. (c) Replace the lumped susceptance with a short-circuited stub. How long should the stub be? (d) What length of open-circuited stub would accomplish the same result?

26. While touring a foreign military installation you note that the single-stub tuning method is used to match a standard $50\text{-}\Omega$ cable ($v_p = 2.4 \times 10^8$ m/s) to a receiver known to have an input impedance of $200\underline{/0°}\ \Omega$. You estimate that the stub is 1.6 m long and is about 3.0 m from the receiver. What are the probable frequencies of operation?

27. An old section of $150\text{-}\Omega$ underground cable is being replaced by a new $50\text{-}\Omega$ cable. At point A the cables will be spliced and buried. At point B the new cable will terminate in a $25\text{-}\Omega$ load inside a relay station. A stub of $50\text{-}\Omega$ cable can be connected at point B, if necessary, but not at point A. We can cut the section AB to any length L within the bounds $60\lambda \leq L \leq 61\lambda$. (a) How can the SWR on the main line ($150\text{-}\Omega$ cable) be reduced to unity? (b) When the main line is matched, what is the SWR on the new section? (c) Is a stub necessary? If so, how long should it be? (d) What is the required length L?

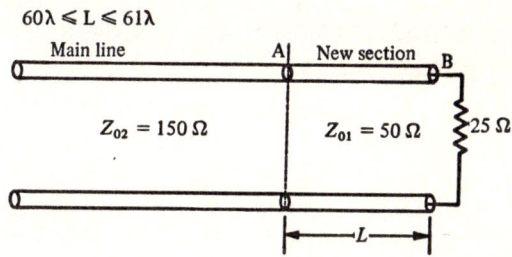

Fig. P.27

28. Show that the normalization procedure used by the function CYIN (where the line input voltage is $1.0 + j0.0$) in SANTLINE yields the correct line voltages when the procedures described in CVLØAD are applied. For a specific case, consider a line with a shorted stub tuner (there are three sections of line).

6
Waves on Dissipative Lines

6.1 INTRODUCTION TO DISSIPATIVE LINES

In the preceding two chapters we derived and discussed many of the properties of waves propagating on lossless transmission lines. However, it is an idealization to neglect losses in many practical transmission lines. For sufficiently short segments of line, we can neglect losses, and then the techniques of Chaps. 4 and 5 can be used. Otherwise, we must carefully take into account the effects of such losses on the waves which propagate on the line. If losses are not considered negligible on a line, the line is said to be *dissipative* or *lossy*.

The equations of the dissipative line were derived and tabulated in Sec. 5.2. Sinusoidal time variations will be assumed throughout this chapter, and the phasor equations for a lossy line are repeated below for convenience.

$$V(z) = V^+(e^{-\gamma z} + \Gamma_r e^{\gamma z}) \tag{1}$$

$$I(z) = I^+(e^{-\gamma z} - \Gamma_r e^{\gamma z}) \tag{2}$$

WAVES ON DISSIPATIVE LINES

$$\gamma = \sqrt{(R + j\omega L)(G + j\omega C)} = \alpha + j\beta \tag{3}$$

$$Z_0 = \sqrt{\frac{R + j\omega L}{G + j\omega C}} = |Z_0|\underline{/\theta_0} \tag{4}$$

The impedance $Z(z)$ at any point on the transmission line is the ratio of phasor voltage to phasor current, given by

$$Z(z) = \frac{Z_0(e^{-\gamma z} + \Gamma_r e^{\gamma z})}{e^{-\gamma z} - \Gamma_r e^{\gamma z}} = \frac{Z_0(1 + \Gamma_r e^{2\gamma z})}{1 - \Gamma_r e^{2\gamma z}} \tag{5}$$

The voltage reflection coefficient at a point, defined as the ratio of the negative-going voltage wave to the positive-going voltage wave, can be obtained from Eq. (1), and is similar to Eq. (5) of Sec. 5.4:

$$\Gamma(z) = \frac{V^- e^{\gamma z}}{V^+ e^{-\gamma z}} = \Gamma_r e^{2\gamma z} \tag{6}$$

For dissipative lines, $\alpha = \operatorname{Re} \gamma \neq 0$, and it can be seen from Eqs. (5) and (6) that as z approaches minus infinity, $Z(z)$ approaches Z_0 and $\Gamma(z)$ approaches 0. The reason is that as z becomes more and more negative, i.e., as we approach the source end, the incident wave becomes larger and the reflected wave becomes smaller. Thus a long section of lossy line effectively isolates the load from the generator and has an input impedance of Z_0. The price paid for this isolation is a generator large enough to supply sufficient power for both the line losses and the load requirements. The greater the line losses with respect to the power requirements of the load, the greater the isolation achieved between the generator and the load.

The characteristic impedance of a lossy line, in general, has both real and reactive components, since $R \neq 0$ or $G \neq 0$, or both. The angle θ_0 of the characteristic impedance, from Eq. (4), is

$$\theta_0 = \tfrac{1}{2}\left(\tan^{-1}\frac{\omega L}{R} - \tan^{-1}\frac{\omega C}{G}\right) \tag{7}$$

and if $L/R = C/G$, the characteristic impedance is real. With many types of cable in use now, however, the conductance of the dielectric material between conductors can be neglected but the resistance of the wires cannot. For transmission lines having small conductance, the angle of the characteristic impedance can be made smaller artificially by increasing the inductance of the line. This can be accomplished by putting many small inductors in series at regularly spaced intervals on the line.

This technique, called *inductive loading*, was commonly used in the early part of this century on open-wire telephone lines as well as cables, in an attempt to obtain distortionless lines. For many low-loss lines, however, the characteristic impedance has a negligibly small angle.

Unlike lossless lines, on a dissipative line it is possible for the magnitude of the reflection coefficient to exceed unity. We now find the maximum magnitude of the reflection coefficient by considering extreme values for the angles of the characteristic impedance Z_0 and the load impedance Z_r. The angle of Z_0 cannot exceed $\pm 45°$ (why?), and the angle of Z_r cannot exceed $\pm 90°$. Thus the angle of the normalized load impedance Z_{nr} must be within the range $\pm 135°$. If $Z_{nr} = N/\underline{\pm 135°}$, the reflection coefficient at the load is

$$\Gamma(0) = \frac{N/\underline{\pm 135°} - 1}{N/\underline{\pm 135°} + 1} = \frac{-N/\sqrt{2} \pm jN/\sqrt{2} - 1}{-N/\sqrt{2} \pm jN/\sqrt{2} + 1} \tag{8}$$

Therefore

$$|\Gamma(0)| = \left(\frac{1 + N^2 + N\sqrt{2}}{1 + N^2 - N\sqrt{2}}\right)^{1/2} = \left(1 + \frac{2\sqrt{2}}{N + 1/N - \sqrt{2}}\right)^{1/2} \tag{9}$$

The reflection coefficient has its greatest magnitude when $N + 1/N$ is a minimum, which occurs when $N = 1$. The resulting maximum value of $|\Gamma(z)|$ is

$$|\Gamma(0)| = 1 + \sqrt{2} \tag{10}$$

For a graphical solution of some problems on dissipative lines, it is necessary to use a special Smith chart extended beyond $|\Gamma(z)| = 1.0$, but extension beyond $|\Gamma(z)| = 1 + \sqrt{2}$ will never be necessary. It should be noted that the reflection coefficient can achieve this extreme magnitude *only at the load* on a dissipative line in which $Z_{nr} = 1/\underline{\pm 135°}$. At all points nearer the source, the magnitude of the reflection coefficient must be less than $1 + \sqrt{2}$ because of the attenuation constant.

It may seem that conservation of energy is violated if the reflection coefficient exceeds unity, but we now show that this is not the case. With voltage and current given by

$$V(z) = V^+(e^{-\alpha z}e^{-j\beta z} + \Gamma_r e^{\alpha z}e^{j\beta z}) \tag{11}$$

and

$$I(z) = Y_0 V^+(e^{-\alpha z}e^{-j\beta z} - \Gamma_r e^{\alpha z}e^{j\beta z}) \tag{12}$$

the total complex power at any point on the line is

$$\langle S(z) \rangle = \tfrac{1}{2} V(z) I^*(z) = \langle P(z) \rangle + jQ(z) \tag{13}$$

WAVES ON DISSIPATIVE LINES

where $I^*(z)$ denotes the complex conjugate of $I(z)$. Then if $\Gamma_r = |\Gamma_r|e^{j\rho}$ and $Y_0 = G_0 + jB_0$, it can be shown by expanding Eq. (13) that the complex power is given by

$$\langle S(z)\rangle = \frac{|V^+|^2}{2}\{G_0(e^{-2\alpha z} - |\Gamma_r|^2 e^{2\alpha z}) + 2B_0|\Gamma_r|\sin(2\beta z + \rho)$$
$$+ j[2G_0|\Gamma_r|\sin(2\beta z + \rho) - B_0(e^{-2\alpha z} - |\Gamma_r|^2 e^{2\alpha z})]\} \quad (14)$$

The power $\langle P(z)\rangle$ is the real part of the complex power $\langle S(z)\rangle$ and is composed of the first three terms of Eq. (14). The first term is the power in the incident wave, the second term is the power in the reflected wave, and the third term is due to interaction of the two waves. This interaction can occur only on transmission lines which have a complex characteristic impedance, a condition also required for the reflection coefficient to exceed unity. On lines like this, total power at a point is *not* the superposition of the power in the positive- and negative-going waves.

6.2 STANDING WAVES

Sinusoidal voltages and currents on a lossy line can be determined by phasor techniques similar to those used for a lossless line. The difference is that the phasor magnitudes change with distance due to the nonzero attenuation constant. The circuit in Fig. 6.2.1 will be used as an example to illustrate phasor techniques for a lossy line and to obtain a standing-wave pattern.

The voltage phasor diagram at any point on the line can be found by substituting the proper value of z into the relations for positive and negative-going phasor voltages given by

$$V^+(z) = V^+ e^{-\alpha z} e^{-j\beta z} \quad (1)$$

and

$$V^-(z) = V^- e^{\alpha z} e^{j\beta z} \quad (2)$$

Since $\Gamma_r = 0.5\underline{/0°}$, the voltage phasor diagram at $z = 0$ is as shown in Fig. 6.2.2, assuming the load voltage is the reference. Figure 6.2.3 shows

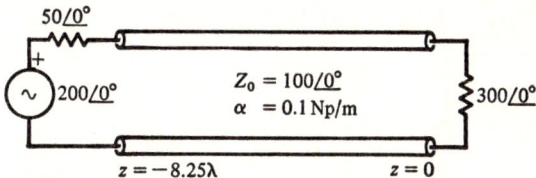

Fig. 6.2.1 The circuit used to illustrate phasors on a dissipative line.

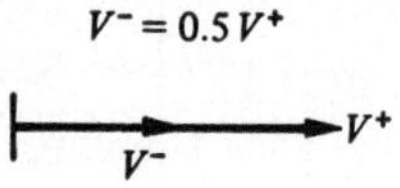

$V^- = 0.5 V^+$

Fig. 6.2.2 Voltage phasor diagram at $z = 0$.

the voltage at $z = -8.25\lambda$, where the load voltage is still the reference. It can be seen that the magnitude of the positive-going phasor voltage is larger, and the magnitude of the negative-going voltage phasor smaller, at the generator end than at the load. The input impedance for this example is found from Eq. (5) of Sec. 6.1:

$$Z(-8.25\lambda) = Z_{in} = \frac{100(1 + 0.5e^{-1.65}e^{-j33\pi})}{1 - 0.5e^{-1.65}e^{-j33\pi}} = 82.5 \, \Omega$$

The sending-end voltage is

$$V(-8.25\lambda) = \frac{82.5}{50 + 82.5}(200\underline{/0°}) = 124.5\underline{/0°}$$

However, from Fig. 6.2.3, the sending-end voltage can also be expressed as $V(-8.25\lambda) = V^+(2.28 - 0.2195)\underline{/90°} = 124.5\underline{/0°}$, which implies that $V^+ = 60.4\underline{/-90°}$. Therefore $V^- = \Gamma_r V^+ = 30.2\underline{/-90°}$ at $z = 0$.

Thus, if the voltages and currents on the line were expressed in terms of the given reference, the generator voltage, it would probably be helpful to redraw the diagrams of Figs. 6.2.2 and 6.2.3 so that V^+ and V^- are at an angle of $-90°$. Then the generator voltage (and the sending-end voltage) would be drawn at an angle of $0°$, as expected. This is equivalent to rotating the diagrams of Figs. 6.2.2 and 6.2.3 clockwise through $90°$. Unfortunately, the angles of V^+ and V^- with respect to the generator voltage cannot be found until the positive- and negative-going phasors are translated from the load to the generator.

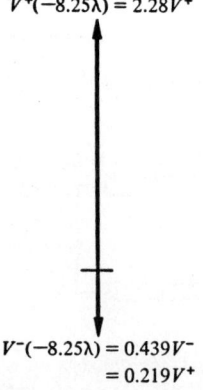

Fig. 6.2.3 Voltage phasor diagram at $z = -8.25\lambda$.

WAVES ON DISSIPATIVE LINES

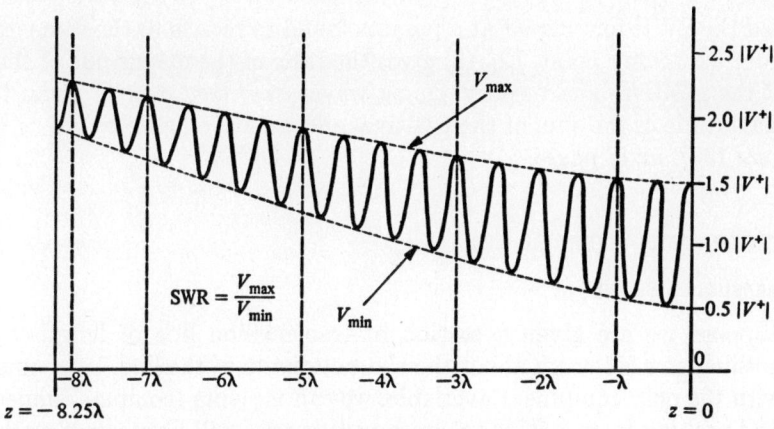

Fig. 6.2.4 Voltage standing-wave pattern for the example.

If voltage or current phasor diagrams are drawn for every fraction of a wavelength on a line and the magnitude of the phasor voltage or current is plotted versus distance on the line, then the voltage or current standing wave is obtained. The voltage standing-wave pattern for this example is shown in Fig. 6.2.4.

In general, the envelope of the maximum phasor voltage is given by

$$V_{\max} = |V^+|e^{-\alpha z} + |V^-|e^{\alpha z} = |V^+|(e^{-\alpha z} + |\Gamma_r|e^{\alpha z}) \tag{3}$$

and the envelope of the minimum phasor voltage is given by

$$V_{\min} = |V^+|e^{-\alpha z} - |V^-|e^{\alpha z} = |V^+|(e^{-\alpha z} - |\Gamma_r|e^{\alpha z}) \tag{4}$$

These envelopes are the dotted lines illustrated in Fig. 6.2.4. From Eqs. (3) and (4) it can be seen that as z approaches minus infinity, the maximum and minimum voltage envelopes approach $|V^+|e^{-\alpha z}$, so that the standing-wave amplitude no longer appears to oscillate but has an exponential form. This is due to the gradual attenuation of the negative-going wave and the ever increasing magnitude of the positive-going wave as the generator is approached on a long lossy line.

For transmission lines in which losses are not negligible, it is convenient to define the SWR as the ratio of the maximum voltage *envelope* to the minimum voltage *envelope* at the point of interest. Thus the SWR at a point is given by

$$\boxed{\mathrm{SWR}(z) = \frac{e^{-\alpha z} + |\Gamma_r|e^{\alpha z}}{e^{-\alpha z} - |\Gamma_r|e^{\alpha z}} = \frac{1 + |\Gamma_r|e^{2\alpha z}}{1 - |\Gamma_r|e^{2\alpha z}}} \tag{5}$$

and the SWR for current at a point is found to reduce to the same expression. At every point, Eq. (5) gives the ratio of the magnitude of the sum of the positive- and negative-going waves *as if they were in phase* to the magnitude of the sum of the positive- and negative-going waves *as if they were 180° out of phase*.

PROGRAMMED EXERCISE 6.2.1

MEASURING Z_0 AND γ

Suppose we are given a section of transmission line of length L, and nothing else is known about the characteristics of the line. Assume that with the only equipment available, we can measure (complex) impedance and nothing more. This programmed exercise will illustrate a method of determining Z_0 and α, as well as a countable set of possible values of β. We will find that additional information is required to determine β uniquely.

We first terminate the line in a known (or measurable) impedance Z_{r_1} and measure the corresponding input impedance $Z_1(-L)$; then we terminate the line in another known impedance $Z_{r_2} \neq Z_{r_1}$ and measure the new input impedance $Z_2(-L)$. The reader should verify each of the following true statements.

1. By definition of $\Gamma(z)$,

$$e^{-2\gamma L} = \frac{\Gamma_1(-L)}{\Gamma_{r_1}} = \frac{\Gamma_2(-L)}{\Gamma_{r_2}} \quad (6)$$

2. Writing Γ_{r_1} and Γ_{r_2} in terms of Z_{r_1} and Z_{r_2} respectively (and Z_0, of course) and writing $\Gamma_1(-L)$ and $\Gamma_2(-L)$ in terms of $Z_1(-L)$ and $Z_2(-L)$ respectively and cross-multiplying twice, we obtain a single equation with only Z_0 unknown

$$Z_0{}^3[Z_{r_1} + Z_2(-L) - Z_{r_2} - Z_1(-L)]$$
$$+ Z_0\{[Z_{r_1} + Z_2(-L)][Z_{r_2}Z_1(-L)]$$
$$- [Z_{r_2} + Z_1(-L)][Z_{r_1}Z_2(-L)]\} = 0 \quad (7)$$

3. The unique nontrivial solution of Eq. (7) is

$$Z_0 = \left\{ \frac{[Z_{r_2} + Z_1(-L)][Z_{r_1}Z_2(-L)] - [Z_{r_1} + Z_2(-L)][Z_{r_2}Z_1(-L)]}{Z_{r_1} + Z_2(-L) - Z_{r_2} - Z_1(-L)} \right\}^{1/2} \quad (8)$$

for *any* pair of terminations Z_{r_1} and Z_{r_2}.

WAVES ON DISSIPATIVE LINES

4. If a perfect short circuit and a perfect open circuit could be obtained for Z_{r_1} and Z_{r_2} respectively, then

$$Z_0 = [Z_1(-L)Z_2(-L)]^{1/2} \tag{9}$$

and Z_0 is obtained rather easily.

5. While it is usually possible to obtain a nearly perfect short circuit, it is often difficult to obtain a very close approximation to an open circuit (why?). It is more realistic, therefore, to set $Z_{r_1} = 0$ and *not* let Z_{r_2} approach infinity, as in deriving Eq. (9). Then

$$Z_0 = \left[\frac{Z_{r_2}Z_1(-L)Z_2(-L)}{Z_{r_2} + Z_1(-L) - Z_2(-L)}\right]^{1/2} \tag{10}$$

In summary, one of the equations (8) to (10) will allow us to find Z_0, providing we can measure impedance.

6. Now that Z_0 is known, we can evaluate and simplify either of the ratios in Eq. (6) to a known complex number $A + jB$

$$e^{-2\gamma L} = \frac{\Gamma_1(-L)}{\Gamma_{r_1}} = A + jB \tag{11}$$

7. Using Euler's identity in expanding the exponential and equating real parts and imaginary parts respectively, we obtain the following two equations in the unknowns α and β:

$$e^{-2\alpha L} \cos 2\beta L = A \tag{12}$$

and

$$e^{-2\alpha L} \sin 2\beta L = -B \tag{13}$$

Squaring both sides of Eqs. (12) and (13) and adding, we find that

$$e^{-4\alpha L} = A^2 + B^2 \tag{14}$$

so that

$$\alpha = \frac{-1}{4L} \ln(A^2 + B^2) \tag{15}$$

Thus the attenuation constant has been found.

8. We can attempt to find β by dividing Eq. (12) by Eq. (13):

$$\cot 2\beta L = -\frac{A}{B} \tag{16}$$

Since $\cot 2\beta L = \cot (2\beta L \pm n\pi)$ for $n = 0, 1, 2, \ldots$, we cannot determine β uniquely. However, β must be one of the following numbers.

$$\beta = \frac{1}{2L}\left[\cot^{-1}\left(-\frac{A}{B}\right) \pm n\pi\right] \quad \text{for } n = 0, 1, 2, \ldots \quad (17)$$

Therefore it is not possible to determine β uniquely from impedance measurements alone.

The interested reader should attempt to answer the following questions:

1. Would additional known terminations, for example, Z_{r_s} yielding the input impedance $Z_3(-L)$, etc., aid in determining β?
2. If not, what sort of experiment must be performed to determine β uniquely?

6.3 GRAPHICAL TECHNIQUES

The Smith chart and the Z-θ chart are useful for finding the admittance and impedance on lossy lines as well as for lossless lines. For lossy lines, however, the magnitude of the reflection coefficient decreases exponentially toward zero as the distance from the load increases. The generalized reflection coefficient for lossy lines was defined in Sec. 6.1, and is given by $\Gamma(z) = \Gamma_r e^{2\alpha z} e^{j2\beta z}$. Thus, if a load impedance is located on the Smith chart or the Z-θ chart, the impedance path to the generator end of the line is not a circle of constant SWR but an exponential spiral approaching the center of the chart, where $\Gamma(z) = 0$ and SWR $= 1.0$.

As an example of graphical techniques for lossy lines, the sending-end impedance and the SWR at each end of the line will be found for the circuit shown in Fig. 6.3.1. The normalized load impedance is

$$Z_n(0) = 2 - j2$$

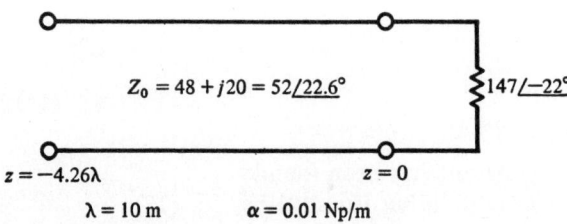

Fig. 6.3.1 The circuit used to demonstrate graphical techniques on a lossy line.

WAVES ON DISSIPATIVE LINES

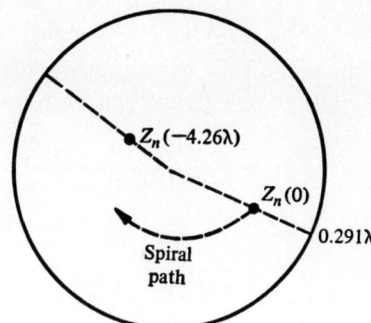

Fig. 6.3.2 Determination of the sending-end impedance on the Smith chart.

and the attenuation constant can be expressed in units of nepers per meter, or in units of nepers per wavelength as follows:

$$\alpha = (0.01 \text{ Np/m})(10 \text{ m/}\lambda) = 0.1 \text{ Np/}\lambda$$

Therefore

$$|\Gamma(-4.26\lambda)| = |\Gamma_r|e^{-2(0.1)(4.26)} = |\Gamma_r|(0.426)$$

Since the SWR at the load is 4.4,

$$|\Gamma_r| = \frac{4.4 - 1}{4.4 + 1} = 0.63$$

and

$$|\Gamma(-4.26\lambda)| = 0.268$$

The SWR at the sending end is therefore

$$\text{SWR}(-4.26\lambda) = \frac{1.268}{0.732} = 1.73$$

and the Smith chart determination of the sending-end impedance is illustrated in Fig. 6.3.2.

On the chart, the path connecting $Z_n(0)$ and $Z_n(-4.26\lambda)$ is a spiral of 8.52 complete turns. It is seen that

$$Z_n(-4.26\lambda) = 0.62 + j0.23 = 0.661\underline{/20.3°}$$

so that the sending-end impedance is

$$Z_s = 52\underline{/22.6°}\,(0.661\underline{/20.3°}) = 34.4\underline{/42.9°}\,\Omega$$

As expected, the sending-end impedance is "closer" to the characteristic impedance than the load impedance, the SWR at the sending end is closer to unity than at the load, and the reflection coefficient at the sending end is closer to zero than at the load.

Since $V(z) = V^+ e^{-\alpha z} e^{-j\beta z}[1 + \Gamma(z)]$ and $I(z) = I^+ e^{-\alpha z} e^{-j\beta z}[1 - \Gamma(z)]$, if the magnitude of voltage or current is known at a point z_1, the magnitude of voltage or current can be found at a point z_2 by using

$$|V^+| = \frac{|V(z_1)|}{e^{-\alpha z_1}|1 + \Gamma(z_1)|} = \frac{|V(z_2)|}{e^{-\alpha z_2}|1 + \Gamma(z_2)|} \tag{1}$$

or

$$|I^+| = \frac{|I(z_1)|}{e^{-\alpha z_1}|1 - \Gamma(z_1)|} = \frac{|I(z_2)|}{e^{-\alpha z_2}|1 - \Gamma(z_2)|} \tag{2}$$

respectively. These equations are similar to equations derived in Sec. 5.6 for a lossless line. The difference is the nonzero attenuation constant, which implies that $\Gamma(z_1)$ and $\Gamma(z_2)$ have different magnitudes and that the exponential terms in Eqs. (1) and (2) have values other than unity.

Thus if $|V(z_1)|$ is known and we want to find $|V(z_2)|$, we can measure the phasors $|1 + \Gamma(z_1)|$ and $|1 + \Gamma(z_2)|$ from the Smith chart (as for lossless lines) and use

$$|V(z_2)| = \frac{|1 + \Gamma(z_2)|}{|1 + \Gamma(z_1)|} e^{-\alpha(z_2 - z_1)} |V(z_1)| \tag{3}$$

to find $|V(z_2)|$. Similarly, we can find the magnitude of current at one point on the line if current is known at another point.

6.4 IMPEDANCE MATCHING

QUARTER-WAVELENGTH TRANSFORMERS

The quarter-wavelength transformer, discussed in Sec. 5.8, can be used to match a lossy line to a load. Practically, however, it is often impossible to find a transmission line having the characteristic impedance necessary for a quarter-wavelength transformer to be placed adjacent to the load. If the lossy transmission line is long enough, the quarter-wavelength transformer can be placed some distance from the load, so that additional choices are available for the characteristic impedance of the transformer, as shown in Fig. 6.4.1.

However, if a lossless line (real characteristic impedance) is to be matched to a load by means of a quarter-wavelength transformer having real characteristic impedance, there are only two choices for the characteristic impedance of the transformer (see Sec. 5.8).

If a lossy line (characteristic impedance Z_0 not necessarily real) is to be matched to a load with a quarter-wavelength section of line having characteristic impedance Z_{01}, the impedance Z'_r at the load end of the transformer must satisfy the relation

$$Z_{01}^2 = Z_0 Z'_r \tag{1}$$

WAVES ON DISSIPATIVE LINES

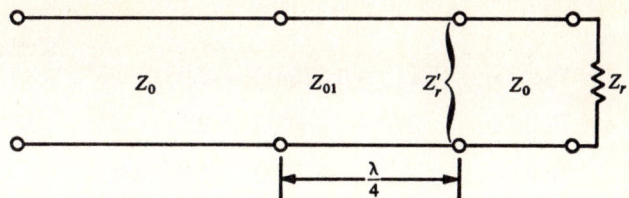

Fig. 6.4.1 A quarter-wavelength transformer.

as shown in Sec. 5.8. Since we would like to select the transformer from commercially available lines, which generally have real characteristic impedances, we often assume that Z_0 is real. However, Z_{01} is real if the product $Z_0 Z'_r$ is real, since $Z_{01}{}^2 = Z_0 Z'_r$ for matched conditions. For the lossy line, we will show that there are as many as four choices for a real Z_{01} per wavelength of line, because of the spiral path due to attenuation, on either the Smith chart or the Z-θ chart. The Z-θ chart is convenient to use, so that Z'_r can be chosen to have an angle with the same magnitude as the angle of Z_0 but opposite sign.

As an example of tuning a lossy line with a quarter-wavelength transformer having real Z_{01}, let $Z_0 = 100/12.5°$, the load impedance $Z_r = 200/42.5°$, and the attenuation constant $\alpha = 0.1$ Np/λ. The spiral impedance path is shown for 1 wavelength, starting at the load, in Fig. 6.4.2. From Eq. (1), Z'_{nr} must have an angle of twice the conjugate of the angle of Z_0, or $-25°$ in this case. (Why?) At each of the four points a, b, c, and d within the first wavelength (there are more points farther from the load) where the angle of the impedance on the line is $-25°$, a real Z_{01} exists for a quarter-wavelength transformer which could be inserted at that point to match the load. For this example, the normalized

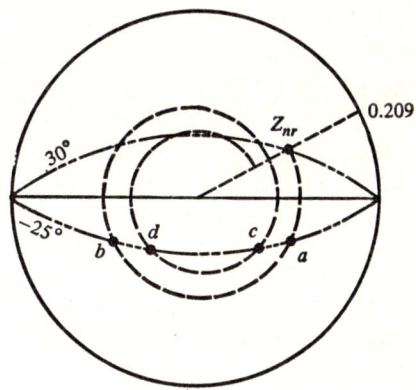

Fig. 6.4.2 An impedance path on the Z-θ chart for the example.

impedances of the four points closest to the load which satisfy this criterion are:

$\quad$ Point a: $\quad Z_{na}(-0.0863\lambda) = 2.08\underline{/-25°}$
$\quad$ Point b: $\quad Z_{nb}(-0.2563\lambda) = 0.5\underline{/-25°}$
$\quad$ Point c: $\quad Z_{nc}(-0.592\lambda) = 1.8\underline{/-25°}$
$\quad$ Point d: $\quad Z_{nd}(-0.7488\lambda) = 0.56\underline{/-25°}$

These points are best found by trial and error with the aid of the Z-θ chart. First we find $|\Gamma(z)| = |\Gamma_r|e^{\alpha z}$ by choosing different values of Z until $\Gamma(z)$ is on the $-25°$ locus. Then the possible values for the characteristic impedance of the quarter-wavelength transformer are found from Eq. (1):

$\quad$ Point a: $\quad Z_{01} = 144\ \Omega$
$\quad$ Point b: $\quad Z_{01} = 70.7\ \Omega$
$\quad$ Point c: $\quad Z_{01} = 136.8\ \Omega$
$\quad$ Point d: $\quad Z_{01} = 74.8\ \Omega$

Thus several choices exist for the characteristic impedance of the section of line used for the transformer. More choices exist, at the rate of four per wavelength, until the locus of impedance has spiraled in so much that it does not cross the proper angle locus (the locus necessary to cancel the angle of the characteristic impedance of the lossy line). For this example, the last locus intersection is approximately where $|\Gamma(z)| = 0.225$, the shortest distance from the origin to the $-25°$ locus. Here, $|\Gamma(z)| = 0.225$ at $z = -3.145\lambda$, so that matching with the quarter-wavelength transformer having real characteristic impedance could not be accomplished farther than about 3.145λ from the load. Tuning with a quarter-wavelength transformer beyond this point could be accomplished only by sections of lossy line having a characteristic impedance with a nonzero angle.

It should be noted that, for some problems, matching with a section of line having real characteristic impedance may not be possible. This occurs when the impedance spiral on the Z-θ chart lies either entirely within the locus of the conjugate of the angle of Z_0 or outside the locus without crossing. The latter case can occur only on short, badly mismatched lines. (When can the former case occur?)

STUB TUNERS

A lossless tuning stub can be conveniently used to match a lossy line if the characteristic impedance of the lossy line is real. If the characteristic impedance is *not* real, the matching procedure is quite difficult. For example, if the admittance on the load side at the stub is $Y'_r = a + jc$ and the characteristic admittance of the lossy line is $Y_0 = a + jb$, the tuning

WAVES ON DISSIPATIVE LINES

stub must have a susceptance of $j(b - c)$ for matching. On the Smith chart, however, the admittance Y'_r is difficult to locate because the real part of the normalized admittance on the load side at the tuning stub is given by

$$\text{Re}\frac{Y'_r}{Y_0} = \text{Re}\frac{a + jc}{a + jb} = \frac{a^2 + bc}{a^2 + b^2} \quad (2)$$

Thus the locus for the normalized admittance at the stub (on the load side) is *not* the locus of $\text{Re } Y_n = 1$ (unless $b = 0$) but is the locus described by Eq. (2). Unfortunately, the susceptance jc is not known until the locus is found, and the locus cannot be found until the susceptance is found. Although the problem could be solved by trial and error, we will consider only lossy lines which have real characteristic impedances.

The normalized admittance at any point is given by

$$Y_n(z) = \frac{1 - |\Gamma_r|e^{2\alpha z}e^{j(2\beta z + \rho)}}{1 + |\Gamma_r|e^{2\alpha z}e^{j(2\beta z + \rho)}} \quad (3)$$

where $\Gamma_r = |\Gamma_r|e^{j\rho}$. Rationalizing, simplifying, and setting the real part of $Y_n(z)$ equal to 1, it is found that at any point z for which

$$\cos(2\beta z + \rho) = -e^{-2\alpha z} \quad (4)$$

a lossless tuning stub could be placed in parallel with the line and adjusted for length to match the line. This is a transcendental equation which must be solved for z to find the distances from the load to points on the line where $\text{Re } Y_n = 1$. This is best done by means of a graphical solution on the Smith chart.

As an example of this technique, for the line shown in Fig. 6.4.3, let

$Z_r = 255 - j150 \, \Omega$

$Z_0 = 100 \, \Omega$

$\alpha = \dfrac{2n}{\lambda}$

Then $Y_{nr} = 0.29 + j0.18$, and the magnitude of the reflection coefficient at the load is 0.565. The point nearest the load where $\text{Re } Y_n = 1$ will be sought. A maximum distance to this first point can be found, as an outer limit, by finding the intersection of the $|\Gamma_r| = 0.565$ locus with the $\text{Re } Y_n = 1.0$ locus. This would be the location of the stub if the attenuation constant were zero. In this example, the maximum distance found in this way is 0.1425λ, illustrated in Fig. 6.4.4. This maximum distance can be substituted for z to find the minimum value of $|\Gamma(z)| = |\Gamma_r|e^{2\alpha z}$. Plotted on the Smith chart, this minimum value of $|\Gamma(z)|$ yields a minimum distance to the point of interest. Here, the minimum value of $|\Gamma(z)|$ is 0.3195, which results in a lower limit for distance of 0.1205λ from the

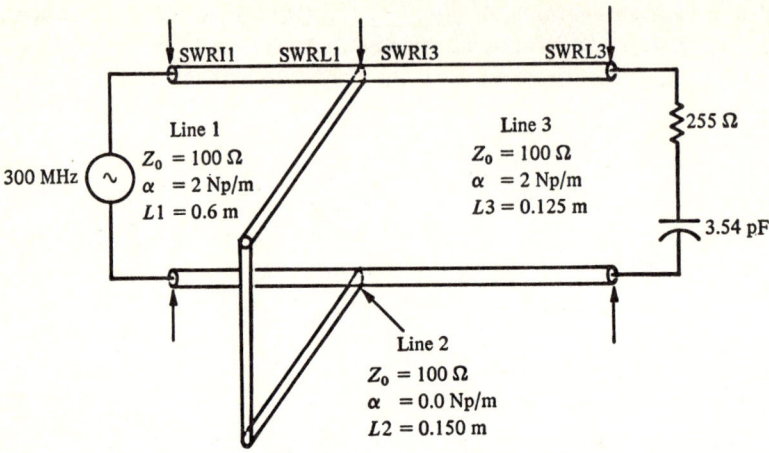

Fig. 6.4.3 An example of single-stub tuning on a lossy line.

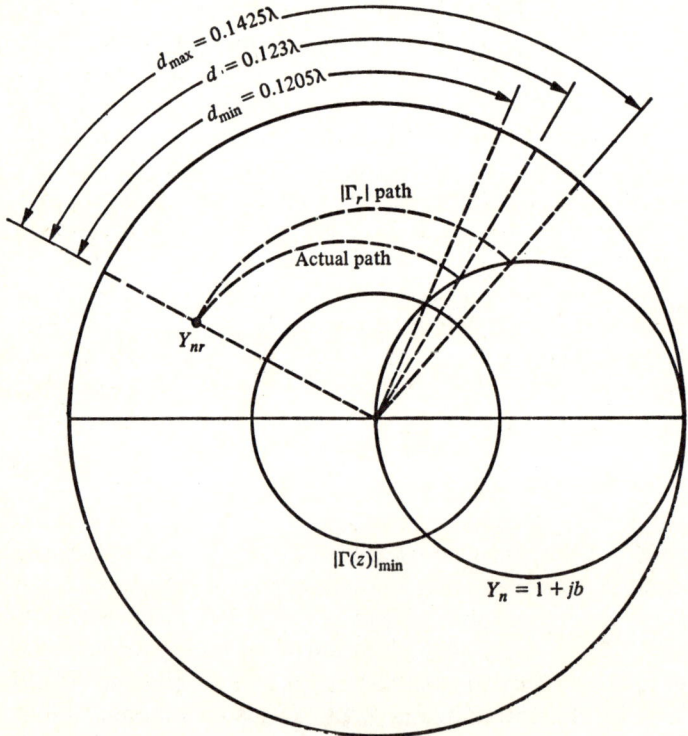

Fig. 6.4.4 Determining where on a lossy line to place a lossless tuning stub.

WAVES ON DISSIPATIVE LINES

Smith chart. The actual value for distance is then somewhere between 0.1425λ and 0.1205λ and is found by the following procedure:

1. Choose a distance d between the maximum and minimum distances.
2. Compute $|\Gamma(-d)|$ for the distance chosen.
3. If the locus of $|\Gamma(-d)|$ and the $z = -d$ ray intersect on the Re $Y_n = 1.0$ locus, then $z = -d$ is the correct distance. If the intersection is inside (outside) the Re $Y_n = 1.0$ locus, then $z = -d$ is too far (not far enough) from the load and another distance must be tried.

For this example, $z = -0.123\lambda$ is the correct point on the line. At that point, $|\Gamma(-0.123\lambda)| = 0.35$, and $Y_n(-d^+) = 1 + j0.75$, so that the tuning stub must have an admittance of $-j0.0075$.

6.5 COMPUTER SYNTHESIS AND ANALYSIS PROBLEMS

As we found in Secs. 6.1 to 6.4, lossy transmission systems are more difficult to analyze than lossless ones. Graphical techniques can be helpful, but tedious, when the attenuation constant is not zero. In many cases we will find that the simplest method is to use the digital computer and SANTLINE. Since the subprograms of SANTLINE are designed for use with lossy lines,† the applications in Sec. 5.10 could be repeated in this chapter. The reader will find it interesting to do so, e.g., to investigate how the bandwidth of the double-stub tuner of Fig. 5.10.3 changes if the main lines are lossy. (Use the same program but let α vary over some range of values.) However, we will use the computer to solve the single-stub system just discussed in the previous section.

Specifically, for the system shown in Fig. 6.4.3, we will use the computer to find FL3 (the line length from load to stub) and the stub length FL2. These lengths (the lengths necessary for matching at 300 MHz) can be compared with the results obtained from the graphical analysis of Sec. 6.4. Then we will allow the frequency to change from 100 to 500 MHz, with the line lengths fixed, and find the following standing-wave ratios (refer to Fig. 6.4.3) as frequency varies:

SWRI1 = SWR at input end of line 1
SWRL1 = SWR at load end of line 1
SWRI3 = SWR at input end of line 3
SWRL3 = SWR at load end of line 3

The stub is assumed to be lossless, and so the SWR is infinite on line 2. Bandwidths at the input and load ends of lines 1 and 3 can be found from graphs of SWR versus frequency.

† The one exception is STUBL. It ignores the actual value of α and uses $\alpha = 0$ to compute the stub length.

The program used for this problem is given in Table 6.5.1, and the results are given in Table 6.5.2. The particular parameter values used for this program are

$F = 3.0E8$ = center frequency, Hz
$A = 17.36 = 17.36$ dB/m attenuation = 2 Np/m
$VF = 1.0$ = velocity factor
$Z0 = 100.0$ = characteristic impedance, Ω
$XTERM(1) = 255.0$ = load resistance, Ω
$XTERM(2) = -150.0$ = load capacitive reactance, Ω at frequency F
$FL1 = 0.6$-length of line l, m

Table 6.5.1 Program STUBBW for Fig. 6.4.3

```
C     THIS PROGRAM IS FOR A SINGLE STUB SYSTEM WITH LOSSY LINES.
C     AT THE CENTER FREQUENCY, IT COMPUTES THE DISTANCE TO THE STUB,
C     AND THE LENGTH OF THE STUB.  IT ASSUMES THE STUB IS LOSSLESS.
C     ALSO, IT COMPUTES THE SWR AT LOAD, STUB, AND INPUT AS A
C     FUNCTION OF FREQUENCY.
C
C     ******** INSERT DECLARATIONS PACKAGE HERE ********
      COMMON/DATA/BDATA(45,3)
      COMPLEX CYIN,CPROP,CYSHRT,GET1C,CYSER,YSEND
      CYSER(F)=1.0/CMPLX(255.0,-1.0/(6.283185*F*3.54E-12))
      F=3.0E8
      FPZ0=F
      Z0=100.0
      P0=CPROP(0.0,1.0,FPZ0)
      CALL ENTER(2)
      P0=CPROP(17.36,1.0,FPZ0)
      CALL ENTER(3)
      LINE=1
      LENGTH=0.6
      CALL STASH
      FL3=FINDL(3,CYSER(F),3)
      FL2=STUBL(2,YLINE-GET1C(3,15))
      WRITE(6,10)F,FL3,FL2
   10 FORMAT(1H1,9X,26HLOSSY STUB TUNING PROBLEM.//10X,29HFREQUENCY FOR
     AMATCH (HERTZ) =,7X,1PE9.2//10X,34HLENGTH FROM LOAD TO STUB (METERS
     B)=,2X,E9.2//10X,25HLENGTH OF STUB (METERS) =,11X,E9.2///10X,25HBAN
     CDWIDTH CALCULATIONS...//16X,4HFREQ,5X,5HSWRI1,5X,5HSWRL1,5X,5HSWRI
     D3,5X,5HSWRL3/)
      F=1.0E8
      DO 30 I=1,21
      YSEND=CYIN(1,CYSHRT(2)+CYIN(3,CYSER(F)))
      SWRL1=GET1R(1,44)
      SWRI1=GET1R(1,43)
      SWRL3=GET1R(3,44)
      SWRI3=GET1R(3,43)
      WRITE(6,20)F,SWRI1,SWRL1,SWRI3,SWRL3
   20 FORMAT(10X,1PE10.3,0P4F10.3)
   30 F=F+2.0E7
      CALL EXIT
      END

C     THERE ARE NO DATA CARDS FOR PROGRAM STUBBW.
```

WAVES ON DISSIPATIVE LINES

Table 6.5.2 Results for the line of Fig. 6.4.3

LOSSY STUB TUNING PROBLEM.

FREQUENCY FOR MATCH (HERTZ) =   3.00E 08

LENGTH FROM LOAD TO STUB (METERS) =   1.23E-01

LENGTH OF STUB (METERS) =   1.50E-01

BANDWIDTH CALCULATIONS...

FREQ	SWRI1	SWRL1	SWRI3	SWRL3
1.000E 08	1.180	21.607	3.062	10.776
1.200E 08	1.168	12.488	2.849	8.327
1.400E 08	1.151	7.757	2.673	6.840
1.600E 08	1.130	5.133	2.530	5.868
1.800E 08	1.108	3.596	2.414	5.196
2.000E 08	1.086	2.651	2.319	4.712
2.200E 08	1.064	2.044	2.241	4.350
2.400E 08	1.045	1.637	2.177	4.073
2.600E 08	1.028	1.354	2.123	3.855
2.800E 08	1.013	1.152	2.078	3.682
3.000E 08	1.000	1.004	2.040	3.540
3.200E 08	1.011	1.124	2.008	3.424
3.400E 08	1.020	1.245	1.980	3.327
3.600E 08	1.028	1.359	1.956	3.246
3.800E 08	1.035	1.465	1.935	3.176
4.000E 08	1.041	1.561	1.917	3.117
4.200E 08	1.045	1.645	1.901	3.065
4.400E 08	1.049	1.718	1.887	3.021
4.600E 08	1.052	1.777	1.875	2.981
4.800E 08	1.054	1.823	1.864	2.947
5.000E 08	1.056	1.854	1.854	2.916

The subroutines FINDL and STUBL are used to find the unknown lengths FL3 and FL2. Then the frequency is changed from 100 to 500 MHz in 20-MHz steps in the DØ loop. However, since the capacitive reactance is a function of frequency, it must be changed each time the frequency is changed. Because the capacitive reactance is known ($-150.0\ \Omega$) at the center frequency, it can be found at any frequency F

$$\text{XTERM}(2) = (3.0\text{E}8)(-150.0)/F$$

Note, however, that this computation is performed in a slightly different fashion by the statement function CYSER(F). Then the load impedance at any frequency is

$$\text{ZLØAD} = \text{XTERM}(1) + j\text{XTERM}(2)$$

which is stored as a complex number in the TLDATA array. The desired SWRs are computed by CYIN for each frequency. To complete the programming, we obtain the SWRs from the BDATA array using the subprogram GET1R.

From Table 6.5.2 we see that the length from the load to the stub (FL3) is 0.123 m, which agrees with the length obtained graphically in Fig. 6.4.4. Looking at the SWR calculations at the load, we see that SWRL3 varies from 10.78 at F = 100 MHz, to 3.54 at 300 MHz, to 2.92 at 500 MHz. At those same frequencies, the SWR at the input end of line 3 is 3.06, 2.04, and 1.85 respectively. This reduction in the SWR is caused by the attenuation term $e^{-\alpha z}$ over the 0.123 m. By the addition of the stub, the SWR is reduced to 1.00 (matched) at 300 MHz, as can be seen from the SWRL1 column in Table 6.5.2. If we plot SWRL1 versus frequency, we find that the bandwidth (for SWR $\leq$ 1.5) is about 133 MHz. However, the attenuation between the stub and the input, a length of 0.6 m, reduces the SWR so that the line is essentially matched over the entire frequency range from 100 to 500 MHz. Longer line lengths will further improve the match, but the loss due to attenuation (17.36 dB/m) also increases.

PROBLEMS

1. Derive Eq. (14) in Sec. 6.1 for the complex power at any point on a dissipative line

2. How must the parameters R, L, G, and C of a dissipative line be restricted so that one can obtain the total power at a point by superposition of the powers in the positive- and negative-going waves? Are these conditions easy or difficult to meet? What is the physical significance of "power at a point"?

3. The line shown has an attenuation constant of 0.1 Np/m, and $\lambda = 10$ m. (a) Plot phasor diagrams at the load and at the sending end. (b) How much power is dissipated in the line? (c) What is the SWR at each end of the line?

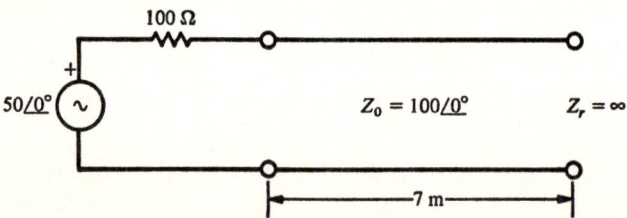

Fig. P.3

4. The lossy line shown has an attenuation of 0.01 Np/λ. (a) Plot a phasor diagram at the load. (b) How much power is dissipated in the line? In the load? (c) What is SWR at each end of the line?

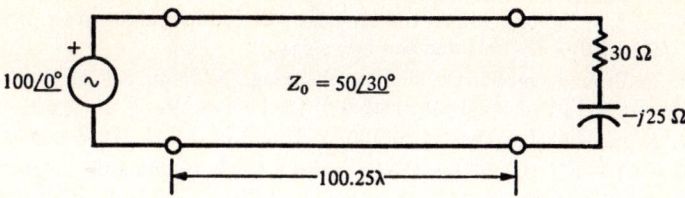

Fig. P.4

5. Show that if a line is terminated in $Z_R \neq Z_0$, then $Z(z) \neq Z_0$ for any finite z.

6. For the line shown, $\alpha = 0.1$ Np/m and $\lambda = 10$ m. (a) Find $Z(-7 \text{ m})$, $\Gamma(0)$, $\Gamma(-7 \text{ m})$, SWR(0), and SWR(-7 m). (b) Find $|I_s|$ and $|V_r|$ by means of the Smith chart.

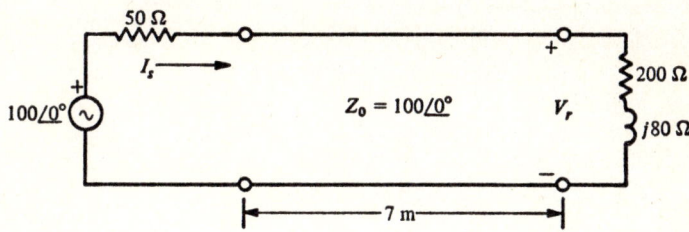

Fig. P.6

7. A transmission line 0.5 m long has $R + j\omega L = 41\underline{/83°}\,\Omega/\text{m}$ and $G + j\omega C = 0.01\underline{/5.6°}$ mho/m and is terminated in $Z_r = 10 - j100\,\Omega$. (a) Find the sending-end impedance by means of phasor diagrams, and check your answer using a Smith or Z-θ chart. (b) A 100-V rms sinusoidal generator with negligible internal impedance is connected to the sending end. Find the power supplied by the generator and the power dissipated in the load and in the line.

8. A section of line 3λ long has $Z_0 = 75\underline{/0°}\,\Omega$ and $\alpha = 0.2$ Np/λ and is terminated in $Z_r = 150\underline{/40°}\,\Omega$. A lossless shorted tuning stub 0.27λ long, with $Z_0 = 50\,\Omega$, is connected across the lossy section of line at a point 1.25λ from the load. Find the input impedance of the lossy section and the SWR on each section of line.

9. In Prob. 8 suppose the shorted stub is not lossless but is composed of the same lossy cable used for the main line. Find the input impedance and the SWR on each section of line.

10. A line 5.4 m long is terminated in $Z_r = 75\underline{/-35°}\,\Omega$ and $Z_0 = 167\underline{/18°}\,\Omega$. The attenuation is 0.06 Np/λ, v_p is 10^8 m/s, and the frequency of operation is 50 MHz. Match this lossy line with a lossless shorted tuning stub having a characteristic impedance $Z_0 = 100$.

11. In Prob. 10, match the lossy line with one lossless quarter-wavelength transformer. If the transformer must be inserted no more than 1 wavelength from the load, what choices are there for the characteristic impedance of the transformer?

12. Perform the derivation required to obtain the (transcendental) equation [Eq. (4) of Sec. 6.4], thereby finding the equation satisfied by points on the intersection of the $Y_n(z) = 1.0 + jb$ circle and the $\Gamma(z)$ spiral.

13. In Prob. 8, match the line by changing the frequency of operation a minimal amount and by changing the length of the lossless stub if necessary.

14. A line 4.8λ long has $Z_0 = 100\underline{/0°}\ \Omega$ and $\alpha = 0.15$ Np/λ and is terminated in $Z_r = 62\underline{/-37°}\ \Omega$. Match the line with a single tuning stub as close to the load as possible. What is the input impedance if frequency is (a) lowered 10 percent and (b) raised 10 percent?

15. A 50-V rms generator with an internal impedance of $20\underline{/0°}\ \Omega$ is attached to the (matched) line of Prob. 10. Find the power dissipated in the line and the load.

16. A 100-mV rms generator with an internal impedance of $200\underline{/0°}\ \Omega$ is connected to the (matched) line of Prob. 11. Is the power dissipated in the load independent of the characteristic impedance of the transformer? Find the power dissipated in the load if the lossless transformer is placed as close to the load as possible.

7
Waveguides

7.1 INTRODUCTION

A transmission line represents a system of conductors capable of guiding electromagnetic waves, whether the line is coaxial cable or parallel wires or has some other geometry. However, the term *waveguide* is normally reserved for a *single* hollow conductor, i.e., no center conductor, usually having either a rectangular or a circular cross section. We will find that electromagnetic energy can be transmitted from one point to another through the interior of a single hollow conductor providing that the wavelength is sufficiently short with respect to dimensions of the cross section of the waveguide. We will also find that, unlike two-conductor transmission lines, a given waveguide has a minimum frequency of operation, usually called the *cutoff frequency*. Thus no waveguide can transmit direct current, and waveguides become impractically large for frequencies lower than a few hundred megahertz.

The waveguides we study have a uniform cross section (either rectangular or circular), and we will not consider the effects of losses† either

---

† These losses are discussed in S. Ramo, J. R. Whinnery, and T. Van Duzer, "Fields and Waves in Communication Electronics," John Wiley & Sons, Inc., New York, 1965.

in the conductor or in the dielectric. Waveguides are useful partly because they can be curved, and the electromagnetic waves will follow a moderate curvature and retain essentially the same characteristics as waves on a straight section of waveguide. We will analyze only straight waveguides, since it is customary to assume that the moderate curvature usually sufficient in practical situations does not change the fields markedly. Measurements of the SWR (of the electric field on curved sections of waveguide) verify that normally a very small amount of energy is reflected due to the curvature of the waveguide; hence there is not a large mismatch.

Problems involving waveguides are more difficult than problems concerning transmission lines since there is no direct analogy to either voltage or current on waveguides. For transmission lines, problems were easily formulated in terms of voltage and current. Thus we could solve these problems without using vectors, although the (vector) fields could be obtained at any time in terms of voltage and current. For waveguides, however, we are often required to work directly with electric and magnetic field vectors.†

Energy transmission through a hollow waveguide, we will find, requires either TE or TM waves, since the propagation of TEM waves requires either a second conductor or free space. TE and TM waves, defined in Sec. 3.1, can often be viewed as TEM waves obliquely incident on the walls of a waveguide. Thus the discussion in Sec. 3.7 of a uniform plane wave obliquely incident on a perfect conductor will be a convenient foundation for the study of TE and TM waves. We first examine propagation between two infinite parallel planes and then we add the "side planes" to obtain a rectangular waveguide.

7.2 INFINITE PARALLEL PLANES

In Sec. 3.7 we found that TEM waves obliquely incident on a perfect conductor are reflected from the conductor, so that there are standing waves in the direction normal to the conductor. There are traveling waves in the $-\mathbf{a}_x$ direction (parallel to the conductor), with the incident fields as shown in either Fig. 3.7.4 or 3.7.5. The direction of energy flow in both cases (E parallel and E normal to the plane of incidence) was $-\mathbf{a}_x$, the direction of travel for the traveling waves. The component of E parallel to the conductor (which, like the other field components, formed standing waves normal to the conductor) was zero on the planes given by Eq. (26) of Sec. 3.7 and repeated here:

$$z = \frac{-m\lambda}{2\cos\theta} \qquad m = 0, 1, 2, \ldots \qquad (1)$$

† For a discussion of some simple cases in which transmission-line techniques can be applied to waveguides, see *ibid.*, p. 439.

WAVEGUIDES

The reader may recall that θ is the angle of incidence of the plane wave on the conductor.

If the purpose of the fields shown in Fig. 3.7.4 is to transport energy in the $-\mathbf{a}_z$ direction, parallel to the conductor, these fields can be thought of as forming a TM wave with respect to the $-\mathbf{a}_z$ direction. That is, the magnetic field (in Fig. 3.7.4) has a component transverse to the direction of energy flow, while the electric field has both transverse and longitudinal components. Similarly, the fields shown in Fig. 3.7.5 form a TE wave with respect to the $-\mathbf{a}_z$ direction (the direction of energy flow). Thus a uniform plane wave, under these circumstances, can be thought of as either a TM or a TE wave, depending upon whether **E** is in the plane of incidence or normal to the plane of incidence. The TM and TE waves propagate parallel to the conductor, while the corresponding TEM wave is obliquely incident on the conductor.

Since the component of **E** parallel to the conductor is zero in the planes given by Eq. (1), we can put a second infinite, plane, perfect conductor at any of the planes given by Eq. (1). Suppose we now have two such parallel perfect conductors, as shown in Fig. 7.2.1. Then the boundary conditions are satisfied, since the component of **E** parallel to the conductors vanishes at each conductor. The fields between conductors are precisely the same as before insertion of the second conductor, and hence these fields satisfy Maxwell's equations. Thus, for a TM wave between parallel conductors, the fields are given by Eqs. (22) to (24) of Sec. 3.7, and for a TE wave, the fields are given by Eqs. (30) to (32) of Sec. 3.7. From Eq. (1), we can see that the standing wave between conductors contains an integral number of half wavelengths, measured normal to the planes of the conductors.

To illustrate some additional interesting properties of TM and TE waves, we use the coordinate system and conductors as shown in Fig.

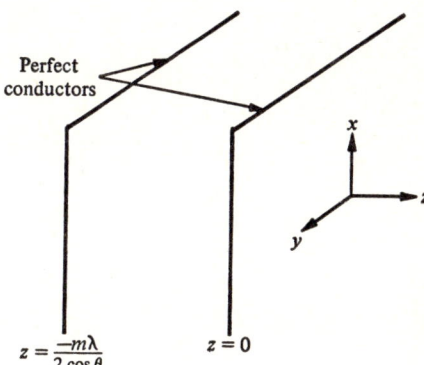

Fig. 7.2.1 Two plane, infinite, parallel, perfect conductors with the same coordinate system used in Sec. 3.7.

$z = \dfrac{-m\lambda}{2\cos\theta}$ $z = 0$

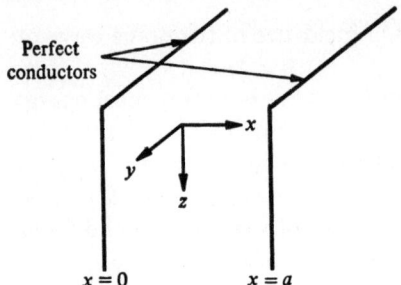

Fig. 7.2.2 Two plane, infinite, parallel, perfect conductors with a new coordinate system.

7.2.2, in which the origin has been located on the left conductor. We first discuss the TM wave, and we can start by rewriting Eqs. (22) to (24) of Sec. 3.7 in terms of the coordinate system shown in Fig. 7.2.2. That is, by replacing z with x and x with $-z$ in Eqs. (22) to (24) of Sec. 3.7 (we must also replace E_x with $-E_z$ and E_z with E_x) we obtain

$$E_z(x,z) = j2E^+ \cos\theta \sin(\beta x \cos\theta) \exp(-j\beta z \sin\theta) \tag{2}$$

$$E_x(x,z) = 2E^+ \sin\theta \cos(\beta x \cos\theta) \exp(-j\beta z \sin\theta) \tag{3}$$

$$H_y(x,z) = \frac{2E^+}{\eta} \cos(\beta x \cos\theta) \exp(-j\beta z \sin\theta) \tag{4}$$

We must apply the boundary condition at the conductor located at $x = a$ (why not the other conductor?), since we did not specifically place this conductor at a plane where the tangential component of **E** is zero. Therefore, from Eq. (2),

$$E_z(a,z) = j2E^+ \cos\theta \sin(\beta a \cos\theta) \exp(-j\beta z \sin\theta) = 0 \tag{5}$$

The only nontrivial way (what are the trivial ways?) for Eq. (5) to be satisfied is if

$$\sin(\beta a \cos\theta) = 0 \tag{6}$$

which will be true if

$$\beta a \cos\theta = m\pi \quad \text{for } m = 0, 1, 2, 3, \ldots \tag{7}$$

where m is the number of half wavelengths in the standing wave between conductors. Therefore,

$$\beta \cos\theta = \frac{m\pi}{a} \tag{8}$$

WAVEGUIDES

For the plane waves discussed in Sec. 3.7, the phase constant is

$$\beta = \frac{\omega}{v} = \omega \sqrt{\mu\varepsilon} \qquad (9)$$

and $\gamma = j\beta$ for the direction of propagation ζ of the plane wave. The TM and TE waves discussed here, however, propagate in the $\mathbf{a}_z$ direction, so that, from Eqs. (2) to (4), the propagation constant for the $\mathbf{a}_z$ direction is given by

$$\gamma' = j\beta' = j\beta \sin\theta = j\beta \sqrt{1 - \cos^2\theta} \qquad (10)$$

From Eq. (8), and substituting $\omega\sqrt{\mu\varepsilon}$ for β, we can rewrite Eq. (10) as

$$\gamma' = j\left[\omega^2\mu\varepsilon - \left(\frac{m\pi}{a}\right)^2\right]^{1/2} \qquad (11)$$

which expresses the propagation constant γ' in terms of the frequency ω, the constants of the medium μ and ε, the *mode number m*, and the separation distance a between conductors. From Chap. 3, the reader can recall that $(\mu\varepsilon)^{-1/2}$ is the velocity of propagation of the TEM wave, which travels a zigzag path between the two conductors. The mode number m is the number of half wavelengths in the standing wave between conductors.

From Eq. (11) we see that γ' can be either real or imaginary, depending upon the frequency. We define the *cutoff frequency* f_c as the frequency at which $\gamma' = 0$ for a particular mode. That is,

$$f_c = \frac{m}{2a\sqrt{\mu\varepsilon}} = \text{cutoff frequency} \qquad (12)$$

Then, from Eq. (11),

$$\gamma' = \alpha' = \frac{m\pi}{a}\left[1 - \left(\frac{f}{f_c}\right)^2\right]^{1/2} \quad \text{for } f < f_c \qquad (13)$$

or

$$\gamma' = j\beta' = j\omega\sqrt{\mu\varepsilon}\left[1 - \left(\frac{f_c}{f}\right)^2\right]^{1/2} \quad \text{for } f > f_c \qquad (14)$$

Since γ' is real if $f < f_c$ for a particular mode, there is no propagation of the fields between conductors at these low frequencies. Instead, the fields simply decay with distance, the phase angle being independent of distance. Conversely, if $f > f_c$ for a given mode, γ' has no real part (we are neglecting losses) and the wave propagates without attenuation.

From Eqs. (8), (9), and (12) we see that

$$\cos\theta = \frac{f_c}{f} \qquad (15)$$

Using $E_0 = j2E^+$, and Eqs. (8), (14), and (15), we can rewrite Eqs. (2) to (4) for a TM wave between parallel planes:

$$E_z(x,z) = E_0 \frac{f_c}{f} \sin \frac{m\pi x}{a} e^{-j\beta' z} \tag{16}$$

$$E_x(x,z) = -jE_0 \left[1 - \left(\frac{f_c}{f}\right)^2\right]^{1/2} \cos \frac{m\pi x}{a} e^{-j\beta' z} \tag{17}$$

$$H_y(x,z) = \frac{-jE_0}{\eta} \cos \frac{m\pi x}{a} e^{-j\beta' z} \tag{18}$$

PHASE AND GROUP VELOCITIES

For the case in which $f > f_c$, so that $\gamma' = j\beta'$, the phase velocity of the TM wave is

$$v_p = \frac{\omega}{\beta'} = \frac{\omega}{\omega \sqrt{\mu\varepsilon}\,[1 - (f_c/f)^2]^{1/2}} = \frac{v}{[1 - (f_c/f)^2]^{1/2}} \tag{19}$$

To find the group velocity it is convenient to find $d\beta'/d\omega$ and then invert, obtaining

$$v_g = \frac{d\omega}{d\beta'} = v \left[1 - \left(\frac{f_c}{f}\right)^2\right]^{1/2} \tag{20}$$

Surprising as it may seem, the phase velocity v_p is always greater than the velocity of light v in the medium if the frequency is finite, while the group velocity v_g is always less than v.

The physical significance of v_p being greater than the velocity of light can be seen from the following alternate derivation. In Fig. 7.2.3, the TEM wave propagates with velocity $v = 1/(\mu\varepsilon)^{1/2}$ in the ζ direction (as it travels from B to C), and its angle of incidence with the conductors is θ.

As the TEM wave travels with velocity v from B to C, its wavefront (the plane of constant phase) progresses from the plane AB to the plane CD. The wavefront, as illustrated in the figure, is perpendicular to the direction of propagation of the TEM wave. Therefore, in the time it takes for the plane wave to travel from B to C, the plane of constant phase has traveled from B to D, in the direction of propagation of the TM or TE wave. This can be written

$$\Delta t_p = \frac{BC}{v} = \frac{BD}{v_p} \tag{21}$$

so that

$$v_p = v \frac{BD}{BC} = \frac{v}{\sin \theta} \tag{22}$$

WAVEGUIDES

From Eqs. (9) and (11), we can write $\sin \theta$ and $\cos \theta$ in terms of the two frequencies, the mode number, and the conductor separation:

$$\sin \theta = \left[1 - \frac{1}{\omega^2 \mu \varepsilon}\left(\frac{m\pi}{a}\right)^2\right]^{1/2} = \left[1 - \left(\frac{f_c}{f}\right)^2\right]^{1/2}$$

and

$$\cos \theta = \frac{f_c}{f} \tag{23}$$

Simplifying v_p in terms of the cutoff frequency f_c given by Eq. (12), we obtain

$$v_p = \frac{v}{\sin \theta} = \frac{v}{[1 - (f_c/f)^2]^{1/2}} \tag{24}$$

which agrees with the previously derived phase velocity for a TM or TE wave.

The velocity of energy transport is the same as the group velocity for the TM and TE waves discussed here, and from Fig. 7.2.3 we see that

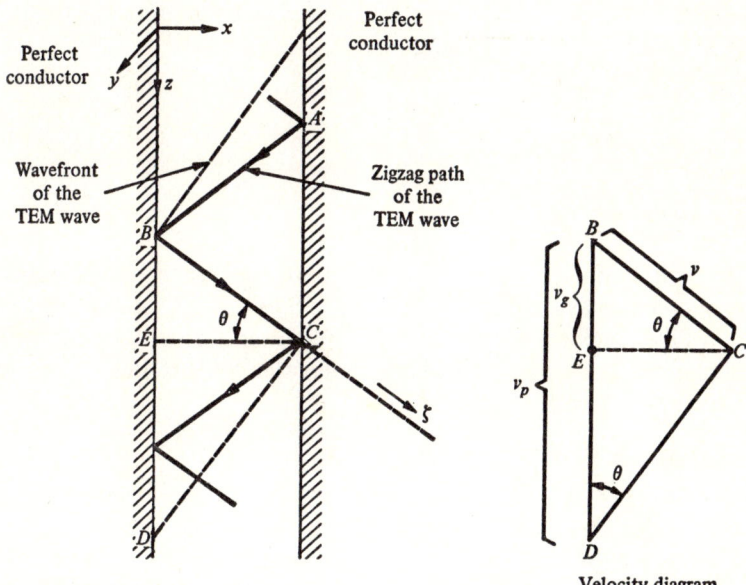

Fig. 7.2.3 As a TEM wave travels between parallel conductors, the phase velocity of the corresponding TM or TE wave exceeds the velocity of light, while the group velocity is less than the velocity of light.

the energy in the wave travels from B to E down the conductor in the time that the TEM wave travels from B to C. Therefore

$$\Delta t_g = \frac{BE}{v_g} = \frac{BC}{v} \tag{25}$$

so that

$$v_g = v \sin \theta = v \left[1 - \left(\frac{f_c}{f}\right)^2\right]^{1/2} \tag{26}$$

which agrees with the group velocity previously derived and given in Eq. (20). Therefore both TM and TE waves travel between parallel conductors with the phase and group velocities given above.

From Fig. 7.2.3 we can see that the group velocity is less than the velocity of light and the phase velocity is greater than the velocity of light for the TM and TE waves. The latter does not contradict the theory of relativity, since the phase velocity does not represent the velocity of either energy or particles. This leads to the proper interpretation of *evanescent* waves, in which $f < f_c$. The variation of such waves with time and distance can be written

$$e^{-\alpha' z} \cos(\omega t + \phi) \tag{27}$$

where α' is given by Eq. (13). For evanescent waves, the phase is independent of distance, so that the phase velocity appears to be infinite. This can be derived by using

$$\gamma' = \alpha' + j\beta'$$

and taking the limit

$$\lim_{\beta' \to 0} v_p = \lim_{\beta' \to 0} \frac{\omega}{\beta'} = \infty$$

We will find later in this section that no power is transmitted in an evanescent wave.

To conclude our discussion of phase and group velocities, Fig. 7.2.4 illustrates the variation of these velocities with frequency, as well as α' versus frequency for $f < f_c$.

WAVELENGTH

The wavelength λ of a TEM wave of a given frequency is fixed, whether it travels a zigzag path between conductors or travels in free space. This wavelength is

$$\lambda = \frac{v}{f} = \frac{2\pi}{\beta} = \frac{2\pi}{\omega \sqrt{\mu \varepsilon}} \tag{28}$$

WAVEGUIDES

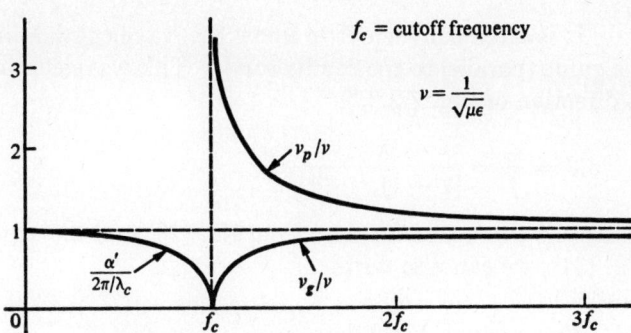

Fig. 7.2.4 Phase and group velocities for $f > f_c$ and α' for $f < f_c$ for TE and TM waves.

where all symbols refer to free-space values and λ is measured in the ζ direction (the direction of propagation of the TEM wave). Substituting this value of λ into Eq. (23) gives

$$\sin \theta = \left[1 - \lambda^2 \left(\frac{m}{2a}\right)^2\right]^{1/2} \qquad (29)$$

Therefore, if the free-space wavelength λ and the mode number m are known, the angle of incidence θ is uniquely determined since the boundary conditions at $x = 0$ and $x = a$ must be satisfied.

There are m half wavelengths in the standing wave between conductors, and for each m, the corresponding cutoff frequency for that mode is given by Eq. (12). From Eqs. (23) and (24) we see that at cutoff the angle of incidence θ is zero. This result should be intuitive, since it represents normal incidence on a conductor, in which no energy is transmitted in a direction parallel to the conductor. Equivalently, the cutoff (free-space) wavelength λ_c is the value of λ for which $\sin \theta = 0$ in Eq. (29). This value is

$$\lambda_c = \frac{2a}{m} \qquad m = 1, 2, 3, \ldots \qquad (30)$$

Rewriting Eq. (29), we obtain

$$\lambda = \lambda_c \cos \theta \qquad (31)$$

which reinforces our earlier conclusion that $f > f_c$, or $\lambda < \lambda_c$, for propagation to occur between the parallel conductors.

It is often convenient to know the wavelength λ' in the direction of the guide (parallel to the conductors). This wavelength, measured in the $\mathbf{a}_z$ direction of Fig. 7.2.3, is

$$\lambda' = \frac{v_p}{f} = \frac{\lambda}{[1 - (f_c/f)^2]^{1/2}} \tag{32}$$

Using the expression for $\sin \theta$ from Eq. (23) and the expression for λ from Eq. (31), we can also write

$$\lambda' = \frac{\lambda}{\sin \theta} = \lambda_c \cot \theta \tag{33}$$

for the wavelength in the direction of propagation of the TE and TM waves. Thus λ' is greater than the free-space wavelength λ, but λ' can be either less than or greater than the cutoff wavelength λ_c for the given mode, depending upon whether $\theta > 45°$ or $\theta < 45°$ respectively. This can be seen from the following example, in which we assume $m = 2$, so that there are two half wavelengths in the standing wave between conductors. Therefore $\lambda_c = a$, and $\lambda' = a \cot \theta$. Figure 7.2.5 shows the zigzag paths and the wavefronts of two TEM waves with mode number

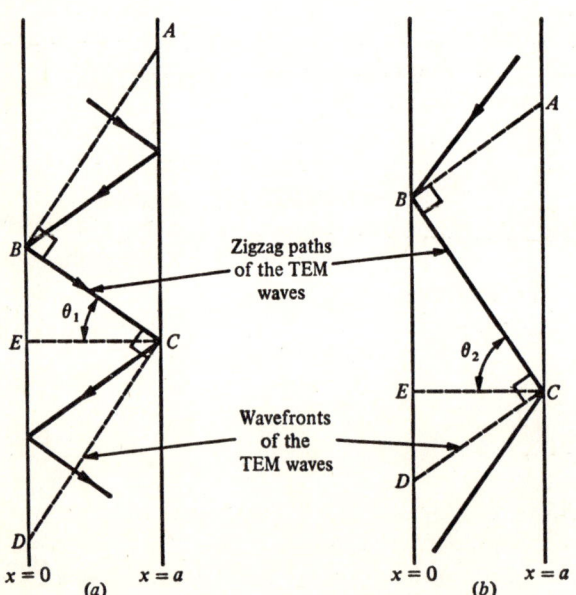

Fig. 7.2.5 The frequency is higher and λ' is smaller (by about one-half) in (b) than in (a). (a) $\theta < 45°$; (b) $\theta > 45°$.

WAVEGUIDES

$m = 2$. The frequency of the wave shown in Fig. 7.2.5b is higher than the frequency of the wave shown in a since $\theta_2 > \theta_1$. In both Fig. 7.2.5a and b

$$\cot \theta = \frac{ED}{EC} = \frac{ED}{a}$$

Since we assumed $m = 2$, so that $\lambda_c = a$,

$$\lambda' = \lambda_c \cot \theta = a \cot \theta = ED$$

Therefore the line ED in either Fig. 7.2.5a or b is the wavelength λ', which is measured in the direction parallel to the conductors. For an arbitrary mode number m, then, we would find that

$$\lambda' = \lambda_c \cot \theta = \frac{2a}{m} \cot \theta = \frac{2}{m} DE$$

In conclusion, for any mode, the length DE is proportional to the wavelength λ' in the guide.

IMPEDANCE CONCEPTS, POWER, AND ENERGY

In Eqs. (3) and (4) of Sec. 3.3, the intrinsic impedance was defined. This impedance is the ratio of the total electric field to the total magnetic field for a plane wave, and was found to be a characteristic of the medium in which the wave travels. Now, for the simple case in which TE or TM waves propagate without reflection, we define the *wave impedance Z* as

$$Z_{\text{TM}} \text{ or } Z_{\text{TE}} = \frac{E_{\text{trans}}}{H_{\text{trans}}} \tag{34}$$

where the *total* transverse components of **E** and **H** must be used. Thus, for the TM wave given by Eqs. (2) to (4), we find that

$$Z_{\text{TM}} = \frac{E_x(x,z)}{H_y(x,z)} = \eta \sin \theta \tag{35}$$

which can be written [from Eqs. (12) and (23)] as

$$Z_{\text{TM}} = \eta \left[1 - \left(\frac{f_c}{f}\right)^2 \right]^{\frac{1}{2}} \tag{36}$$

From Eq. (36), it can be seen that Z_{TM} is real for frequencies above cutoff ($f > f_c$) but that Z_{TM} is imaginary for frequencies too low to propagate between the two conductors. It can be shown that these two cases have the following interpretations:

1. If Z_{TM} (or Z_{TE}) is real, there is a time-average *real* power flow in the direction of propagation.

2. If Z_{TM} (or Z_{TE}) is imaginary, only time-average *reactive* power flows in the direction of propagation and energy is stored in the volume occupied by the fields.

Thus, for the positive-going TM wave given by Eqs. (16) to (18), the Poynting vector is

$$\langle S \rangle = \langle P \rangle + jQ = \tfrac{1}{2}[(E_x \mathbf{a}_x + E_z \mathbf{a}_z) \times (H_y^* \mathbf{a}_y)]$$
$$= \frac{E_0^2}{2\eta} \left\{ \left[1 - \left(\frac{f_c}{f}\right)^2\right]^{1/2} \cos^2 \frac{m\pi x}{a} \mathbf{a}_z \right.$$
$$\left. - j \frac{f_c}{f} \sin \frac{m\pi x}{a} \cos \frac{m\pi x}{a} \mathbf{a}_x \right\} \quad (37)$$

Using Eq. (36) and the fact that $E_0 = \eta H_0$ for this TM wave, we can rewrite Eq. (37) as

$$\langle S \rangle = \frac{H_0^2}{2} \left(Z_{TM} \cos^2 \frac{m\pi x}{a} \mathbf{a}_z - j\eta \frac{f_c}{f} \sin \frac{m\pi x}{a} \cos \frac{m\pi x}{a} \mathbf{a}_x \right) \quad (38)$$

To find the total power transfer down the guide, we integrate the axial component of the Poynting vector over the transverse area of the guide. Since we assumed the conducting planes were infinite, we shall have to be satisfied with finding the power transferred for a width of 1 m in the y direction.

$$\langle W_T \rangle = \int_0^1 \int_0^a \langle S_z \rangle \, dx \, dy = \int_0^1 \int_0^a \langle P_z \rangle \, dx \, dy \quad (39)$$

Using the power density given in Eq. (38), we obtain

$$\langle W_T \rangle = \int_0^1 \int_0^a \frac{H_0^2}{2} \left(Z_{TM} \cos^2 \frac{m\pi x}{a} \right) dx \, dy = \frac{a H_0^2 Z_{TM}}{4} \quad W \quad (40)$$

which is the power flow in the $\mathbf{a}_z$ direction for a width of 1 m.

It is often very important in waveguide systems to be able to determine the power transferred down the guide. The method illustrated above (for a particular case) is easily applied to waveguides of any shape and is valid for any type of electromagnetic wave. The procedure is to (1) find the time-average Poynting vector $\langle S \rangle$, and (2) integrate the axial component of $\langle S \rangle$ throughout the cross-sectional area of the guide.

In the next section, we will study the characteristics of an interesting class of waves which can propagate in a rectangular waveguide.

7.3 TE_{m0} WAVES IN RECTANGULAR WAVEGUIDES

The infinite parallel planes discussed in the preceding section are obviously not a practical waveguide. However, we can easily construct a practical

WAVEGUIDES

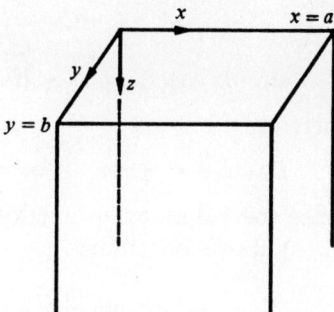

Fig. 7.3.1 The rectangular waveguide is constructed by adding conducting planes at $y = 0$ and $y = b$ to the infinite conducting planes used previously.

waveguide by considering a width b of the infinite planes and enclosing this width in two parallel planes normal to the first two planes. The result is a waveguide of rectangular cross section, as shown in Fig. 7.3.1.

The subject of this section is to determine the simplest types of waves which can propagate in rectangular waveguides and to understand some of the characteristics of these waves. We start with a class of waves which is easy to derive from a family of waves capable of propagating between infinite parallel planes.

THE TE$_{m0}$ WAVES

We begin the derivation by finding the TE waves which can exist between infinite parallel conducting planes. Starting with Eqs. (30) to (32) of Sec. 3.7, for the oblique incidence of a wave on a perfect conductor with **H** in the plane of incidence, we translate to the coordinate system of Fig. 7.2.2. We therefore replace z with x and x with $-z$. The components of **H** must also be relabeled, so that $-H_z$ replaces H_x and H_x replaces H_z. Thus Eq. (30) of Sec. 3.7 becomes

$$E_y = -j2E^+ \sin(\beta x \cos \theta) \exp(-j\beta z \sin \theta) \tag{1}$$

and Eqs. (32) and (31) of Sec. 3.7 become

$$H_x = j\frac{2E^+}{\eta} \sin \theta \sin(\beta x \cos \theta) \exp(-j\beta z \sin \theta) \tag{2}$$

and

$$H_z = \frac{2E^+}{\eta} \cos \theta \cos(\beta x \cos \theta) \exp(-j\beta z \sin \theta) \tag{3}$$

respectively. Applying the boundary conditions for the planes at $x = 0$ and $x = a$, we see that

$$E_y = 0 \quad \text{at } x = 0 \text{ and } x = a$$

requires that
$$\sin(\beta x \cos \theta) = 0 \quad \text{if } x = 0 \text{ or } x = a$$
therefore
$$\beta a \cos \theta = \pm m\pi \qquad m = 1, 2, 3, \ldots \tag{4}$$

Since $\cos \theta$ is an even function, we replace $\beta \cos \theta$ with $m\pi/a$ in Eqs. (1) to (3) above, obtaining

$$E_y = -j2E^+ \sin \frac{m\pi x}{a} \exp(-j\beta z \sin \theta) \tag{5}$$

$$H_x = j \frac{2E^+}{\eta} \sin \theta \sin \frac{m\pi x}{a} \exp(-j\beta z \sin \theta) \tag{6}$$

$$H_z = \frac{2E^+}{\eta} \cos \theta \cos \frac{m\pi x}{a} \exp(-j\beta z \sin \theta) \tag{7}$$

Relations from Secs. 3.7 and 7.2 which we have used previously are

$$\beta = \frac{\omega}{v_p} \quad \text{free-space values}$$

$$\beta \sin \theta = \beta \left[1 - \left(\frac{f_c}{f}\right)^2 \right]^{1/2} = \beta'$$

$$\cos \theta = \frac{f_c}{f}$$

where the cutoff frequency† is

$$f_c = \frac{m}{2a\sqrt{\mu\varepsilon}} \tag{8}$$

The wave impedance for a wave traveling in the $+z$ direction is

$$Z_{TE} = \frac{-E_y}{H_x} = \frac{\eta}{[1 - (f_c/f)^2]^{1/2}} \tag{9}$$

Using these relations and the substitution of E_0 for $-j2E^+$, we can rewrite Eqs. (5) to (7):

$$E_y = E_0 \sin \frac{m\pi x}{a} e^{-j\beta' z} \tag{10}$$

$$H_x = \frac{-E_0}{Z_{TE}} \sin \frac{m\pi x}{a} e^{-j\beta' z} \tag{11}$$

$$H_z = j \frac{E_0 f_c}{\eta\ f} \cos \frac{m\pi x}{a} e^{-j\beta' z} \tag{12}$$

† Note that the cutoff frequency for TE_{m0} waves is the same as for TE or TM waves between parallel planes. Why?

WAVEGUIDES

We now have a simple set of equations for a TE_{m0} wave propagating between infinite planes, and this wave has the same frequency-cutoff characteristics as the TM wave discussed in Sec. 7.2 and given in Eq. (8).

The TE_{m0} waves are the simplest waves which can propagate in a rectangular waveguide, and the TE_{10} mode, in particular, is widely used. Probably the primary reason for the popularity of the TE_{10} mode is that the cutoff frequency for this mode is lower than the cutoff frequency of any other mode which will propagate in a rectangular guide, assuming $a > b$. For a given frequency, using the TE_{10} mode allows one to use the smallest rectangular waveguide possible. In the centimeter-wavelength portions of the spectrum, smaller waveguides are cheaper because less metal is used. But when the cutoff wavelength of the guide is much less than a centimeter or so, the smaller waveguides become quite expensive because of increased machining difficulties.

In the next section, we will investigate some of the other, more complicated modes which can propagate in a rectangular guide.

7.4 A PROGRAMMED EXAMPLE ON RECTANGULAR WAVEGUIDES

This programmed example introduces some of the more complex modes which propagate in rectangular waveguides. The cases considered will probably seem less tedious than our previous ones, however, because we work just with the coordinates of the waveguides (and not with ξ coordinates). Also, by omitting some of the derivational details, we can focus more attention on treating the waveguide as a boundary-value problem.

The basic differential equations for the fields of any waveguide (charge-free inside the guide) are the vector wave equations (see Sec. 2.3)

$$\nabla^2 \mathbf{E} = \mu\sigma \frac{\partial \mathbf{E}}{\partial t} + \mu\varepsilon \frac{\partial^2 \mathbf{E}}{\partial t^2} \tag{1}$$

$$\nabla^2 \mathbf{H} = \mu\sigma \frac{\partial \mathbf{H}}{\partial t} + \mu\varepsilon \frac{\partial^2 \mathbf{H}}{\partial t^2} \tag{2}$$

An infinite number of solutions are possible, but a properly given set of boundary conditions will determine the solution uniquely. Also, in a given waveguide $\mathbf{E}$ and $\mathbf{H}$ differ, even though they both satisfy the vector wave equation, because they must satisfy different boundary conditions.

We will use the method of solving partial differential equations called *separation of variables* and assume that the solution can be written as a product of simpler functions. Substitution of the product into the original partial differential equation reduces it to simpler differential equations which can be solved, subject to the boundary conditions.

PART 1 HELMHOLTZ' EQUATION

Electrical engineers are especially interested in the sinusoidal time function, and so we will assume that the electric field intensity $\mathbf{E}(x,y,z,t)$ in Eq. (1) can be replaced by the product $\mathbf{E}(x,y,z)e^{j\omega t}$. Show that this substitution reduces the vector wave equation (1) to

$$\nabla^2 \mathbf{E}(x,y,z) + \beta^2 \left(1 - j\frac{\sigma}{\omega\varepsilon}\right) \mathbf{E}(x,y,z) = 0 \qquad (3)$$

called the *vector Helmholtz equation*. Of course, Eq. (2) can be similarly reduced, but we will concentrate on the $\mathbf{E}$ field. Now explain why the following approximation is valid for waveguides:

$$\gamma^2 = \beta^2 \left(1 - j\frac{\sigma}{\omega\varepsilon}\right) \approx \beta^2 \qquad (4)$$

In Chap. 1 it was noted that the vector laplacian separates in rectangular coordinates (see Table 1.2.2). Show that for waveguides, the vector Helmholtz equation reduces to three scalar Helmholtz equations

$$\nabla^2 E_x + \beta^2 E_x = 0 \qquad (5)$$
$$\nabla^2 E_y + \beta^2 E_y = 0 \qquad (6)$$
$$\nabla^2 E_z + \beta^2 E_z = 0 \qquad (7)$$

In general, the components E_x, E_y, and E_z of the vector $\mathbf{E}(x,y,z)$ are functions of all three variables x, y, and z. Although it might appear that our problem is compounding, since we now have three partial differential equations to solve rather than a single vector wave equation, this is not the case. Boundary conditions can be easily applied in many cases to Eqs. (5) to (7), whereas Eq. (1) is normally difficult to solve.

PART 2 UNIFORM PLANE WAVES

The simplest electromagnetic wave is the uniform plane wave in free space, discussed extensively in Chap. 3. Assume a linearly polarized wave of the form

$$\mathbf{E}(x,y,z) = E_x(x,y,z)\mathbf{a}_x \qquad (8)$$

so that $E_y(x,y,z) = 0 = E_z(x,y,z)$. Since the wave is uniform, show that

$$\frac{\partial E_x(x,y,z)}{\partial x} = 0 = \frac{\partial E_x(x,y,z)}{\partial y} \qquad (9)$$

Now prove that the three scalar Helmholtz equations reduce to the simple form

$$\frac{d^2 E_x}{dz^2} + \beta^2 E_x = 0 \qquad (10)$$

WAVEGUIDES

and that this equation has a solution

$$E_x = E^+ e^{-j\beta z} + E^- e^{+j\beta z} \tag{11}$$

where E^+ and E^- are arbitrary constants with units of volts per meter. Prove that Eq. (11) is composed of two traveling waves which carry energy in the $\pm z$ direction.

PART 3 NONUNIFORM PLANE WAVES

Suppose that we have a uniform TEM wave propagating in free space and we erect two large parallel plates in the field, oriented as shown in Fig. 7.4.1a. We want to know whether the wave will propagate in the region between the plates.

The boundary conditions at either plate require that

$$E_y = 0 = E_z \quad \text{at } x = 0 \text{ and } x = a \tag{12}$$

for perfect conductors (why?). In addition, we have assumed a uniform TEM wave, and so the boundary conditions of Eq. (9) apply. Thus the boundary conditions are the same as for the uniform plane wave of part 2, and the solution is given by Eq. (11). We conclude that the wave will propagate through the region between the plates as if they were not there (neglecting any edge effects). We could obtain this same result from the more general solution given in Sec. 7.2. Specifically, prove that the positive-going wave of Eq. (11) can be obtained from Eqs. (2) and (3) of Sec. 7.2. Note that for the case at hand, $\theta = \pi/2$.

The situation is significantly different if we erect the parallel plates as shown in Fig. 7.4.1b, where E_x is parallel to the plates. We want to

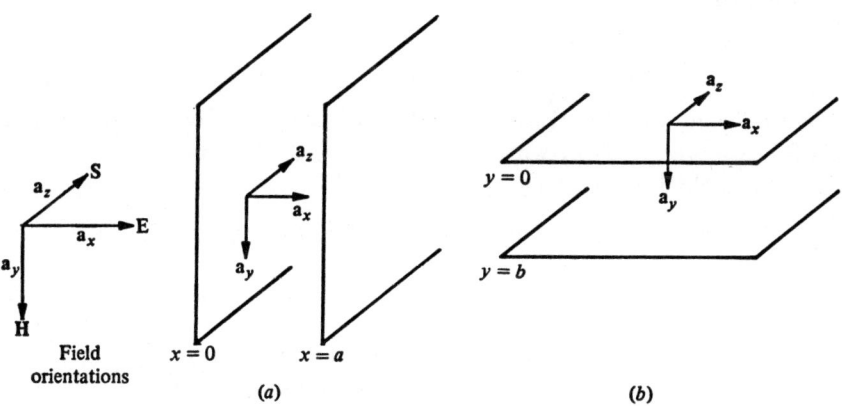

Fig. 7.4.1 The type of wave propagation between parallel plates depends upon the orientation of the plates with respect to the polarized wave.

determine whether the given uniform plane wave can propagate in the region between the plates. For this TEM wave, $E_y = 0 = E_z$, as before, so that only one scalar Helmholtz equation (5) is useful here. Show that the boundary conditions are

$$E_x = 0 \quad \text{at } y = 0 \text{ and } y = b \tag{13}$$

and consequently the E field cannot be uniform in the region between the plates (unless it is zero).

To investigate the possibility of a nonuniform wave, we assume that E_x can vary in the y and z directions but not in the x direction. Show why this assumption is reasonable. We will also assume that the field $E_x(y,z)$ can be written as the product

$$E_x(y,z) = V(y)W(z) \tag{14}$$

where V and W are unknown functions to be determined. By substituting Eq. (14) into the scalar Helmholtz equation, show that we get the following two differential equations (in which the variables are separated):

$$\frac{\partial^2 V}{\partial y^2} + \xi^2 V = 0 \tag{15}$$

$$\frac{\partial^2 W}{\partial z^2} + (\beta^2 - \xi^2)W = 0 \tag{16}$$

Here ξ^2 is an arbitrary constant, called the *separation constant*, to be determined by the boundary conditions. Prove that solutions of Eqs. (15) and (16) are, respectively,

$$V(y) = \sin \xi y \tag{17}$$

and

$$W(z) = E_0^{\pm} \exp(\mp j \sqrt{\beta^2 - \xi^2}\, z) \tag{18}$$

where E_0^+ and E_0^- are arbitrary constants determined by the strength of the field sources. (They will be left arbitrary in this example.)

When the boundary conditions of Eq. (13) are applied to Eq. (17), show that

$$\xi = \frac{n\pi}{b} \qquad n = 1, 2, 3, \ldots \tag{19}$$

This is an important step, essential for an understanding of fields in waveguides. We obtain an infinite number of possible solutions, one for each value of the mode number n. The frequency of the source will determine which one (or more) mode will actually exist. The E fields for three modes are sketched in Fig. 7.4.2.

WAVEGUIDES

In conclusion, we can obtain the solutions for the **E** field from Eqs. (14), (17), and (18):

$$E_x(y,z) = E_0^+ \sin\frac{n\pi y}{b} e^{-j\beta'z} + E_0^- \sin\frac{n\pi y}{b} e^{+j\beta'z} \tag{20}$$

where

$$\beta' = \sqrt{\beta^2 - \left(\frac{\pi n}{b}\right)^2} \tag{21}$$

Use Eqs. (20) and (21) to prove that wave propagation does not occur below the cutoff frequency. Prove that the cutoff frequency for this example is given by Eq. (12) of Sec. 7.2. Compare Eq. (20) with Eq. (1) of Sec. 7.3 and explain any differences that occur. Can you show that it is possible for multiple modes to exist simultaneously between the plates?

Once the **E** field has been determined, the **H** field can be obtained from Maxwell's curl equation, rather than from the scalar Helmholtz equation. Using the assumption that the time variation of **H** is sinusoidal, show that

$$\nabla \times \mathbf{E} = -j\omega\mu\mathbf{H} \tag{22}$$

Now find $\nabla \times \mathbf{E}$ from Eq. (20) and show that **H** must have both y and z components

$$\mathbf{H} = H_y(y,z)\mathbf{a}_y + H_z(y,z)\mathbf{a}_z \tag{23}$$

From this derivation, find the expressions for $H_y(y,z)$ and $H_z(y,z)$ for this particular case. Sketch these components of **H** as was done for **E** in Fig. 7.4.2.

We find that the Poynting vector

$$\mathbf{E} \times \mathbf{H} = -E_x H_z \mathbf{a}_y + E_x H_y \mathbf{a}_z \tag{24}$$

is not along the z axis but has a component in the $\mathbf{a}_y$ direction. Consequently, the TEM wave will follow a zigzag path between the plates,

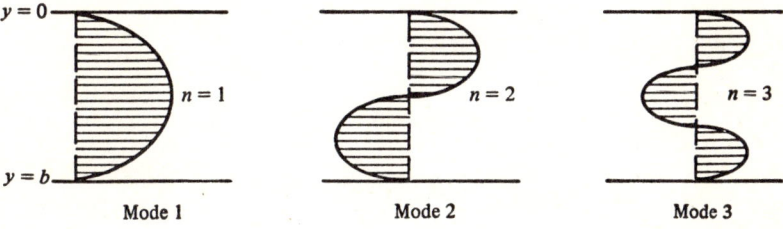

Fig. 7.4.2 The nonuniform E field for three modes between the plates.

but the net power flow is in the $\mathbf{a}_z$ direction (from which came the designation TE mode).

PART 4 RECTANGULAR WAVEGUIDES

Suppose the parallel plates of Fig. 7.4.1b are replaced with the rectangular waveguide oriented as shown in Fig. 7.4.3. As before, the basic equations are the scalar Helmholtz equations (5) to (7), under the assumption of perfectly conducting walls and a charge-free dielectric (or free space) inside the waveguide. To use the method of separation of variables again, assume that

$$E_x(x,y,z) = U_x(x)V_x(y)W_x(z)$$
$$E_y(x,y,z) = U_y(x)V_y(y)W_y(z) \quad (25)$$
$$E_z(x,y,z) = U_z(x)V_z(y)W_z(z)$$

where the U's, V's, and W's are nine unknown functions. Determine these unknowns by substituting Eqs. (25) into the scalar Helmholtz equations, obtaining second-order differential equations, like Eqs. (15) and (16), with separation constants.

To evaluate the separation constants, you must use the boundary conditions that apply to the given waveguide

$$E_y = 0 = E_z \quad \text{at } x = 0 \text{ and } x = a \quad (26)$$
$$E_x = 0 \quad \text{at } y = 0 \text{ and } y = b \quad (27)$$

Show that the separation constants are

$$\xi_1 = \frac{m\pi}{a} \quad m = 0, 1, 2, 3, \ldots \quad (28)$$

$$\xi_2 = \frac{n\pi}{b} \quad n = 0, 1, 2, 3, \ldots \quad (29)$$

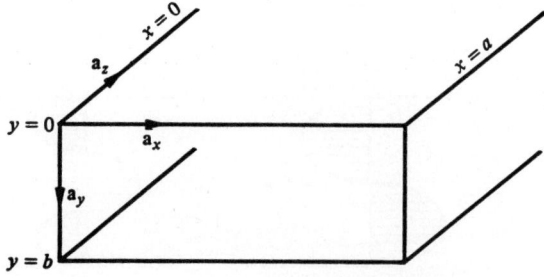

Fig. 7.4.3 TE or TM waves can propagate through the rectangular waveguide in several modes.

WAVEGUIDES

and the phase constant is

$$\beta' = \sqrt{\beta^2 - \left(\frac{m\pi}{a}\right)^2 - \left(\frac{n\pi}{b}\right)^2} \tag{30}$$

Finally, prove that the solutions for the positive-going waves are

$$E_x = E_{x0}{}^+ \cos\frac{m\pi x}{a} \sin\frac{n\pi y}{b} e^{-j\beta' z} \tag{31}$$

$$E_y = E_{y0}{}^+ \sin\frac{m\pi x}{a} \cos\frac{n\pi y}{b} e^{-j\beta' z} \tag{32}$$

$$E_z = E_{z0}{}^+ \sin\frac{m\pi x}{a} \sin\frac{n\pi y}{b} e^{-j\beta' z} \tag{33}$$

where $E_{x0}{}^+$, $E_{y0}{}^+$, and $E_{z0}{}^+$ are arbitrary constants with units of volts per meter.

In the case of the transverse electric wave there can be no $\mathbf{a}_z$ component, and so the constant $E_{z0}{}^+$ must be zero. This situation is denoted TE$_{mn}$, where m is the mode number in the $\mathbf{a}_x$ direction (usually the longer dimension of the rectangle) and n is the mode number in the $\mathbf{a}_y$ direction. The simplest mode that can exist is TE$_{10}$, called the *dominant mode*. Obtain the equations for this mode from Eqs. (31) to (33).

The equations for the **H** field can be obtained from the curl equation (22). Show that the solutions corresponding to Eqs. (31) to (33) are as follows:

$$H_x = \frac{E_y}{\eta\beta}\left(-\beta' + j\frac{n\pi}{b}\frac{E_{z0}{}^+}{E_{y0}{}^+}\right) \tag{34}$$

$$H_y = \frac{E_x}{\eta\beta}\left(\beta' - j\frac{m\pi}{a}\frac{E_{z0}{}^+}{E_{x0}{}^+}\right) \tag{35}$$

$$H_z = \frac{j}{\eta\beta}\left(\frac{m\pi E_{y0}{}^+}{a} - \frac{n\pi E_{x0}{}^+}{b}\right)\cos\frac{m\pi x}{a}\cos\frac{n\pi y}{b} e^{-j\beta' z} \tag{36}$$

The **H** field for any TE$_{mn}$ mode can be obtained from these equations by substitution of $E_{z0}{}^+ = 0$ and the given constants for m and n.

The transverse magnetic mode TM$_{mn}$ can be found by setting H_z equal to zero. Choose the constants so that

$$\frac{mE_{y0}{}^+}{a} = \frac{nE_{x0}{}^+}{b} \tag{37}$$

or

$$E_{y0}{}^+ = \frac{na}{mb}E_{x0}{}^+ \tag{38}$$

Prove that the simplest TM mode is TM_{11} since all the fields must be zero if either $m = 0$ or $n = 0$. Sketch the **E** and **H** fields for the TM_{11} mode.

In conclusion, Eqs. (31) to (36) describe the fields which can exist in rectangular waveguides. Either TE or TM modes can exist, and since superposition of these modes is possible, the resulting waves in the guide can be quite complicated.

7.5 WAVEGUIDE BOUNDARY-VALUE PROBLEMS

Electromagnetic waves will propagate in waveguides of almost any cross-sectional shape, although analysis of the boundary-value problem can be very difficult in odd-shaped guides. In practice, the most commonly used guides have rectangular, circular, or elliptic cross sections. The advantages of these guides are that (1) analysis is relatively simple, (2) desired modes are easily excited, and (3) construction and assembly are not complicated. We have studied the rectangular guide in detail because a thorough understanding of it will help us understand other guides.

Because of our study of rectangular guides we are now in a position to summarize the most commonly used method of solving such boundary-value problems. The problem is, given a waveguide of known cross section, to find the various modes of the **E** and **H** fields that can propagate. The procedure is as follows:

1. Choose a coordinate system that is most natural for the cross section, i.e., one in which the equations for the boundaries are simply expressed.
2. The vector Helmholtz equations

$$\nabla^2 \mathbf{E} + \beta^2 \mathbf{E} = 0 \tag{1}$$

and

$$\nabla^2 \mathbf{H} + \beta^2 \mathbf{H} = 0 \tag{2}$$

apply in any charge-free low-loss waveguide. Express the vector laplacian in the chosen coordinate system. Note that

$$\nabla^2 \mathbf{E} = -\nabla \times \nabla \times \mathbf{E} \tag{3}$$

and

$$\nabla^2 \mathbf{H} = -\nabla \times \nabla \times \mathbf{H} \tag{4}$$

because the divergence of **E** and **H** is zero.

WAVEGUIDES

3. From Eq. (1) or (2) obtain three partial differential equations by equating each vector component to zero. In general, these three equations will be coupled, involving more than one of the dependent variables E_u, E_v, and E_w [or H_u, H_v, and H_w if Eq. (2) is used], where u, v, and w are the coordinates of the chosen system. Ideally, one or more of the three partial differential equations will involve just one of the variables. For example, in the rectangular system, we obtained Eqs. (5) to (7) of Sec. 7.4, each in a single dependent variable.
4. Assume separation of variables and write the dependent variables as products of functions of u, v, and w. See Eq. (14) of Sec. 7.4 for an example using two terms. Substitute these functions into the partial differential equations, attacking the simplest one first. If this method is to be used for the given problem, these substitutions must yield differential equations in single variables, related only by separation constants. Often the problem can be simplified by searching only for the TE or TM modes, so that longitudinal components of the transverse wave can be assumed to be zero. If none of the equations separate, some other method must be used (see any text on partial differential equations).
5. Apply the boundary conditions to the ordinary differential equations obtained in step 4 and solve in terms of the mode numbers and arbitrary constants.
6. After a few components are found, it is generally easier to use Maxwell's curl equations for $\nabla \times \mathbf{E}$ and $\nabla \times \mathbf{H}$ to obtain other components than to continue with the separation of variables. Final solutions will be of the form

$$\mathbf{E} = E_u \mathbf{a}_u + E_v \mathbf{a}_v + E_w \mathbf{a}_w$$
$$\mathbf{H} = H_u \mathbf{a}_u + H_v \mathbf{a}_v + H_w \mathbf{a}_w \tag{5}$$

where each component is (in general) a product of functions of u, v, and w.

7.6 CIRCULAR WAVEGUIDES

The six-step procedure given in Sec. 7.5 will now be applied to circular waveguides like that shown in Fig. 7.6.1.

Step 1 The cylindrical coordinate system (r,ϕ,z) is a natural one for this geometry. The tangential $\mathbf{E}$ field must be zero on the perfectly conducting cylinder, and so the boundary condition is

$$E_\phi(r_0,\phi,z) = 0 \tag{1}$$

where r_0 is the radius of the cylinder.

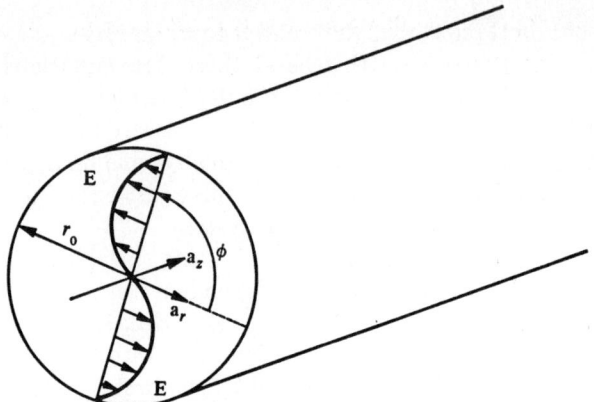

Fig. 7.6.1 A cross section of cylindrical waveguide showing the E field for the TE_{01} mode.

Steps 2 and 3 For the vector Helmholtz equations we write

$$-\nabla \times \nabla \times \mathbf{E} + \beta^2 \mathbf{E} = 0 \qquad (2)$$

$$-\nabla \times \nabla \times \mathbf{H} + \beta^2 \mathbf{H} = 0 \qquad (3)$$

Now consider Eq. (2) and show that the vector components are as follows:
The $\mathbf{a}_r$ component:

$$\frac{\partial^2 E_r}{\partial r^2} + \frac{1}{r}\frac{\partial E_r}{\partial r} - \frac{E_r}{r^2} + \frac{1}{r^2}\frac{\partial^2 E_r}{\partial \phi^2} - \frac{2}{r^2}\frac{\partial E_\phi}{\partial \phi} + \frac{\partial^2 E_r}{\partial z^2} + \beta^2 E_r = 0 \qquad (4)$$

The $\mathbf{a}_\phi$ component:

$$\frac{\partial^2 E_\phi}{\partial r^2} + \frac{1}{r}\frac{\partial E_\phi}{\partial r} - \frac{E_\phi}{r^2} + \frac{1}{r^2}\frac{\partial^2 E_\phi}{\partial \phi^2} + \frac{2}{r^2}\frac{\partial E_r}{\partial \phi} + \frac{\partial^2 E_\phi}{\partial z^2} + \beta^2 E_\phi = 0 \qquad (5)$$

The $\mathbf{a}_z$ component:

$$\frac{\partial^2 E_z}{\partial r^2} + \frac{1}{r}\frac{\partial E_z}{\partial r} + \frac{1}{r^2}\frac{\partial^2 E_z}{\partial \phi^2} + \frac{\partial^2 E_z}{\partial z^2} + \beta^2 E_z = 0 \qquad (6)$$

Although these three equations appear rather formidable, note that Eqs. (4) and (5) are of the same form except for the sign of the fifth term. Also, these two equations are said to be *coupled* because of the occurrence of the variables E_ϕ and E_r respectively in the fifth terms. However, Eq. (6) is not coupled to the others because E_z is the one and only variable. These same equations could be obtained for the magnetic field components of Eq. (3) with the E-field variables replaced by the corresponding H-field variables (E_r replaced by H_r, etc.).

WAVEGUIDES

Step 4 Assume that the component $E_z(r,\phi,z)$ of Eq. (6) can be separated into a product of three functions

$$E_z(r,\phi,z) = R(r)P(\phi)e^{-j\gamma z} \tag{7}$$

The particular function $e^{-j\gamma z}$ was chosen since we are interested in waves that propagate in the $\mathbf{a}_z$ direction. The time function $e^{j\omega t}$ has been omitted, as was customary with transmission-line analysis (see Sec. 5.2). The propagation constant γ is unknown but will be found. $R(r)$ and $P(\phi)$ are unknown functions of single variables, and the immediate object is to find them. Substituting Eq. (7) into Eq. (6), we can show that

$$r^2 \frac{R''}{R} + r\frac{R'}{R} + r^2(\beta^2 - \gamma^2) + \frac{P''}{P} = 0 \tag{8}$$

where

$$R' = \frac{\partial R}{\partial r} \qquad R'' = \frac{\partial^2 R}{\partial r^2} \qquad P'' = \frac{\partial^2 P}{\partial \phi^2} \tag{9}$$

Since P''/P is not a function of r and the first three terms of Eq. (8) are not functions of ϕ, the equation consists of two independent parts which must add to zero for all values of r and ϕ. Thus there exists some separation constant ξ^2 such that

$$\frac{P''}{P} = -\xi^2 \tag{10}$$

$$r^2 \frac{R''}{R} + r\frac{R'}{R} + r^2(\beta^2 - \gamma^2) = \xi^2 \tag{11}$$

Step 5 These ordinary differential equations can be solved. From Eq. (10), show that

$$P(\phi) = A_\phi e^{j\xi\phi} + B_\phi e^{-j\xi\phi} \tag{12}$$

where A_ϕ and B_ϕ are arbitrary complex constants. Since the **E** field exists in a cylindrical waveguide, the field must repeat at least once each revolution as ϕ increases. That is,

$$P(0) = P(2\pi m) \qquad m = 0, 1, 2, \ldots \tag{13}$$

Therefore you can show that ξ must be an integer

$$\xi = m = 0, 1, 2, \ldots \tag{14}$$

It is a *mode number* describing the number of repetitions of the field pattern as ϕ changes from zero to 2π. Further, by selecting the zero for

ϕ we can express the solution in terms of a sine or cosine function. For simplicity, we let

$$P(\phi) = A_\phi e^{\pm jm\phi} \qquad (15)$$

where the sign is chosen by any boundary conditions imposed on $P(\phi)$.

Equation (11) can be written in a standard form called *Bessel's equation*

$$R'' + \frac{1}{r}R' + \left(\beta^2 - \gamma^2 - \frac{m^2}{r^2}\right)R = 0 \qquad (16)$$

Solutions of this equation, called *Bessel functions*, can be obtained by the Frobenius method with series solutions.† The Bessel functions cannot be expressed in terms of simpler functions. In this respect the Bessel functions are similar to the sine and cosine functions, which are fundamentally defined by infinite series. The solutions of Eq. (16) of interest here are called *Bessel functions of the first kind*‡ and are given by

$$J_m(qr) = \frac{(qr)^m}{2^m m!}\left[1 - \frac{(qr)^2}{2(2m+2)} + \frac{(qr)^4}{(2)(4)(2m+2)(2m+4)} \cdots\right] \qquad (17)$$

For our particular problem we let

$$q^2 = \beta^2 - \gamma^2 \qquad (18)$$

where the unknown constant γ is yet to be determined. Since these Bessel functions have been extensively tabulated, one seldom resorts to actual evaluation of the series. (The same is true for the trigonometric functions, of course.) In conclusion, the solution of Eq. (16) is

$$R(r) = A_2 J_m(qr) \qquad (19)$$

where A_2 is an arbitrary complex constant.

The E_z component can now be written as a product of known functions by substitution of Eqs. (15) and (19) into (7)

$$E_z(r,\phi,z) = A_0 J_m(qr) e^{\pm jm\phi} e^{-j\gamma z} \qquad (20)$$

---

† Most texts on differential equations cover series solutions and the Frobenius method. Also many references are available on Bessel functions, e.g., Frank Bowman, "Introduction to Bessel Functions," Dover Publications, Inc., New York, 1958; N. W. McLachlan, "Bessel Functions for Engineers," Oxford University Press, London, 1955; G. N. Watson, "A Treatise on the Theory of Bessel Functions," Cambridge University Press, Cambridge, 1944 (Dover Publications, Inc., New York, 1966).

‡ Other Bessel series solutions exist; hence "of the first kind" is used to distinguish this particular series.

WAVEGUIDES

where A_0 is an arbitrary complex constant which must be selected to satisfy boundary conditions.

Since the magnetic field must satisfy the same partial differential equations, but with different boundary conditions, we can show that the solution for H_z is

$$H_z(r,\phi,z) = B_0 J_m(qr) e^{\mp j(m\phi - \pi/2)} e^{-j\gamma z} \tag{21}$$

where B_0 is the arbitrary constant.

Step 6 We will not solve Eqs. (4) and (5) because of their complexity, but we can obtain the remaining components from the curl equations

$$\nabla \times \mathbf{E} = -j\omega\mu\mathbf{H} \tag{22}$$

$$\nabla \times \mathbf{H} = j\omega\varepsilon\mathbf{E} \tag{23}$$

From the components of these equations, and using the known functions for E_z and H_z, we can obtain the following results:

$$E_r = \frac{-j}{q^2}\left[\frac{m\omega\mu B_0}{r} J_m(qr) e^{\mp jm\phi} + \gamma A_0 J'_m(qr) e^{\pm jm\phi}\right] e^{-j\gamma z} \tag{24}$$

$$E_\phi = \frac{j}{q^2}\left[\frac{m\gamma A_0}{r} J_m(qr) e^{\pm j(m\phi - \pi/2)} + \omega\mu B_0 J'_m(qr) e^{j\mp(m\phi - \pi/2)}\right] e^{-j\gamma z} \tag{25}$$

$$H_r = \frac{-1}{\omega\mu}\left[\gamma E_\phi + \frac{jmA_0}{r} J_m(qr) e^{\pm j(m\phi - \pi/2)}\right] e^{-j\gamma z} \tag{26}$$

$$H_\phi = \frac{1}{\omega\mu}[\gamma E_r - jA_0 J'(qr) e^{\mp j(m\phi - \pi/2)}] e^{-j\gamma z} \tag{27}$$

where

$$J'_m(qr) = \frac{d}{dr} J_m(qr) \equiv \frac{m}{r} J_m(qr) - q J_{m+1}(qr) \tag{28}$$

The only remaining unknown is the propagation constant γ. It can be evaluated by substituting the boundary condition of Eq. (1) into Eq. (25), which requires that $E_\phi = 0$ at $r = r_0$. Consequently γ must be chosen so that the bracketed term in Eq. (25) is zero for $r = r_0$.

$$\frac{m\gamma A_0}{r_0} J_m(qr_0) + \omega\mu B_0 J'_m(qr_0) = 0 \tag{29}$$

This is a transcendental equation which cannot be solved directly for γ. Fortunately in most cases of interest, either $A_0 = 0$ or $B_0 = 0$, and so the problem reduces to finding the roots, or zeros, of either $J_m(qr_0)$ or $J'_m(qr_0)$. Extensive tables† of the values of qr_0 such that

$$J_m(qr_0) = 0 \quad \text{and} \quad J'_m(qr_0) = 0 \tag{30}$$

are available. Abbreviated forms of these tables are given. We will find

Zeros $q_n r_0$ of the Bessel functions

Root no.	$J_0 = 0$	$J_1 = 0$	$J_2 = 0$	$J_3 = 0$
$n = 1$	2.405	3.832	5.136	6.380
2	5.520	7.016	8.417	9.761
3	8.654	10.173	11.620	13.015
4	11.792	13.324	14.796	16.223

Zeros $q_n r_0$ of the Bessel function derivatives

Root no.	$J'_0 = 0$	$J'_1 = 0$	$J'_2 = 0$	$J'_3 = 0$
$n = 1$	3.832	1.841	3.054	4.201
2	7.016	5.331	6.706	8.015
3	10.173	8.536	9.969	11.346
4	13.324	11.706	13.170	14.586

that the first table is used for TM waves and the second for TE waves.

As an example of the use of these tables, suppose $B_0 = 0$, $A_0 \neq 0$ in Eq. (29), and let $m = 0$. Then Eq. (29) is satisfied by the values of qr_0 from Eq. (30) such that

$$J_0(qr_0) = 0 \tag{31}$$

From the first table, the zeros are

$$\begin{aligned} q_1 r_0 &= 2.405 \\ q_2 r_0 &= 5.520 \\ q_3 r_0 &= 8.654 \\ q_4 r_0 &= 11.792 \end{aligned} \tag{32}$$

where the subscript n of q_n denotes the root number.

† "Handbook of Mathematical Functions," National Bureau of Standards AMS-55, Washington, 1964.

WAVEGUIDES

This subscript is used in the notation TE_{mn} and TM_{mn} to identify the root, or zero. From Eq. (18)

$$\gamma = \sqrt{\beta^2 - q_n^2} \tag{33}$$

If for example, the second root is used ($n = 2$), then

$$\gamma = \sqrt{\beta^2 - \left(\frac{5.520}{r_0}\right)^2} \tag{34}$$

and hence γ can be evaluated for any given β and r_0.

The field components obtained in Eqs. (20), (21), and (24) to (27) are the general solutions of the circular waveguide problem. They can be simplified considerably if we restrict our solutions to either TE_{mn} or TM_{mn} modes. For example, TE waves are obtained when $E_z = 0$, which requires that A_0 in Eq. (20) be zero, and consequently Eqs. (24) to (27) can be simplified. Thus for TE_{mn} waves

$$E_r = \frac{j\omega\mu m B_0}{q_n^2 r} J_m(q_n r) \sin m\phi \, e^{-j\gamma z} \tag{35}$$

$$E_\phi = \frac{j\omega\mu B_0}{q_n^2} J'_m(q_n r) \cos m\phi \, e^{-j\gamma z} \tag{36}$$

$$E_z = 0 \tag{37}$$

$$H_r = -\frac{\gamma E_\phi}{\omega\mu} \tag{38}$$

$$H_\phi = \frac{\gamma E_r}{\omega\mu} \tag{39}$$

$$H_z = B_0 J_m(q_n r) \cos m\phi \, e^{-j\gamma z} \tag{40}$$

Similarly for TM waves, the H_z component must be zero, and so $B_0 = 0$. Thus for TM_{mn} waves

$$E_r = \frac{-j\gamma A_0}{q_n^2} J'_m(q_n r) \cos m\phi \, e^{-j\gamma z} \tag{41}$$

$$E_\phi = \frac{jm\gamma A_0}{q_n^2 r} J_m(q_n r) \sin m\phi \, e^{-j\gamma z} \tag{42}$$

$$E_z = A_0 J_m(q_n r) \cos m\phi \, e^{-j\gamma z} \tag{43}$$

$$H_r = \frac{-\omega\varepsilon E_\phi}{\gamma} \tag{44}$$

$$H_\phi = \frac{\omega\varepsilon E_r}{\gamma} \tag{45}$$

$$H_z = 0 \tag{46}$$

The simplest mode is the TM_{01} wave ($m = 0$, $n = 1$) where three of the six field components are zero, namely E_ϕ, H_r, and H_z. In this case

$$\gamma^2 = \beta^2 - \left(\frac{2.405}{r_0}\right)^2 \tag{47}$$

The TE_{01} wave also has a simple structure since $E_r = E_z = H_\phi = 0$ and

$$\gamma^2 = \beta^2 - \left(\frac{3.832}{r_0}\right)^2 \tag{48}$$

However, complete solutions of the wave equation can be obtained from superposition of simpler modes, and many different waves can propagate simultaneously. In most cases, most of the energy will be carried by a dominant mode which is determined by the method of excitation.

Any circular waveguide has a lower cutoff frequency, below which energy will not propagate in the guide. Note that γ must not be imaginary in the term $e^{-j\gamma z}$ if propagation is to occur, so that from Eq. (33)

$$\beta^2 > q_n^2 \tag{49}$$

The critical case occurs when $\beta^2 = q_n^2$. Substituting

$$\beta = \omega\sqrt{\mu\varepsilon} \tag{50}$$

we find that the cutoff frequency is

$$f_c = \frac{q_n}{2\pi\sqrt{\mu\varepsilon}} \tag{51}$$

where q_n is obtained from the tables for the zeros of $J_m(q_n r_0)$ and $J'_m(q_n r_0)$. For example, for the TM_{01} mode,

$$f_c = \frac{2.405}{2\pi r_0 \sqrt{\mu\varepsilon}} \tag{52}$$

and this mode has the lowest cutoff frequency of the TM_{mn} modes [since 2.405 is the smallest number in the table of zeros of $J_m(q_0 r_0)$]. Similarly, cutoff frequencies for the TE_{mn} modes can be obtained from the zeros of $J'_m(q_n r_0)$. The smallest of all the zeros, and hence the lowest cutoff frequency, occurs for the TE_{11} mode, for which

$$f_c = \frac{1.841}{2\pi r_0 \sqrt{\mu\varepsilon}} \tag{53}$$

The radius r_0 of the guide always appears in the denominator of the equation for cutoff frequency, and so we see that low-frequency propagation requires large-diameter waveguides.

WAVEGUIDES

In conclusion, the field components for circular waveguides can be found using the separation-of-variables method summarized in Sec. 7.5. In the steady-state condition, circular guides will propagate either TE or TM waves but not TEM waves. For either TE or TM modes, there are cutoff frequencies, determined by the diameter of the guide, below which any given mode will not propagate.

7.7 RESONANT CAVITIES

If a transmission line is shorted at the receiving end with a perfect conductor, standing waves develop and both the **E** and **H** fields have null points at periodic intervals along the line. Similar phenomena can occur if any waveguide is shorted: standing waves develop, and periodic zeros exist in the steady-state fields. Suppose we form a cavity by using perfectly conducting sheets to block each end of a section of waveguide and then insert a small probe to inject electromagnetic energy into the cavity. If there were no losses in the walls of the waveguide, standing waves would develop and continue to exist, forever trapped inside the cavity. If there are losses (and there always are), the fields decay exponentially with time unless additional energy is supplied.

More familiar examples of energy in cavities occur when sound waves are trapped inside reflecting enclosures, such as empty rooms or tunnels. In many musical instruments, the design of the resonant cavity can make the difference between high and poor quality.

In electromagnetic systems, cavities are used as resonant elements, similar to LC filters in circuits. The frequencies and modes at which the cavity is resonant are determined by the geometry of the cavity. Theoretically, a cavity has an infinite number of resonant frequencies, just as there are an infinite number of frequencies that will propagate above the lowest cutoff frequency in the waveguide. Cavities of almost any shape will serve as resonators, but mathematical analysis is generally confined to relatively simple cases.

Energy loss in a cavity is a function of the conductivity of the metal walls, and so minimum losses occur when highly conducting materials such as silver or copper are used. Any dielectric material inside the guide will increase the losses. Losses in cavities are described by a quality factor Q, which is the ratio of the energy stored to the energy lost per cycle (as in circuit theory). The Q of a well-designed cavity is on the order of tens of thousands, in contrast with a few hundred obtainable with LC circuits. Since the energy is stored in the fields inside the cavity but the losses occur primarily on the surface, the Q is approximately proportional to the ratio of the volume enclosed to the inner surface area. Thus large Q's are obtained with large volumes and small surface areas. The sphere is an ideal shape for a cavity, but it is more difficult to construct than a rectangular or cylindrical resonator. Its analysis is undertaken in more advanced texts.† For our analysis, we will consider the rectangular cavity only.

Given a rectangular waveguide oriented as shown in Fig. 7.4.3, we assume that perfectly conducting plates seal the ends at $z = 0$ and $z = c$. Then the boundary conditions are

$$E_y = 0 \quad E_z = 0 \quad \text{at } x = 0 \text{ and } x = a \tag{1}$$

$$E_x = 0 \quad E_z = 0 \quad \text{at } y = 0 \text{ and } y = b \tag{2}$$

$$E_x = 0 \quad E_y = 0 \quad \text{at } z = 0 \text{ and } z = c \tag{3}$$

We can solve the three Helmholtz equations

$$\nabla^2 E_i + \beta^2 E_i = 0 \quad i = x, y, z \tag{4}$$

subject to the boundary conditions, using the separation-of-variables technique. The results are very similar to those obtained in Sec. 7.4 for rectangular guides. For example, if we assume

$$E_x(x,y,z) = U_x(x)V_x(y)W_x(z) \tag{5}$$

the Helmholtz equation for E_x becomes

$$\frac{U_x''(x)}{U_x(x)} + \frac{V_x''(y)}{V_x(y)} + \frac{W_x''(z)}{W_x(z)} + \beta^2 = 0 \tag{6}$$

where the double primes denote second derivatives with respect to the given independent variables of each function. To solve for $W_x(z)$, we replace the functions of x and y with constants of separation

$$-\xi_x^2 - \xi_y^2 + \frac{W_x''(z)}{W_x(z)} + \beta^2 = 0 \tag{7}$$

and let

$$\xi_z^2 = \beta^2 - \xi_x^2 - \xi_y^2 \tag{8}$$

Then Eq. (7) reduces to

$$W_x'' + \xi_z^2 W_x = 0$$

and from the boundary condition of Eq. (3) we see that

$$W_x(z) = A_1 \sin \xi_z z \tag{9}$$

$$\xi_z = \frac{p\pi}{c} \quad p = 0, 1, 2, \ldots \tag{10}$$

† J. A. Stratton, "Electromagnetic Theory," McGraw-Hill Book Company, New York, 1941.

WAVEGUIDES

Similarly we can solve for $V_x(y)$, using the boundary condition of Eq. (2),

$$V_x(y) = A_2 \sin \xi_y y \tag{11}$$

$$\xi_y = \frac{n\pi}{b} \qquad n = 0, 1, 2, \ldots \tag{12}$$

When we attempt to evaluate $U_z(x)$, we find that none of the boundary conditions apply; we also find that, in general, $U_z \neq 0$ at $x = 0$ and at $x = a$. However, we can write the solution in the form

$$U_z(x) = A_3 e^{-j\xi_x x} + B_3 e^{+j\xi_x x} \tag{13}$$

and when we evaluate the separation constants from the Helmholtz equations for E_y or E_z, we find

$$\xi_x = \frac{m\pi}{a} \qquad m = 0, 1, 2, \ldots \tag{14}$$

The constants A_3 and B_3 can be evaluated for any given field intensity, as determined by the field source. The simplest solution is the cosine wave. In that case

$$E_x = A \cos\left(\frac{m\pi}{a} x\right) \sin\left(\frac{n\pi}{b} y\right) \sin\left(\frac{p\pi}{c} z\right) \tag{15}$$

where the constant A is the product of the previous constants A_1, A_2, A_3, selected to satisfy the E_x component of the field intensity.

The phase constant can be written in terms of the cavity dimensions from Eqs. (8), (10), (12), and (14)

$$\beta^2 = \left(\frac{m\pi}{a}\right)^2 + \left(\frac{n\pi}{b}\right)^2 + \left(\frac{p\pi}{c}\right)^2 \tag{16}$$

This equation is valid for all three of the Helmholtz equations given in Eq. (4). The solutions of the other two Helmholtz equations, found in a similar manner, are

$$E_y = B \sin\left(\frac{m\pi}{a} x\right) \cos\left(\frac{n\pi}{b} y\right) \sin\left(\frac{p\pi}{c} z\right) \tag{17}$$

$$E_z = C \sin\left(\frac{m\pi}{a} x\right) \sin\left(\frac{n\pi}{b} y\right) \cos\left(\frac{p\pi}{c} z\right) \tag{18}$$

The constants A, B, and C must be selected so that the rather complex excitation conditions on the vector E field are satisfied. Again, simplifications were made to obtain the cosine terms. In the transverse electric mode TE_{mnp} the E_z component is zero, and so in that case the constant C is zero. The notation TE_{mnp} is a direct extension of the waveguide notation. The integer m indicates the number of half wavelengths of the standing wave along the x direction; n, along the y direction; and p, along the z direction. Thus a TE_{101} mode has one half wavelength on each of the x and z axes and none on the y axis. Since there is no unique orientation of the cavity (such as direction of power flow in a waveguide), the notation depends upon the arbitrary selection of the axes with respect to the cavity.

Equations (15) to (18) for the electric fields and β should be compared with Eqs. (30) to (33) of Sec. 7.4. The similarity of the cavity fields and the waveguide solutions becomes apparent.

The **H** fields for the rectangular cavity can be obtained from the curl equation

$$\nabla \times \mathbf{E} = -j\omega\mu\mathbf{H} \tag{19}$$

by equating components

$$H_x = \frac{j}{\omega\mu}\left(\frac{\partial E_z}{\partial y} - \frac{\partial E_y}{\partial z}\right) \tag{20}$$

$$H_y = \frac{j}{\omega\mu}\left(\frac{\partial E_x}{\partial z} - \frac{\partial E_z}{\partial x}\right) \tag{21}$$

$$H_z = \frac{j}{\omega\mu}\left(\frac{\partial E_y}{\partial x} - \frac{\partial E_x}{\partial y}\right) \tag{22}$$

Since the electric fields are known, the magnetic fields can be obtained. If the TM_{mnp} mode is to be obtained, the H_z component must be zero and the constants must be determined to fulfill this condition.

The resonant frequency of the cavity can be found from Eq. (16)

$$f_r = \frac{v_p \beta}{2\pi} = \frac{1}{2\pi\sqrt{\mu\varepsilon}}\left[\left(\frac{m\pi}{a}\right)^2 + \left(\frac{n\pi}{b}\right)^2 + \left(\frac{p\pi}{c}\right)^2\right]^{\frac{1}{2}} \tag{23}$$

from which one can find the resonant wavelength

$$\lambda_r = 2\left(\frac{m^2}{a^2} + \frac{n^2}{b^2} + \frac{p^2}{c^2}\right)^{-\frac{1}{2}} \tag{24}$$

Equations (23) and (24) are valid for both TE and TM modes, which can exist simultaneously in the cavity. Since the integers m, n, and p represent the half-wave periodicity in the x, y, and z directions respectively, the lowest resonant frequency (dominant mode) occurs when the integer associated with the smallest dimension is zero and the other two are unity. For example, these modes are TE_{011}, TE_{101}, or TE_{110}. Note that TM_{0np} or other TM modes where one integer is zero cannot exist.

WAVEGUIDES

In conclusion, we have found that the analytical methods used to find the fields in waveguides also apply to resonant cavities. For the simplest geometric shapes, such as rectangular, cylindrical, and spherical cavities, the fields can be described analytically. Analysis of odd-shaped cavities is difficult, but essential concepts can be obtained from the simpler cases and used as models for more difficult problems.

PROBLEMS

1. Two infinite, parallel, perfect conductors are separated by an air dielectric 3 cm thick. For each of the lowest three modes, find the range of frequencies in which a TM wave can propagate between the conductors.
2. Show that Eqs. (16) to (18) in Sec. 7.2 satisfy the boundary conditions shown in Fig. 7.2.2.
3. From Eqs. (23) and (24) in Sec. 7.2 it is evident that $\sin \theta$ need not be a real number. What is the physical significance of this phenomenon?
4. Write the equations for an evanescent TM wave propagating in the z direction between the parallel planes shown in Fig. 7.2.3. Show that the time-average real power transmitted in the z direction is zero.
5. Show that the wave impedance Z_{TM} appears to be a capacitive reactance and the wave impedance Z_{TE} appears to be an inductive reactance for frequencies below cutoff. Find the resulting Poynting vectors for each case.
6. For a TM wave propagating between perfectly conducting parallel planes, show that (a) if Z_{TM} is real, there is a time-average real power flow in the direction of propagation and (b) if Z_{TM} is imaginary, there is a time-average reactive power flow in the direction of propagation.
7. Let $m = 1$, so that a TE_{10} wave is described by Eqs. (10) to (12) of Sec. 7.3, and calculate the power flow in the z direction. The waveguide dimensions are $a = 5$ cm and $b = 3$ cm, and the waveguide is filled with air.
8. Given the vector wave equation for the magnetic field intensity

$$\nabla^2 \mathbf{H} = \mu\sigma \frac{\partial \mathbf{H}}{\partial t} + \mu\varepsilon \frac{\partial^2 \mathbf{H}}{\partial t^2}$$

assume that $\mathbf{H}$ is sinusoidal and show that the propagation constant is given by

$$\gamma^2 = \beta^2 \left(1 - j\frac{\sigma}{\omega\varepsilon}\right)$$

as was obtained for the electric field.

9. Show that the vector Helmholtz equation

$$\nabla^2 \mathbf{H} + \beta^2 \mathbf{H} = 0$$

in rectangular coordinates separates into three scalar Helmholtz equations in H_x, H_y, and H_z.

10. Find the boundary conditions at the perfectly conducting plates in Fig. 7.4.1a and b for the given magnetic field.

11. Find the solution of the scalar Helmholtz equations of Prob. 9 for the boundary conditions of Prob. 10. Show that the same solution is obtained from the curl equation (22) of Sec. 7.4.

12. Sketch the magnetic fields accompanying the electric fields of the three modes illustrated in Fig. 7.4.2.

13. Derive the phase constant β' of Eq. (30) of Sec. 7.4 for the rectangular waveguide, and show that the solutions for the positive-going waves E_x, E_y, and E_z are given by Eqs. (31) to (33) of the same section.

14. Prove that the simplest TM mode in a rectangular waveguide is the TM_{11} mode, and sketch the E and H fields.

15. For a rectangular waveguide with dimensions $a = 2b$, show that the TE_{10} mode has the lowest cutoff frequency of all possible modes. Show that the modes can be arranged by increasing order of cutoff frequencies as follows:

$$TE_{10}, \; (TE_{01}, TE_{20}), \; (TE_{11}, TM_{11}), \; \ldots$$

16. For the waveguide of Prob. 15, show that the cutoff frequencies for the TE_{01} and the TE_{20} modes are the same and equal to twice the cutoff frequency of the TE_{10} mode. Is this true in a square waveguide (where $a = b$)?

17. If the waveguide of Prob. 15 were square ($a = b$), what would be the order of the cutoff frequencies for the first 10 modes?

18. Separate the vector Helmholtz equation

$$\nabla \times \nabla \times \mathbf{E} - \beta^2 \mathbf{E} = 0$$

into its vector components in the cylindrical coordinate system and thereby show that Eqs. (4) to (6) of Sec. 7.6 are obtained.

19. Arrange the first six modes (both TE and TM) of a circular waveguide in order of increasing cutoff frequency, as in Prob. 15.

20. Sketch the E and H fields in the following modes for a circular waveguide: TM_{01}, TE_{01}, TE_{21}.

21. In the rectangular cavity of Sec. 7.7, show that the same propagation constant β^2 of Eq. (16) is obtained for the E_y and the E_z components of the field.

22. Prove that the resonant wavelength of the rectangular cavity is given by Eq. (24) of Sec. 7.7, assuming Eq. (23).

23. Assume that TE_{011} is the dominant mode of a rectangular cavity with a being the smallest dimension. Show that the resonant wavelength is

$$\lambda_r = \frac{2bc}{\sqrt{b^2 + c^2}}$$

24. For a cubical resonant cavity where $a = b = c$, show that the resonant wavelength equals the length of the diagonal on one face of the cube, or $\lambda_r = \sqrt{2}\,a$.

appendix A

Tables for Reference

Table 1.1.1 Dimensions and units
The International System (or the rationalized mks system) of Units is used throughout. The fundamental dimensions are mass (M) in kilograms, length (L) in meters, time (T) in seconds, and electric charge (Q) in coulombs. Some of the more commonly used quantities are tabulated below.

Symbol	Name and unit	Abbreviation	Dimension
α	attenuation constant (neper/meter)	Np/m	L^{-1}
C	capacitance (farad)	F	$M^{-1}L^{-2}T^2Q^2$
σ	conductivity (mho/meter)	℧/m	$M^{-1}L^{-3}TQ^{-2}$
I	current (ampere)	A	$T^{-1}Q$
E	electric field intensity (volt/meter)	V/m	$MLT^{-2}Q^{-1}$
Ψ	electric flux (coulomb)	C	Q
D	electric flux density (coulomb/meter²)	C/m²	$L^{-2}Q$
F	force (newton)	N	MLT^{-2}
f	frequency (hertz)	Hz	T^{-1}
L	inductance (henry)	H	ML^2Q^{-2}
H	magnetic field intensity (ampere/meter)	A/m	$L^{-1}T^{-1}Q$
Φ	magnetic flux (weber)	Wb	$ML^2T^{-1}Q^{-1}$
B	magnetic flux density (weber/meter² or tesla)	Wb/m² or T	$MT^{-1}Q^{-1}$
μ	permeability (henry/meter)	H/m	MLQ^{-2}
ϵ	permittivity (farad/meter)	F/m	$M^{-1}L^{-3}T^2Q^2$
β	phase constant (radian/meter)	rad/m	L^{-1}
V	potential (volt)	V	$ML^2T^{-2}Q^{-1}$
P	power (watt)	W	ML^2T^{-3}
ω	radian frequency (radian/second)	rad/s	T^{-1}
R	resistance (ohm)	Ω	$ML^2T^{-1}Q^{-2}$
A	vector magnetic potential (weber/meter)	Wb/m	$MLT^{-1}Q^{-1}$
λ	wavelength (meter)	m	L

Table 1.2.1 Vector transformations between coordinate systems
Given a vector $\mathbf{R}(u_1,v_1,w_1)$ with components R_{u_1}, R_{v_1}, and R_{w_1} in an "old" system, the transformed vector $\mathbf{R}(u_2,v_2,w_2)$ in the "new" system is given by

$$\mathbf{R}(u_2,v_2,w_2) = [R_{u_1}\ \ R_{v_1}\ \ R_{w_1}][T_{12}]\begin{bmatrix} a_{u_2} \\ a_{v_2} \\ a_{w_2} \end{bmatrix}$$

where $[T_{12}]$ is the transformation matrix from system 1 to system 2. The components R_{u_1}, R_{v_1}, and R_{w_1} must be expressed in terms of the new variables u_2, v_2, and w_2.

Variable substitutions and transformation matrices

Rectangular to cylindrical	Cylindrical to rectangular
$x = r\cos\phi,\ y = r\sin\phi,\ z = z$	$r = (x^2 + y^2)^{1/2},\ \phi = \tan^{-1}\left(\dfrac{y}{x}\right),\ z = z$
$[T_{RC}] = \begin{bmatrix} \cos\phi & -\sin\phi & 0 \\ \sin\phi & \cos\phi & 0 \\ 0 & 0 & 1 \end{bmatrix}$	$[T_{CR}] = \begin{bmatrix} \dfrac{x}{r} & \dfrac{y}{r} & 0 \\ -\dfrac{y}{r} & \dfrac{x}{r} & 0 \\ 0 & 0 & 1 \end{bmatrix}$

Rectangular to spherical	Spherical to rectangular
$x = r\sin\theta\cos\phi$ $y = r\sin\theta\sin\phi$ $z = r\cos\theta$	$r = (x^2+y^2+z^2)^{1/2}$ $\theta = \cos^{-1}\dfrac{z}{r}$ $\phi = \tan^{-1}\dfrac{y}{x}$
$[T_{RS}] = \begin{bmatrix} \sin\theta\cos\phi & \cos\theta\cos\phi & -\sin\phi \\ \sin\theta\sin\phi & \cos\theta\sin\phi & \cos\phi \\ \cos\theta & -\sin\theta & 0 \end{bmatrix}$	$[T_{SR}] = \begin{bmatrix} \dfrac{x}{r} & \dfrac{y}{r} & \dfrac{z}{r} \\ \dfrac{xz}{r\alpha} & \dfrac{yz}{r\alpha} & -\dfrac{\alpha}{r} \\ -\dfrac{y}{\alpha} & \dfrac{x}{\alpha} & 0 \end{bmatrix}$ where $\alpha = (x^2+y^2)^{1/2}$

Cylindrical to spherical	Spherical to cylindrical
r_c = radial variable in cylindrical system	r_s = radial variable in spherical system
$r_s = (r_c^2 + z^2)^{1/2}$	$r_c = r_s \sin\theta$
$\theta = \cos^{-1}\dfrac{z}{r_s}$	$\phi = \phi$
$\phi = \phi$	$z = r_s \cos\theta$
$[T_{CS}] = \begin{bmatrix} \sin\theta & \cos\theta & 0 \\ 0 & 0 & 1 \\ \cos\theta & -\sin\theta & 0 \end{bmatrix}$	$[T_{SC}] = \begin{bmatrix} \dfrac{r_c}{r_s} & 0 & \dfrac{z}{r_s} \\ \dfrac{z}{r_s} & 0 & -\dfrac{r_c}{r_s} \\ 0 & 1 & 0 \end{bmatrix}$

Table 1.2.2 Divergence, gradient, and curl in curvilinear coordinates

Coordinate system	Variables			Unit vectors			Scale factors		
	u	v	w	$\mathbf{a}_u$	$\mathbf{a}_v$	$\mathbf{a}_w$	h_1	h_2	h_3
Cartesian	x	y	z	$\mathbf{a}_x$	$\mathbf{a}_y$	$\mathbf{a}_z$	1	1	1
Cylindrical	r	ϕ	z	$\mathbf{a}_r$	$\mathbf{a}_\phi$	$\mathbf{a}_z$	1	r	1
Spherical	r	θ	ϕ	$\mathbf{a}_r$	$\mathbf{a}_\theta$	$\mathbf{a}_\phi$	1	r	$r \sin \theta$

Vector: $\mathbf{A} = A_u \mathbf{a}_u + A_v \mathbf{a}_v + A_w \mathbf{a}_w$ \hfill (1)

Differential vector: $d\mathbf{l} = h_1\, du\, \mathbf{a}_u + h_2\, dv\, \mathbf{a}_v + h_3\, dw\, \mathbf{a}_w$ \hfill (2)

Differential volume: $dV = h_1 h_2 h_3\, du\, dv\, dw$ \hfill (3)

$$\text{grad}\, \psi = \nabla \psi = \frac{1}{h_1} \frac{\partial \psi}{\partial u} \mathbf{a}_u + \frac{1}{h_2} \frac{\partial \psi}{\partial v} \mathbf{a}_v + \frac{1}{h_3} \frac{\partial \psi}{\partial w} \mathbf{a}_w \quad (4)$$

$$\text{div}\, \mathbf{A} = \nabla \cdot \mathbf{A} = \frac{1}{h_1 h_2 h_3} \left(\frac{\partial}{\partial u} h_2 h_3 A_u + \frac{\partial}{\partial v} h_3 h_1 A_v + \frac{\partial}{\partial w} h_1 h_2 A_w \right) \quad (5)$$

$$\text{curl}\, \mathbf{A} = \nabla \times \mathbf{A} = \frac{1}{h_1 h_2 h_3} \begin{vmatrix} h_1 \mathbf{a}_u & h_2 \mathbf{a}_v & h_3 \mathbf{a}_w \\ \dfrac{\partial}{\partial u} & \dfrac{\partial}{\partial v} & \dfrac{\partial}{\partial w} \\ h_1 A_u & h_2 A_v & h_3 A_w \end{vmatrix} \quad (6)$$

Scalar laplacian:

$$\nabla^2 \psi = \frac{1}{h_1 h_2 h_3} \left[\frac{\partial}{\partial u} \left(\frac{h_2 h_3}{h_1} \frac{\partial \psi}{\partial u} \right) + \frac{\partial}{\partial v} \left(\frac{h_3 h_1}{h_2} \frac{\partial \psi}{\partial v} \right) + \frac{\partial}{\partial w} \left(\frac{h_1 h_2}{h_3} \frac{\partial \psi}{\partial w} \right) \right] \quad (7)$$

Vector laplacian:

$\nabla^2 \mathbf{A} = \nabla(\nabla \cdot \mathbf{A}) - \nabla \times \nabla \times \mathbf{A}$ \quad any coordinate system \hfill (8)

$\nabla^2 \mathbf{A} = (\nabla^2 A_x)\mathbf{a}_x + (\nabla^2 A_y)\mathbf{a}_y + (\nabla^2 A_z)\mathbf{a}_z$ \quad rectangular coordinates only \hfill (9)

Table 1.2.3   Vector identities
A, B, C = vectors; a_i = unit vectors; a_n = normal unit vector; Ψ = scalar

The divergence theorem $\quad \int_v \nabla \cdot \mathbf{A} \, dv = \oint_s \mathbf{A} \cdot \mathbf{a}_n \, ds \quad$ (1)

Stokes' theorem $\quad \oint_C \mathbf{A} \cdot d\mathbf{l} = \int_s \nabla \times \mathbf{A} \cdot \mathbf{a}_n \, ds \quad$ (2)

The following equations are valid only in the rectangular coordinate system:

$$\nabla \equiv \mathbf{a}_x \frac{\partial}{\partial x} + \mathbf{a}_y \frac{\partial}{\partial y} + \mathbf{a}_z \frac{\partial}{\partial z} \tag{3}$$

$$\nabla \cdot \mathbf{A} = \mathbf{a}_x \frac{\partial A_x}{\partial x} + \mathbf{a}_y \frac{\partial A_y}{\partial y} + \mathbf{a}_z \frac{\partial A_z}{\partial z} \tag{4}$$

$$\nabla \times \mathbf{A} = \left(\frac{\partial A_z}{\partial y} - \frac{\partial A_y}{\partial z}\right)\mathbf{a}_x + \left(\frac{\partial A_x}{\partial z} - \frac{\partial A_z}{\partial x}\right)\mathbf{a}_y + \left(\frac{\partial A_y}{\partial x} - \frac{\partial A_x}{\partial y}\right)\mathbf{a}_z \tag{5}$$

$$\nabla \Psi = \frac{\partial \Psi}{\partial x}\mathbf{a}_x + \frac{\partial \Psi}{\partial y}\mathbf{a}_y + \frac{\partial \Psi}{\partial z}\mathbf{a}_z \tag{6}$$

$$\nabla^2 \Psi = \frac{\partial^2 \Psi}{\partial x^2} + \frac{\partial^2 \Psi}{\partial y^2} + \frac{\partial^2 \Psi}{\partial z^2} \tag{7}$$

$$\nabla^2 \mathbf{A} = (\nabla^2 A_x)\mathbf{a}_x + (\nabla^2 A_y)\mathbf{a}_y + (\nabla^2 A_z)\mathbf{a}_z \tag{8}$$

The following equations are independent of the coordinate system:

$\nabla \cdot \Psi \mathbf{A} = \nabla \Psi \cdot \mathbf{A} + \Psi \nabla \cdot \mathbf{A}$ (9)
$\nabla \times \Psi \mathbf{A} = \nabla \Psi \times \mathbf{A} + \Psi \nabla \times \mathbf{A}$ (10)
$\nabla \times \nabla \Psi = 0$ (11)
$\nabla \cdot (\nabla \times \mathbf{A}) = 0$ (12)
$\nabla \times \nabla \times \mathbf{A} = \nabla(\nabla \cdot \mathbf{A}) - \nabla^2 \mathbf{A}$ (13)
$\nabla \times (\mathbf{A} \times \mathbf{B}) = \mathbf{A}(\nabla \cdot \mathbf{B}) - \mathbf{B}(\nabla \cdot \mathbf{A}) - (\mathbf{A} \cdot \nabla)\mathbf{B} + (\mathbf{B} \cdot \nabla)\mathbf{A}$ (14)
$\mathbf{A} \cdot \mathbf{B} \times \mathbf{C} = \mathbf{C} \cdot \mathbf{A} \times \mathbf{B} = \mathbf{B} \cdot \mathbf{C} \times \mathbf{A}$ (15)
$\mathbf{A} \times (\mathbf{B} \times \mathbf{C}) = (\mathbf{A} \cdot \mathbf{C})\mathbf{B} - (\mathbf{A} \cdot \mathbf{B})\mathbf{C}$ (16)
$(\mathbf{A} \times \mathbf{B}) \times \mathbf{C} = (\mathbf{A} \cdot \mathbf{C})\mathbf{B} - (\mathbf{B} \cdot \mathbf{C})\mathbf{A}$ (17)

TABLES FOR REFERENCE

Table 1.7.1 Gradient, divergence, curl, and scalar laplacian in the common coordinate systems

Vector: $\mathbf{A} = A_u\mathbf{a}_u + A_v\mathbf{a}_v + A_w\mathbf{a}_w$
Scalar: Ψ

Gradient

Rectangular: $\nabla \Psi = \dfrac{\partial \Psi}{\partial x}\mathbf{a}_x + \dfrac{\partial \Psi}{\partial y}\mathbf{a}_y + \dfrac{\partial \Psi}{\partial z}\mathbf{a}_z$ \hfill (1)

Cylindrical: $\nabla \Psi = \dfrac{\partial \Psi}{\partial r}\mathbf{a}_r + \dfrac{1}{r}\dfrac{\partial \Psi}{\partial \phi}\mathbf{a}_\phi + \dfrac{\partial \Psi}{\partial z}\mathbf{a}_z$ \hfill (2)

Spherical: $\nabla \Psi = \dfrac{\partial \Psi}{\partial r}\mathbf{a}_r + \dfrac{1}{r}\dfrac{\partial \Psi}{\partial \theta}\mathbf{a}_\theta + \dfrac{1}{r \sin \theta}\dfrac{\partial \Psi}{\partial \phi}\mathbf{a}_\phi$ \hfill (3)

Divergence

Rectangular: $\nabla \cdot \mathbf{A} = \dfrac{\partial A_x}{\partial x} + \dfrac{\partial A_y}{\partial y} + \dfrac{\partial A_z}{\partial z}$ \hfill (4)

Cylindrical: $\nabla \cdot \mathbf{A} = \dfrac{1}{r}\dfrac{\partial r A_r}{\partial r} + \dfrac{1}{r}\dfrac{\partial A_\phi}{\partial \phi} + \dfrac{\partial A_z}{\partial z}$ \hfill (5)

Spherical: $\nabla \cdot \mathbf{A} = \dfrac{1}{r^2}\dfrac{\partial r^2 A_r}{\partial r} + \dfrac{1}{r \sin \theta}\left(\dfrac{\partial \sin \theta\, A_\theta}{\partial \theta} + \dfrac{\partial A_\phi}{\partial \phi} \right)$ \hfill (6)

Curl

Rectangular: $\nabla \times \mathbf{A} = \left(\dfrac{\partial A_z}{\partial y} - \dfrac{\partial A_y}{\partial z} \right)\mathbf{a}_x + \left(\dfrac{\partial A_x}{\partial z} - \dfrac{\partial A_z}{\partial x} \right)\mathbf{a}_y + \left(\dfrac{\partial A_y}{\partial x} - \dfrac{\partial A_x}{\partial y} \right)\mathbf{a}_z$ \hfill (7)

Cylindrical: $\nabla \times \mathbf{A} = \left(\dfrac{1}{r}\dfrac{\partial A_z}{\partial \phi} - \dfrac{\partial A_\phi}{\partial z} \right)\mathbf{a}_r + \left(\dfrac{\partial A_r}{\partial z} - \dfrac{\partial A_z}{\partial r} \right)\mathbf{a}_\phi + \dfrac{1}{r}\left(\dfrac{\partial r A_\phi}{\partial r} - \dfrac{\partial A_r}{\partial \phi} \right)\mathbf{a}_z$ \hfill (8)

Spherical: $\nabla \times \mathbf{A} = \dfrac{1}{r \sin \theta}\left(\dfrac{\partial \sin \theta\, A_\phi}{\partial \theta} - \dfrac{\partial A_\theta}{\partial \phi} \right)\mathbf{a}_r + \dfrac{1}{r}\left(\dfrac{1}{\sin \theta}\dfrac{\partial A_r}{\partial \phi} - \dfrac{\partial r A_\phi}{\partial r} \right)\mathbf{a}_\theta$
$\qquad + \dfrac{1}{r}\left(\dfrac{\partial r A_\theta}{\partial r} - \dfrac{\partial A_r}{\partial \theta} \right)\mathbf{a}_\phi$ \hfill (9)

Scalar laplacian

Rectangular: $\nabla^2 \Psi = \dfrac{\partial^2 \Psi}{\partial x^2} + \dfrac{\partial^2 \Psi}{\partial y^2} + \dfrac{\partial^2 \Psi}{\partial z^2}$ \hfill (10)

Cylindrical: $\nabla^2 \Psi = \dfrac{1}{r}\dfrac{\partial}{\partial r}\left(r \dfrac{\partial \Psi}{\partial r} \right) + \dfrac{1}{r^2}\dfrac{\partial^2 \Psi}{\partial \phi^2} + \dfrac{\partial^2 \Psi}{\partial z^2}$ \hfill (11)

Spherical: $\nabla^2 \Psi = \dfrac{1}{r^2}\dfrac{\partial}{\partial r}\left(r^2 \dfrac{\partial \Psi}{\partial r} \right) + \dfrac{1}{r^2 \sin \theta}\dfrac{\partial}{\partial \theta}\left(\sin \theta \dfrac{\partial \Psi}{\partial \theta} \right) + \dfrac{1}{r^2 \sin^2 \theta}\dfrac{\partial^2 \Psi}{\partial \phi^2}$ \hfill (12)

Table 2.7.1 Summary of the elementary field equations

Basic equations

Equation	Point form	Integral form	
Lorentz force equation	$\mathbf{F} = q\mathbf{E} + q(\mathbf{v} \times \mathbf{B})$		(1)
Conservation of charge	$\nabla \cdot \mathbf{J} + \dfrac{\partial \rho}{\partial t} = 0$	$I = \oint_s \mathbf{J} \cdot \mathbf{a}_n \, ds = -\dfrac{\partial q}{\partial t}$	(2)
Gauss' law	$\nabla \cdot \mathbf{B} = 0$	$\oint_s \mathbf{B} \cdot \mathbf{a}_n \, ds = 0$	(3)
Gauss' law	$\nabla \cdot \mathbf{D} = \rho$	$\oint_s \mathbf{D} \cdot \mathbf{a}_n \, ds = q = \int_v \rho \, dv$	(4)
Faraday's law	$\nabla \times \mathbf{E} = -\dfrac{\partial \mathbf{B}}{\partial t}$	$\oint_C \mathbf{E} \cdot d\mathbf{l} = -\int_s \dfrac{\partial \mathbf{B}}{\partial t} \cdot \mathbf{a}_n \, ds = \dfrac{-d\phi}{dt}$	(5)
Ampere's law	$\nabla \times \mathbf{H} = \mathbf{J} + \dfrac{\partial \mathbf{D}}{\partial t}$	$\oint_C \mathbf{H} \cdot d\mathbf{l} = I + \int_s \dfrac{\partial \mathbf{D}}{\partial t} \cdot \mathbf{a}_n \, ds$	(6)

Equations of a homogeneous isotropic medium

$\mathbf{J} = \sigma \mathbf{E}$ $\sigma = 0$ in free space   (7)

$\mathbf{D} = \varepsilon \mathbf{E}$ $\varepsilon_0 = 8.854 \times 10^{-12} \approx \dfrac{10^{-9}}{36\pi}$ F/m in free space   (8)

$\mathbf{B} = \mu \mathbf{H}$ $\mu_0 = 4\pi \times 10^{-7}$ H/m in free space   (9)

Poynting's theorem (for conservation of energy)

$$P_{\text{in}} = P_d + \dfrac{\partial}{\partial t}(W_E + W_M) + \int_s \mathbf{E} \times \mathbf{H} \cdot \mathbf{a}_n \, ds \tag{10}$$

Boundary conditions

$D_{2n} - D_{1n} = \rho_s$ $B_{2n} - B_{1n} = 0$ $E_{2t} - E_{1t} = 0$ $H_{2t} - H_{1t} = J_s$ (11)

The wave equations

$$\nabla^2 \mathbf{A} = \mu\sigma \dfrac{\partial \mathbf{A}}{\partial t} + \mu\varepsilon \dfrac{\partial^2 \mathbf{A}}{\partial t^2} \tag{12}$$

In free space:

$\sigma = 0$ $c = 3 \times 10^8$ m/s $\nabla^2 \mathbf{A} = \dfrac{1}{c^2} \dfrac{\partial^2 \mathbf{A}}{\partial t^2}$ (13)

TABLES FOR REFERENCE

Table 5.9.1 Summary of the steady-state equations

Dissipative lines

$$\gamma = \sqrt{(R + j\omega L)(G + j\omega C)} = \alpha + j\beta \tag{1}$$

$$Y_0 = \frac{1}{Z_0} = \sqrt{\frac{G + j\omega C}{R + j\omega L}} \tag{2}$$

$$v = \frac{\omega}{\beta} \tag{3}$$

$$V(z,t) = \operatorname{Re}\left[(V^+ e^{-\gamma z} + V^- e^{+\gamma z})e^{j\omega t}\right] \tag{4}$$

$$I(z,t) = \operatorname{Re}\left[Y_0(V^+ e^{-\gamma z} - V^- e^{+\gamma z})e^{j\omega t}\right] \tag{5}$$

$$\langle S(z) \rangle = \tfrac{1}{2} V(z) I^*(z) = \langle P(z) \rangle + jQ(z) \tag{6}$$

$$V^+ = \frac{V(-L)}{[1 + \Gamma(-L)]e^{-\gamma(-L)}} \tag{7}$$

$$\Gamma(z) = \frac{V^- e^{+\gamma z}}{V^+ e^{-\gamma z}} = \Gamma_r e^{2\gamma z} = \frac{Z_r - Z_0}{Z_r + Z_0} e^{2\gamma z} \tag{8}$$

$$Y(z) = Y_0 \frac{1 - \Gamma(z)}{1 + \Gamma(z)} \tag{9}$$

$$\mathrm{SWR}(z) = \frac{1 + |\Gamma(z)|}{1 - |\Gamma(z)|} = \frac{1 + |\Gamma_r| e^{2\alpha z}}{1 - |\Gamma_r| e^{2\alpha z}} \tag{10}$$

Lossless lines

Each of the above equations is valid for lossless lines, if we set

$$R = 0 \quad G = 0 \tag{11}$$

so that

$$\alpha = 0 \tag{12}$$

and

$$\gamma = j\beta \tag{13}$$

appendix B

Description of the TLDATA Array and the SANTLINE Computer Package

```
        ******** DECLARATIONS PACKAGE ********

             THIS DECLARATIONS PACKAGE MUST BE INSERTED
             INTO THE SUBPROGRAMS WHERE THE FOLLOWING
             DESIGNATION APPEARS.

        ******** INSERT DECLARATIONS PACKAGE HERE ********

C       ********* BEGINNING OF DECLARATIONS PACKAGE *********
        COMPLEX ZO,PO,ZLINE,PLINE,YLOAD,RCFL,RCFI,YIN,VPLUS,
       * IPLUS,VLOAD,ILOAD,VIN,IIN,PLOAD,PIN,YLINE
        REAL LENGTH,L,IPLUSX,IPLUSQ,ILOADX,ILOADQ,IINX,IINQ,LINEID
        DIMENSION LINEID(2)
        EQUIVALENCE (TLDATA(1),LINEID),(TLDATA(3),ZO,ZOX),
       A (TLDATA(4),ZOQ),(TLDATA(5),PO,POX),(TLDATA(6),POQ),
       B (TLDATA(7),LENGTH),(TLDATA(8),FPZO),
       C (TLDATA(9),YLINE,YLINEX),(TLDATA(10),YLINEQ),
       D (TLDATA(11),VPLUS,VPLUSX),(TLDATA(12),VPLUSQ),
       E (TLDATA(13),IPLUS,IPLUSX),(TLDATA(14),IPLUSQ)
        EQUIVALENCE (TLDATA(15),YIN,YINX),(TLDATA(16),YINQ),
       F (TLDATA(17),RCFI,RCFIX), (TLDATA(18),RCFIQ),
       G (TLDATA(19),VIN,VINX), (TLDATA(20),VINQ),
       H (TLDATA(21),IIN,IINX),(TLDATA(22),IINQ),
       I (TLDATA(23),PIN,PINX),(TLDATA(24),PINQ),
       J (TLDATA(25),YLOAD,YLOADX),(TLDATA(26),YLOADQ)
        EQUIVALENCE (TLDATA(27),RCFL,RCFLX),(TLDATA(28),RCFLQ),
       K (TLDATA(29),VLOAD,VLOADX),(TLDATA(30),VLOADQ),
       L (TLDATA(31),ILOAD,ILOADX),(TLDATA(32),ILOADQ),
       M (TLDATA(33),PLOAD,PLOADX),(TLDATA(34),PLOADQ),
       N (TLDATA(35),ZLINE,ZLINEX),(TLDATA(36),ZLINEQ),
       O (TLDATA(37),PLINE,ALPHA),(TLDATA(38),BETA),(TLDATA(39),R),
       P (TLDATA(40),L),(TLDATA(41),G),(TLDATA(42),C),(TLDATA(43),SWRI),
       Q (TLDATA(44),SWRL),(TLDATA(45),FCOMP)
        COMMON/TLDATA/TLDATA(45),F,LINE
C       ************ END OF DECLARATIONS PACKAGE ************
```

DESCRIPTION OF THE TLDATA ARRAY AND THE SANTLINE COMPUTER PACKAGE   431

```
C         COMPLEX FUNCTION CYIN(NLINE,YTERM)
C
C         CYIN..........GIVEN THE TERMINATING ADMITTANCE, CYIN COMPUTES THE
C                       INPUT ADMITTANCE OF A SEGMENT OF TRANSMISSION LINE
C                       WHOSE PARAMETERS HAVE BEEN PREVIOUSLY STORED IN THE
C                       STORAGE ARRAY BDATA.  THE REFLECTION COEFFICIENTS
C                       AND STANDING WAVE RATIOS AT BOTH ENDS ARE COMPUTED.
C                       ASSUMING AN INPUT VOLTAGE OF 1.0+J(0.0), THE
C                       VOLTAGES, CURRENTS AND POWERS AT EACH END ARE ALSO
C                       COMPUTED.  THESE ARE THEN PLACED IN THE STORAGE
C                       ARRAY, BDATA, AND THE FUNCTION EXITS.  THE FREQUENCY
C                       AT WHICH THE CALCULATIONS WERE MADE (CONTAINED IN
C                       THE COMMON VARIABLE F) IS ALSO SAVED.
C
C         NLINE.........THE LINE NUMBER OF THE LINE IN THE ARRAY BDATA
C                       WHICH IS TO BE USED IN THE CALCULATIONS.
C         IF NLINE IS LESS THAN 1, ONLY THE DATA IN THE COMMON ARRAY TLDATA
C         WILL BE AFFECTED.  NOTHING WILL BE COPIED INTO OR FROM BDATA.
C
C         YTERM.........THE TERMINATING ADMITTANCE IN MHOS.
C
C         ******** INSERT DECLARATIONS PACKAGE HERE ********
C
          COMPLEX YTERM,E2PL,EPL,CEXP,CABS,CONJG
          LINE=NLINE
          CALL FETCH
          YLOAD=YTERM
          RCFL=(YLINE-YLOAD)/(YLINE+YLOAD)
          EPL=CEXP(PLINE*LENGTH)
          E2PL=EPL*EPL
          RCFI=RCFL/E2PL
          YIN=(1.0-RCFI)/(1.0+RCFI)*YLINE
          CYIN=YIN
          VPLUS=1.0/((1.0+RCFI)*EPL)
          IPLUS=VPLUS*YLINE
          VLOAD=(1.0+RCFL)*VPLUS
          ILOAD=(1.0-RCFL)*IPLUS
          VIN=1.0
          IIN=(1.0-RCFI)*IPLUS*EPL
          PLOAD=CONJG(ILOAD)*VLOAD
          PIN=CONJG(IIN)
          RCM=CABS(RCFL)
          T=1.0-RCM
          IF(ABS(T).GE.2.0E-10)GO TO 10
          SWRL=1.0E75
          GO TO 20
   10     SWRL=(1.0+RCM)/T
   20     RCM=CABS(RCFI)
          T=1.0-RCM
          IF(ABS(T).GE.2.0E-10)GO TO 30
          SWRI=1.0E75
          GO TO 40
   30     SWRI=(1.0+RCM)/T
   40     CALL STASH
          RETURN
          END
```

```
      COMPLEX FUNCTION CYSHRT(NLINE)
C
C
C
C     CYSHRT........CYSHRT COMPUTES THE INPUT ADMITTANCE OF A LINE
C                   SEGMENT ASSUMING THAT IT IS SHORTED.  IN ADDITION,
C                   IT PERFORMS ALL OTHER FUNCTIONS OF CYIN.
C
C     NLINE.........THE LINE NUMBER OF THE LINE IN THE ARRAY BDATA
C                   WHICH IS TO BE USED IN THE CALCULATIONS.
C
C     IF NLINE IS LESS THAN -1, ONLY THE DATA IN THE COMMON ARRAY TLDATA
C     WILL BE AFFECTED.  NOTHING WILL BE COPIED INTO OR FROM BDATA.
C
C     SEE FUNCTION CYIN FOR VARIABLE ASSIGNMENTS FOR THE COMMON ARRAY
C     TLDATA.
C
C     ******** INSERT DECLARATIONS PACKAGE HERE ********
C
      COMPLEX E2PL,EPL,CEXP,CABS,CONJG
      LINE=NLINE
      CALL FETCH
      YLOAD=1.0E75
      EPL=CEXP(PLINE*LENGTH)
      E2PL=EPL*EPL
      RCFL=-1.0
      RCFI=-1.0/E2PL
      YIN=(1.0-RCFI)/(1.0+RCFI)*YLINE
      CYSHRT=YIN
      VPLUS=1.0/((1.0+RCFI)*EPL)
      IPLUS=VPLUS*YLINE
      VLOAD=0.0
      ILOAD=2.0*IPLUS
      VIN=1.0
      IIN=(1.0-RCFI)*IPLUS*EPL
      PLOAD=0.0
      PIN=CONJG(IIN)
      SWRL=1.0E75
      RCM=CABS(RCFI)
      T=1.0-RCM
      IF(ABS(T).GE.2.0E-10)GO TO 10
      SWRI=1.0E75
      GO TO 20
   10 SWRI=(1.0+RCM)/T
   20 CALL STASH
      RETURN
      END
```

DESCRIPTION OF THE TLDATA ARRAY AND THE SANTLINE COMPUTER PACKAGE

```
      COMPLEX FUNCTION CVLOAD(NLINE,CVIN)
C
C
C     CVLOAD.........ONCE CYIN OR CYSHRT HAS CALCULATED THE VOLTAGES,
C                    CURRENTS AND POWERS AT THE LOAD AND INPUT ENDS, AND
C                    THE VALUES OF VPLUS AND IPLUS ASSUMING THAT
C                    VIN=1.0+J(0.0), CVLOAD ADJUSTS THESE VALUES TO
C                    REFLECT THE ACTUAL VALUE OF VIN.
C
C     NLINE..........THE LINE NUMBER OF THE LINE IN THE ARRAY BDATA
C                    WHICH IS TO BE MODIFIED.
C     CVIN...........THE ACTUAL VALUE OF THE INPUT VOLTAGE IN VOLTS.
C
C     IF LINE IS LESS THAN 1, ONLY THE DATA IN THE COMMON ARRAY TLDATA
C     WILL BE AFFECTED, NOTHING WILL BE CHANGED IN BDATA.
C
C     SEE FUNCTION CYIN FOR VARIABLE ASSIGNMENTS FOR THE COMMON ARRAY
C     TLDATA.
C
      ******** INSERT DECLARATIONS PACKAGE HERE ********
C
      COMPLEX CVIN
      LINE=NLINE
      CALL FETCH
      VIN=CVIN
      VIN2=VINX*VINX+VINQ*VINQ
      IIN=IIN*VIN
      VPLUS=VPLUS*VIN
      IPLUS=IPLUS*VIN
      VLOAD=VLOAD*VIN
      ILOAD=ILOAD*VIN
      PIN=PIN*VIN2
      PLOAD=PLOAD*VIN2
      CALL STASH
      CVLOAD=VLOAD
      RETURN
      END
```

```
      FUNCTION FINDL(NLINE,YTERM,ITER)
C
C
C     FINDL.........FINDL DETERMINES THE SHORTEST LENGTH OF A PIECE OF
C                   TRANSMISSION LINE SUCH THAT THE REAL PART OF THE
C                   INPUT ADMITTANCE IS APPROXIMATELY EQUAL TO THE REAL
C                   PART OF THE CHARACTERISTIC ADMITTANCE.
C
C     NLINE.........THE NUMBER OF THE LINE SEGMENT WHOSE LENGTH IS TO
C                   BE DETERMINED.
C     YTERM.........THE COMPLEX LOAD ADMITTANCE.
C     ITER..........ITER CONTROLS THE DEGREE OF APPROXIMATION.  IF ITER
C                   IS 1, THE DISTANCE WILL BE FOUND TO THE NEAREST 0.05
C                   WAVELENGTH, IF 2, TO THE NEAREST 0.005 WAVELENGTH,
C                   IF 3, TO THE NEAREST .0005 WAVELENGTH, ETC.
C                   I.E.  THE TOLERANCE IS 0.5*10.0**(-ITER)
C                   WAVELENGTHS.
C
C     SEE FUNCTION CYIN FOR VARIABLE ASSIGNMENTS FOR THE COMMON ARRAY
C     TLDATA.
C
      ******** INSERT DECLARATIONS PACKAGE HERE ********
C
      COMPLEX CYIN,YTERM
      LINE=NLINE
      CALL FETCH
      DELTAL=3.141593/BETA
      LENGTH=0.0
      A=REAL(YTERM)-YLINEX
      DO 20 I=1,ITER
      DELTAL=DELTAL*0.1
      DO 10 J=1,9
      LENGTH=LENGTH+DELTAL
      B=REAL(CYIN(0,YTERM))-YLINEX
      IF(SIGN(1.0,A).NE.SIGN(1.0,B))GO TO 20
   10 A=B
   20 LENGTH=LENGTH-DELTAL
      FINDL=LENGTH
      LINE=NLINE
      CALL STASH
      RETURN
      END
```

DESCRIPTION OF THE TLDATA ARRAY AND THE SANTLINE COMPUTER PACKAGE

```
      FUNCTION STUBL(NLINE,YINST)
C
C
C
C     STUBL.........THIS FUNCTION COMPUTES THE LENGTH OF A SECTION OF
C                   SHORTED, LOSSLESS TRANSMISSION LINE NEEDED TO GIVE
C                   AN INPUT SUSCEPTANCE EQUAL TO THE IMAGINARY PART OF
C                   YINST.  ALTHOUGH THE LINE TO BE USED MAY NOT
C                   ACTUALLY BE LOSSLESS, IT IS TREATED AS SUCH.  ANY
C                   ERRORS INTRODUCED BY THIS PROCEDURE MAY BE ADJUSTED
C                   FOR IN THE FIELD.
C
C     NLINE.........THE LINE NUMBER OF THE LINE SEGMENT TO BE USED FOR
C                   THE STUB.
C     YINST.........THE COMPLEX ADMITTANCE WHOSE IMAGINARY PART IS TO BE
C                   PRODUCED BY THE SHORTED STUB.
C
C     IF NLINE IS LESS THAN 1, THE LINE SEGMENT WHOSE DATA IS IN THE
C     COMMON ARRAY TLDATA WILL BE USED FOR THE STUB.
C
C     SEE FUNCTION CYIN FOR VARIABLE ASSIGNMENTS FOR THE COMMON ARRAY
C     TLDATA.
C
      ******** INSERT DECLARATIONS PACKAGE HERE ********
      COMPLEX YINST
      LINE=NLINE
      CALL FETCH
      Y0=SQRT(C/L)
      THETAP=ATAN(Y0/AIMAG(YINST))
      IF(THETAP.GT.0.0)THETAP=THETAP-3.141593
      LENGTH=-THETAP/(Y0*F*L)
      STUBL=LENGTH
      CALL STASH
      RETURN
      END
```

```
      SUBROUTINE ENTER(NLINE)
C
C
C     ENTER.........ONCE THE PROGRAMMER HAS INITIALIZED THE REQUIRED
C                   ELEMENTS IN THE COMMON ARRAY TLDATA, ENTER COMPUTES
C                   THE VALUES OF R,L,G AND C (L AND C ARE MULTIPLIED
C                   BY 2*PI) AND THEN THE CONTENTS OF TLDATA ARE COPIED
C                   INTO THE ARRAY BDATA IN COMMON BLOCK DATA.
C
C     NLINE.........THE LINE NUMBER OF THE LINE FOR WHICH DATA IS TO BE
C                   COPIED INTO BDATA FROM TLDATA.
C
C     IF NLINE IS LESS THAN 1, ONLY THE DATA IN THE COMMON ARRAY
C     TLDATA IS AFFECTED.  NOTHING WILL BE COPIED INTO BDATA.
C
C     SEE FUNCTION CYIN FOR VARIABLE ASSIGNMENTS FOR THE COMMON ARRAY
C     TLDATA.
C
      ******** INSERT DECLARATIONS PACKAGE HERE ********
      COMPLEX RL,GC
      EQUIVALENCE (R,RL),(G,GC)
      LINE=NLINE
      LENGTH=ABS(LENGTH)
      RL=PO*ZO
      GC=PO/ZO
      L=L/FPZO
      C=C/FPZO
      CALL STASH
      RETURN
C
      ENTRY RESTOR(NLINE)
C
C     RESTOR........RESTOR PLACES ALL DATA FOR THE LINE SEGMENT
C                   SPECIFIED BY THE PARAMETER NLINE INTO THE COMMON
C                   ARRAY TLDATA.  IN ADDITION, THE VALUES OF L AND
C                   C CONTAINED IN BDATA ARE DIVIDED BY 2*PI TO REFLECT
C                   THEIR TRUE VALUES.
C
C     NLINE.........THE LINE NUMBER OF THE LINE WHOSE DATA IS TO BE
C                   COPIED INTO THE COMMON ARRAY TLDATA FROM BDATA.
C
C     IF LINE IS LESS THAN 1, ONLY THE DATA IN THE COMMON ARRAY TLDATA
C     IS AFFECTED.  NOTHING WILL BE COPIED FROM BDATA.
C
      LINE=NLINE
      CALL BRING
      L=L/6.283185
      C=C/6.283185
      RETURN
      END
```

```
      SUBROUTINE STASH
C
C     STASH.........STASH COPIES THE CONTENTS OF THE COMMON ARRAY TLDATA
C                   INTO THE LINE IN BDATA SPECIFIED BY THE COMMON
C                   VARIABLE LINE.
C
C     IF LINE IS LESS THAN 1, THE CALL IS IGNORED.
C
C     SEE FUNCTION CYIN FOR VARIABLE ASSIGNMENTS FOR THE COMMON ARRAY
C     TLDATA.
C
      ******** INSERT DECLARATIONS PACKAGE HERE ********
C
      COMMON/DATA/BDATA(45,1)
      COMPLEX CSQRT,CMPLX
      FCOMP=F
      IF(LINE.LT.1)RETURN
      DO 10 I=1,45
10    BDATA(I,LINE)=TLDATA(I)
      RETURN
C
      ENTRY BRING
C     BRING.........BRING COPIES THE CONTENTS OF THE LINE OF BDATA,
C                   SPECIFIED BY THE COMMON VARIABLE LINE, INTO THE
C                   COMMON ARRAY TLDATA.
C
C     IF THE VALUE OF LINE IS LESS THAN 1, THE CALL IS IGNORED.
C
      IF(LINE.LT.1)RETURN
      DO 20 I=1,45
20    TLDATA(I)=BDATA(I,LINE)
      RETURN
C
      ENTRY FETCH
C     FETCH.........FETCH PERFORMS THE SAME FUNCTION AS BRING EXCEPT
C                   THAT NEW VALUES FOR ZLINE, PLINE AND YLINE ARE
C                   COMPUTED AT THE FREQUENCY OF OPERATION, F.
C
C     IF LINE IS LESS THAN 1, ONLY THE DATA IN TLDATA IS AFFECTED.
C
      IF(LINE.LT.1)GO TO 40
      DO 30 I=1,45
30    TLDATA(I)=BDATA(I,LINE)
40    IF(F.NE.FPZO)GO TO 50
      ZLINE=ZO
      PLINE=PO
      GO TO 60
50    YLINE=CMPLX(R,F*L)
      PLINE=CMPLX(G,F*C)
      ZLINE=CSQRT(YLINE/PLINE)
      PLINE=ZLINE*PLINE
60    YLINE=1.0/ZLINE
      RETURN
      END
```

```
      SUBROUTINE MODIFY(NLINE,NUMBER,VALUE,MODE)
C
C
C     MODIFY........MODIFY ALLOWS THE PROGRAMMER TO CHANGE THE VALUES OF
C                   SELECTED PARAMETERS OF ANY LINE SEGMENT WITHOUT
C                   CHANGING ANY OTHERS (EXCEPT FOR CHANGING R, L, G AND
C                   C TO REFLECT CHANGES IN ZO, PO OR FPZO).
C
C     NLINE.........THE LINE NUMBER OF THE LINE IN BDATA TO BE
C                   AFFECTED.
C     NUMBER........SPECIFIES THE ELEMENT IN THE ARRAY BDATA WHICH IS
C                   TO BE MODIFIED OR THE FIRST OF SEVERAL ELEMENTS TO
C                   BE MODIFIED. THE VALUE OF NUMBER IS USED AS A
C                   SUBSCRIPT FOR ACCESSING THE ARRAY BDATA.
C     VALUE.........THE NEW VALUE TO BE INSERTED INTO
C                   BDATA(NUMBER,NLINE), OR IF MORE THAN ONE VALUE IS
C                   TO BE INSERTED, THE NAME OF THE ARRAY CONTAINING
C                   THE NEW VALUES. (COMPLEX VARIABLES ARE
C                   CONSIDERED AS TWO ELEMENT REAL ARRAYS BY MODIFY).
C     MODE..........SPECIFIES THE LENGTH OF THE REAL ARRAY VALUE. (1 FOR
C                   A SIMPLE REAL VARIABLE, 2 FOR A COMPLEX VARIABLE)
C
C     NOTE          IF THE COMBINATION OF NUMBER AND MODE WOULD ALLOW
C                   CHANGING DATA NOT CONCERNED WITH THE DESIRED LINE
C                   SEGMENT, NO MODIFICATION IS ALLOWED.
C
C     IF NLINE IS LESS THAN 1, THE CALL IS IGNORED, NOTHING IS DONE.
C
      COMMON/DATA/BDATA(45,1)
      DIMENSION VALUE(2),TEMP(8)
      REAL L
      COMPLEX PO,ZO,RL,GC
      EQUIVALENCE (TEMP(1),ZO),(TEMP(3),PO),(TEMP(5),RL),(TEMP(6),L),
     *(TEMP(7),GC),(TEMP(8),C)
      NO=NUMBER-1
      IF(NO+MODE.GT.45.OR.NUMBER.LT.1.OR.MODE.LT.1.OR.NLINE.LT.1)RETURN
      DO 10 I=1,MODE
   10 BDATA(NO+I,NLINE)=VALUE(I)
      IF((NUMBER.LT.3.OR.NUMBER.GT.6).AND.NUMBER.NE.8) RETURN
      DO 20 I=1,4
   20 TEMP(I)=BDATA(I+2,NLINE)
      RL=PO*ZO
      GC=PO/ZO
      L=L/BDATA(8,NLINE)
      C=C/BDATA(8,NLINE)
      DO 30 I=5,8
   30 BDATA(I+34,NLINE)=TEMP(I)
      RETURN
      END
```

DESCRIPTION OF THE TLDATA ARRAY AND THE SANTLINE COMPUTER PACKAGE 439

```
      COMPLEX FUNCTION GET1C(NLINE,NUMBER)
C
C
C     GET1C.........THIS COMPLEX FUNCTION RETURNS ANY COMPLEX DATA ITEM
C                   FOR THE LINE SEGMENT SPECIFIED BY THE PARAMETER
C                   NLINE.
C
C     NLINE.........THE LINE NUMBER OF THE LINE IN THE ARRAY BDATA
C                   WHICH IS TO BE ACCESSED.
C     NUMBER........SPECIFIES THE ELEMENT IN THE COMMON ARRAY TLDATA
C                   CORRESPONDING TO THE VARIABLE WHOSE VALUE IS TO BE
C                   RETURNED.  NUMBER IS USED AS A SUBSCRIPT FOR
C                   ACCESSING THE ARRAY BDATA.
C
C     NOTE          NO CHECK IS MADE TO ASSURE THAT NUMBER ADDRESSES AN
C                   ACTUAL COMPLEX DATA ITEM.  THIS IS LEFT TO THE
C                   PROGRAMMER.  IF NUMBER IS LESS THAN 1 OR EXCEEDS 44
C                   OR NLINE IS LESS THAN 1, A VALUE OF 0.0+J(0.0) IS
C                   RETURNED.
C
      COMPLEX CMPLX
      COMMON/DATA/BDATA(45,1)
      IF(NUMBER.LE.44.AND.NUMBER.GE.1.AND.NLINE.GE.1)GO TO 10
      GET1C=0.0
      RETURN
10    GET1C=CMPLX(BDATA(NUMBER,NLINE),BDATA(NUMBER+1,NLINE))
      RETURN
      END




      FUNCTION GET1R(NLINE,NUMBER)
C
C
C     GET1R.........THIS REAL FUNCTION RETURNS ANY REAL DATA ITEM FOR
C                   THE LINE SEGMENT SPECIFIED BY THE PARAMETER
C                   NLINE.
C
C     NLINE.........THE LINE NUMBER OF THE LINE IN THE ARRAY BDATA
C                   WHICH IS TO BE ACCESSED.
C     NUMBER........SPECIFIES THE ELEMENT IN THE COMMON ARRAY TLDATA
C                   CORRESPONDING TO THE VARIABLE WHOSE VALUE IS TO BE
C                   RETURNED.  NUMBER IS USED AS A SUBSCRIPT FOR
C                   ACCESSING THE ARRAY BDATA.
C
C     NOTE          IF NUMBER IS LESS THAN 1 OR EXCEEDS 45 OR NLINE IS
C                   LESS THAN 1, A VALUE OF 0.0 IS RETURNED.
C
      COMMON/DATA/BDATA(45,1)
      IF(NUMBER.LE.45.AND.NUMBER.GE.1.AND.NLINE.GE.1)GO TO 10
      GET1R=0.0
      RETURN
10    GET1R=BDATA(NUMBER,NLINE)
      RETURN
      END
```

```
      COMPLEX FUNCTION CPROP(A,VF,F)
C
C
C     CPROP.........THIS COMPLEX FUNCTION COMPUTES THE VALUE FOR P0
C                   REQUIRED FOR A LINE SEGMENT FROM THE POWER
C                   ATTENUATION IN DB/METER, THE VELOCITY FACTOR (RATIO
C                   OF THE VELOCITY OF PROPAGATION TO THE SPEED OF LIGHT
C                   IN A VACUUM) AND THE FREQUENCY AT WHICH THESE TWO
C                   ARE KNOWN.  THIS IS USUALLY THE WAY SUCH
C                   INFORMATION IS SUPPLIED IN CATALOGS.
C
C     A.............THE POWER ATTENUATION IN DB PER METER AT FREQUENCY
C                   F.
C     VF............THE VELOCITY FACTOR.
C     F.............THE FREQUENCY AT WHICH BOTH A AND VF ARE KNOWN.
C
C     HINT          ONE FOOT IS APPROXIMATELY 0.3048 METERS
C
      COMPLEX CMPLX
      CPROP=CMPLX(A*0.115179,F/VF*2.094395E-8)
      RETURN
      END




      FUNCTION CNVRTM(WAVEL,VF,F)
C
C
C     CNVRTM........THIS FUNCTION COMPUTES THE LENGTH OF A LINE SEGMENT
C                   IN METERS FROM THE LENGTH IN WAVELENGTHS, THE
C                   VELOCITY FACTOR (RATIO OF THE VELOCITY OF
C                   PROPAGATION TO THE SPEED OF LIGHT IN A VACUUM) AND
C                   THE FREQUENCY AT WHICH THESE ARE KNOWN.
C
C     WAVEL.........LENGTH OF THE LINE SEGMENT IN WAVELENGTHS AT
C                   FREQUENCY F.
C     VF............VELOCITY FACTOR AT FREQUENCY F.
C     F.............FREQUENCY AT WHICH WAVEL AND VF ARE KNOWN
C
      CNVRTM=WAVEL*3.0E8*VF/F
      RETURN
C
      ENTRY CNVRTW(LENGTH,VF,F)
C
C     CNVRTW........THIS FUNCTION ENTRY CONVERTS A LENGTH IN METERS TO
C                   A LENGTH IN WAVELENGTHS, GIVEN THE VELOCITY
C                   CONSTANT AND THE FREQUENCY AT WHICH THE LENGTH IN
C                   WAVELENGTHS IS DESIRED.
C
C     LENGTH        THE LENGTH OF THE LINE SEGMENT IN METERS.
C
C     ALL OTHER PARAMETERS ARE AS FOR CNVRTM.
C
      REAL LENGTH
      CNVRTW=LENGTH*F/(3.0E8*VF)
      RETURN
      END
```

DESCRIPTION OF THE TLDATA ARRAY AND THE SANTLINE COMPUTER PACKAGE

```
      SUBROUTINE ENTER(NLINE)
C
C
C     ENTER.........ONCE THE PROGRAMMER HAS INITIALIZED THE REQUIRED
C                   ELEMENTS IN THE COMMON ARRAY TLDATA, ENTER COMPUTES
C                   THE VALUES OF R,L,G AND C (L AND C ARE MULTIPLIED
C                   BY 2*PI) AND THEN THE CONTENTS OF TLDATA ARE COPIED
C                   INTO THE ARRAY BDATA IN COMMON BLOCK DATA.
C
C     NLINE.........THE LINE NUMBER OF THE LINE FOR WHICH DATA IS TO BE
C                   COPIED INTO BDATA FROM TLDATA.
C
C     IF NLINE IS LESS THAN 1, ONLY THE DATA IN THE COMMON ARRAY
C     TLDATA IS AFFECTED.  NOTHING WILL BE COPIED INTO BDATA.
C
C     SEE FUNCTION CYIN FOR VARIABLE ASSIGNMENTS FOR THE COMMON ARRAY
C     TLDATA.
C
C
      ******** INSERT DECLARATIONS PACKAGE HERE ********
      COMPLEX RL,GC
      EQUIVALENCE (R,RL),(G,GC)
      LINE=NLINE
      LENGTH=ABS(LENGTH)
      RL=PO*ZO
      GC=PO/ZO
      L=L/FPZO
      C=C/FPZO
      CALL STASH
      RETURN
C
      ENTRY RESTOR(NLINE)
C
C     RESTOR........RESTOR PLACES ALL DATA FOR THE LINE SEGMENT
C                   SPECIFIED BY THE PARAMETER NLINE INTO THE COMMON
C                   ARRAY TLDATA.  IN ADDITION, THE VALUES OF L AND
C                   C CONTAINED IN BDATA ARE DIVIDED BY 2*PI TO REFLECT
C                   THEIR TRUE VALUES.
C
C     NLINE.........THE LINE NUMBER OF THE LINE WHOSE DATA IS TO BE
C                   COPIED INTO THE COMMON ARRAY TLDATA FROM BDATA.
C
C     IF LINE IS LESS THAN 1, ONLY THE DATA IN THE COMMON ARRAY TLDATA
C     IS AFFECTED.  NOTHING WILL BE COPIED FROM BDATA.
C
      LINE=NLINE
      CALL BRING
      L=L/6.283185
      C=C/6.283185
      RETURN
      END
```

appendix C
Charts

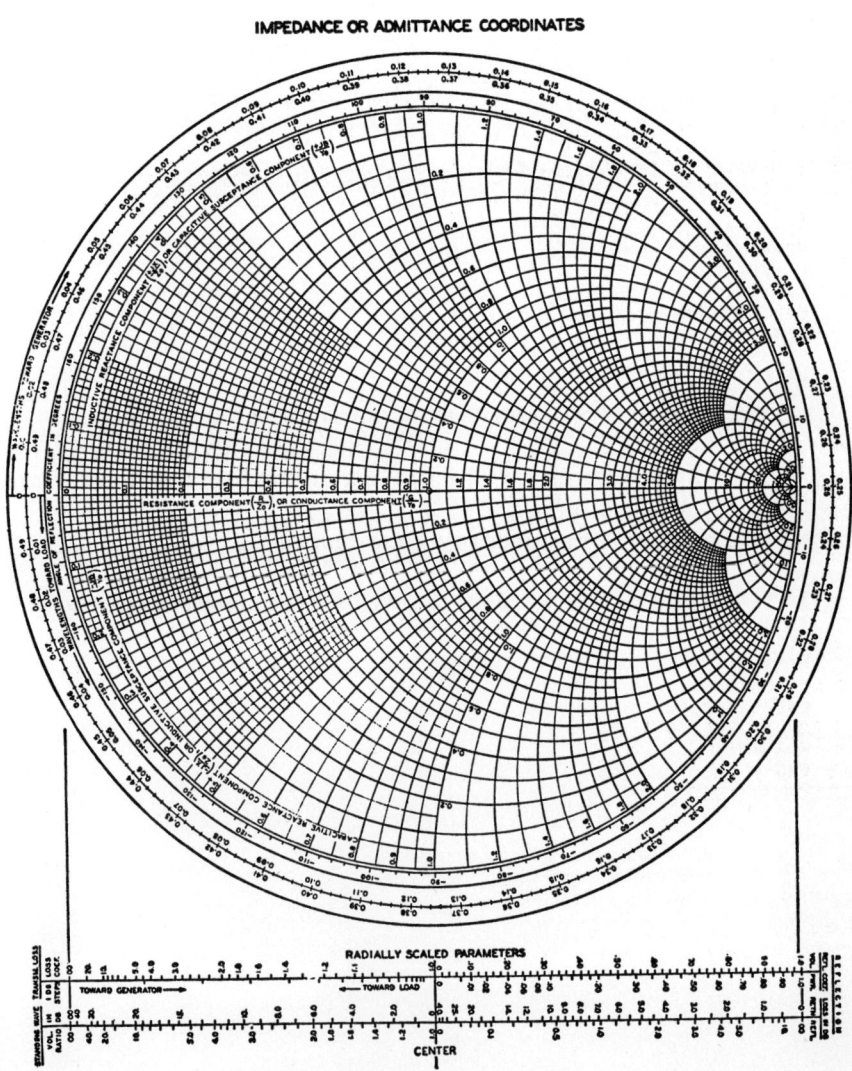

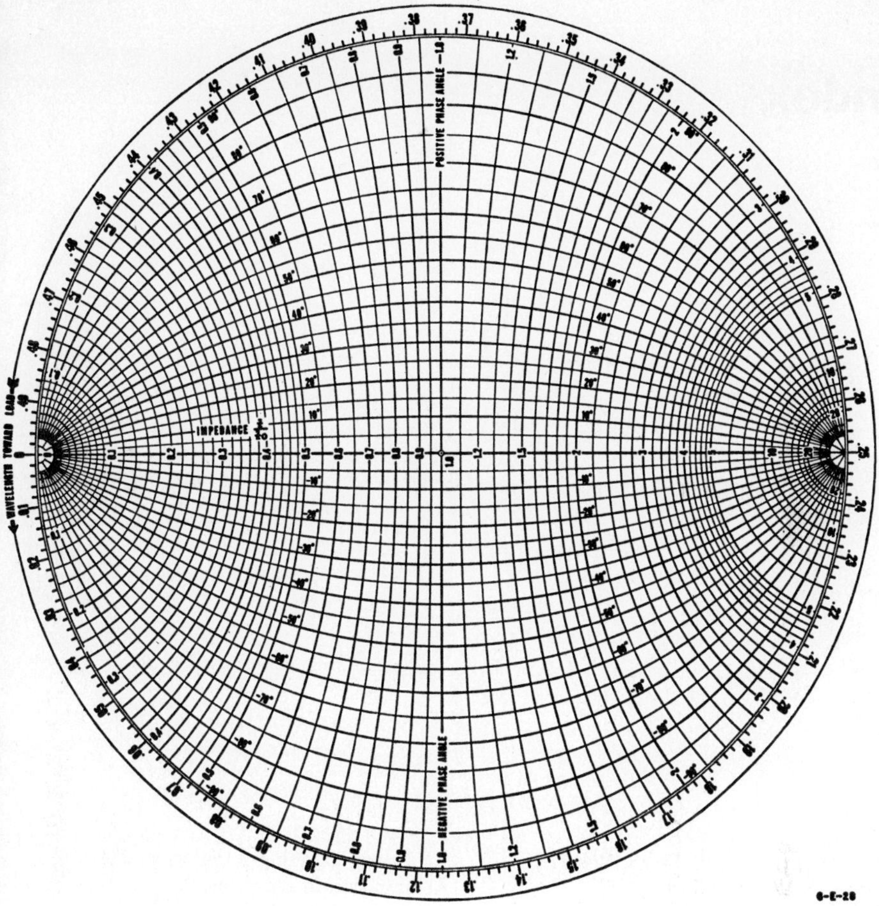

CHARTS

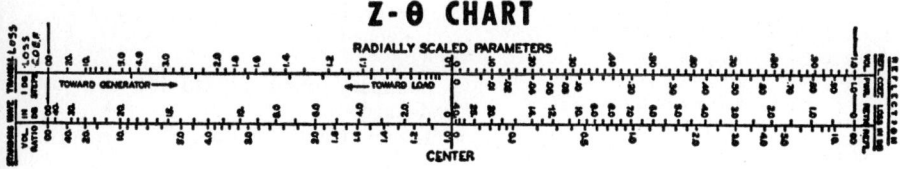

$Z_0 = 1\,\Omega$

Z-θ CHART

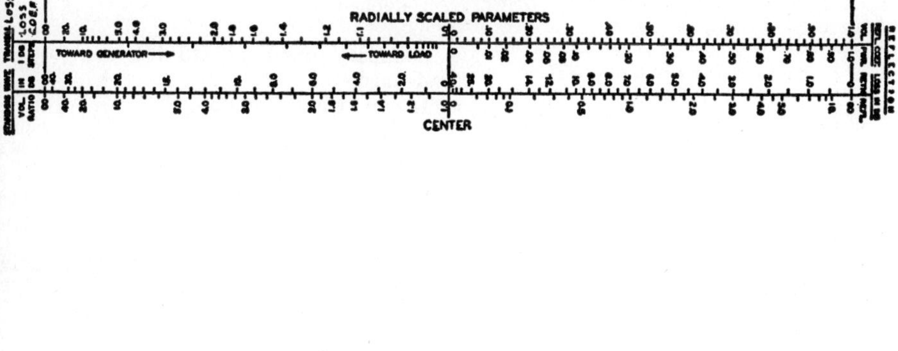

Index

A, vector potential, 76–78, 124–128, 133–134, 140–145
Absolute potential, 38
Adler, R. B., 179, 209
Admittance, characteristic, 221, 271, 272
AM, 184, 349
Ampere's law, 67, 72, 111, 117, 123
Amplitude modulation, 184, 349
Andreasen, M. G., 24
Angle:
   of characteristic impedance, 367
   of incidence, 191, 198, 395
   of reflection, 191, 198
Anisotropic medium, 147
Antenna:
   efficiency of, 145
   point source, 141–145
   power, 145
Antiferromagnetic medium, 148
Apparent time, 91
Arbitrary impedance terminations, 284–299

Arbitrary traveling waveforms, 234–237
Attenuation in seawater, 202
Attenuation constant, 171, 269, 373
Axial vectors, 59

B, vector, 77–78, 102–103
   defined, 102
B field:
   boundary conditions, 148–154
   defined, 102
   Lorentz transformation, 103
   static, 83
Band of frequencies, 349
Bandwidth:
   B, 353
   comparison of double and single transformers, 357–361
   discussion of, 349–354
   of double-stub tuner (example), 354
   optimum, 357

INDEX

Bandwidth:
  of quarter-wave transformer (example), 355–357
  and SWR, 353
BDATA array, 336, 384
  array description, 339–340
Bessel functions, 412
Bessel's equation, 412
Bilinear transformation, 300
Binomial series, 232
Bohr, N., 83
Born, Max, 83, 147
Bounce diagram, 230–234
Boundary conditions:
  at a dielectric, 193–195
  normal components, 148–150
  perfect conductor, 152–153
  resonant cavity, 418
  summary, 154
  tangential components, 150–152
  waveguides, 389–390, 395, 403, 406, 409
Boundary-value problem, 401
  waveguide, 408–409
Bowman, Frank, 412
Brewster angle, 206
Brillouin, L., 185

C, transmission line capacitance, 212–222, 366–367
Cable, coaxial (see Coaxial transmission line)
Capacitance:
  of coaxial capacitor, 212
  of transmission lines, 212–222
Capacitive termination, 254–255
Card deck for SANTLINE, 335
Carson, J., 218
Cavity:
  energy loss, 417–418
  resonant frequency of, 420
Center frequency, 349
Chain rule, 106
Characteristic admittance, 221, 271, 272
Characteristic impedance, 218–220, 270–272
  exponential section, 361
  measurement, 372–374
  quarter-wave transformers, 357
Characteristic and intrinsic impedances, 220
Charge:
  electric, 17–22, 50
  invariance under Lorentz transformation, 100–101
  Q, 17–22, 50

Charge density, 21–22
  surface, 22
  volume, 21
Child-Langmuir law, 47
Chu, L. J., 179, 209
Churchill, R. V., 300
Circular waveguides, 409–417
Circularly polarized waves, 182
Circulation of a field, 58
Circulation integral, 59
Closed path, 36
Closed surface, 40
CNVRTM, subroutine, 344–345
Coaxial transmission line:
  capacitance, 212
  characteristic impedance, 220
  charged, 238–244
  conductance, 217
  description, 207–208
  E field, 212
  external inductance, 213
  of finite length, 222–226
  H field, 211
  internal inductance, 259
  resistance, 218
Common data block, 332
COMMON statement, 335
Complete standing waves, 277
Complex power, 368–369
Components of a vector, 3
Compressibility constant, 362
Computational subprograms, 342
Computer:
  applications, 345–361
  data array: BDATA, 336, 339–340, 384
    TLDATA, 333, 336
  declaration package, 333, 345
    (See also Appendix B, 430–441)
  description of subprograms, 329–345
    (See also Appendix B, 430–441)
  solutions of typical problems, 294–296
  synthesis, of lossy lines, 381
    (See also Function)
Conduction current, 124
Conductivity, 39, 117, 173–175
Conservation of charge, 117
Conservative field, 38, 76
Constants, separation, 404, 406, 411, 418
Continuity equation, 75, 111, 117
  of circuit theory, 118
Contraction factor, 87–88
Convection current, 124
Coordinate systems, 2–17
  cylindrical, 3–10, 409–417
  orthogonal curvilinear, 13–17

Coordinate systems:
  rectangular, 2–3
  spherical, 10–13
Coordinate transformations:
  rectangular-cylindrical, 4–10
  spherical-rectangular, 10–13
  table of, 14
Coulomb gauge, 127–128
Coulomb's law, 17–24, 49–53, 100
Coupled equations, 410
CPROP, subroutine, 345
Curl, 40, 75
  definition, 60
  determinant, 63
  and energy, 64
  examples of, 65–68
  and orthogonal fields, 113
  properties of, 60–61
  table of, 44
  vector components of, 61–63
Current:
  conduction, 124
  continuity of, 75, 111, 117
  convection, 124
  density vector, 38–39, 76–81, 104–105, 118, 123, 136, 151
  displacement, 111, 123–124, 209
  point source, 141–145
  surface current, 151
Curvilinear coordinates, orthogonal, 13–17
Cutoff frequency, 387, 391, 393, 405
  circular guides, 416–417
  rectangular guide, 400
  tapered line, 329
Cutoff wavelength, 395
CVIN, subroutine, 343
CVLOAD, subroutine, 333, 343
CYIN, subroutine, 332, 333, 342, 383
Cylindrical coordinate system, 3–10, 409–417
CYSER, subroutine, 383

D, flux density defined, 49–53
Damped-wave equation, 131, 218, 369
Data array:
  BDATA, 336, 339–340, 384
  TLDATA, 333, 336
Data-transfer subprograms, 340
Declaration package, 333, 345
  (*See also* Appendix B, 430–441)
Delta function, 26–29, 42, 66, 70
∇ operators, 32, 75–76
∇V, 31
Depth, skin, 175, 202
Determinant, curl, 63
Diamagnetic medium, 148

Dibner, Bern, 83
Dielectric boundary conditions, 148–150
Dielectric constant:
  of free space, 19
  relative, 49, 147
Dielectric tensor, 145–148
Difference, potential, 37, 132–135
Differential operators, 161
Digital computer (*see* Computer)
Dimensions and units, 2
  (*See also* Appendix A, 423–429)
Diode, ideal, 47–49, 237–238
Dirac delta function, 26–29, 42, 66, 70
Directional derivative, 32
Discontinuities on a transmission line, 245–252
Dispersion, 185
Dispersive medium, 183
Displacement current, 123–124
  on coax, 209
  density, 111
Dissipative lines, 266–386
Dissipative media, 169–173
Distance quadrature of voltage and current, 278–279
Distortion of distance and time (relativistic), 88–91
Divergence, 39–49, 75
  definition, 40, 46
  table, 44
  theorem, 54–55, 118
Dominant mode, 407
Double-stub tuner, 318–323
  SWR and bandwidth of, 355
Drift velocity, 39
Driving-point impedance, 280–282
Duality relationship, 284

$\mathbf{E}$, electric intensity defined, 22
E^+ and E^-, 165–170, 190
$\mathbf{E}$ field, 22–24, 34
  boundary conditions, 148–154
  on coax, 212
  energy density, 138–139
  Lorentz transformation, 100–102
  normal to plane of incidence, 192–193
  in plane of incidence, 189–192
  of point charge, 23, 34
  potential solution of, 125
$\mathbf{E} \times \mathbf{H}$ (Poynting vector), 137–138
$E = mc^2$, 95
Earth, moon attraction, 52
Effective line lengths and bandwidth, 350
Efficiency of the antenna, 145
Einstein, Albert, 83, 94
Electric charge, 17–19

INDEX

Electric field intensity (*see* **E** field)
Electric intensity (*see* **E** field)
Electrical lengths, 351
Electromotive force (emf), 122, 132–135
Electron, 20–21
Electron spin, 147–148
Electron spin moment, 148
Electrostatic force, 19, 53, 100
Elliot, R. S., 19, 94, 98
Elliptically polarized waves, 180–182
Emf, 122, 132–135
Energy, 135–140, 397–398
  law of conservation, 137
  Lorentz transformation, 94–96
  stored in capacitor, 140
    in coaxial cable, 140
Energy density:
  electric field, 138–139
  magnetic field, 138–139
ENTER, subroutine, 340
Envelope, voltage phasor, 371
Equipotential surface, 24, 29
Equivalent circuits of transmission lines, 216–222
Evanescent waves, 394
Exponential section, 360–361
Exponential spiral, 374
Exponentially tapered line, 327–329, 360–361
External inductance of a coaxial cable, 213

Fano, R. M., 179, 209
Far field radiation, 144
Faraday's law, 108, 117, 121
Ferranti effect, 282–284
Ferromagnetic medium, 148
Feshbach, H., 79, 80
FETCH, subroutine, 340
Field equations:
  static, summary, 81, 82
  summary, table, 154
Field point, 22
Fields:
  on a coaxial line, 210–212
  conservative, 82
  E (*see* **E** field)
  electric (*see* **E** field)
  flux density (*see* **B**; **D**)
  H (*see* **H** field)
  intensity (*see* **E** field; **H** field)
  irrotational, 73, 82
  magnetic (*see* **H** field)
  nonconservative, 57–59
  between plates, 388–398
  of point source, 143

Fields:
  rectangular cavity, 419–420
  reflected, from dielectric, 193–202
  scalar, 24–25
  solenoidal, 74, 82
  static, summary, 81–82
  summary, table, 154
  transmitted into dielectric, 193–202
  variables, list of, 116
  vector, 25–26
  of any waveguide, 401
FINDL, subroutine, 343–344, 383
Flow lines, 25
Flux density field **B**, 77–78, 102–103
Flux density field **D**, 49–53
FM, 349
Forbidden region for stub tuning, 323
Force:
  Coulomb, 17–24, 49–53, 100
  electromotive, 122, 132–135
  electrostatic, 19, 53, 100
  Lorentz, 102, 117, 133
  Lorentz transformation, 96–99
  magnetic, 102
Free space:
  permeability of, 116
  permittivity of, 116
Frequency:
  center, 349
  computer designations, 337–339
  F, FPZO, and FCOMP, 337–339
  response of two quarter-wave sections (example), 357–361
  sensitivity of the quarter wave transformer, 325–326
Frizeau's experiment, 84
Frobenius method, 412
FUNCTION:
  CNVRTM, 344–345
  CPROP, 345
  CVLOAD, 343
  CYIN, 342
  CYSHRT, 342
  FINDL, 343–344
  GET1C, 341
  GET1R, 341, 384
  STUBL, 344
  (*See also* Computer)

G, transmission line conductance, 217–218, 366–367
Γ (*see* Reflection coefficient)
Gamma phasor, 297
Gamma plane, 296–299
Gauge transformation, 127–128
Gauss, K. F., 54

Gaussian surface, 50
Gauss' law, 49–53, 109, 117
　for B field, 107
　for electric fields, 119
　for magnetic fields, 120
Gauss' theorem (see Gauss' law)
Geometric mean, 324
GET1C, subroutine, 341
GET1R, subroutine, 341, 384
Grad V, 31
Gradient, 29–39, 75
　table, 44
Graphical techniques, lossy line, 374–376
Grimes, O. M., 20
Group velocity, 183–185, 392–394

H, vector, defined, 67, 110
H field:
　boundary conditions, 148–154
　on coaxial cable, 211
　of long wire, 67
　Lorentz transformation, 102–103
　static fields, 67–83
Hall effect, 135
Hall voltage, 133, 135
"Handbook of Mathematical Functions," 414n.
Hayt, W. H., 77
Helmholtz' equations, 128, 402, 408, 410, 418
　scalar, 402
　vector, 402
Helmholtz' theorem, 75–83
Homogeneous medium, 146
Horizontally polarized wave, 180

I (see Current)
I^+ and I^-:
　steady state, 269–272
　traveling waves, 219–222
Ideal diode, 237–238
Identities, vector, table, 18, 426
Impedance:
　characteristic, 218–220, 270–272
　　exponential section, 361
　　measurement, 372–374
　　quarter-wave transformers, 357
　driving point, 280–282
　as a function of $\Gamma(z)$, 286
　intrinsic, 103, 144, 167, 173, 193
　lossless transmission line, 285
　normalized, 300
　wave, 397
　$Z(z)$, 280–282, 306, 367

Impedance matching, 314–323
　double stub, 318–323, 355
　lossy lines, 376–381
　quarter wave, 323–327, 355–361, 376–378
　single stub, 315–318
Impedance spiral, 377–378
Incidence:
　normal, 186–189, 194–196
　oblique, 186, 189–193, 196–202
　plane of, 189
　plane wave, 185–193
Independence of path, 37
Inductance:
　external, 213
　internal, 259
　of transmission lines, 213–222
Inductive loading, 368
Inductive termination, 253–254
Inhomogeneous medium, 146
Inhomogeneous wave equation, 127
Integral theorems, 68
Intensity (see E field; H field)
Internal inductance, 259
International system of units, 423
Intrinsic impedance, 103, 144, 167, 173, 193
Inverse square-law field (see Coulomb's law)
Irrotational field, 37, 73, 76, 80
Isolated singularity, 66
Isotropic medium, 147
ITER, subroutine, 344

J, current density vector, 38–39, 76–81, 104–105
Johnson, Walter C., 329

KODE, 346
Korn, G. A., 29
Korn, T. M., 29
Krauss, H. L., 356

L, transmission line inductance, 213–222, 259, 366–367
Laplace's equation, 55, 76, 82
Laplacian:
　scalar, 55, 76
　vector, 76, 78
Law of conservation of energy, 137
Lawton, W. C., 19
LC filters, 417
LENGTH, 341
Lenz's law, 122, 134

INDEX

Light, velocity of, 84
Line length, effective, 352
Linear medium, fields of, 146
Linearly polarized wave, 179–180
LINEID, 347
Load impedance and bandwidth, 350
Loci:
   of constant amplitude, 309
   of constant angle, 309
   of constant reactance, 303
   of constant resistance, 302
Lorentz force, 102, 117, 133
Lorentz gauge, 127–128, 140
Lorentz transformation, 83–99
   of B field, 103
   of charge, 100–101
   definition, 87
   of divergence, 108–109
   of electric field E, 101
   of force, 96–99
   of mass and energy, 94–99
   of static sources q and J, 104–105
   of velocity, 91–94
Lossless lines, 331
Lossy lines, 366–386
   characteristic impedance of, 367

McLachlan, N. W., 412
McQuistan, R. B., 64, 78
Magnetic field intensity (*see* H field)
Magnetic flux density B, 77–78, 102–103
Magnetic materials, 148
Magnetic vector potential A, 76–78, 124–128, 133–134, 140–145
Main program, TLINE, 330, 333
Mass, Lorentz transformation, 96–97
Mass density for seawater, 362
Matched load, 230
Matching (*see* Impedance matching)
Material properties, 145–154, 164–165
Matrix:
   multiplication, 5
   permeability, 147
   permittivity, 146
   transformation, 5–17
Maxwell, James Clerk, 83
Maxwell's equations:
   physical interpretations of, 118–124
   potential solutions of, 124–128
   for static fields, 81, 83
   summary, 115–118
Medium, 145–148
   anisotropic, 147
   diamagnetic, 148
   dielectric, 145–148
   distortionless, 164–165

Medium:
   free space, 147
   isotropic, 147
   polarized, 148
Metric coefficients, 15
Michelson-Morley experiment, 84
mks units, 2
   (*See also* Appendix A, 423–429)
Mobility constant, 39
Mode, 341
Mode number, 391, 404, 411
MODIFY, subroutine, 341
Momentum, 97
Moon, P., 13, 24, 61
Moon, earth attraction, 52
Moore, Richard K., 202
Morse, P. M., 79, 80
Multiplication, matrix, 5

National Bureau of Standards AMS-55, 414
Near-field radiation, 144
Negative-going wave, 270
Neper (Np), 172
NLINE, 341
Nonuniform plane wave, 162, 403–406
Nonzero initial conditions on coax lines, 238–244
Nonzero initial current on a coax line, 242
Normal components:
   at a conductor, 150–153
   at dielectric, 148–150
Normal incidence:
   on a dielectric, 194–196
   on a perfect conductor, 186–189
Normal vector, 29–31
Normalized impedance $Z_n(z)$:
   definition, 300
   maximum and minimum, 306
Normalized voltage phasor, 298
Nuclear spin, 147–148
Nuclear-spin moment, 148
Null points, 277–278, 417
NUMBER, 341

Oblique incidence:
   on a dielectric, 196–202
   on a perfect conductor, 186, 189–193
Ohm's law, 39, 117, 227
Open-circuited line, 223–226, 275–280
Operator, ∇, 32, 75–76
Orbital moment, 148
Ordung, P. F., 356

Page, Leigh, 83
Parallel plane waveguides, 388–392
Paramagnetic medium, 148
Parameters, transmission line, 216–222, 366–367
Parametric equations of an ellipse, 181
Path, closed, 3–6
Penetration, depth of, 175
Permanent store, 332, 335
Permeability μ, 77, 110, 117, 145–153
 free space, 116
 matrix, 147
Permittivity ϵ, 19, 116, 145–153
 in crystals, 146
 free space, 116
 relative, 49
 tensor, 145–153
Phase constant β, 171–172, 269, 272, 274–275, 374, 391, 407, 419
Phase velocity, 164, 182–183, 392–394
Phasor, 169–172
 current, 270–271
 definition, 171
 on a dissipative line, 369–372
 rotation, 370
 voltage, 270–271
Physical interpretations of Maxwell's equations, 118–124
Plane of incidence, 189
Plane-polarized waves, 179–180
Plane wave:
 in dissipative media, 169–173
 incident: on a perfect conductor, 185–193
  on a perfect dielectric, 193–202
 nonuniform, 162
 uniform, 162, 166–169
Plateau voltages, 232
Plimpton, S. J., 19
Point source:
 of current, 141–145
 fields, 143
 radiation, 142–145
Poisson's equation, 55–56, 76–80, 82
 scalar, 55–56, 76–80
 solutions, 78–79
 vector, 78–80
Polarization of plane waves, 179–182
 conventional notation, 180
Polarized medium, 148
Position vector, 6–7
Positive-going wave, 270
Potential, absolute, 38
Potential difference, 37, 132–135
Potential field, 29–39
Potential gradient, 29–32

Power, 135–140, 397–398
 attenuation, 345
 complex, 175–179
 density, 136–137
  (*See also* Poynting vector)
 factor angle, 177, 313
 maximum, 364
 $\langle P \rangle$, time-average, 280
 transfer, 398
 traveling on line, 229
Poynting vector, 137–138, 144, 169, 188, 398, 405
 complex, 175–179
 time-average, 178–179
Poynting's theorem, 135–140, 154
Pressure wave in seawater, 204
Product, matrix, 5
Programmed example on rectangular waveguides, 401–408
Programmed exercises:
 divergence of the E field in a diode, 47–49
 examples of velocity transformations, 93–94
 exponentially tapered line, 322–329
 Ferranti effect, 282–284
 ideal diode on a line, 237–238
 measuring Z_0 and γ, 372–374
 pulse generator, 244–246
 radiation from a point source, 142–145
 undersea communication, 202–204
Propagation constant, 171, 269, 272, 345, 391
 circular guides, 413–415
 measurement, 372–374
Proper length, 89–90
Proper time, 91
Properties of materials, 145–153
Pulse generator, 244–246

Q:
 charge, 17–22, 50
 quality factor, 417
Quadrature of E and H fields, 188–189
Quality factor Q, 417
Quarter-wavelength transformer, 323–327, 376–378
 bandwidth, 355–357
 SWR and bandwidth, 357–361
 two sections, 357–361
Quasipotential, 61
QWTX, program, 356
QWTX2, program, 359

R, transmission line resistance, 218, 366–367

INDEX

Radiation, 140–145
 far field, 141
 near field, 144
 pattern, 144
 point source, 142–145
 resistance, 145
Ramo, S., 169, 387, 388
Rationalized mks system of units, 423
Reactive terminations, 252–259
Receiver impedance, 284
Rectangular cavity, 418–421
Rectangular coordinate system, 2–3
Rectangular waveguides, 398–408
 E field, 407
 H field, 407
"Reference Data for Radio Engineers," 216, 220, 359
Reflected fields on a shorted coax, 223
Reflected wave, 187
Reflection at a dielectric boundary, 193–202
Reflection coefficient, 195–196, 367
 and characteristic impedance, 285
 at discontinuities, 247
 for an electric field, 195, 201
 exceeds unity, 368
 $\Gamma(z)$, 285
 Γ_r, 227–228
 generalized, 300
 for magnetic field, 196, 199
 at the sending end, 229
Reich, H. J., 356
Relative dielectric constant ϵ_r, 49, 147
Relative velocity, 91
Relativistic distortion:
 distance, 88–91
 time, 88–91
Relativity:
 E and M fields, 99–104
 Maxwell's equations, 105–112
 principle of, 84
 special theory of, 84
Resistance:
 radiation, 145
 transmission line, 218, 366–367
Resistive terminations on coax lines, 226–230
Resistivity:
 skin effect, 173–175
 transmission lines, 217–218
Resonant cavities, 417–421
Resonant elements, 417
Resonant frequency of cavity, 420
Resonant wavelength of cavity, 420
Rest mass, 95
RESTOR, subroutine, 340
Rogers, W. R., 24

Rot E, 61
Rotating phasor, 276
Rotation, 61
 of phasor diagrams, 370

S (see Poynting vector)
SANTLINE:
 card deck, 335
 computer package, 333–345, 430–443
 computer program, 293, 330–345
 lossy lines, 381
 major parts, 334
 summary, 333–340
Scalar field, 24–25, 39
Scalar Laplacian, 44, 55, 76
 table, 44
Scalar Poisson equation, 55–56, 76–80
Scalar potential, 24, 34, 35, 77, 124
Scalar source, 75, 76
Scale factor, 15
Schelkunoff, S. A., 218
Schwartz, Laurent, 26
Seawater:
 attenuation, 202
 communication through, 202–204
 mass density, 362
 skin depth, 202
 sonar, 362
Separation of variables, 401, 406, 409, 418
Separation constant, 404, 406, 411, 418
Series solutions, 412
Shorted transmission line, 222–224, 280
Single-stub tuning, 315–318
Singularity, 33
 isolated, 66
Sink of field, 40
Sinusoidal traveling waves, 273–275
Skalnik, J. G., 356
Skin depth, 175
 seawater, 202
Skin effect, 173–175
Smith, P. H., 293
Smith chart, 293, 301–308
 admittance, 306–307
 complex power, 313–314
 current, 309–314
 current ratios, 310–314
 derivation of, 301–308
 power, 309–314
 voltage, 309–314
 voltage ratios, 310–314
Snell's law, 198
Solenoidal field, 74, 80
 of B and H, 120
Sonar, 362

Source of field, 22, 38–40, 76–81, 104–105, 141–145
Source point, 22
Spangenberg, K. R., 47
Spectrum of frequencies, 349
Spencer, D., 13, 24, 61
Spherical coordinate system, 10–13
Spin:
  electron, 147–148
  nuclear, 147–148
Spiral, exponential, 374
Spiral impedance path, 377–378
Standing-wave ratio (SWR) (*see* SWR)
Standing waves, 188, 275–282, 369–372, 390
  definition, 276
  at dielectric boundary, 199
  at a perfect conductor, 188
STASH, subroutine, 340
Static fields, summary of, 75–83
Steady-state solution:
  from transient analysis, 266–268
  of the wave equation, 268–273
Steinmetz, Charles P., 84, 96
Step-function notation, 223
Stoke's theorem, 68–75
  examples, 71–73
Stratton, J. A., 185, 418
Stub tuners (*see* Impedance matching)
STUBL, subroutine, 344, 383
Submarine communication, 202–204
SUBROUTINES (*see* FUNCTION)
Surface, closed, 40
Surface current, 151
Susceptance of shorted stub, 315
SWR, 299, 304
  and bandwidth, 352–353
  definition, 299
  and Γ_r, 305
  lossy lines, 371–372
  periodic function of frequency, 359–360
System of units, 2
  (*See also* Appendix A, 423–429)

Tables:
  definitions of words and element assignments for common array TLDATA, 337–338
    (*See also* Appendix B, 430–441)
  description of subprograms of SANTLINE and glossary of terms, 346–350
    (*See also* Appendix B, 430–441)
  dimensions and units, 423
  divergence, gradient, and curl in curvilinear coordinates, 16, 425

Tables:
  frequency response of quarter-wave transformer, 357
  gradient, divergence, curl, and scalar laplacian in common coordinate systems, 44, 427
  lossless line: with complex load, 294
    with imaginary load, 295
    with no load, 296
    with real load, 293
  mathematical postulates of electromagnetic theory, 117
  program double-stub bandwidth, 354
  program QWTX, 356
  program QWTX2, 359
  program STUBBW, 382
  program TLINE1, 351
  single-stub tuning on lossy line, results for, 383
  summary of elementary field equations, 154, 428
  summary of steady-state equations, 272, 331, 429
  vector identities, 17, 426
  vector transformations between coordinate systems, 14, 424
  voltage, current, and power for line, results for, 352
Tangential components:
  at a conductor, 150–153
  at a dielectric, 148–150
Tanner, R. L., 24
Tapered line, 327–329, 360–361
Taylor series, 45
TE waves, 161, 165, 388–421
TE_{mn} waves, 407
  circular guides, 415
TE_{mnp} mode, 419
TE_{m0} waves, 399–401
TE_{01} mode, 419
Telegrapher's equations, 213, 216, 270, 328
  for a lossy line, 217–218
TEM wave, 161, 165, 208–209
Temporary store, 335
Tensor, dielectric, 145–148
Teslas, 102
Thévenin equivalent circuit, 228–229
Time-average power, lossless line, 287–288
Time constant, 257
TLDATA array, 333, 336, 383, 430–443
TLINE main program, 330
TLINE1 simple test program, 345–349
TM waves, 161, 165, 388–421
TM_{mn} waves, 407–408
  circular guides, 415

INDEX

453

TM$_{mnp}$ mode, 420
Total differential, 211
Transcendental equation, 413
  for stub tuning, 379
Transformation:
  of coordinate systems, 14, 424
  Lorentz (*see* Lorentz transformation)
Transformation matrix, 5-17
Translation of vectors, 8
Transmission:
  coefficient, 196, 205
  at dielectric boundary, 193-202
  at discontinuities, 247
  for electric field, 201
  for magnetic field, 199
Transmission line equivalent circuit, 216-222
Transmission line geometries, 215-216
Transmission line parameters, 216-222, 366-367
Transverse components of fields, 161
Transverse electric (TE) waves, 161, 388-421
Transverse electromagnetic (TEM) wave, 161, 165, 208-209
Transverse magnetic (TM) waves, 161, 388-421
Traveling waves, 163-166
  energy storage, 168
  functional solution, 164
  on matched line, 274
  Poynting vector, 169
  vectors, 165-166
Triangular waveform on coax, 234-237
Tuning stubs (*see* Impedance matching)
TV, 349

Uniform plane wave, 162, 186, 193, 389, 402-403
Unit impulse function (*see* Delta function)
Unit step function, 27, 221-226
Unit vector, 2, 23
Units, system of, 2
  (*See also* Appendix A, 423-429)
Utility subprograms, 333, 344

V:
  scalar potential, 34-37
  in time varying fields, 132-135
V^+ and V^-:
  evaluation, 286
  steady state, 269-272
  traveling waves, 219-222, 267-268

VALUE, 341
Van Der Ziel, Albert, 148
Van Duzer, T., 169, 387, 388
Vector:
  axial, 59
  complex, 176
  components, 410
  fields, 25-26, 39, 68-75
    properties of, 73-75
  identities, table, 18, 426
  (*See also* Fields)
  laplacian, 76, 78, 402, 408
  normal, 29-31
  Poisson equation, 78-81
  potential **A**, 76-78, 124-128, 133-134, 140-145
  source, 75, 76
  spaces, 24
  translation, 8-10
  wave equation, 127
Velocity:
  group, 183-185, 392-394
  of light, 84
  phase, 164, 182-183, 392-394
  of propagation, 164, 182-185, 214-215, 272, 274-275
  of sinusoid, 172, 270-271
Velocity factor, 345
Vertically polarized wave, 180
Voltage, 37, 132-135
  drop, 133
  phasor, normalized, 298
  phasor envelope, 371
  plateau, 232
Vortex, 73, 76
Vorticity lines, 74

Watson, G. N., 412
Wave equation, 128-132, 154
  analytic function solution, 162
  complex variable, solution, 169-173
  for the current, 214-216
  damped, 131, 169-173
  inhomogeneous, 127
  phasor solution, 169-173
  sinusoidal solution, 169-173, 266-273
  solution of, 159-163
  steady-state solution of, 268-273
  vector, 127
  for voltage, 214-216
Wave impedance, 397, 400
Waveguide boundary-value problems, 408-409
Waveguides, 387-421
  circular, 411-417

Waveguides:
  parallel planes, 388–398
  rectangular, 398–408
WAVEL, subroutine, 344–345
Wavelength, 273–275, 394–397
  and bandwidth, 350
Waves:
  evanescent, 394
  reflected, 187

Waves:
  standing, 369–372
  TE, 161, 388–421
  TEM, 161, 208–209
  TM, 161, 388–421
Weeks, W. L., 145, 169
Wells, James C,. 333
Whinnery, J. R., 169, 387, 388
Wolf, Emil, 147